FERN ECOL

MW00836925

Ferns are an integral part of the world's flora, appreciated for their beauty as ornamentals, problematic as invaders, and endangered by human interference. They often dominate forest understories but also colonize open areas, invade waterways, and survive in nutrient-poor wastelands and eroded pastures. This is the first comprehensive summary of fern ecology, with worldwide examples from Siberia to Hawaii. Topics include a brief history of the ecological study of ferns, their biogeography and population dynamics, their role in ecosystem nutrient cycles and adaptations to xeric environments, and their responses to disturbance and interactions with other organisms. Fully illustrated concepts provide a framework for students and professionals in ecology, conservation, and land management, and a wealth of information for anyone interested in ferns.

KLAUS MEHLTRETER is a Researcher and Professor at the Instituto de Ecología, A. C., Xalapa, Mexico. His research focuses on the ecology of tropical ferns in cloud forests, coastal mangroves, and seasonally dry forests, and he is a fellow of the Mexican research system (Sistema Nacional de Investigadores).

LAWRENCE R. WALKER is a Professor of plant ecology at the University of Nevada, Las Vegas. His research focuses on the mechanisms that drive plant succession, particularly primary succession on volcanoes, landslides, glacial moraines, floodplains, dunes, mine tailings, and abandoned roads. Walker received a Fulbright Award in 2004 for work in Iceland and has also received three Distinguished Researcher awards from the University of Nevada.

JOANNE M. SHARPE is an independent Consultant specializing in fern ecology, long-term studies, the role of amateurs in ecological research, and education. She is currently a Research Associate with the NSF-funded Luquillo Long-Term Ecological Research (LTER) Program in Puerto Rico, where for the past 18 years she has monitored fern growth and spore production as part of rain forest productivity, hurricane impact, and demographic research projects.

FERN ECOLOGY

Edited by

KLAUS MEHLTRETER
Instituto de Ecología, A. C., Xalapa, Mexico

LAWRENCE R. WALKER
University of Nevada, Las Vegas

JOANNE M. SHARPE
Sharplex Services, Edgecomb, Maine

CAMBRIDGE
UNIVERSITY PRESS

CAMBRIDGE
UNIVERSITY PRESS

University Printing House, Cambridge CB2 8BS, United Kingdom

Published in the United States of America by Cambridge University Press, New York

Cambridge University Press is part of the University of Cambridge.

It furthers the University's mission by disseminating knowledge in the pursuit of education, learning and research at the highest international levels of excellence.

www.cambridge.org
Information on this title: www.cambridge.org/9780521728201

© Cambridge University Press 2010

First published 2010

A catalogue record for this publication is available from the British Library

ISBN 978-0-521-89940-6 Hardback
ISBN 978-0-521-72820-1 Paperback

Contents

Klaus Mehltreter

Roderick C. Robinson, Elizabeth Sheffield and Joanne M. Sharpe

Color plates are between pages 176 and 177.

Contributors

Peter Hietz Institute of Botany, Department of Integrative Biology, University of Natural Resources and Applied Life Sciences (BOKU), Gregor Mendel-Str. 33, 1180 Vienna, Austria

Michael Kessler Institute of Systematic Botany and Botanical Garden, University of Zürich, Zollikerstrasse 107, CH-8008 Zürich, Switzerland

Klaus Mehltreter Instituto de Ecología, A. C., Departamento de Ecología Funcional, km 2.5 antigua carretera a Coatepec No. 351, Congregación El Haya, Xalapa 91070, Veracruz, Mexico

Sarah J. Richardson Landcare Research, PO Box 40, Lincoln 7640, New Zealand

Roderick Robinson Landward Consultancy, Shinglebeck, Leavening, Malton, N.Yorks, YO17 9SG, UK

Joanne M. Sharpe Sharplex Services, PO Box 499, Edgecomb, Maine 04556, USA

Elizabeth Sheffield Faculty of Life Sciences, University of Manchester, G30B Stopford Building, Oxford Road, Manchester, M13 9PT, UK

Alan R. Smith University Herbarium, University of California, Berkeley, California 94720–2465, USA

Lawrence R. Walker School of Life Sciences, University of Nevada, Las Vegas, 4505 Maryland Parkway, Las Vegas, Nevada 89154–4004, USA

Preface

There are many ways to appreciate ferns and lycophytes. We admire their shapes, from tiny, filmy ferns on tree trunks, to lacy maidenhair ferns in rock crevices, to sturdy tree ferns with their huge, dissected leaves. We wonder at the beauty of their leaves that vary from subtle shades of green to gray, pale yellow, reddish or even iridescent blue. As we become more acquainted with ferns, the fascination deepens and the questions begin. Some of the first questions are about fern habitats. How do they survive in the deep shade of forest understories? On flooded banks of streams? On wind-swept mountain tops? On hot, vertical rock faces? Why are they so abundant on tropical mountains and oceanic islands? Questions then arise about fern growth. How do they survive drought or freezing temperatures? How do they reproduce? How old are they? What is a spore and that always elusive little "gametophyte"? We also ponder how the presence of ferns impacts other organisms. Do ferns compete with seed plants? Do they get eaten by herbivores? Finally, how do humans interact with ferns? Which species are edible or have medicinal qualities? Why are some ferns a nuisance to us? What makes them weedy ferns? We, the editors of this book, have each pondered these questions, both as people who are fascinated by the beauty and variety of ferns and as scientists whose job it is to question how the natural world is assembled, collate information about it and synthesize what is known.

Most fern research in the past has focused on fern morphology, taxonomy and phylogenetics and the practical applications of ferns for horticulture. Published studies of fern ecology have increased rapidly in recent years, but have not been previously collated. This book was produced in order to provide a synthesis of fern ecology and to answer some of that long list of questions – answers that we have found from the literature, convinced colleagues to address or discovered in our own research. Of course, we end with a new list of questions to help direct future studies of fern ecology!

Each of us has approached the world of ferns from different perspectives. Klaus was fascinated by the ferns on Costa Rican mountain slopes, and wondered for how long their leaves grew under these extreme climatic conditions. He marveled about the enormous fern diversity of the cloud forests and was concerned to see how this habitat was rapidly being lost. These concerns led to studies of the growth and seasonality of ferns, their adaptations to various tropical environments, interactions with herbivores and their conservation. Lawrence was first dazzled by tree ferns in a greenhouse in chilly Vermont (USA) but only began working on tree ferns and thicket-forming ferns when they were critical species in the recovery process (succession) on severely disturbed tropical landscapes (following volcanoes and landslides). His interest is thus in the ecological role that ferns have in changing plant communities. Joanne was amazed at the diversity of fern species and habitats in tropical Trinidad and learned the names of 100 species while on vacation from a management consulting job in upstate New York (USA). Frustrated by the lack of ecological and demographic information about these species in the local university library, she changed careers. Despite these different perspectives about ferns, we all eagerly undertook this task to describe and synthesize what is known about fern ecology. Joanne and Lawrence decided, over lunch in fern-rich Puerto Rico, that such a book needed to happen. Approaching Klaus with our idea we were delighted to hear that Cambridge University Press had recently asked Klaus to consider just such a project. The editorial team was established.

We are indebted to the early pioneers of fern ecology, including Hermann Christ, Michael Cousens, Adrian Dyer, Richard Holttum, Clive Jermy, Karl Kramer, Chris Page, Rolla and Alice Tryon, Herb and Florence Wagner, and Alexander Watt for seeing ferns as not just taxonomic oddities, but plants with a dynamic relationship to their environment. The recent book *Biology and Evolution of Ferns and Lycophytes*, edited by Tom Ranker and Christopher Haufler, provided us with an excellent and up-to-date reference work and the opportunity to complement their opus with our own, more ecological perspective. Alan Smith graciously provided his most recent checklist of genera of the world (see Appendix B). We gratefully acknowledge the assistance of external reviewers who provided us with invaluable feedback, shared their enthusiasm and expertise, and encouraged us to integrate details we would have otherwise overlooked. These generous individuals include: Victor Amoroso, Nan Crystal Arens, Mike Barker, Piet Bremer, Tim Brodribb, Beverley Clarkson, Bruce Clarkson, David Coomes, Adrian Dyer, Paul Hanson, Ruth Kirkpatrick, Rosario Medel, Sue Olsen, Elizabeth Powell, Tom Ranker, Gillian L. Rapson, Ann Russell, Harald Schneider, Alan R. Smith, Tamara Ticktin, James Volin, Margery Walker, James Watkins Jr., Paulo Windisch, and George Yatskievych. In addition, most chapter authors contributed reviews of one or more chapters. Arturo Piña amended images with graphical software. Jacqueline Garget at Cambridge University Press encouraged us to undertake this book project, Charlotte Broom,

Denise Cheuk, Chris Hudson, Rachel Eley and Dominic Lewis guided us through the publication process.

Klaus thanks his early teachers of botany and plant ecology Wilfried Bennert, Hermann Schraudolf, Focko Weberling, and especially Sieghard Winkler for directing him toward ferns; his parents and Ludwig Müller for encouraging him to follow his convictions even when they required distant travels; his students for asking all those supposedly easy questions about ferns that he tries to answer in this book; his wife, Sandra, for her unconditional love and support; and his son Lukas for holding out patiently when he would prefer to play with Dad. Klaus is grateful for the rewarding and inspirational collaboration and friendship of his coeditors. Klaus was supported by the Instituto de Ecología, A. C., Xalapa (902–11–796), CONACYT-SEMARNAT (2002-C01–0194) and CONACYT-SEP (2003-C02–43082), Mexico.

Lawrence humbly thanks his coeditors for reminding him how much botany he has forgotten and reawakening his love of detail and wonder about plant form and function in general and that of ferns in particular. This cooperative effort has helped him see outside his plant succession and disturbance ecology boxes. Lawrence was partially supported by the Luquillo Long-Term Ecological Research Program in Puerto Rico funded by the U.S. National Science Foundation.

Joanne thanks John Mickel who encouraged her to study ferns; Judy Jernstedt who was willing to guide a graduate student interested in ferns; her colleagues at the Luquillo Ecological Research site and the Holt Forest in Maine for their willingness to include ferns in their research programs; her coeditors for their unflagging enthusiasm and patience; and most of all her husband, Henry Minot, who has patiently encouraged and supported her quest. She dedicates this work to the memory of Joe Copeland who kindly identified her first ferns in Trinidad; Herb Wagner who asked and answered fern questions in Costa Rica; and Bruce Haines who was an inspiration throughout her graduate and professional career. Joanne was partially supported by the Luquillo Long-Term Ecological Research Program in Puerto Rico funded by the U.S. National Science Foundation.

Photo credits

The following people are thanked for providing photographs. Other line drawings or half-tones are from the corresponding chapter authors or are accompanied by the acknowledgement, reference and copyright permission within the figure legend.

Scott Bauer, USDA Agricultural Research Service, Bugwood.org: Plate 7B.

Chris Evans, River to River, CWMA, Bugwood.org: Plate 8C.

Peggy Greb, USDA Agricultural Research Service, Bugwood.org: Plate 8D.

Victoria Hernández: Fig. 1.6.

Victoria Hernández and Tiburcio Láez: Fig. 1.3a–f.

Peter Hietz: Fig. 5.1a–b, d–e.

Kensington Aviation: Fig. 8.12.

Keysotyo, commons.wikimedia.org: Fig. 8.8.

May and Baker Ltd.: Fig. 8.3.

Klaus Mehltreter: Figs. 5.1c, 7.1, 7.2, 9.1a–b, 9.3a–b, Plates 1C, 2A–B, 3A–D, 4A–D, 5A–D, 6A–D, 7C–D, 8A–B, front cover, back cover.

Henry Minot: Fig. 3.9b.

Chris Morse: Fig. 4.3.

Sarah J. Richardson: Figs. 4.1, 4.6.

Roderick Robinson: Figs. 8.2, 8.4, 8.5.

Peter Room, University of Queensland, Bugwood.org: Fig. 8.10a–b.

Joanne M. Sharpe: Figs. 3.1b–c, 3.3a–d, 3.4, 3.5, 3.9a, 6.5, Plate 7A.

Forest and Kim Starr: Fig. 8.6.

Lawrence R. Walker: Figs. 4.2b, 6.1, 6.2, 6.7a–f, 6.8, 6.9a–b, 6.10.

Janet Wilmshorst: Fig. 4.2a.

1

Ecological importance of ferns

JOANNE M. SHARPE, KLAUS MEHLTRETER
AND LAWRENCE R. WALKER

1.1 Introduction

Ferns immediately capture the imagination of all who are fortunate enough to notice them. With their large, highly dissected and shiny green leaves, ferns are so visually appealing that many are sold as ornamentals. Most moist woodlands will have a number of fern species blanketing the understory with their pungent foliage. In tropical woodlands, ferns are often at eye level or above, providing an aesthetic and delicate subcanopy. Even in arid lands or on newly exposed surfaces such as burns, clear-cuts or landslides, ferns can be present and sometimes dominant, catching your full attention as you push through fern thickets or get snagged by their spines. Beyond their immediate visual appeal, ferns are curious objects. How do plants of such ancient origin persist in the modern world? How can something so fragile survive trampling, burning, logging or grazing? Ferns and lycophytes were long considered as mystical plants, because people did not understand how they could reproduce without ever producing a flower, a fruit or a seed (Moran, 2004). In this book, we address the mystique that surrounds ferns by exploring fern ecology, or how ferns relate to their environment. Throughout the world, whenever ferns are the focus of ecological research, important and often surprising findings emerge.

We present four approaches to fern ecology. First, we provide a conceptual synthesis of the rapidly expanding field of fern ecology in order to establish a framework for future research and to encourage interdisciplinary approaches to studies of ferns. For example, modern molecular and genetic tools are used to probe the linkages of extant ferns to their fossil progenitors while evolutionary ecologists explore how ferns have coexisted with seed plants for so long. Second, we highlight key aspects of the rapidly expanding literature on fern ecology in order to provide a solid background for both researchers and teachers who inspire the next generation. We address the ecological underpinnings of such questions as how ferns reproduce, grow and successfully invade such a diversity of habitats. Third, we hope to broaden the nascent appreciation

Fern Ecology, ed. Klaus Mehltreter, Lawrence R. Walker and Joanne M. Sharpe. Published by Cambridge University Press. © Cambridge University Press 2010.

of the ecological importance of ferns among natural resource professionals such as ecologists, conservationists and land-use managers. The remarkable adaptations of ferns to various disturbances, including their abilities to accumulate toxins in their environment, suggest an important role for ferns in conservation and restoration. Fourth, we provide an in-depth focus on ecological processes with examples and details to further educate and intrigue anyone who is already fascinated by ferns.

Humans use ferns for food, medicine, agriculture and horticulture. For example, the unfurled croziers (immature leaves) of *Matteuccia struthiopteris* (ostrich fern) are eaten in the USA. In medicine, the lycophyte *Huperzia serrata* (Chinese club-moss) provides an alkaloid that has been important for the control of epilepsy. Agricultural uses of ferns include the aquatic nitrogen-fixing ferns in the genus *Azolla* (mosquito fern) that are used in parts of Asia as green manure for rice fields and in India where they are fed to cows to increase milk production. Leaves of the South African fern *Rumohra adianti-formis* (leatherleaf fern) are grown commercially for their ornamental value for flower arrangements. Horticultural research has resulted in hundreds of fern species being grown as ornamentals and successfully cultivated in greenhouses and fields throughout the world (Jones, 1987; Rickard, 2000; Hoshizaki and Moran, 2001; Mickel, 2003; Olsen, 2007) but less is known about how ferns interact with their environment. Recent human activities have also enhanced the long distance dispersal capabilities of ferns that are weedy colonizers. Introduced fern species have displaced native plants in some parts of the world. Successful control measures for invading species, as well as the conservation of rare ferns, have an important role in many natural ecosystem management strategies and require a thorough understanding of fern ecology.

Novel insights can be gained about whole ecosystem interactions and species links to abiotic components by focusing on the spore-producing ferns and lyco-phytes as a group (see Section 1.4.3) that has a unique reproductive system but an ecology that is both similar and distinct from that of the better known plants for which reproduction depends on seeds. In this introductory chapter, we highlight some of the trends in the study of fern ecology and provide a brief overview of the topics included in the remaining chapters of this book. Because ferns (Plates 4–6, 8) and lycophytes (Plate 3) may be groups of plants that are unfamiliar to some of our readers, we then discuss (1) important differences between the elements of the fern life cycle and those of the more commonly studied seed plants, (2) the history of ferns on a geologic timescale and (3) recent changes in phylogenetics and systema-tics that have redefined the classification of ferns and lycophytes.

1.2 Advances in the study of fern ecology

Early ecological studies rarely included ferns, instead concentrating on seed plant ecology. However, a survey of the recent fern literature (International Association of

Pteridologists, 1994–2007) shows a threefold increase in the number of publications devoted to fern ecology over the 14-year period. The study of fern ecology began with taxonomy. Early collectors in the tropics sent fern specimens with field notes to taxonomists at temperate herbaria for examination and comparison. Using such annotated herbarium collections in Switzerland, Christ (1910) wrote the first bio-geographical treatment of ferns, even though he had never traveled further than to the Canary Islands. It became clear that there were many more species of ferns and lycophytes in tropical than in northern temperate climates (Plates 1, 2). Today, most fern species are found in the humid tropics where their ancestors originated, although some groups have adapted to cooler climates of tropical mountains or temperate latitudes. Furthermore, the tropics provided a wider range of habitats in which ferns developed a variety of growth forms such as tree ferns, water ferns, epiphytes, hemiepiphytes, and climbers (Plates 6, 8). Notable centers of fern diversity include mountainous tropical islands.

Holttum, a tropical fern taxonomist who went to Malaya in 1922 and directed the Singapore Botanical Gardens for several years, described fern habitats, their light conditions and associated seed plants based on his own observations (Holttum, 1938). These descriptions were a major contribution to tropical fern ecology that is still recognized today (Price, 1996). Following in this tradition, Tryon and Tryon (1982) integrated extensive ecological field notes into their treatment of the genera of ferns and lycophytes in the neotropics. The focus of fern ecology shifted from simple habitat descriptions to more detailed observations of plant growth and population dynamics when Watt (1940) published the first of his many ecological studies of *Pteridium aquilinum* (bracken) in Britain (see Watt, 1976 for a review). *Pteridium* is an extremely successful colonizer worldwide and can form dense, nearly impenetrable stands. *Pteridium* generally contains several biochemical compounds such as high levels of carcinogens that make it a health hazard when eaten by livestock or humans although the early Maori of New Zealand used *Pteridium esculentum* rhizomes as a crop. Where it is unwanted (i.e., considered an invasive weed), *Pteridium* is difficult to eradicate. *Pteridium* therefore continues to be an important subject for ecological research (Robinson, 2007). Studies of *Pteridium* and other abundant ferns have helped to develop an understanding of the complex relationships between ferns and their environments.

An early assessment of a forest understory in a northern New England watershed emphasized the potential ecological significance of ferns. Siccama *et al.* (1970) reported that ferns comprised 37% of annual understory biomass production (shrubs and herbaceous plants) and that species of *Dryopteris* (wood fern) contributed 70% of the biomass of the herbaceous layer. One difference between ferns and seed plants was highlighted in the 1960s during an experiment in which a section of a Puerto Rican rain forest was irradiated with cesium (Odum and Pigeon, 1970). This study was sponsored by the United States Atomic Energy Commission in order to understand the consequences of nuclear warfare or major reactor accidents on rain forests. Many plant species were negatively affected by the radiation, but ferns and lycophytes were unexpectedly resistant (Sorsa, 1970). More recently, ecologists have turned their attention to the high fern diversity of the tropics.

The temporal and spatial dimensions of fern growth have attracted interest recently. For example, there were unexpectedly strong seasonal patterns detected in mangrove ferns (Mehltreter and Palacios-Rios, 2003), climbing ferns (Mehltreter, 2006) and tree ferns in Mexico (Mehltreter and García-Franco, 2008). Also unexpected was the discovery of year-to-year variation in spore production in a rheophytic (i.e., flood-adapted) fern in the rivers of Puerto Rico (Sharpe, 1997). The spatial dynamics of tropical ferns have been examined in the Amazonian lowlands of Peru and Ecuador (Tuomisto *et al.*, 1995; Tuomisto and Ruokolainen, 2005). These studies have led to a better understanding of the intricate pattern of edaphic variability in a rain forest that was thought to be quite homogeneous because of its more uniform tree canopy composition. Portugal Loayza (2005) detected very distinctive fern abundance patterns in a 16-ha grid at a long-term research area in a Puerto Rican rain forest and linked specific fern species to areas of the forest where coffee had been grown more than 70 years earlier. These studies have shown that unexpected spatial and temporal patterns can be explained by evaluating fern growth and distribution.

It has not escaped the attention of some ecophysiologists that ferns can be found in habitats featuring extremes of temperature, light or humidity. Bannister and Wildish (1982) found that native fern species in New Zealand were less likely to develop frost resistance than the non-native male wood fern *Dryopteris filix-mas*. They also showed that the shade tolerance in forest ferns in New Zealand is greater than had been reported for forest floor seed plants. With a positive gravitropic response facilitated by the low light environment of the rain forest floor, the fertile leaf of *Danaea wendlandii* stays flat on the ground during early development (Sharpe and Jernstedt, 1990). Growth of the leaf then dramatically reverses to the vertical direction during the very last stages of development, allowing spore dispersal from above the low layer of sterile leaves that had shaded it. Early work with fern water relations of a Colorado desert fern, *Notholaena parryi* by Nobel (1978) demonstrated that, in contrast to many other temperate ferns, only minimal growth occurs during the summer months of June, July and August when soil water potential is lowest and temperature is highest. More recently, Milius (2007) has discussed high tolerance for low humidity among the tiny green gametophytes of epiphytic ferns of the rain forest. There are, no doubt, many other unusual aspects of fern ecophysiology to be discovered.

Fern interactions with other species and their impacts on community dynamics are also beginning to be explored. On tropical landslides, Walker (1994) found that dense thickets of scrambling ferns such as *Gleichenella pectinata* (syn. *Dicranopteris pectinata*) inhibited tree seedling growth early in succession but were likely to facilitate later stages of succession by stabilizing slopes and improving soil conditions. In a study of a temperate hardwood forest in the northeastern USA, George and Bazzaz (2003) discovered that a dense fern layer dominated by *Dennstaedtia punctilobula* (hay-scented fern) limited the growth of some species of tree seedlings but not others, thus controlling the ultimate structure of the upper canopy.

Human alteration of natural habitat, often combined with the appearance of fern species introduced from other parts of the world, is presenting new challenges to ecologists and land managers. *Lygodium microphyllum* (small-leaf climbing fern) from the Old World was introduced into gardens in the southeastern USA in the 1950s, escaped, and has now become a serious pest in vast areas where water regimes modified earlier by human disturbance are being restored (Hutchinson and Langeland, 2006). Researchers have had to conduct ecological studies of *L. microphyllum* (Plate 8D) in both its native and invaded habitats in order to understand how to control its spread. Tu and Ma (2005) recently reported that the hyperaccumulating *Pteris vittata* (Chinese brake fern, Plate 8B) can remove quantities of otherwise inaccessible arsenic from mined wastelands. While this plant process could be of great benefit, understanding the ecology of *P. vittata* before remediation is now essential because *P. vittata* has been identified as an invasive fern (Palmer, 2003).

The field of fern ecology has advanced from simple observation of fern habitat characteristics to long-term studies of their complex roles in nutrient cycling and successional dynamics of natural ecosystems. Recent human activities have required that the ecological role of ferns be better understood as ferns become increasingly important in the horticultural trade and as invaders of disturbed habitats. Issues regarding conservation of rare ferns and management of invasive ferns require a well-grounded understanding of fern dispersal, colonization and growth. Throughout the world, whenever ferns are the focus of ecological research, surprising and important findings emerge. While pioneering fern ecologists (e.g., Wagner, 1973; Page, 1979b) have highlighted the need for research in many areas of fern and lycophyte ecology, it is only in the last 20 years that these calls have been answered by ecologists. A productive trend in fern and lycophyte ecology is clearly emerging and we encourage readers with a wide variety of perspectives to contribute to the growth of this exciting field of research.

1.3 Fern ecology topics in this book

In order to encourage the inclusion of ferns in future ecological research, this book provides a comprehensive and updated review of fern ecology. This review integrates a broad range of topics and highlights questions that present new research opportunities and interdisciplinary approaches. Selected topics in fern ecology have been addressed in edited volumes about ferns (Verdoorn, 1938; Dyer, 1979; Dyer and Page, 1985; Kramer *et al.*, 1995, Camus *et al.*, 1996; Chandra and Srivastava, 2003; Ranker and Haufler, 2008) but the last comprehensive reviews on fern ecology were published nearly 30 years ago (Page, 1979a, 1979b). Since that time, there have been considerable advances and remarkable changes in the emphasis of ecological studies of ferns from the individual species and its habitat to studies at the population, community and ecosystem levels. In addition to the pteridologists who have traditionally conducted fern ecology studies in the past, general ecologists with interests in a wider variety of topics such as nutrient cycling, edaphic influences, spatial distribution and succession have begun to discover the importance of ferns. The scope of ecological studies of ferns is also changing, widening to include large-scale spatial and temporal dynamics as well as social and economic influences. While research on the individual species provides the fundamentals, it is within the larger context of global ecology that the role of ferns may prove vital to our future, given their long and successful history of adaptation to environmental change.

Michael Kessler (see Chapter 2) presents a world map and an analysis of latitudinal and altitudinal gradients of fern distribution to explain the observed patterns and the reasons for their success throughout the world (Plate 1A, B). Joanne Sharpe and Klaus Mehltreter (see Chapter 3) discuss how elements of the life cycle and demography of individuals and populations respond to environmental triggers and document a wide array of phenological patterns in temperate and tropical ferns. Sarah Richardson and Lawrence Walker (see Chapter 4) describe fern responses to and impacts on ecosystem nutrient dynamics, while Peter Hietz (see Chapter 5) highlights the surprising success of ferns in xeric environments. Lawrence Walker and Joanne Sharpe (see Chapter 6) examine colonization and the negative and positive influences of ferns on successional processes that follow disturbance at several spatial and temporal scales. Klaus Mehltreter (see Chapter 7) highlights some of the strategies that ferns have developed to interact successfully with mutualistic and antagonistic fungi as well as animals.

Humans also have an impact on the role of ferns in many ecosystems. Roderick Robinson, Elizabeth Sheffield and Joanne Sharpe (see Chapter 8) present some of the most widespread but aggressive ferns that have a negative influence for humans as they invade natural ecosystems and impact agriculture. Fern and lycophyte lineages have survived millions of years by adapting successfully to a multitude of natural, sometimes even catastrophic changes, but the challenges are even greater today because of human alterations of the environment. As Klaus Mehltreter explains (see Chapter 9), a well-integrated approach to fern conservation requries a more complete understanding of their ecology. Finally, Lawrence Walker, Klaus

Mehltreter and Joanne Sharpe (see Chapter 10) summarize the ecological role of ferns, identify central themes in current research and suggest avenues for future research. We hope that advances in fern ecology will not only increase our knowledge of ferns as model organisms to investigate general and applied ecological questions but also ensure the future of the diversity of ferns and lycophytes as important components of our world.

1.4 Fern structure, life cycles, evolution and classification

A general grounding in the basics of fern structure can be helpful in reading this book and can be found in most elementary botany books. Lellinger (2002) has written a multilingual (English, French, Portuguese and Spanish) glossary that includes many of the terms specific to ferns and this book includes a short glossary of terms relating not only to ferns, but to ecological topics as well. We will discuss elements of the fern life cycle and then compare the alternation of generations in ferns with that found in seed plants. The ancient history of extant fern lineages provides a perspective on several aspects of ecology including, for example, alliances between herbivores and their host plants. Recent changes in the understanding of the phylogeny of ferns and lycophytes (Fig. 1.1) have resulted in a modified classification requiring several nomenclatural changes at the generic level. A current list of fern and lycophyte taxa, including brief descriptions of their key characteristics (Appendix A) may prove helpful for their distinction. Alan Smith has provided an updated alphabetical list of genera with synonyms (Appendix B) that can be invaluable in following the most recent taxonomical and nomenclatural changes.

1.4.1 Life cycle and alternation of generations

Background information on ferns and lycophytes compared with the much more diverse and numerous seed plant groups, particularly the angiosperms (i.e., flowering plants), is essential to a complete understanding of the ecological importance of ferns. Functionally, ferns are no different from other green plants that capture the energy needed for photosynthesis. But differences in structure, life cycle and dispersal can result in unique roles for ferns.

Fern sporophytes are common and very distinctive plants in the vegetation of many parts of the world while a gametophyte is quite inconspicuous. The general sexual life cycle of ferns (Fig. 1.2) is characterized by the alternation of two generations consisting of (1) a prominent sporophyte plant and (2) a much smaller but independent plant, the gametophyte. The sexual life cycle requires (1) that the sporophyte leaf produces asexual spores (Fig. 1.3, Plate 5B) that can germinate into

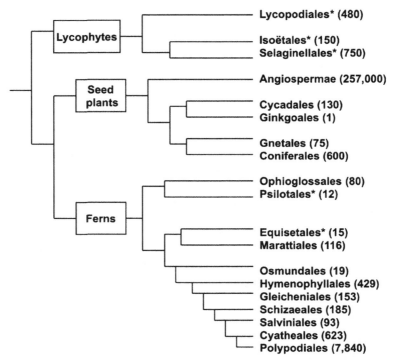

Fig. 1.1 Diagram of the vascular plant lineages (Schuettpelz and Pryer, 2008), including species counts, of the orders of lycophytes and ferns (Appendix A; Smith *et al.*, 2008) as well as seed plants (Judd *et al.*, 2008). Note that all flowering plant orders are included in Angiospermae, whereas the four gymnosperm orders are shown. Plant groups formerly called "fern allies" are denoted by an asterisk (*).

a gametophyte and (2) that the gametophyte reproduces sexually (see Box 1.2 for details) to develop into a sporophyte. The differences between the fern and seed plant life cycles are discussed in detail in Box 1.1. Here we focus on a comparison of ferns and seed plants with respect to (1) the more conspicuous, usually leafy green sporophyte generation and (2) dispersal units (sperm cells and spores for ferns and pollen and seeds for seed plants). Characteristics of both of these elements of the life cycle have profound ecological consequences.

Sporophytes

There are several obvious morphological characteristics that differentiate fern sporophytes from seed plant sporophytes such as the angiosperms (Table 1.1). Ferns and lycophytes do not produce secondary woody tissue as do woody seed plants. Fern sporophytes do not have long-lived primary roots as do most seed plants. Instead, the root system of most ferns consists of thin adventitious secondary roots that are fibrous and surface foraging, although in the Marattiales and

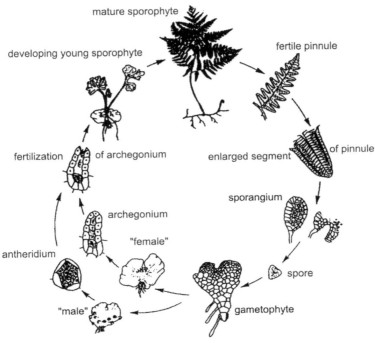

Fig. 1.2 General life cycle for homosporous ferns. Note the separate drawing of male and female gametophytes to imply outcrossing, which is more common than self-fertilization that is usually shown in life cycle diagrams for ferns. (Adapted with permission from Sheffield, 2008.)

Ophioglossales (Fig. 1.1) they can be thick and fleshy. Fern leaves exhibit three distinctive characteristics. First, with few exceptions (e.g., Ophioglossales, adder's tongues; Psilotales, whisk-ferns) emergent fern leaves unroll as they develop from a coiled, or occasionally hooked, leaf structure called a crozier or fiddlehead. The crozier exhibits circinate vernation in that the leaf is rolled from apex to base (Fig. 3.9a), a feature lacking in lycophytes and very rare in seed plants (e.g., the Cycadales and a few angiosperms, Fig. 1.1). Second, mature fern leaves can bear sporangia aggregated into sori that are often arrayed in various patterns, mostly on the lower leaf surface (Plate 5B) or the leaf margin or on specialized fertile, spore-bearing leaves (Plate 5C). In contrast, lycophytes bear one or more sporangia on the upper leaf surface. Third, free leaf venation (i.e., branched but not netted veins) for minor veins or partly netted (i.e., aerolate) venation occurs in many ferns (Plate 5B), whereas completely netted or parallel venation is characteristic of leaves of most seed plants. When fully expanded, mature leaves of ferns take on a variety of simple to highly dissected complex shapes (Plates 4–6). Fern leaves are mostly 10–200 cm in length, but can range from <1 cm to >10 m long. Croziers and leaves (also called

Box 1.1 Alternation of generations: ferns and seed plants

The reproductive life cycles of ferns and seed plants are similar, because in both groups the most common pattern is for a large usually diploid (2n) sporophyte generation to produce usually haploid (1n) spores. These spores develop into small haploid gametophytes (in ferns, sometimes called a prothallus) that produce gametes which are joined by fertilization to produce a diploid (2n) sporophyte. We describe the details of a general fern life cycle of using the more common homosporous ferns as an example (Fig. 1.4a) and its important differences from the seed plant life cycle using angiosperms as an example (Fig. 1.4b). We start with the diploid sporophyte. Sporophytes of both ferns and angiosperms may reproduce asexually (off the right side of the diagrams, Fig. 1.4a, b) by vegetative growth processes such as budding, rhizome division or by producing runners. During the sexual life cycle (bold circle of diagrams, Fig. 1.4a, b) the number of sets of chromosomes changes twice from one generation to another, first through meiosis and then through fertilization. In homosporous ferns meiosis occurs in sporangia on the fertile leaves of a diploid fern sporophyte and results in haploid spores. Meiosis in angiosperms occurs in flowers where haploid male microspores are produced in the anthers and haploid female megaspores are produced in the ovary, making them heterosporous. Fern spores are released from the sporangia, usually dispersed by wind, and then develop into independent haploid gametophytic plants (see Box 1.2), while the alternate generation in seed plants is completely dependent on the sporophyte. In some fern species, the fern gametophytes are capable of vegetative propagation by producing gemmae that are multicellular propagules which break off, disperse and grow into mature gametophytes (left portion of the diagram, Fig. 1.4a).

In our angiosperm example (Fig. 1.4b) the microspore develops into a male haploid gametophyte (pollen grain) that is released and dispersed (by wind, water or animals) to the stigma of a flower while the female megaspore remains attached to the mother plant where it develops in the ovary of the flower into a haploid female gametophyte (embryo sac). Fern gametophytes mostly produce either motile sperm or immobile egg cells although some produce both types of gametes at the same time. Fertilization occurs on the gametophyte (in its habitat, e.g., the forest floor, rock crevice or tree bark) and the diploid zygote then develops directly into a new diploid sporophyte plant. In contrast, the angiosperm gametophyte is located within the flower where the egg cell is fertilized by a sperm nucleus that is delivered by a pollen grain that is dispersed to a stigma, germinates and forms a pollen tube that grows down the style to the ovule. After fertilization, the diploid zygote develops into a diploid embryo that together with the nutritive tissue (endosperm) and a protective wall (testa) constitute a seed. The mature seed is released from the parent seed plant and after dispersal (by wind, water or animals) to suitable habitat, germinates into a new diploid sporophyte plant (Fig. 1.4b). Some ferns may have an asexual apogamic cycle (inner central life cycle, Fig. 1.4a) in which all stages (sporophytes, spores and gametophytes) have the same set of chromosomes (usually two, but sometimes three or more). In the apogamic life cycle, the sporophyte is produced from vegetative cells of the gametophyte and the gametophyte germinates from a diplospore (i.e., spores that develop without undergoing meiosis; Fig. 1.3e).

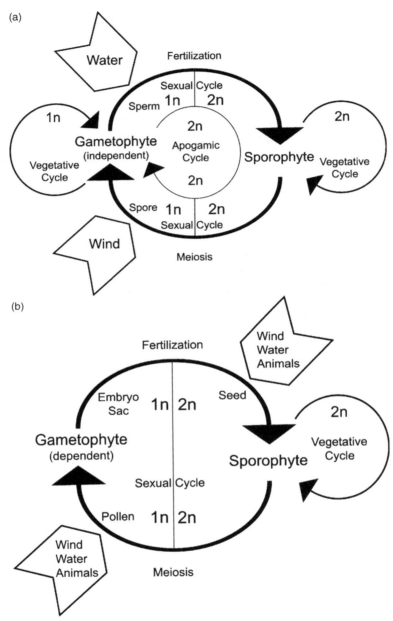

Fig. 1.4 Diagram of alternation of generations for (a) homosporous ferns and (b) angiosperms. The haplophase and diplophase are represented by 1n and 2n, respectively. See Box 1.1 for detailed explanation.

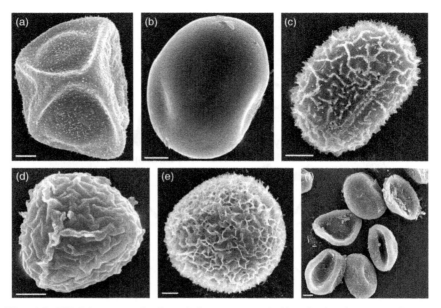

Fig. 1.3 SEM images of selected fern spores. (a–b) *Elaphoglossum peltatum*, monolete spore (a) with and (b) without the outer spore wall layer (i.e., perispore), (c) *Woodsia mexicana*, monolete spore with one linear aperture through which spore germination occurs, (d) *Pellaea cordifolia*, trilete spore with a three-part aperture. (e) *Cheilanthes* sp., possible triploid apogamic species with globose spores, (f) *Blechnum* sp., sterile hybrid with aborted, deformed spores. Scales 10 μm.

fronds or megaphylls) with relatively large blades are diagnostic for the ferns but the small leaves (i.e., microphylls) of most lycophytes are quite different (Plate 3). The lycophyte leaf has a single vein, no rachis, does not uncoil as it matures and those that are fertile are often grouped in a strobilus (i.e., cone, Plate 3A). Most fern leaves are photosynthetic, but they also perform the critical function of producing spores.

Spores

Fern spores are light in weight (<0.01 mg; Westoby *et al.*, 1990). Spores are about the same size as pollen grains of seed plants, but both are smaller than orchid seeds, the smallest seeds of any seed plant group. Spore colors vary from green to white, cream, yellow, orange, gray, and brown to black. Green spores are found in several fern families (e.g., Onocleaceae, sensitive ferns; Hymenophyllaceae, filmy ferns) and lose their viability within 2–6 weeks because their chlorophyll pigment is quickly catabolized. Spores of other colors have a mean viability of several months to years. Fern spores have two basic shapes and an often ornate surface structure that can take on an almost infinite variety of forms (Fig. 1.3; Tryon and Lugardon, 1990).

Box 1.2 Gametophytes

The gametophyte is the sexual generation in the life cycle of ferns and lycophytes and is photosynthetic, mycotrophic (i.e., obtains nutrients through a fungal relationship) or endosporic (lives on its stored nutrients and nearly entirely enclosed by the spore wall). Although there are several types of nonphotosynthetic subterranean gametophytes, most are green, flat, one to several cell layers thick and develop aboveground. These photosynthetic gametophytes are generally anchored to the substrate by rhizoids and exhibit a variety of growth forms. Many green gametophytes are short lived and cordiform or "heart shaped" (Figs. 1.2, 1.5 I, 1.6). In contrast, gametophytes of some species (mostly epiphytes) are perennials that grow slowly and can branch repeatedly (Fig. 1.5 II–V, Farrar *et al.*, 2008).

Once the gametophyte is established, male and female gametangia are produced on the side of the gametophyte that is not exposed to light. An antheridium (Fig. 1.2) produces motile sperm cells and an archegonium (Fig. 1.2) produces one egg cell. Gametophytes may be male or female, or may produce both types of gametangia. For fertilization, the sperm cell must swim through water to an egg cell (Fig. 1.2). Most fern species cross-fertilize (i.e., sperm fertilizes an egg cell from a different gametophyte), but the gametophytes are potentially bisexual. If the gametophyte has simultaneously functioning archegonia and antheridia it may self-fertilize (i.e., sperm fertilizes an egg cell from the same gametophyte), which is of advantage after long distance dispersal.

Many of the perennial gametophytes produce gemmae (Fig. 1.5 IV, V), vegetative buds that break off and disperse. The gemmae allow these gametophytes to form large populations in habitats that may be free of sporophytes of their species. Longevity and the ability to propagate vegetatively are of advantage for long-distant colonization, because they increase the chances that, over time, one gametophyte will germinate near another gametophyte so that cross-fertilization may occur (for a review on gametophyte ecology, see Farrar *et al.*, 2008).

Species of the Ophioglossales, Psilotales and some Lycopodiales (see Fig. 1.1) have gametophytes that develop and mature belowground. Without photosynthesis, these gametophytes are nutritionally dependent on an association with fungi that feed on decaying organic matter. These gametophytes usually require several years to mature and, since they do not orient toward the light, many can produce gametangia on all surfaces. Although aboveground gametophytes can produce only one sporophyte, multiple fertilization events can occur on subterranean gametophytes resulting in the production of several young sporophytes.

Monolete spores are bean shaped (Fig. 1.3a–c) and are more commonly found in the most recently evolved order of ferns (Fig. 1.1, e.g., Polypodiales). Trilete spores are tetrahedral (Fig. 1.3d, e), and are commonly found in early fossil records and in some of the older extant fern families (e.g., Cyatheaceae, tree ferns). Spores of both

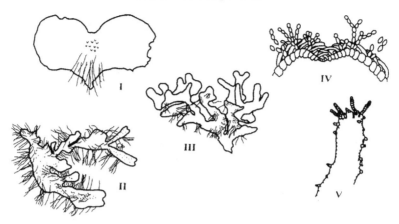

Fig. 1.5 Types of photosynthetic gametophytes of the fern class Polypodiopsida (Appendix A). I, cordiform; II, strap; III, ribbon; IV, gemmiferous strap (detail of budding along the edge of a branched proliferation of the gametophyte); V, gemmiferous ribbon (detail of budding along the edge of a branch of the gametophyte). Cordiform gametophytes are short lived, while types II–V are long-lived perennials. (Adapted with permission from Farrar *et al.*, 2008.)

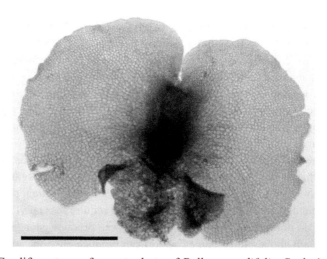

Fig. 1.6 Cordiform type of gametophyte of *Pellaea cordifolia*. Scale 1 mm.

shapes can be found in some families (e.g., Gleicheniaceae, scrambling ferns) and spores of hybrids are often deformed (Fig. 1.3f).

The ecological consequences of spore size, form and surface structure (Fig. 1.3) are unknown. Most species of ferns and lycophytes are homosporous, producing only a single size of spore; spore size ranges in diameter from 15 to 150 μm but most are 30–50 μm in diameter (Tryon and Lugardon, 1990). However, the lycophyte

Table 1.1 *Comparison of most common characteristics of the biology of spore-bearing vascular plants (i.e., ferns and lycophytes) and flowering seed plants (i.e., angiosperms). Gymnosperms have an intermediate position between ferns and angiosperms. Note that there are uncommon exceptions to most of these categories.*

Characteristic	Ferns and lycophytes	Angiosperms
Alternation of generations	Two independent generations: gametophyte and sporophyte	Two generations: gametophyte embedded in sporophyte
Gametophyte nutrition	Photosynthetic or mycotrophic	"Parasitic" on sporophyte
Gametophyte development	Exosporic (outside the spore wall), endosporic heterosporous groups	Endosporic (within the pollen grain or embryo sac)
Fertilization	On gametophyte, sperm cell swims through water	Within ovule, sperm nuclei transported within the pollen tube, no water required
	Simple fertilization	Double fertilization
Sporophyte	Homosporous (36 families) or heterosporous (4 families)	Heterosporous
Xylem	Mostly tracheids, some vessels	Mostly vessels, some tracheids
Secondary growth	Only in *Isoëtes* and *Botrychium*	Present in woody species
Leaves	Megaphylls (ferns) with usually circinate vernation or microphylls (lycophytes)	Megaphylls, although sometimes of small size
Dispersal	Haploid spores Mostly by wind; water	Diploid seeds Mostly by animals, wind or water

families Selaginellaceae (spike-mosses) and Isoëtaceae (quillworts) and the fern families Marsileaceae (water-clover) and Salviniaceae (floating ferns) are heterosporous, producing both large female megaspores ranging in size from 200 to 1000 μm and smaller male microspores measuring 10 to 60 μm, which are often even smaller than spores of homosporous ferns (Tryon and Lugardon, 1990).

Dispersal limitations in ferns and seed plants

The very small, usually wind-dispersed spores (Figs. 1.3, 1.4a), the potential bisexuality of most fern gametophytes with possible self-fertilization and the vegetative reproduction of some fern gametophytes all allow ferns to reproduce successfully from a single spore and to establish on distant oceanic islands. Once germinated, the gametophyte produces male and female structures (i.e., antheridia and archegonia) that form male and female gametes (i.e., sperm and egg). A motile sperm must swim to and enter into the archegonium (see Box 1.2) to fertilize the egg cell. The zygote is the product of fertilization and develops into the sporophyte. Thus, for ferns, the completion of the life cycle requires first, the wind dispersal of the spore(s) and second, the movement of the sperm through water to the egg cell of the same or a different gametophyte (Fig. 1.2); these two journeys differ greatly in the distance traveled.

There are two stages of dispersal in the seed plant life cycle but pollen grain and seed dispersal can be facilitated by animals as well as by wind or water (Fig. 1.4b). First, a usually haploid pollen grain (i.e., immature male gametophyte) that is about the same size as a fern spore must travel to reach an ovule of the sporophyte of the same species, a very specific and relatively small target. Mutualistic relationships with animal pollinators often facilitate this difficult task. Many angiosperms with larger seeds that cannot be wind dispersed rely on animal seed dispersers, making the species dependent on these animals and even vulnerable if the pollinator becomes rare or extinct. Although ferns and seed plants have somewhat different dispersal challenges, many aspects of the ecology of their sporophytes are comparable once they are established within an ecosystem.

1.4.2 Fern evolution

The ferns and lycophytes (Fig. 1.1) originated in ancient tropical habitats and have been successful in many types of environments for thousands of millennia (Plates 1C, 2A, B). Ferns (and fern-like plants) first appeared in the fossil record during the Middle Devonian 390 mya (million years ago; see Appendix C for a geologic time-table) according to Rothwell and Stockey (2008). Some of these groups had representatives such as *Calamites* (related to the extant order Equisetales, horsetails) and lepidodendrids (related to the extant order Isoëtales), that grew as tall as 10–35 m and were dominant elements of the swamps of the Carboniferous (355–290 mya) that we now harvest in the form of coal and oil. Most early fern groups vanished during two large extinction events at the end of the Carboniferous (300 mya) and at the end of the Permian (240 mya), the exceptions being ancestors of today's orders (Fig. 1.1) of lycophytes (Isoëtales; Lycopodiales, club-mosses; and Sellaginellales, spike-mosses), and ferns (Ophioglossales, Psilotales, Equisetales and Marattiales). After these mass extinctions, the seed plant species (gymnosperms in the Mesozoic, 205–65 mya, and angiosperms in the Cenozoic, 65 mya to present) increased greatly in number and took on a dominant role in many ecosystems. Most extant fern and lycophyte lineages began to diversify at the same time as did seed plants (Schneider *et al.*, 2004). A new group called leptosporangiate ferns, characterized by a one cell thick sporangium wall (in contrast to the two or more cell thick sporangium walls of eusporangiate ferns), appeared at about the same time as the gymnosperms and increased dramatically in species richness concurrently with the angiosperms (Rothwell and Stockey, 2008). These modern ferns are included in the orders Osmundales, Hymenophyllales (filmy ferns), Gleicheniales, Schizaeales, Salviniales (heterosporous ferns), Cyatheales (tree ferns) and the Polypodiales which includes the majority of fern species (Fig. 1.1). The leptosporangiate ferns adapted very successfully to new environments in the forests dominated by seed plants, both as epiphytes on trunks and branches and as terrestrial plants under low-light conditions created by the forest canopy. In some environments (e.g., forest understories, oceanic islands, or disturbed surfaces), ferns are still important or even dominant components of the vegetation (Plates 1C, 2B), and some fern lineages, especially in the Pteridaceae (brake fern) family, have adapted to drier regions and now survive in semi-arid environments (Plate 2A).

1.4.3 Current systematic classification of ferns and lycophytes

Ferns and lycophytes have continued to evolve and diversify into about 11 000 extant species worldwide (Fig. 1.1). Recent changes in molecular phylogenies of lycophytes, ferns and seed plants have been providing new insights into the past history of the distribution and interrelationships among fern assemblages on Earth today. This work has changed our understanding of the natural relationships of ferns and lycophytes and resulted in some important changes to their classification. The most recent phylogenetic work recognizes 40 families and approximately 300 genera of ferns and lycophytes (Appendices A, B; Smith *et al.*, 2008). The variety of characteristics used to describe and delimit some of these taxa (Appendix A) only begins to suggest their potential for diversity in habitat specialization and ecological function. Systematists recently have reinterpreted the phylogeny of vascular plants in light of new molecular findings (Fig. 1.1; Schuettpelz and Pryer, 2008) and found an unexpectedly deep division and diverse affinities in the group formerly called "fern allies". Extant vascular plants are now divided into two major groups with the oldest lineage, the lycophytes (about 1400 species; Plate 3), separated from the other two vascular plant lineages, the ferns (Plates 4–6) and the seed plants (Fig. 1.1). The formerly used category of fern allies had included the current lycophytes (Lycopodiales, club-mosses; Selaginellales, spike-mosses) as well as Psilotales (whisk-ferns), and Equisetales (horsetails). The last two orders are now thought to be more closely related to the ferns than to modern lycophytes (Schuettpelz and Pryer, 2008) because of strong molecular evidence (e.g., nuclear and plastid DNA sequences), and because their morphological similarities to lycophytes, such as the extreme reduction of their leaves, appear to be a consequence of ecological habitat specialization rather than of shared phylogenetic origin.

Like ferns, lycophytes are vascular plants that disperse and reproduce by spores and form gametophytes that exist independently from the sporophyte. Because lycophytes encounter many of the same ecological challenges as the ferns, in this book on fern ecology we include both of these groups even though they are not currently considered to be monophyletic. We will usually refer to all seedless vascular plants as ferns in the broadest sense (ferns *sensu lato*), which includes ferns (*sensu stricto*) and lycophytes. Each general statement about ferns will therefore also refer to lycophytes but when necessary we will specifically identify the appropriate group.

1.5 Summary

Ferns and lycophytes represent only about 4% of all vascular plant species on Earth, yet they are widely distributed and inhabit nearly all types of tropical, subtropical, temperate and boreal ecosystems. Ferns occur in rain forests, high montane cloud forests, temperate forests, mangroves and even floating or submerged in lakes. Ferns are weedy colonizers of disturbed landscapes and are also found scattered among the rocks of semi-arid deserts, savannas, coasts and high alpine mountains where they resist drought, fire or cold temperatures. A number of herbivorous insects feed on ferns, and ferns have responded with a diverse array of defense strategies to protect their rhizomes, roots and leaves. Humans have moved ferns around the Earth, sometimes with unintended consequences that result in huge expenditures to eradicate them from their new habitats. Finally, humans realize that many ferns are becoming less common because of their activities, and only modern conservation approaches can reverse this trend. It is only by understanding the ecology of this diverse group of 11 000 species that humans can appreciate and manage the role of ferns and lycophytes.

Acknowledgements

Joanne Sharpe and Lawrence Walker each acknowledge support from the Luquillo Long-Term Ecological Research Program in Puerto Rico funded by the U.S. National Science Foundation. Klaus Mehltreter is grateful for support from the Instituto de Ecología, A. C., Xalapa, Veracruz, Mexico, and Victoria Hernández and Tiburcio Láez for providing microscopic photographs. We thank Beverley and Bruce Clarkson, Adrian Dyer, Elizabeth Powell, Gillian L. Rapson, Margery Walker and George Yatskievych for very helpful reviews of this manuscript.

References

Bannister, P. and Wildish, K. L. (1982). Light compensation points and specific leaf areas in some New Zealand ferns. *New Zealand Journal of Botany*, **20**, 421–4.
Camus, J. M., Gibby, M. and Johns, R. J. (eds.) (1996). *Pteridology in Perspective.* Whitstable, UK: Whitstable Litho. Ltd.
Chandra, S. and Srivastava, M. (eds.) (2003). *Pteridology in the New Millennium.* Dordrecht, Netherlands: Kluwer Academic Publishers.

Christ, H. (1910). *Die Geographie der Farne*. Leipzig, Germany: Gustav Fischer.

Dyer, A. F. (ed.) (1979). *The Experimental Biology of Ferns*. London: Academic Press.

Dyer, A. F. and Page, C. N. (eds.) (1985). Biology of pteridophytes. *Proceedings of the Royal Society of Edinburgh, Section B (Biological Sciences)*, **86**, 1–474.

Farrar, D. R., Dassler, C., Watkins, J. E., Jr. and Skelton, C. (2008). Gametophyte ecology. In *Biology and Evolution of Ferns and Lycophytes*, ed. T. A. Ranker and C. H. Haufler. Cambridge, UK: Cambridge University Press, pp. 222–56.

George, L. O. and Bazzaz, F. A. (2003). The herbaceous layer as a filter determining spatial pattern in forest tree regeneration. In *The Herbaceous Layer in Forests of Eastern North America*, ed. F. S. Gilliam and M. R. Roberts. New York: Oxford University Press, pp. 265–82.

Holttum, R. E. (1938). The ecology of tropical pteridophytes. In *Manual of Pteridology*, ed. F. Verdoorn. The Hague: Nijhoff, pp. 420–50.

Hoshizaki, B. J. and Moran, R. M. (2001). *Fern Grower's Manual*. Portland, OR, USA: Timber Press.

Hutchinson, J. T. and Langeland, K. A. (2006). Survey of control measures of old world climbing fern (*Lygodium microphyllum*) in southern Florida. *Florida Scientist*, **69**, 217–23.

International Association of Pteridologists (1994–2007). *Annual Review of Pteridological Research*. Volumes 8–21. Edgecomb, ME, USA.

Jones, D. L. (1987). *Encyclopedia of Ferns*. Portland, OR, USA: Timber Press.

Judd, W. S., Campbell, C. S., Kellogg, E. A., Stevens, P. F. and Donaghue, M. J. (2008). *Plant Systematics, a Phylogenetic Approach*. Sunderland MA, USA: Sinauer Associates.

Kramer, K. U., Schneller, J. J. and Wollenweber, E. (1995). *Farne und Farnverwandte*. Stuttgart, Germany: Georg Thieme Verlag.

Lellinger, D. L. (2002). A modern multilingual glossary for taxonomic pteridology. *Pteridologia*, **3**, 1–263.

Mehltreter, K. (2006). Leaf phenology of the climbing fern *Lygodium venustum* in a semideciduous lowland forest on the Gulf of Mexico. *American Fern Journal*, **96**, 21–30.

Mehltreter, K. and García-Franco, J. G. (2008). Leaf phenology and trunk growth of the deciduous tree fern *Alsophila firma* (Baker) D. S. Conant in a lower montane Mexican forest. *American Fern Journal*, **98**, 1–13.

Mehltreter, K. and Palacios-Rios, M. (2003). Phenological studies of *Acrostichum danaeifolium* (Pteridaceae, Pteridophyta) at a mangrove site on the Gulf of Mexico. *Journal of Tropical Ecology*, **19**, 155–62.

Mickel, J. T. (2003). *Ferns for American Gardens*. Portland, OR, USA: Timber Press.

Milius, S. (2007). Tough frills: ferns' wimp stage aces survival test. *Science News*, **172**, 307.

Moran, R. C. (2004). *A Natural History of Ferns*. Portland, OR, USA: Timber Press.

Nobel, P. S. (1978). Microhabitat, water relations, and photosynthesis of a desert fern, *Notholaena parryi.Oecologia*, **31**, 293–308.

Odum, H. T. and Pigeon, R. F. (eds.) (1970). *A Tropical Rainforest: a Study of Irradiation and Ecology at El Verde, Puerto Rico*. Washington, D.C.: Division of Technical Information, U.S. Atomic Energy Commission.

Olsen, S. (2007). *Encyclopedia of Garden Ferns*. Portland, OR, USA: Timber Press.

Page, C. N. (1979a). The diversity of ferns. An ecological perspective. In *The Experimental Biology of Ferns*, ed. A. F. Dyer. London: Academic Press, pp. 10–57.

Page, C. N. (1979b). Experimental aspects of fern ecology. In *The Experimental Biology of Ferns*, ed.. A. F. Dyer. London: Academic Press, pp. 552–89.

Palmer, D. D. (2003). *Hawai'i's Ferns and Fern Allies*. Honolulu, HI, USA: University of Hawai'i Press.

Portugal Loayza, A. B. D. R. (2005). Effects of environmental factors and past land use on pteridophyte distributions in tropical forests. Masters thesis, University of Puerto Rico.

Price, M. G. (1996). Holttum and ferns. In *Pteridology in Perspective*, ed. J. M. Camus, M. Gibby and R. J. Johns. Whitstable, UK: Whitstable Litho Ltd., pp. 13–26.

Ranker, T. A. and Haufler, C. H. (eds.) (2008). *Biology and Evolution of Ferns and Lycophytes*. Cambridge, UK: Cambridge University Press.

Rickard, M. (2000). *The Plantfinder's Guide to Garden Ferns*. Portland, OR, USA: Timber Press.

Robinson, R. C. (2007). Steps to more effective bracken management. *Aspects of Applied Biology*, **82**, 143–55.

Rothwell, G. W. and Stockey, R. A. (2008). Phylogeny and evolution of ferns: a paleontological perspective. In *Biology and Evolution of Ferns and Lycophytes*, ed. T. A. Ranker and C. H. Haufler. Cambridge, UK: Cambridge University Press, pp. 332–66.

Schneider, H., Schuettpelz, E., Pryer, K. M., *et al.* (2004). Ferns diversified in the shadow of angiosperms. *Nature*, **428**, 553–7.

Schuettpelz, E. and Pryer, K. M. (2008). Fern phylogeny. In *Biology and Evolution of Ferns and Lycophytes*, ed. T. A. Ranker and C. H. Haufler. Cambridge, UK: Cambridge University Press, pp. 395–416.

Sharpe, J. M. (1997). Leaf growth and demography of the rheophytic fern *Thelypteris angustifolia* (Willdenow) Proctor in a Puerto Rican rainforest. *Plant Ecology*, **130**, 203–12.

Sharpe, J. M. and Jernstedt, J. J. (1990). Tropic responses controlling leaf orientation in the fern *Danaea wendlandii* (Marattiaceae). *American Journal of Botany*, **77**, 1050–9.

Sheffield, E. (2008). Alternation of generations. *In Biology and Evolution of Ferns and Lycophytes*, ed. T. A. Ranker and C. H. Haufler. Cambridge, UK: Cambridge University Press, pp. 49–74.

Siccama, T. G., Bormann, F. H. and Likens, G. E. (1970). The Hubbard Brook ecosystem study: productivity, nutrients and phytosociology of the herbaceous layer. *Ecological Monographs*, **40**, 389–402.

Smith, A. R., Pryer, K. M., Schuettpelz, E., *et al.* (2008). Fern classification. In *Biology and Evolution of Ferns and Lycophytes*, ed. T. A. Ranker and C. H. Haufler. Cambridge, UK: Cambridge University Press, pp. 417–67.

Sorsa, V. (1970). Fern cytology and the radiation field. In *A Tropical Rainforest: a Study of Irradiation and Ecology at El Verde, Puerto Rico*, ed. H. T. Odum and R. F. Pigeon. Washington, D.C.: Division of Technical Information, U.S. Atomic Energy Commission.

Tryon, A. and Lugardon, B. (1990). *Spores of the Pteridophyta*. New York: Springer-Verlag.

Tryon, R. M. and Tryon, A. F. (1982). *Ferns and Allied Plants with Special Reference to Tropical America*. New York: Springer-Verlag.

Tu, C. and Ma, L. Q. (2005). Effects of arsenic on concentration and distribution of nutrients in the fronds of the arsenic hyperaccumulator *Pteris vittata* L. *Environmental Pollution*, **135**, 330–40.

Tuomisto, H. and Ruokolainen, K. (2005). Environmental heterogeneity and the diversity of pteridophytes and Melastomataceae in western Amazonia. *Biologiske Skrifter*, **55**, 37–56.

Tuomisto, H., Ruokolainen, K., Kalliola, R., *et al.* (1995). Dissecting Amazon biodiversity. *Science*, **269**, 63–6.

Verdoorn, F. ed. (1938). *Manual of Pteridology*. The Hague: Nijhoff.

Wagner, W. H., Jr. (1973). Some future challenges of fern systematics and phylogeny. In *The Phylogeny and Classification of the Ferns*, ed. A. C. Jermy, J. A. Crabbe and B. A. Thomas. New York: Academic Press, pp. 245–56.

Walker, L. R. (1994). Effects of fern thickets on woodland development on landslides in Puerto Rico. *Journal of Vegetation Science*, **5**, 525–32.

Watt, A. S. (1940). Contributions to the ecology of bracken (*Pteridium aquilinum*). I. The rhizome. *New Phytologist*, **39**, 401–22.

Watt, A. S. (1976). The ecological status of bracken. *Botanical Journal of the Linnean Society*, **73**, 217–39.

Westoby, M., Rice, B. and Howell, J. (1990). Seed size and growth form as factors in dispersal spectra. *Ecology*, **71**, 1307–15.

2

Biogeography of ferns

MICHAEL KESSLER

Key points

1. Biogeographical patterns of ferns and angiosperms are the result of a combination of vicariance and long distance dispersal, but due to their more effective dispersal via spores, the latter is more frequent among ferns. Therefore, fern species tend to have wider ranges and the relative number of fern species compared with seed plants is highest on remote, mountainous tropical islands such as Hawaii and the Mascarenes. Also, fern communities on different continents are more similar compositionally than those of seed plants.
2. Despite their potential for long distance spore dispersal, many fern species have localized ranges as a result of low frequency of successful long distance dispersal, habitat specialization, geographical isolation and competitive interactions between species.
3. Species richness of ferns follows a latitudinal gradient that peaks in the tropics, where ferns are especially diverse and abundant in wet habitats with moderate temperatures at elevations of about 1000–2500 m. On average, species in tropical mountains have elevational amplitudes of about 1000 m. The peak of endemism is located at higher elevations than that of species richness.

2.1 Introduction

Biogeography deals with the distribution patterns of species and communities, and their causal relationships with factors such as climate, soil and evolutionary history (Humboldt, 1805; Lomolino *et al.*, 2006). Specific topics addressed by biogeographers include the sizes of geographical ranges and their spatial placement, the way individual species attain their distribution ranges (dispersal, extinction and vicariance), the distribution of species numbers (alpha diversity), changes in species composition (beta diversity) and the spatial distribution of species traits (macroecology). These aspects can be addressed at different spatial and temporal scales, ranging from global to local and focusing on the present-day situation as well as on past conditions.

Fern Ecology, ed. Klaus Mehltreter, Lawrence R. Walker and Joanne M. Sharpe. Published by Cambridge University Press. © Cambridge University Press 2010.

Ferns have long been the subject of biogeographical studies (e.g., Christ, 1910), but many of these topics have only been dealt with cursorily or within a specific geographical or ecological context that has not previously permitted general conclusions. From a biogeographical point of view, ferns are primarily distinguished from flowering plants by their spore dispersal and their separate gametophytic and sporophytic generations. Accordingly, they are independent of biotic pollination and distribution vectors (with few exceptions such as aquatic ferns transported by waterfowl), eliminating two of the aspects that strongly influence flowering plant biogeography, resulting in a closer biogeographic relationship to climate and substrate (Barrington, 1993; Given, 1993). In addition, ferns are particularly suitable organisms for biogeographic studies because of their worldwide distribution, as well as their moderately high but manageable species richness that allows for both quantitative sampling and statistical inference (Tuomisto, 1994; Kessler and Bach, 1999). Additionally, the occurrence of independent gametophytic and sporophytic generations in ferns (see Chapter 3) allows a direct comparison of "moss-like" and "angiosperm-like" forms within the same taxon and hence might serve as models on how land plants have evolved and adapted to their new environment.

2.2 Dispersal and vicariance

2.2.1 Long distance dispersal

Dispersal of fern (and lycophyte) species takes place primarily via dust-like spores, although some species produce gametophytic gemmae also suitable for medium distance dispersal (Dassler and Farrar, 2001), or vegetative buds (Plate 5A) and rhizomes from sporophytes as well as sporocarps in aquatic species that mainly serve for short distance dispersal. Single fern individuals may produce millions of spores, and while most of them only disperse a few meters (Conant, 1978; Wolf *et al.*, 1991; Sheffield, 1996), dispersal over thousands of kilometers is possible (Tryon, 1985; Barrington, 1993; Smith, 1993; Schneller and Liebst, 2007). The latter presumably takes place mainly via air currents at intermediate (e.g., trade winds, storm systems) to high (e.g., jet streams) altitudes (Erdtman, 1937; Polunin, 1951; Gressitt *et al.*, 1961; Punetha, 1991; Caulton *et al.*, 2000), although direct experimental evidence for this is still lacking. However, the ability of bryophyte and fern spores to survive extreme environmental conditions in the high atmosphere (e.g., low temperatures, high ultraviolet radiation) has been experimentally demonstrated (Gradstein and van Zanten, 2001).

   Long distance dispersal events are rare but disproportionately important in determining the distribution patterns of fern species (Renner, 2005; Nathan, 2006). Factors influencing the potential for long distance dispersal among ferns have never really been addressed by specific studies. Small spore size (20–100 μm in diameter), and the ability of a number of species to be apogamous or to produce bisexual gametophytes capable of self-fertilization (see Chapter 3), certainly play an important role (Tryon, 1970, 1976; Smith, 1972, 1993; Wolf *et al.*, 2001). Spores vary in size, shape, and surface ornamentation (see Chapter 1; Tryon and Lugardon, 1990), but these traits may be unimportant in dispersal because at the scale of spore size the laws of fluid mechanics apply. A different relationship is presented by epiphytic species with long-lived, colonial, gemmiferous gametophytes (e.g., grammitid, filmy and vittarioid ferns, Dassler and Farrar, 1997, 2001). While there is no direct evidence for long distance dispersal by gemmae, because all these species also produce spores, ferns with such gametophytes are relatively more abundant on islands than on continental land areas (Dassler and Farrar, 2001), and are over-represented among species pairs or putative species pairs shared between the neotropics and Africa/Madagascar (Moran and Smith, 2001; Janssen *et al.*, 2007). Whether this is due to a higher probability of arrival or a higher probability of successful establishment remains to be explored.

Yet another comparison related to dispersal can be made between spores with and without chlorophyll. Some fern groups such as Equisetaceae (Plate 4C), grammitid ferns, Hymenophyllaceae (Plate 6A, B), Onocleaceae, and Osmundaceae, have green spores that are presumably more vulnerable to climatic extremes (e.g., drought, frost) and are shorter lived (48 days on average) than spores without developed chloroplasts that may survive for months to years (see Chapter 1; Lloyd and Klekowski, 1970). Accordingly, such species may be less capable of long distance dispersal and hence they should have smaller ranges. While they should therefore be underrepresented on islands, such a tendency has been observed to be weak (Kessler, 2002b). This distribution pattern suggests that even short-lived, chlorophyllous spores are viable long enough to achieve very broad ranges (e.g., *Melpomene flabelliformis* ranging from the Andes to the Mascarenes; *Hymenophyllum polyanthos*, a filmy fern of pantropical distribution). A lack of correlation between range size and presence of chlorophyll in spores has also been found among tropical liverworts, which have comparable spore sizes (van Zanten and Gradstein, 1988).

2.2.2 *Dispersal to oceanic islands*

The potential for long distance dispersal among ferns is best illustrated by oceanic islands that have never been connected to the mainland and thus can only be colonized via long distance dispersal (Tryon, 1985; Barrington, 1993; Smith, 1993). The best studied island system is Hawaii, located approximately 4000 km from the nearest continent. This island chain is about 85 million years old, but erosion and subsidence of older islands imply that the current terrestrial biota could not have colonized the islands prior to about 23 million years ago (mya; Price and Clague, 2002). The Hawaiian Islands are inhabited by 188 species of native ferns, of which 77% are endemic (Palmer, 2003). Taxonomic and phylogenetic studies suggest that these species are the result of at least 140 independent colonization events (Wagner *et al.*, 1990; Wagner, 1995), although this number is likely to be an underestimate, because one species alone, *Asplenium adiantum-nigrum* (black spleenwort), may have colonized the Hawaiian Islands at least three times and perhaps as many as 17 times (Ranker *et al.*, 1994; but see Vogel *et al.* 1999a). In a review of phylogenetic studies of six Hawaiian fern genera, Geiger *et al.* (2007) showed that at least 11 independent colonization events have occurred. Six or seven of these lineages are related to southern Pacific or Southeast Asian clades and probably arrived in Hawaii via the subtropical jet stream, a high-altitude wind current. Other lineages arrived from the New World either via the trade winds or in storm systems acting at lower altitudes. Once established on Hawaii, several of these lineages underwent adaptive radiations such as the endemic genus *Adenophorus*, related to the widespread *Grammitis*, which has evolved into 11 species (Ranker *et al.*, 2003, 2004). In the genus *Diellia*, there is evidence for island chain hopping (i.e., the colonization of newly formed volcanic islands from successively older ones; Schneider *et al.*, 2005).

On the Galápagos Islands, another remote volcanic island system that is located about 1000 km west of South America, a preliminary study by Adersen (1988) suggests that the species composition of fern and lycophyte communities on the islands corresponds to patterns expected from random dispersal events, and that the number of species on the islands is primarily driven by island size, as predicted by the theory of island biogeography (MacArthur and Wilson, 1967). However, detailed phylogenetic analyses are lacking to support this statement.

The capacity of long distance dispersal between continents can be exemplified by the relationships between the neotropics and Africa/Madagascar. Moran and Smith (2001) list 27 species and 87 putative species pairs (i.e., assumed sister species) shared between both continents separated by the Atlantic. With the exception of one example from an old evolutionary lineage (*Anemia*), these putative disjunctions are best explained by long distance dispersal, although superficial similarities between some pairs have proven to be the result of convergence (e.g., Rouhan *et al.*, 2004). Among the derived fern family Polypodiaceae, the mainland African representatives mostly arrived by long distance dispersal from either the neotropics or Asia, with some subsequent speciation (Janssen *et al.*, 2007). Except for perhaps the genus *Platycerium* (staghorn ferns), none of the genera studied by Janssen *et al.* (2007) originated in their host region, Africa.

2.2.3 Vicariance

Long distance dispersal is not the only way in which fern species may attain large and disjunct distribution ranges (Kato, 1993) – that is, ranges that consist of geographically separate parts, either on different islands or continents, or in habitat patches within a geographical region (Fig. 2.1). Such distribution patterns can be explained by three fundamentally different processes: dispersal (see Sections 2.2.1 and 2.2.2), partial extinction from a prior dispersal episode, or vicariance, by which originally contiguous populations are split by geological or climatic changes into separate populations.

In contrast to the numerous examples of long distance dispersal among ferns, undisputed examples for vicariance are rare. This is due to the fact that vicariance by continental drift took place such a long time ago that the resulting biogeographic patterns have often been obscured by speciation, extinction and dispersal events (Wolf *et al.*, 2001). The degree to which vicariance has contributed to the currently observed distribution patterns has long been debated, but the increasing availability of dated molecular phylogenies now allows a more comprehensive overview of the relative importance of vicariance versus dispersal. At the level of species or species pairs, a number of intercontinental disjunctions that have formerly been attributed to vicariance events have now been shown by molecular studies to be so recent that long distance dispersal is more likely (e.g., Moran and Smith, 2001; Kreier and Schneider, 2006; Janssen *et al.*, 2007). In contrast, deep phylogenetic splits may be as old as or even older than the breakups between the continents. For example, the separation of the

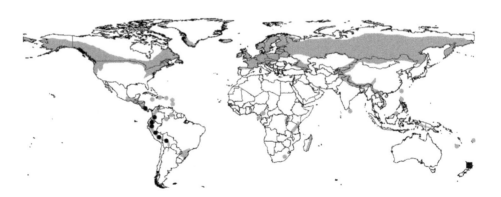

Fig. 2.1 Distributions of *Lycopodium clavatum* (gray shading), one of the most widespread lycophyte species (albeit with several subspecies and varieties, some of which may be independent species), as well as the fern species *Loxoma cunninghamii* (black square) and *Loxsomopsis pearcei* (black circles), the only two members of the highly disjunct and localized family Loxomataceae.

lycophyte genus *Huperzia* (fir-moss, Plate 3C) into a New World and an Old World clade is estimated to have taken place before the separation of Africa from South America some 95 mya (Wikström and Kenrick, 1997, 2000, 2001), followed by more recent dispersal events of younger subclades between the continents (Wikström *et al.*, 1999). Similar cases of deep phylogenetic splits between Old and New World clades that may be attributed to vicariance are found in Blechnaceae (Cranfill and Kato, 2003), Cyatheaceae (Conant *et al.*, 1995; Korall *et al.*, 2006), Isoëtaceae (Hoot *et al.*, 2006), Hymenophyllaceae (Dubuisson *et al.*, 2003), Polypodiaceae (Ranker *et al.*, 2004; Schneider *et al.*, 2004a) and *Polystichum* (sword fern; Little and Barrington, 2003). As a result of these deep phylogenetic splits, and despite numerous cases of long distance dispersal, much of the fern diversity of the New and Old World tropics has evolved independently of each other. For example, the fern floras of Mount Kinabalu in Borneo and Carrasco National Park in Bolivia, each with about 600 species, probably share no more than 60 ancestors between both areas (Kessler *et al.*, 2001).

The fern floras of the four major vegetated landmasses of the southern hemisphere provide other case studies for the relative roles of dispersal and vicariance. At least 103 species occur on two or more of the southern continental regions, including 13 present in South America as well as Australia (including New Zealand; Parris, 2001). New Zealand was separated from other mainland areas about 80 mya; and because some major fern lineages are much older than this, it was proposed that vicariance played an important role in determining the composition of its fern flora (Copeland, 1939; Lovis, 1959). However, recent fossil, distributional and cytological data now favor dispersal (Pole, 1994; Brownsey, 2001). Molecular studies of 31 genera have shown that, in the majority of cases, divergence times from the most closely related relatives studied were very short, supporting long distance dispersal into or out of New Zealand (Perrie and Brownsey, 2007). However, determination of the age of phylogenetic lineages is often problematic, because of the difficulty of finding reliably dated fossils or geological events, and because of varying rates of molecular substitution (Bromham and Penny, 2003). In the case of New Zealand, in three to six cases (depending on the parameters chosen for the analysis) divergence times were long enough to be in accordance with both vicariance and dispersal.

At a different spatial and temporal scale, vicariance can also be caused by climatic shifts that influence the distribution of vegetation types. In particular, glacial events have shaped distribution patterns of many species, especially in temperate and high montane regions. In New Zealand, for example, *Asplenium hookerianum* is currently widespread on both main islands, but genetic studies suggest that during the ice ages this species was distributed in scattered populations, often close to the ice shields (Shepherd *et al.*, 2007). In Europe, diploid populations of some species of *Asplenium* (spleenwort) are restricted to the southern Alps, presumably corresponding to refugial populations formed during glacial periods, whereas the tetraploid forms are found over large areas of Central Europe, but colonized these regions only in the last *c.* 10 000 years (Trewick *et al.*, 2002).

2.3 Range size: variability and spatial distribution

2.3.1 *Range size and its correlates*

Fern and lycophyte species, genera and families are on average more geographically widespread than flowering plants (Tryon, 1970, 1986; Smith, 1972). This is most likely due to their efficient spore dispersal, but other interpretations have also been proposed (see Section 2.2.3; Smith, 1993; Wolf *et al.*, 2001). It was long believed that fern and lycophyte taxa are older, with some of them predating the major continental drift events. This is certainly the case for the older fern families such as Dicksoniaceae, Gleicheniaceae, Hymenophyllaceae, Osmundaceae and Schizaeaceae whose lineages are 150–300 million years old, but most modern families have arisen at the same time as, or even after, many angiosperm families (50–100 mya, Schneider *et al.*, 2004a). It is also conceivable that ferns may evolve more slowly than angiosperms due to differences in reproductive biology and ecology (Smith, 1972).

Distributional ranges of individual fern species vary by many orders of magnitude. Among the most widespread fern species are *Cystopteris fragilis* (brittle fern) and *Lycopodium clavatum* (common club-moss) that occur almost throughout temperate regions and tropical montane ecosystems (Fig. 2.1). At the other end of the scale are species known from single populations with few individuals. Examples include *Isoëtes tennesseensis* (quillwort), known only from two small rivers in southeastern Tennessee (Luebke and Budke, 2003), and *Asplenium tunquiniense* which is locally common in about 1 km² of cloud forest in Bolivia, but absent in nearby areas (Kessler and Smith, 2006).

The range of a species is determined by its age, its ecological requirements, the availability of suitable habitats and the ability of the species to reach these habitats and survive there (Lester *et al.*, 2007). Ranges of ferns are usually considered to be more strongly determined by habitat availability than by their, generally high, dispersal capability (Tryon, 1970, 1986; Smith, 1972, 1993; Richard *et al.*, 2000; Guo *et al.*, 2003; Tuomisto *et al.*, 2003a, 2003b; Jones *et al.*, 2006). The efficiency of spore dispersal can be illustrated by the uniformity of haplotypes among some ferns in the northern hemisphere (e.g., *Asplenium viride*, green spleenwort, James *et al.*, 2008). However, there is also some evidence for limitations on dispersal such as for several calcicolous ferns (i.e., species that thrive on calcareous soils) in Canada that are absent from what are apparently environmentally suitable microsites (Wild and Gagnon, 2005).

On a different spatial scale, Moran (1995, 1996) estimated that nearly 25% of the fern species in Colombia and Ecuador on the western side of the Andes do not occur on the eastern side, suggesting that the mountains have acted as a barrier to migration, although environmental differences between the slopes may also play a role. Limitations on dispersal are more likely to play a role in species with highly specific habitat requirements and small, isolated patches of suitable habitat (Peck *et al.*, 1990).

Range size may also be correlated with other reproductive and morphological traits. Among Japanese ferns, for example, ranges are larger in species with multiple reproductive modes (sexual + vegetative or apogamous) than in species with a single mode (Guo *et al.*, 2003). The same study showed that while diploid species display wide variability in range sizes, polyploid species uniformly had small ranges, possibly as a result of the relatively young age of species derived from polyploidization. Interestingly, these findings contrast with the observations on *Asplenium* in Europe where diploid taxa often are glacial refugees (Trewick *et al.*, 2002). Similarly, in *Pellaea* (cliff brake), polyploid derivatives have been assumed to start out with small ranges but often eventually outcompeted their diploid progenitors (Tryon, 1957). Further, in Japan seasonally green species have larger ranges than evergreen species, presumably due to limited subtropical habitat suitable for evergreen species (Guo *et al.*, 2003). In Bolivia, latitudinally widespread fern species also tend to have broader elevational ranges, implying that they are ecologically more adaptable than localized species (Kessler 2002b). On the other hand, in a Bolivian cloud forest, fern species with restricted ranges tend to be locally more frequent than widespread species, which may be due to more specific adaptations of localized species to specific habitats, although a sampling bias (widespread species have more distribution gaps and are therefore on average less frequent) cannot be excluded (Kessler, 2002b). Contrary to the Japanese study (Guo *et al.*, 2003), range size of Bolivian ferns was neither correlated to life form nor to the studied reproductive aspects (sexual versus asexual reproduction, spores with chlorophyll versus spores without chlorophyll; Kessler, 2002b).

2.3.2 Endemism

Endemic plant species (i.e., species with restricted ranges) are not randomly distributed. They are especially well represented on oceanic islands (Carlquist, 1974), in isolated habitats such as mountain ranges (Humphries, 1979; Gentry, 1986), on localized geological substrates (Kruckeberg and Rabinowitz, 1985; Cowling *et al.*, 1994) and in tropical and subtropical regions (Gentry, 1986; Stevens, 1989). Considering ferns, geographically isolated islands are well known for their high levels of endemism as exemplified by the 77% endemic fern species in Hawaii (Palmer, 2003). This can be explained by rare colonization events resulting in initially small, genetically isolated populations that can diverge quickly from their source populations through genetic drift, bottleneck effects or adaptation and selection.

Along elevational gradients, patterns of endemism are less easily explained. Concentrations of endemic fern species are generally found at high elevations, typically above the peak of total species richness (Fig. 2.2; Tryon 1972; Kessler, 2000b, 2002b; Kluge and Kessler, 2006). This distribution is probably influenced by a more subtle kind of geographical isolation. Mountains are typically steepest at high elevations (Körner, 2000). As a result, species with given elevational

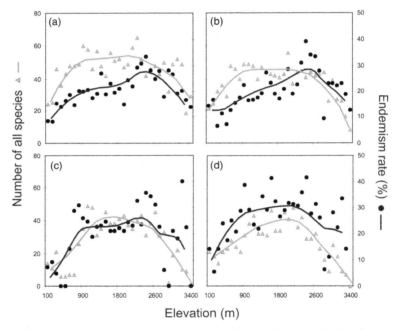

Fig. 2.2 Patterns of species richness and rates of endemism (% of endemic species) of (a) all fern species, (b) terrestrial species, (c) trunk epiphytes and (d) high-canopy epiphytes in plots of 400 m^2 along an elevational gradient in Costa Rica (modified after Kluge and Kessler, 2006). For this analysis, species are considered to be endemics if they are restricted to Costa Rica and Panama (24% of all fern species recorded in this study). Note that while overall the peak of endemism is located at higher elevations than that of species richness, this pattern is primarily caused by the terrestrial species. The trend lines in this figure were drawn by distance-weighted least-squares smoothing and are intended to guide the eye.

amplitudes have narrower (and steeper) ranges at high than at low elevations. For example, a species occurring in cloud forest at 2500–3500 m on the eastern Andean slope from Venezuela to Bolivia will have a range that is 600 times longer than wide, whereas at 500–1500 m this ratio will be less than 100 (Graves, 1988). Such narrowly linear ranges are not conducive to gene flow, resulting in genetic fragmentation between populations, and eventual speciation (Kessler, 2002a). However, the story does not end here. In Costa Rica, fern endemism is not highest at the topographically steepest and most fragmented elevations, but rather at somewhat lower elevations with higher humidity, suggesting that species formation and survival may be favored there (Kluge and Kessler, 2006). Even more interestingly, species of different life forms and habitats show distinct patterns (Fig. 2.2). For example, maximum endemism of crown epiphytes is found at lower elevations than for trunk epiphytes and terrestrial species (Kluge and Kessler, 2006). The causes for these distinct patterns are unknown.

Species with small ranges are generally considered to be of particular conservation concern because their small ranges render them particularly susceptible to habitat loss (Balmford and Long, 1994) and to human-induced habitat disturbance (Moolman and Cowling, 1994; Samways, 1994; Andersen *et al.*, 1997). Comparing natural and strongly degraded habitats (Kessler, 2001a; Paciencia and Prado, 2005), this is certainly true for ferns. However, two studies in Bolivia suggest that endemic ferns may profit, in terms of diversity, from slight habitat disturbance. In one study, the relative abundance of ferns at 16 sites was higher in forests with a low level of human disturbance (through logging or cattle grazing) than in nearby undisturbed forests (Fig. 2.3a; Kessler, 2001a). A second study showed that along a chronosequence of vegetation succession on landslides, fern communities in senescent (i.e., decaying) forests were less species-rich and more strongly dominated by widespread species than communities in mature (i.e., climax) forests (Fig. 2.3b; Kessler 1999). The relationship between range size and competitive ability may be explained by the difficulty less competitive species have in expanding their ranges, whereas strong competitors have a higher probability of establishing new populations following medium to long distance dispersal. Thus, species with small ranges might be competitively inferior to widespread species. Further, localized species may depend on a certain level of habitat disturbance that disrupts competitive interactions and thus limits their competitive exclusion by widespread species. Under natural conditions, disturbance regimes may be high enough to allow the persistence of the localized species but perhaps below their optimum level. In contrast, high levels of human disturbance often lead to massive losses of species (Kessler, 2001a). Clearly, intriguing questions for future research include the relationship between range size and dispersal, competitive ability and taxonomic affinity of ferns.

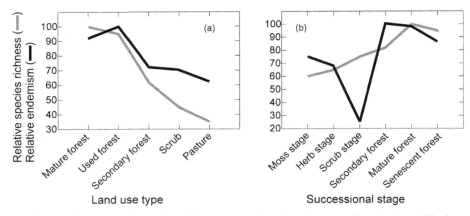

Fig. 2.3 Patterns of species richness (gray lines) and rates of endemism (black lines) of ferns in plots of 400 m² along (a) a gradient of human vegetation degradation and (b) a gradient of vegetation succession on landslides (modified after Kessler, 1999 and 2001a). Note that endemism is highest in forest used by humans for timber extraction and cattle grazing as well as at intermediate stages of vegetation succession, suggesting that localized species are competitively rather weak.

2.4 Alpha diversity

2.4.1 Patterns of species richness

The total number of known ferns is estimated at about 9600 species worldwide (see Appendix A; Smith *et al.*, 2006), plus about 1400 lycophytes. However, about 100 "new" species are being described yearly and the total number of species may be closer to 15 000 (Roos, 1996; Chapman, 2006). "New" species are recognized both as a result of intensive collection activity, especially in remote tropical regions and through detailed taxonomic studies. For example, about 110 fern species, 9% of the currently known Bolivian fern and lycophyte flora, have been described from Bolivia in the last ten years (e.g., Lehnert, 2006; Kessler and Smith, 2007). Of these, roughly two thirds are based on specimens first gathered in the last two decades, whereas the other third represents species already collected and present in herbaria but not recognized as distinct new taxa until recently. The number of Bolivian fern and lycophyte collections and known species increased exponentially from 881 collections and 404 species by 1970 to 4207 collections and 743 species by 1990, and to 23 221 collections and 1165 species by 2006. When this collecting activity is mapped using grid cells, it emerges that 95% of all collections come from less than 5% of the Bolivian territory. However, even the most densely sampled area in Bolivia, with 2270 samples representing 401 species in about 700 km², is still incompletely known, and statistically we infer that at least

560 species should be expected there. Importantly, the degree of undersampling is more pronounced in the species-rich grid cells, even if the number of collections is higher (Soria-Auza and Kessler, 2008). We can conclude that our knowledge of the spatial distribution of ferns at global and regional scales is still far from complete and that it is biased toward the better studied, species-poor areas because of greater undersampling in species-rich regions. Of course, this does not only apply to Bolivia but also to many other tropical regions such as the Andes, Madagascar and Southeast Asia, which probably have the least well known fern floras.

Despite problems in obtaining accurate species counts, the diversity of ferns varies to such a degree among regions and habitats that broad patterns of diversity are clearly discernible. Fern communities are richest in wet tropical regions, particularly in cloud forests. A global analysis of fern diversity, based on data from 205 mainland and 117 island floras, has only recently been conducted (Plate 1A; H. Kreft, unpublished data). The analysis confirmed that fern richness is highest in regions with high potential evapotranspiration (i.e., a measure of how much water vaporizes or is transpired per year), a high number of rainy days and marked topographical relief (i.e., mountains). This largely corresponds to results obtained from regional studies, most of which find that the species richness of ferns is primarily determined by the level of rainfall and topographic complexity (Rwanda: Dzwonko and Kornaś, 1994; Bolivia: Kessler, 2000b; Iberian Peninsula: Pausas and Sáez, 2000, Ferrer-Castán and Vetaas, 2005; New Zealand: Lehmann *et al.*, 2002; Australia: Bickford and Laffan, 2006). The only known exception is Uganda, where fern diversity at a local scale is primarily determined by soil fertility and the distance from putative Pleistocene forest refugia (i.e., forest-covered areas during the ice ages; Lwanga *et al.*, 1998). Whether it is a natural phenomenon that climate is more important globally and soils at a local scale, or whether there is a methodological bias remains to be explored. Data for soil nutrients have been rarely considered, because they are more difficult to obtain for large geographic areas than climate data.

While the overall diversity pattern of ferns coincides with that of seed plants (Barthlott *et al.*, 2005; Kreft and Jetz, 2007), there are some marked differences (Kramer, 1993; H. Kreft, personal communication). Whereas seed plants are highly diverse in regions with Mediterranean-type climates (hot, dry summers and cool, wet winters), and moderately diverse in arid regions, ferns are generally species-poor in these habitats, although some groups such as *Selaginella* (spike-moss, Plate 3B), cheilanthoid ferns (Fig. 5.1c), and the *Asplenium aethiopicum* complex are represented by numerous species. Furthermore, the latitudinal increase of species numbers from the poles toward the equator is 2–3 times steeper for ferns than for seed plants (H. Kreft, personal communication). Clearly, the range of climatic conditions under which ferns can occur and could diversify is more limited than the range for seed plants, with the exception of epiphytic ferns. These extend farther into cold habitats than any epiphytic flowering plant group both in temperate regions (e.g., *Polypodium*, Zotz, 2005) and in high tropical mountains (e.g., *Melpomene*, Krömer *et al.*, 2005). Also, along with some bromeliads, ferns of the genus *Pleopeltis* (Fig. 5.1e, Plate 7C) are among the vascular epiphytes reaching farthest into dry habitats in the neotropics. Interestingly, these three genera all belong to the family Polypodiaceae, suggesting that familial traits within this phylogenetically derived family support resistance to extreme climatic conditions.

In addition to patterns of absolute species numbers, one may focus on the relative contribution of ferns to the vascular plant flora in different geographical regions or habitats. Worldwide, ferns comprise an average of 3.6% of different vascular plant floras (H. Kreft, personal communication.). However, in tropical forests and in montane habitats they can represent up to 13% of the local flora, whereas in desert regions they may be completely absent. On islands, ferns on average represent about 15% of the local vascular plant species, and in exceptional cases up to 70% on islands such as Easter Island, Palau, Saint Helena and Tristan da Cunha (Plate 1B). The proportion of fern taxa typically increases with the distance of the islands from the mainland, undoubtedly as a result of the higher dispersal ability of ferns compared with most seed plants. However, this relationship is not perfect, and some remote islands have fewer ferns than expected. Examples are the cold, windswept Falkland Islands, and especially flat and mountainless tropical atolls such as the Maldives that lack habitats suitable for most ferns.

Within the global framework of fern diversity, mainland Africa is particularly poor in fern species. Whereas the neotropics and Southeast Asia both have in the order of 3000–4500 species, mainland Africa has only about 630 species (Tryon, 1986). Even Madagascar, which is 40 times smaller than mainland Africa, has about 560 species. This pattern has been explained by the limited surface area of humid tropical mountains in Africa (Moran, 1995). However, in the global analysis (H. Kreft, personal communication), Africa has fewer species than expected under present-day conditions, even when all climatic and landscape factors (including area and topographic relief) are considered. This low diversity of African ferns may thus be best explained by Pleistocene climatic oscillations, during which rain forests in Africa were restricted to small refugia, while most of the continent was covered with drier vegetation types, presumably leading to widespread extinctions of ferns (Kornaś, 1993). A detailed analysis of distribution patterns of African ferns has shown that these can be separated into three main groups that correspond to refugial areas (Aldasoro *et al.*, 2004): Guinea–Congolian thermophilous species inhabit the lowland rain forests of West and Central Africa and had their refuge around the Gulf of Guinea; the cold-tolerant, Afro-montane taxa survived the dry periods mainly in the East African mountains; and the southern drought-tolerant elements had their refuge in the Cape Region.

2.4.2 Mountains as hot spots of fern diversity

Tropical mountains are clearly the hot spots of fern diversity. A number of studies have shown that on high tropical mountains, species richness shows a hump-shaped elevational richness pattern, with highest diversity at mid-elevations and decreasing diversity toward both high and low elevations (Fig. 2.4; Jacobsen and Jacobsen, 1989; Parris *et al.*, 1992; Kessler, 2000b, 2001b, 2001c, 2002a; Kessler *et al.*, 2001; Hemp, 2002; Bhattarai *et al.*, 2004; Kluge *et al.*, 2006; Watkins *et al.*, 2006). The elevation at which maximum fern diversity occurs differs somewhat among mountain ranges, with maxima at about 1800 m in both Costa Rica (maximum mountain height 3820 m, Kluge *et al.*, 2006) and on Mount Kinabalu, Borneo (4000 m, Kessler *et al.*, 2001), at 2000 m in Bolivia (4500 m, Kessler, 2000b), and at 2400 m on Mount Kilimanjaro, Tanzania

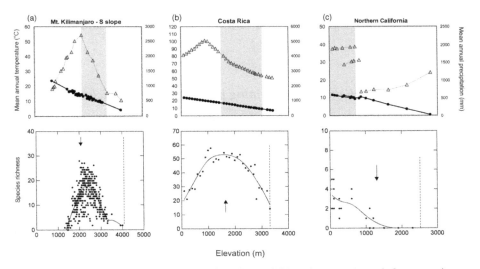

Fig. 2.4 Elevational patterns of climatic variables (top row) and fern species richness (bottom row) in (a) Tanzania (after Hemp, 2002), (b) Costa Rica (after Kluge *et al.*, 2006) and (c) California (M. Kessler, unpublished data). In the climate figures, left axes and dotted lines show values of mean annual temperature; right axes and triangles mean annual precipitation; elevational steps with extensive occurrences of fog ("cloudbelts") are shaded gray. In the species richness figures, dashed vertical lines show the natural upper limits of the study gradients (timberline or mountain peaks), and arrows the elevational midpoints of the gradients. Note different scales of axes. The trend lines in this figure were drawn by distance-weighted least-squares smoothing and are intended to guide the eye.

(5850 m, this volcano is placed on a plateau at about 1000 m, Hemp 2002). In temperate regions such as New Zealand (Ohlemüller and Wilson, 2000) or North America (Fig. 2.4), richness may decline continuously with elevation or remain roughly constant. Climatically, these gradients correspond to the upper parts of tropical gradients, where richness also declines, and indeed species numbers are comparable between these data sets at similar values of mean annual temperature (M. Kessler and J. Kluge, unpublished data). The maxima of fern richness have often been interpreted as reflecting the elevation of maximum humidity (Kessler, 2001a; Hemp, 2002) or an optimal combination of humidity and mild temperatures (Bhattarai *et al.*, 2004; Kluge *et al.*, 2006). For example, in South America a number of fern species that typically occur along the lower slopes of the Andes extend into the wettest part of the Ecuadorian Amazon basin, suggesting that the lower elevational limit of these species is determined by environmental humidity (Kessler 2001b).

At an even finer spatial scale, fern richness typically is higher in humid micro-habitats (Richard *et al.*, 2000), in ravines, and along streams rather than on slopes, and lower on mountain ridges than on slopes (Kessler, 2002b; Kluge and Kessler, 2006). Again, this has been linked to the higher and more constant humidity close to water bodies, although higher soil fertility on lower slopes and near streams may also play a role.

2.4.3 Mechanisms determining patterns of species richness

Area dependence of species richness

Area is a crucial parameter determining biodiversity patterns for all taxa, because larger areas can support higher numbers of individuals and hence more viable populations of a larger number of species, reducing extinction risks (Rahbek, 1995; Rosenzweig and Ziv, 1999; Lomolino, 2001). Furthermore, larger areas typically contain a wider range of habitats. While the influence of area on fern richness is obvious when we compare countries (Plate 1B) or islands (Roos *et al.*, 2004) of different sizes, it is much less clear-cut on mountains, where area per altitudinal belt typically declines with elevation (Körner, 2000), but where fern richness peaks at mid-elevations. On mountains, other factors, in particular climate, play a primary role.

Climate dependence of species richness

It has frequently been noted that most ferns are particularly water-dependent plants that require high humidity for their growth and their reproduction (e.g., Tryon, 1976; Barrington, 1993); this is because gametophytes depend on water for the transport of gametes, and because some of the conspicuous adaptations to water stress among flowering plants (e.g., succulence, annual life cycle) are rare or absent among ferns. This reasoning may involve circular argumentation, however, because the "optimal" conditions are defined by where the largest number of species is found rather than on where environmental conditions are truly optimal for fern development, both relative to environmental conditions and to interactions with other plants. Furthermore, ferns are not totally absent in many relatively inhospitable habitats, and many fern and lycophyte lineages have evolved morphological adaptations to deal with water stress, low temperatures or both (see Chapter 5; Pickett, 1931; Gaff, 1977; Kramer *et al.*, 1995; Page, 2002; Kessler *et al.*, 2007). If these drought-adapted species can survive under these relatively adverse conditions, why are there so few species?

There are two fundamentally different ways in which the number of coexisting species in a given habitat can be limited: by actual local environmental conditions such as energy availability or by historical factors.

Local limitations of species richness

Favorable climatic and soil conditions allow the growth and persistence of a larger number of individuals of a given higher taxon, in our case ferns. Accordingly, the energy available through primary production to the fern communities for growth and reproduction is higher under favorable conditions. The species-energy theory states that the more energy that is available in an ecosystem, the more species will be able to coexist (Wright, 1983). Indeed, indirect measures of energy availability at the ecosystem level such as AET (actual evapotranspiration) or PET (potential evapotranspiration) are often positively related to plant species richness (Kreft and Jetz, 2007; H. Kreft, personal communication). However, there are two challenging aspects in addressing the species–energy relationship. The first is to get the appropriate measure for energy availability. The theoretical relationship of species richness to energy availability is based on net primary productivity (NPP, i.e., the amount of energy fixed by plants through photosynthesis and made available to the plants themselves and to the consumers). Yet, because NPP is difficult to measure, field studies usually use surrogates of NPP such as AET, PET, vegetation biomass, or tree growth (Waide *et al.*, 1999). However, none of these measures gives a reliable estimate of NPP, and Waide *et al.* (1999) have concluded that there are no studies "relating diversity directly to productivity". Secondly, measures of NPP at the ecosystem level may be quite meaningless for specific study groups such as ferns. In a forest ecosystem, most of NPP is contributed by the trees, whereas ferns only contribute a small and variable percentage. Clearly, in order to address the species–energy relationship for ferns, we first need measurements of the productivity of fern species. Such data are completely lacking at present.

Moreover, even less is known about the mechanisms that may answer the question "how can increasing energy materialize into more species?" (Willig *et al.*, 2003; Mönkkönen *et al.*, 2006). Until recently, literature on species richness–energy relationships referred only in very few cases to such causal mechanisms, probably because of the belief that they were understood. For example, Rosenzweig and Abramsky (1993) stated that the relationship between richness and energy "troubles no one": a poor environment does not provide many resources for rare species, and they become extinct. This simplistic view has proven to be misleading, as the few studies addressing this relationship report conflicting results (Willig *et al.*, 2003; Evans *et al.*, 2005). Most recently, the set of potential mechanisms that may explain the causality of the species–energy relationship has been reviewed by Evans *et al.* (2005). They evaluated nine hypotheses, three of which may apply directly to ferns and are discussed below to provide an impression of their theoretical background and the data needed to evaluate them.

The *population size hypothesis* basically postulates that increasing energy may support more individuals and thus more species. This idea is analogous to the Theory of Island Biogeography, in which species richness increases with area because larger areas support more resources (but only in total, not necessarily per unit area) and thus more individuals, enabling species to maintain higher population sizes and to reduce extinction risks (MacArthur and Wilson, 1967). Replacing "area" by "energy availability", indicates that high energy areas support more species by shifting population sizes above the threshold of viability (Kaspari *et al.*, 2000; Hurlbert, 2004; Pautasso and Gaston, 2005, 2006). This hypothesis predicts that along a gradient of species richness, the number of individuals per species, or at least the minimum number of individuals per species needed to maintain viable populations, should remain roughly constant, if species assemblages were saturated. In contrast to this assumption, the number of fern individuals per species increases in Costa Rica with elevation in survey plots of constant size (Fig. 2.5; J. Kluge and M. Kessler, unpublished data). This may suggest that at high elevations, fern communities do not reach their maximum species diversity because the occurring species are overrepresented by more individuals than needed to maintain their populations, leaving no room for the immigration of additional species. However, the vegetative size of fern individuals also declines with elevation (Kessler and Siorak, 2007; Kessler *et al.*, 2007; Kluge and Kessler, 2007), and it may simply be the case that at lower elevations the density of individuals is limited by their spatial requirements. And finally, we know nothing about the potential competition of ferns with seed plants at different elevations. Clearly, we have not even started to properly address these topics.

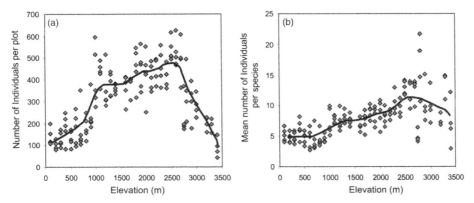

Fig. 2.5 (a) Total number of fern individuals and (b) mean number of individuals per species along an elevational gradient in Costa Rica (J. Kluge and M. Kessler, unpublished data). Compared with the pattern of species richness (Fig. 2.2A), the peak of number of individuals is located at higher elevations, resulting in an overall increase of individuals per species. The trend lines in this figure were drawn by distance-weighted least-squares smoothing and are intended to guide the eye.

The *niche position hypothesis* proposes that niche position specialists use relatively specific, rare resource types that may be too scarce in low energy areas to allow the survival of viable populations (Abrams, 1995). Thus, the niche position hypothesis provides a causal link between energy and habitat heterogeneity (Hurlbert and Haskell, 2003; Hurlbert, 2006). The central prediction is that energy input determines the variety and abundance of resources and thus population sizes of species specialized on certain resource types; in high energy areas, rare resource types are more likely to support viable populations than in low energy areas. This hypothesis has never been tested for ferns.

In contrast, the *niche breadth hypothesis* assumes that when resource abundance increases, species may switch from a wide array of resources to their preferred ones, leading to reduced niche breadth in high-energy areas. This restriction may lead to reduced niche overlap and hence lower competition, promoting coexistence. The diagnostic prediction of the niche breadth hypothesis is that along gradients of energy input, species have their widest niche breadth at the end of lowest energy. For example, along an elevational transect in Costa Rica, epiphytic ferns at lower elevations usually occur only in specific parts of trees (e.g., trunk, thick branches, outer branches), whereas close to the tree line they usually grew on most parts of the trees (J. Kluge and M. Kessler, unpublished data). This may suggest that niche breadth increases, but also that the environmental conditions on tree trunks and in the canopy are much more similar to each other in stunted forests close to timberline than in a 40 m tall forest in the lowlands. Microenvironmental measurements at the actual growth sites of the species are needed to evaluate the niche breadth hypothesis for ferns.

These few examples should suffice to give an idea how local energy availability may limit the richness of fern assemblages. We can conclude that while the theoretical framework is reasonably well established, empirical studies do not yet exist, and that the methodological hurdles in testing these hypotheses will be high.

Historical limitations of species richness

Gradients of richness are not static, and present-day patterns can be seen as a snapshot of a dynamic history on a very large temporal scale (e.g., Wiens and Donoghue, 2004; Ricklefs, 2005; Roy and Goldberg, 2007). A classic example is the low diversity of African ferns that is attributed to adverse climatic conditions in the past, which have led to extinction of ferns or reduced possibilities for their diversification. On other continents, the diversity of ferns may have experienced similar historical limitations or perhaps reached a maximum determined by the carrying capacity of the ecosystems. Given enough time and suitable environmental conditions for future diversification, fern richness may continue to increase differentially over evolutionary time and result in diversity patterns different from those observed today.

There are several ways in which historical factors may influence patterns of diversity (Wiens and Donoghue, 2004; Allen and Gillooly, 2006; Mittelbach *et al.*, 2007; Ricklefs, 2007). For example, the tropics may harbor a higher diversity than temperate regions because the taxa under consideration, in our case the major phylogenetic lineages of ferns, have originated there, and adaptations necessary to disperse and persist in cold and climatically seasonal regions have evolved only in some taxa and at a later point in time (niche conservatism *sensu* Wiens and Donoghue, 2004). Humboldt (1808) proposed that high latitudes support fewer species because they have to tolerate low temperatures, a line of thinking leading to Currie *et al.*'s (2004) "climatic tolerance" hypothesis. Tropical regions had a greater extent in the historical past (Behrensmeyer *et al.*, 1992), and many extant clades are originally tropical, leading to greater time and space availability for speciation under tropical conditions. Alternatively, regions with lower present-day diversity may have suffered from more pronounced extinction events, as suggested above for Africa. These two hypotheses are based on the assumption that rates of diversification are similar in all geographical regions and that the differences in species richness are the consequence of the time available for diversification and differing rates of extinction (Mittelbach *et al.*, 2007; Ricklefs, 2007). However, rates of diversification may differ between regions or habitats, because of higher mutation rates and shorter generation times in high energy areas (Evans *et al.*, 2005). This mechanism predicts that mutation rates are directly linked to solar energy by temperature and ultraviolet radiation and indirectly by reduced generation time, although few studies have shown such a positive relationship (e.g., Cardillo, 1999).

These potential historical explanations have not yet been explicitly tested for ferns or for most major groups of organisms (Wiens and Donoghue, 2004; Ricklefs, 2005; Harrison and Cornell, 2007). It has been proposed that ferns have lower diversification rates than flowering plants, either due to the lack of many of the isolating mechanisms found in angiosperms, or due to higher levels of gene flow between populations (Tryon, 1970; Smith, 1972; Ranker *et al.*, 1994, 2004). Within the ferns, several recent molecular studies have shown that different clades often have distinct rates of molecular diversification (Pryer *et al.*, 2001; Schuettpelz and Pryer, 2006), perhaps as a result of alterations in life cycles and reproductive biology linked to ecological adaptations (Haufler *et al.*, 2000; Bromham and Penny, 2003), or due to genome-wide shifts in the rates of nucleotide substitutions (Schuettpelz and Pryer, 2006), although there is no direct evidence for either of these explanations. Schneider *et al.* (2004b) found that the deeper branches of the asplenioid fern diversification correspond to a division into tropical and temperate clades, with temperate clades evolving on up to six different occasions from tropical ones. Although no statistical analyses have been conducted to ascertain if the lower diversity of asplenioid ferns in temperate regions may be attributed to later or slower diversification or to higher extinction, the overall pattern found by Schneider *et al.* (2004b) certainly suggests that evolutionary processes played a role in shaping present-day diversity patterns in this fern lineage. Comparing radiations of several genera of Polypodiaceae, Haufler *et al.* (2000) suggested that in temperate regions fern speciation may be primarily driven by allopatry (i.e., divergence of geographically separate populations), whereas in the tropics, ecological specialization and sympatric speciation may predominate. Similar thoughts were developed by Yatabe *et al.* (2001) for the diversification of Asian species of *Asplenium*. This interesting hypothesis remains to be tested more thoroughly.

On a different timescale, Schneider *et al.* (2004a) have shown that most of the contemporary fern diversity evolved after the main diversification of angiosperms in the Late Cretaceous and Early Tertiary. They proposed that flowering plants profoundly changed the habitat conditions, especially in forest ecosystems, from fern- (and gymnosperm-) dominated forests with an open vegetation structure and much light reaching the ground level to denser and darker angiosperm-dominated forests. Fern groups that have managed to adapt to these new conditions, e.g., through the development of more sensitive photoreceptors that allow sufficient photosynthesis even under low light conditions (Schneider-Pötsch *et al.*, 1998; Kawai *et al.*, 2003), could then diversify in the new niche. The radiation of fern lineages may also be linked to ecological specialization as in Pteridaceae (Schuettpelz *et al.*, 2007). In this family, distinct clades are characterized by their aquatic (*Acrostichum*, leather fern and *Ceratopteris*, antler fern), epiphytic (vittarioid ferns, Plate 6D) or xeric habitats (cheilanthoid ferns, see Chapter 5), suggesting that ecological innovation was responsible for the initial diversification of the family.

Dispersal limitation

Species richness may not only be influenced by deterministic processes as outlined above, but also by stochastic factors and dispersal limitation. For example, the range of placements of species within geographically or ecologically constrained domains may independently create patterns of species richness (Colwell and Hurtt, 1994). This mid-domain effect (MDE), a geometrical null model in which species ranges are randomly placed along a geographical domain with hard outer boundaries resulting in hump-shaped richness patterns, has been invoked as an explanation for richness patterns along elevational and latitudinal gradients (see review in Colwell *et al.*, 2004). While MDE models closely correlate with diversity patterns (especially along some elevational gradients), both for ferns (Kessler, 2000b; Watkins *et al.*, 2006) and other taxa (Colwell *et al.*, 2004; McCain, 2005), the meaning and implications of such geometric constraint models are hotly debated (Bokma and Mönkkönen, 2000; Colwell and Lees, 2000; Jetz and Rahbek, 2001; Hawkins and Diniz-Filho, 2002; Grytnes, 2003a, 2003b; Colwell *et al.*, 2004; Herzog *et al.*, 2005). The main criticism is that the frequency distribution of range sizes, which generates hump-shaped patterns in the MDE null model, exists only because of the presence of environmental gradients (Hawkins *et al.*, 2005).

On a different spatial scale, Hubbell (2001) has suggested that the occurrence and abundance of species at given sites within a larger ecoregion or habitat is influenced by dispersal limitation. This "neutral theory of biodiversity and biogeography" considers variations in the dispersal ability of species, their population size, and their immigration rate as the crucial factors determining the composition and distribution of species abundances of communities. The relative influence of these three factors over ecological factors varies with scale and study group; local scales within ecologically more homogeneous areas show a stronger effect of dispersal limitation (e.g., Dalling *et al.*, 2002; Potts *et al.*, 2002; Svenning *et al.*, 2004; Chust *et al.*, 2006), whereas at regional scales and across strong ecological gradients, the niche assembly model appears to be more applicable (e.g., Condit *et al.*, 2002; Tuomisto *et al.*, 2003c; Jones *et al.*, 2006, 2007; Ruokolainen *et al.*, 2007).

Dispersal may also modify patterns of species richness due to mass or source–sink effects (Shmida and Wilson, 1985; Pulliam, 1988). This occurs when propagules of a species are dispersed to suboptimal habitats, where they may survive but are unable to produce enough offspring to maintain self-sustaining populations. Although such sink populations have been documented in numerous individual species (Gilpin and Hanski, 1991; Wilson, 1992; Leibold *et al.*, 2004), the extent of such populations at the community level and therefore their influence on observed richness patterns remains largely unexplored. For example, hump-shaped patterns of species richness along elevational gradients have been hypothe-sized to be due to dispersal of species from lower and higher elevations, resulting in highest overlap of such sink populations at mid-elevations, whereas commu-nities at the extremes of the gradient can receive immigrants from only one direction (Rahbek, 1997; Kessler, 2000b; Lomolino, 2001; Grytnes and Vetaas, 2002; Grytnes, 2003a, 2003b; Kattan and Franco, 2004). Grytnes *et al.* (2008) assessed elevational patterns of species richness of vascular plants along four elevational gradients in Norway. They compared patterns of species richness using only fertile populations with patterns obtained when only sterile populations were considered, assuming that species recorded only as sterile individuals represent sink populations. In several cases, they found that "sterile-only" richness patterns were more strongly hump shaped than "fertile-only" patterns. They interpreted this as an accumulation of sterile sink populations at mid-elevation sourced from species mainly inhabiting either higher or lower elevations. In contrast, the extremes of the gradient can only receive sink propagules from one direction and hence have lower diversity. This supports the hypothesis that source–sink effects do indeed modify elevational richness patterns. Comparable studies have not yet been published on ferns, but recent analyses from an elevational gradient in Costa Rica clearly show that separating sterile and fertile records indeed results in distinct patterns and that dispersal-mediated population processes appear to modify patterns of species richness (Fig. 2.6; J. Kluge and M. Kessler, unpub-lished data).

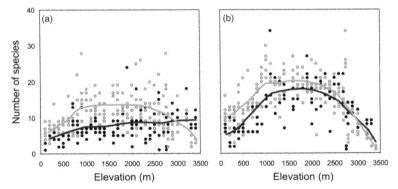

Fig. 2.6 Elevational patterns of the number of fertile (black symbols) and sterile (gray symbols) (a) terrestrial and (b) epiphytic ferns in plots of 400 m^2 along an elevational gradient in Costa Rica (J. Kluge and M. Kessler, unpublished data). Note that the hump-shaped pattern of the diversity of terrestrial ferns is caused by the accumulation of sterile species at mid-elevations. The trend lines in this figure were drawn by distance-weighted least-squares smoothing and are intended to guide the eye.

2.5 Beta diversity

Beta diversity is the change of plant or animal communities along ecological gradients. While these gradients can cover a wide range of spatial, temporal and ecological aspects, studies on the beta diversity of ferns have mostly focused on changes in geological substrates and soils as well as in elevation.

2.5.1 Soil gradients

It is well known that many species of ferns occur preferentially on specific rock or soil types. In Europe, for example, many species of *Asplenium* are restricted to either limestone, serpentine or acidic rock (e.g., basalt, granite; Vogel *et al.*, 1999b; James *et al.*, 2008). However, in most cases it is not known if their substrate specificity is based on physiological requirements, or whether it is driven by competitive interactions among species.

In the tropics, the most extensive study on the relationship of fern community composition to soil factors has been conducted by Hanna Tuomisto and colleagues in western Amazonia. They have shown that in this climatically fairly homogeneous region, differences in topography and soil conditions explain many of the changes in species composition of communities of ferns and other plant groups (Tuomisto and Ruokolainen, 1994; Tuomisto and Poulsen, 1996, 2000; Tuomisto *et al.*, 2002; Poulsen *et al.* 2006; see also Young and León, 1989). These local environmental factors appear to be overriding the effect of dispersal limitation at local (Jones *et al.*, 2006, in Costa Rica) and at regional (Tuomisto *et al.*, 2003a, 2003b, 2003c, in Amazonia) scales. Detailed studies of *Adiantum* (maidenhair fern; Tuomisto *et al.*, 1998) and *Polybotrya* (Tuomisto, 2006) in Amazonia, and of tree ferns in Costa Rica (Jones *et al.*, 2008), have likewise shown that individual species have distinct "preferences" for specific soil conditions. The specificity of fern assemblages to soil conditions renders them suitable as indicators for forest types (Ruokolainen *et al.*, 1997; Salovaara *et al.*, 2004) and for distribution patterns of other plant groups (Vormisto *et al.*, 2000). Combining these floristic studies with satellite imagery has allowed the distinction of more than 200 forest types in Peruvian Amazonia alone (Tuomisto, 1998; Tuomisto *et al.*, 1995, 2003c).

2.5.2 Elevational gradients

Along elevational gradients, the composition of fern communities changes considerably. While some major families such as Aspleniaceae and Dryopteridaceae are found from the lowest to the highest elevations, many others show distinct elevational preferences (Mehltreter, 1995). For example, within the Hymenophyllaceae, the trichmanoid ferns (Plate 6A, B) typically occur at low and middle elevations, whereas the hymenophylloid ferns are mainly found at mid to high elevations (Dubuisson *et al.*, 2003). At the species level, the restriction to specific elevational zones becomes even more apparent. Species in tropical mountains have, on average, elevational amplitudes of about 1000 m, although many occur only within a few hundred meters of altitude, and some species have wider elevational ranges of up to 3000 m (Kessler, 2002b; Kluge *et al.*, 2006; Jácome *et al.*, 2007).

While it is logical that individual fern species have specific elevational distributions determined by their ecological requirements, a more complex question is whether these elevational limits apply simultaneously to a large number of species. The latter scenario would result in a specific elevational zonation of fern assemblages that would reflect the elevational vegetation belts typically distinguished in mountains (e.g., Holdridge *et al.*, 1971; Grubb, 1974; Ellenberg, 1975; Frahm and Gradstein, 1991). Recently, a number of studies have addressed this question using statistical tests to assess if the turnover of communities at specific elevations is higher than expected by chance (Kessler, 2000a; Bach, 2004; Hemp, 2006; Bach *et al.*, 2007; Kluge *et al.*, 2008). These studies have obtained mixed results. In some cases, no distinct boundaries were found over elevational amplitudes of more than 2000 m, but in other cases, several distinct elevational zones were

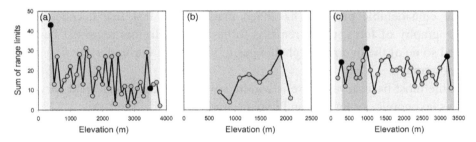

Fig. 2.7 Elevational zonation (different shades of gray) of fern assemblages along elevational gradients in the (a) very wet Carrasco National Park, Bolivia, (b) semihumid Masicurí valley, Bolivia, and (c) Costa Rica (modified after Kessler, 2000a, Kluge *et al.*, 2008). The symbols show the number of species that reach their upper or lower distributional limits at a given elevation, with the black symbols denoting zones of statistically significant elevated rates of species turnover. In (a) Carrasco and (c) Costa Rica, the uppermost and lowermost two limits correspond to the transition from the lowlands to foothills and to tree line, respectively. In addition, the limit at *c.* 1000 m in (c) Costa Rica denotes the lower limit of the regular cloud condensation layer. In (b) Masicurí, the only limit shows the transition from mixed evergreen forest to forest dominated by the conifer *Podocarpus*.

discernible (Fig. 2.7). In most cases, boundaries correspond to zones where other taxa experience abrupt ecological shifts such as the upper timberline, the lower limit of the cloud condensation belt, changes in forest type (e.g., evergreen to deciduous) or the transition from the steep mountains to the flat lowlands. In many cases, floristic boundaries are determined only by accumulations of either upper or lower elevational limits of species rather than by both. For example, the abrupt shift in community composition at the lower limit of the cloud condensation belt is determined mainly by lower elevational limits of species restricted to the cloud condensation zone (Bach *et al.*, 2007; Kluge *et al.*, 2008). In the other direction, species growing at lower elevation extend to various degrees into the cloud condensation zone without presenting abrupt changes. This pattern is certainly determined by the degree to which the environmental conditions are more stressful on one side of the ecological boundary. For many fern species of the cloud condensation belt such as those of the Hymenophyllaceae, the dry conditions below it are limiting, whereas the humid conditions within the belt may not limit the range extension of any fern species of lower elevations. In conclusion, it appears that where ecological conditions change gradually, fern communities also change gradually, whereas abrupt ecological shifts lead to a distinct zonation of fern communities, with most distributional limits being directed toward the more stressful environmental conditions.

2.6 Questions for the future

While considerable progress has been made in the last few decades on the biogeography of ferns, much remains to be learned. In this review, I have discussed some of the overriding challenges to improving our understanding of fern biogeography.

At the most basic level, there is continued need for additional field surveys as well as taxonomic, systematic and floristic studies (Smith, 2006). About 15% of all fern and lycophyte species may not yet be known to science (Chapman, 2006), and for many of those already known, information on distribution and ecology is rudimentary. Unfortunately, extensive field surveys have been restricted to the few remaining, mostly tropical ecosystems that are still unaltered. Further, climatic shifts, both as a result of global climate change (e.g., Pounds *et al.*, 1999; Still *et al.*, 1999) and of forest clearance (Lawton *et al.*, 2001), may already have disrupted the natural climate–fern community relationships. Ultimately, an understanding of the factors determining the present-day composition and diversity of fern communities will depend on linking phylogenies at species and population levels with detailed field- and laboratory-based ecological and ecophysiological measurements.

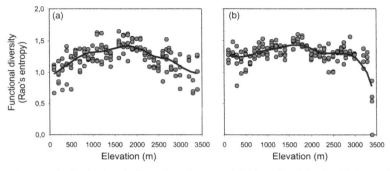

Fig. 2.8 Morphological trait diversity of terrestrial (a) and epiphytic (b) ferns along an elevational gradient in Costa Rica (J. Kluge and M. Kessler, unpublished data). The diversity index Rao's entropy gives a measure of the variability of morphological traits within a community, independently of the species richness. Note that both patterns of morphological diversity are less hump shaped than the patterns of species richness (Fig. 2.2), and that the morphological diversity of epiphytic ferns is almost constant, except above tree line, where there are only a handful of epiphytes living in scattered bushes. The trend lines in this figure were drawn by distance-weighted least-squares smoothing and are intended to guide the eye.

A deeper understanding of the population biology and genetics of ferns is indispensable in order to understand the phylogeography of many clades, and the evolution and maintenance of contemporary fern communities (Haufler, 2007). Especially for tropical taxa, questions of how the populations are spatially distributed and what levels of gene flow exist between them remain largely unexplored (James *et al.*, 2008).

In the context of community diversity, one may not only focus on species numbers, but also on the diversity of the morphological traits of the species. Thus, a community consisting of three similar species from the same genus may be morphologically less diverse than a community of three species belonging to different families and looking very different from each other (Magurran, 2004). No such studies have yet been published on ferns, but preliminary analyses along an elevational gradient in Costa Rica suggest that interesting patterns may be found (Fig. 2.8).

Finally, very little is known about the ecology and biogeography of gametophytes. Although the haploid gametophyte is a fundamentally different organism than the diploid sporophyte, there are only a few studies on the ecology of gametophytes of selected species (Cousens, 1981, 1988; Cousens *et al.*, 1985; Greer and McCarthy, 1999; Watkins *et al.*, 2007). In some of these species in which gametophytes inhabit temperate regions inhospitable to the respective sporophytes (Farrar, 1967; Peck *et al.*, 1990), the gametophytes are more stress tolerant than their sporophytes (Sato and Sakai, 1980, 1981).

Acknowledgements

I thank Jürgen Kluge for sharing his unpublished data and for preparing some of the figures; Klaus Mehltreter, Harald Schneider and two anonymous reviewers for painstakingly reviewing and greatly improving the manuscript; and the German Research Foundation (DFG) for funding numerous fern-related projects over the years.

References

Abrams, P. A. (1995). Monotonic or unimodal diversity-productivity gradients: what does competition theory predict? *Ecology*, **76**, 2019–27.

Adersen, H. (1988). Null hypotheses and species composition in the Galápagos Islands. In *Diversity and Patterns in Plant Communities*, ed. H. J. During, M. J. A. Werger and J. H. Willems. The Hague: Academic Publishing, pp. 37–46.

Aldasoro, J. J., Cabezas, F. and Aedo, C. (2004). Diversity and distribution of ferns in sub-Saharan Africa, Madagascar and some islands of the South Atlantic. *Journal of Biogeography*, **31**, 1579–604.

Allen, A. P. and Gillooly, J. F. (2006). Assessing latitudinal gradients in speciation rates and biodiversity at the global scale. *Ecology Letters*, **9**, 947–54.

Andersen, M., Thornhill, A. and Koopowitz, H. (1997). Tropical forest disruption and stochastic biodiversity losses. In *Tropical Forest Remnants: Ecology, Management, and Conservation of Fragmented Communities*, ed. W. F. Laurance and R. O. Bierregaard. Chicago, IL, USA: University of Chicago Press, pp. 281–91.

Bach, K. (2004). *Vegetationskundliche Untersuchungen zur Höhenzonierung tropischer Bergwälder in den Anden Boliviens*. Marburg, Germany: Verlag Görich and Weiershäuser.

Bach, K., Kessler, M. and Gradstein, S. R. (2007). A simulation approach to determine statistical significance of species turnover peaks in a species-rich tropical cloud forest. *Diversity and Distributions*, **13**, 863–70.

Balmford, A. and Long, A. (1994). Avian endemism and forest loss. *Nature*, **372**, 623–4.

Barrington, D. S. (1993). Ecological and historical factors in fern biogeography. *Journal of Biogeography*, **20**, 275–80.

Barthlott, W., Mutke, J., Rafiqpoor, M. D., Kier, G. and Kreft, H. (2005). Global centres of vascular plant diversity. *Nova Acta Leopoldina*, **92**, 61–83.

Behrensmeyer, A. K., Damuth, J. D., DiMichele, W. A. and Potts, R. (1992). *Terrestrial Ecosystems through Time: Evolutionary Paleoecology of Terrestrial Plants and Animals*. Chicago, IL, USA: University of Chicago Press.

Bhattarai, K. R., Vetaas, O. R. and Grytnes, J. A. (2004). Fern species richness along a central Himalayan elevational gradient, Nepal. *Journal of Biogeography*, **31**, 389–400.

Bickford, S. A. and Laffan, S. W. (2006). Multi-extent analysis of the relationship between pteridophyte species richness and climate. *Global Ecology and Biogeography*, **15**, 588–601.

Bokma, F. and Mönkkönen, M. (2000). The mid-domain effect and the longitudinal dimension of continents. *Trends in Ecology and Evolution*, **15**, 288–9.

Bromham, L. and Penny, D. (2003). The modern molecular clock. *Nature Reviews Genetics*, **4**, 216–24.

Brownsey, P. J. (2001). New Zealand's pteridophyte flora: plants of ancient lineage but recent arrival? *Brittonia*, **53**, 284–303.

Cardillo, M. (1999). Latitude and rates of diversification in birds and butterflies. *Proceedings of the Royal Society of London, Series B*, **266**, 1221–5.

Carlquist, S. (1974). *Island Biology*. New York: Columbia University Press.

Caulton, E., Keddie, S., Carmichael, R. and Sales, J. (2000). A ten year study of the incidence of spores of bracken (*Pteridium aquilinum* (L.) Kuhn) in an urban rooftop airstream in south-east Scotland. *Aerobiologia*, **16**, 29–33.

Chapman, A. D. (2006). *Numbers of Living Species in Australia and the World*. Report for the Department of the Environment and Heritage, Canberra, Australia.

Christ, H. (1910). *Die Geographie der Farne*. Leipzig, Germany: Gustav Fischer Verlag.

Chust, G., Chave, J., Condit, R., *et al.* (2006). Determinants and spatial modeling of tree β-diversity in a tropical forest landscape in Panama. *Journal of Vegetation Science*, **17**, 83–92.

Colwell, R. K. and Hurtt, G. C. (1994). Nonbiological gradients in species richness and a spurious Rapoport effect. *American Naturalist*, **144**, 570–95.

Colwell, R. K. and Lees, D. C. (2000). The mid-domain effect, geometric constraints on the geography of species richness. *Trends in Ecology and Evolution*, **15**, 70–76.

Colwell, R. K., Rahbek, C. and Gotelli N. J. (2004). The mid-domain effect and species richness patterns, what have we learned so far? *American Naturalist*, **163**, E000–E023.

Conant, D. S. (1978). A radioisotope technique to measure spore dispersal of the tree fern *Cyathea arborea* Sm. *Pollen et Spores*, **20**, 583–93.

Conant, D. S., Raubeson, L. A., Attwood, D. K. and Stein, D. B. (1995). The relationships of Papuasian Cyatheaceae to New World tree ferns. *American Fern Journal*, **85**, 328–40.

Condit, R., Pitman, N., Leigh, E. G., *et al.* (2002). Beta-diversity in tropical forest trees. *Science*, **295**, 666–9.

Copeland, E. B. (1939). Fern evolution in Antarctica. *Philippine Journal of Science*, **70**, 157–88.

Cousens, M. I. (1981). *Blechnum spicant*, habitat and vigor of optimal, marginal, and disjunct populations and field observations of gametophytes. *Botanical Gazette*, **142**, 251–8.

Cousens, M. I. (1988). Reproductive strategies in pteridophytes. In *Plant Reproductive Ecology, Patterns and Strategies*, ed. J. L. Doust and L. L. Doust. New York: Oxford University Press, pp. 307–28.

Cousens, M. I., Lacey, D. G. and Scheller, J. M. (1985). Life-history studies of ferns: a consideration of perspective. *Proceedings of the Royal Society of Edinburgh, Series B*, **86**, 371–80.

Cowling, R. M., Witkowski, E. T. F., Milewski, A. V. and Newbey, K. R. (1994). Taxonomic, edaphic and biological aspects of narrow plant endemism on matched sites in mediterranean South Africa and Australia. *Journal of Biogeography*, **21**, 651–64.

Cranfill, R. and Kato, M. (2003). Phylogenetics, biogeography, and classification of the woodwardioid ferns (Blechnaceae). In *Pteridology in the New Millennium*, ed. S. Chandra and M. Srivastava. Dordrecht, The Netherlands: Kluwer Academic Publishers, pp. 25–48.

Currie, D. J., Mittelbach, G. G., Cornell, H. V., *et al.* (2004). Predictions and tests of climate-based hypotheses of broad-scale variation in taxonomic richness. *Ecology Letters*, **7**, 1121–34.

Dalling, J. W., Muller-Landau, H. C., Wright, S. J. and Hubbell, S. P. (2002). Role of dispersal in the recruitment limitation of neotropical pioneer species. *Journal of Ecology*, **90**, 714–27.

Dassler, C. L. and Farrar, D. R. (1997). Significance of form in fern gametophytes, clonal, gemmiferous gametophytes of *Callistopteris baueriana* (Hymenophyllaceae). *International Journal of Plant Sciences*, **158**, 622–39.

Dassler, C. L. and Farrar, D. R. (2001). Significance of gametophyte form in long-distance colonization by tropical, epiphytic ferns. *Brittonia*, **53**, 325–69.

Dubuisson, J.-Y., Hennequin, S., Rakotondrainibe, F. and Schneider, H. (2003). Ecological diversity and adaptive tendencies in the tropical fern *Trichomanes* L. (Hymenophyllaceae) with special reference to epiphytic and climbing habits. *Botanical Journal of the Linnean Society*, **142**, 41–63.

Dzwonko, Z. and Kornaś, J. (1994). Patterns of species richness and distribution of pteridophytes in Rwanda (Central Africa), a numerical approach. *Journal of Biogeography*, **21**, 491–501.

Ellenberg, H. (1975). Vegetationsstufen in perhumiden bis perariden Bereichen der tropischen Anden. *Phytocoenologia*, **2**, 368–78.

Erdtman, G. (1937). Pollen grains recovered from the atmosphere over the Atlantic. *Acta Horti Gotoburgensis*, **12**, 185–96.

Evans, K. L., Warren, P. H. and Gaston, K. J. (2005). Species-energy relationships at the macroecological scale, a review of mechanisms. *Biological Review*, **80**, 1–25.

Farrar, D. R. (1967). Gametophytes of four tropical fern genera reproducing independently of their sporophytes in the southern Appalachians. *Science*, **155**, 1266–7.

Ferrer-Castán, D. and Vetaas, O. R. (2005). Pteridophyte richness, climate and topography in the Iberian Peninsula, comparing spatial and nonspatial models of richness patterns. *Global Ecology and Biogeography*, **14**, 155–65.

Frahm, J.-P. and Gradstein, S. R. (1991). An altitudinal zonation of tropical rain forests using bryophytes. *Journal of Biogeography*, **18**, 669–78.

Gaff, D. F. (1977). Desiccation-tolerant vascular plants of Southern Africa. *Oecologia*, **31**, 95–109.

Geiger, J. M. O., Ranker, T. A., Neale, J. M. R. and Klimas, S. T. (2007). Molecular biogeography and origins of the Hawaiian fern flora. *Brittonia*, **59**, 142–58.

Gentry, A. H. (1986). Endemism in tropical vs temperate plant communities. In *Conservation Biology: The Science of Scarcity and Diversity*, ed. M. Soulé. Sunderland, MA. USA: Sinauer Associates, pp. 153–81.

Gilpin, M. E. and Hanski, I. A. (1991). *Metapopulation Dynamics: Empirical and Theoretical Investigations*. London: Academic Press.

Given, D. R. (1993). Changing aspects of endemism and endangerment in Pteridophyta. *Journal of Biogeography*, **20**, 293–302.

Gradstein, S. R. and van Zanten, B. O. (2001). High altitude dispersal of spores, an experimental approach. *XVI International Botanical Congress, St. Louis*. Abstract Number 15.14.13.

Graves, G. L. (1988). Linearity of geographic range and its possible effect on the population structure of Andean birds. *Auk*, **105**, 47–52.

Greer, G. K. and McCarthy, B. C. (1999). Gametophytic plasticity among four species of ferns with contrasting ecological distributions. *International Journal of Plant Sciences*, **160**, 879–86.

Gressitt, J. L., Sedlacek, J., Wise, K. A. J. and Yoshimoto, C. M. (1961). A high speed airplane trap for air-borne organisms. *Pacific Insects*, **3**, 549–55.

Grubb, P. J. (1974). Factors controlling the distribution of forest types on tropical mountains, new facts and new perspective. In *Altitudinal Zonation in Malaysia*, ed. J. R. Flenley. Trans. 3rd Aberdeen–Hull Symposium on Malaysian Ecology. University of Hull, Department of Geography, Miscellaneous Series, 16, pp. 13–46.

Grytnes, J. A. (2003a). Species-richness patterns of vascular plants along seven altitudinal transects in Norway. *Ecography*, **26**, 291–300.

Grytnes, J. A. (2003b). Ecological interpretations of the mid-domain effect. *Ecology Letters*, **6**, 883–8.

Grytnes, J. A. and Vetaas, O. R. (2002). Species richness and altitude, a comparison between null models and interpolated plant species richness along the Himalayan altitudinal gradient, Nepal. *American Naturalist*, **159**, 294–304.

Grytnes, J. A., Heegaard, E. and Romdal, T. S. (2008). Can the mass effect explain the mid-altitudinal peak in vascular species richness? *Basic and Applied Ecology*, **9**, 371–82.

Guo, Q., Kato, M. and Ricklefs, R. E. (2003). Life history, diversity and distribution, a study of Japanese pteridophytes. *Ecography*, **26**, 129–38.

Harrison, S. and Cornell, H. V. (2007). Introduction, merging evolutionary and ecological approaches to understanding geographic gradients in species richness. *American Naturalist*, **170**, S1–S4.

Haufler, C. H. (2007). Genetics, phylogenetics, and biogeography: Considering how shifting paradigms and continents influence fern diversity. *Brittonia*, **59**, 108–14.

Haufler, C. H., Hooper, E. A. and Therrien, J. P. (2000). Modes and mechanisms of speciation in pteridophytes: Implications of contrasting patterns in ferns representing temperate and tropical habitats. *Plant Species Biology*, **15**, 223–36.

Hawkins, B. A. and Diniz-Filho, J. A. F. (2002). The mid-domain effect cannot explain the diversity gradient of Nearctic birds. *Global Ecology and Biogeography*, **11**, 419–26.

Hawkins, B. A., Diniz-Filho, J. A. F. and Weis, A. E. (2005). The mid-domain effect and diversity gradients, is there anything to learn? *American Naturalist*, **166**, E140–E143.

Hemp, A. (2002). Ecology of the pteridophytes on the southern slopes of Mt. Kilimanjaro. I. Altitudinal distribution. *Plant Ecology*, **159**, 211–39.

Hemp, A. (2006). Continuum or zonation? Altitudinal diversity patterns in the forests on Mt. Kilimanjaro. *Plant Ecology*, **184**, 27–42.

Herzog, S. K., Kessler, M. and Bach, K. (2005). The elevational gradient in Andean bird species richness at the local scale, a foothill peak and a high-elevation plateau. *Ecography*, **28**, 209–22.

Holdridge, L. R., Grenke, W. C., Hatheway, W. H., Liang, T. and Tosi, J. A. (1971). *Forest Environments in Tropical Life Zones, a Pilot Study*. Oxford, UK: Pergamon Press.

Hoot, S. B., Taylor, W. C. and Napier, N. S. (2006). Phylogeny and biogeography of *Isoëtes* (Isoëtaceae) based on nuclear and chloroplast DNA sequence data. *Systematic Botany*, **31**, 449–60.

Hubbell, S. P. (2001). *The Unified Neutral Theory of Biodiversity and Biogeography*. Princeton, NJ, USA: Princeton University Press.

Humboldt, A. von (1805). *Essai sur la géographie des plantes accompagné d'un tableau physique des régions équinoxiales, fondé sur des mesures exécutées, depuis le dixième degré de latitude boréale jusqu'au dixième degré de latitude australe, pendant les années 1799, 1800, 1801, 1802 et 1803*. Paris: Levrault Schoell.

Humboldt, A., von (1808). *Ansichten der Natur mit wissenschaftlichen Erläuterungen*. Tübingen, Germany: J. G. Gotta.

Humphries, C. J. (1979). Endemism and evolution in Macaronesia. In *Plants and Islands*, ed. D. Bramwell. London: Academic Press, pp. 171–99.

Hurlbert, A. H. (2004). Species-energy relationships and habitat complexity in bird communities. *Ecology Letters*, **7**, 714–20.

Hurlbert, A. H. (2006). Linking species-area and species-energy relationships in *Drosophila* microcosms. *Ecology Letters*, **9**, 287–94.

Hurlbert, A. H. and Haskell, J. P. (2003). The effect of energy and seasonality on avian species richness and community composition. *American Naturalist*, **161**, 83–97.

Jacobsen, W. B. G. and Jacobsen, N. H. G. (1989). Comparison of the pteridophyte floras of southern and eastern Africa, with special reference to high-altitude species. *Bulletin du Jardin Botanique National de Belgique*, **59**, 261–317.

Jácome, J., Kessler, M. and Smith, A. R. (2007). A human-induced skewed elevational frequency distribution of ferns in the Andes. *Global Ecology and Biogeography*, **16**, 313–8.

James, K. E., Schneider, H., Ansell, S. E., *et al.* (2008). Diversity arrays technology (DArT) for pan-genomic evolutionary studies of non-model organisms. *PLoS ONE*, 3, e 1682, 1–11.

Janssen, T., Kreier, H.-P. and Schneider, H. (2007). Origin and diversification of African ferns with special emphasis on Polypodiaceae. *Brittonia*, **59**, 159–81.

Jetz, W. and Rahbek, C. (2001). Geometric constraints explain much of the species richness pattern in African birds. *Proceedings of the National Academy of Sciences, USA*, **98**, 5661–6.

Jones, M. M., Tuomisto, H., Clark, D. B. and Olivas, P. (2006). Effects of mesoscale environmental heterogeneity and dispersal limitation on floristic variation in rain forest ferns. *Journal of Ecology*, **94**, 181–95.

Jones, M. M., Olivas-Rojas, P., Tuomisto, H. and Clark, D. B. (2007). Environmental and neighbourhood effects on tree fern distributions in a neotropical lowland rain forest. *Journal of Vegetation Science*, **18**, 13–24.

Jones, M. M., Tuomisto, H., Clark, D. B. and Olivas, P. (2008). Differences in the degree of environmental control of large and small tropical plants, just a sampling effect? *Journal of Ecology*, **96**, 367–77.

Kaspari, M., O'Donnell, S. and Kercher, J. R. (2000). Energy, density, and constraints to species richness, ant assemblages along a productivity gradient. *American Naturalist*, **155**, 280–93.

Kattan, G. H. and Franco, P. (2004). Bird diversity along elevational gradients in the Andes of Colombia, area and mass effects. *Global Ecology and Biogeography*, **13**, 451–8.

Kato, M. (1993). Biogeography of ferns; dispersal and vicariance. *Journal of Biogeography*, **20**, 265–74.

Kawai, H., Kanegae, T., Christensen, S., *et al.* (2003). Responses of ferns to red light are mediated by an unconventional photoreceptor. *Nature*, **421**, 287–90.

Kessler, M. (1999). Plant species richness and endemism during natural landslide succession in a perhumid montane forest in the Bolivian Andes. *Ecotropica*, **5**, 123–36.

Kessler, M. (2000a). Altitudinal zonation of Andean cryptogam communities. *Journal of Biogeography*, **27**, 275–82.

Kessler, M. (2000b). Elevational gradients in species richness and endemism of selected plant groups in the central Bolivian Andes. *Plant Ecology*, **149**, 181–93.

Kessler, M. (2001a). Maximum plant community endemism at intermediate intensities of anthropogenic disturbance in Bolivian montane forests. *Conservation Biology*, **15**, 634–41.

Kessler, M. (2001b). Pteridophyte species richness in Andean forests in Bolivia. *Biodiversity and Conservation*, **10**, 1473–95.

Kessler, M. (2001c). Patterns of diversity and range size of selected plant groups along an elevational transect in the Bolivian Andes. *Biodiversity and Conservation*, **10**, 1897–1920.

Kessler, M. (2002a). The elevational gradient of Andean plant endemism, varying
influences of taxon-specific traits and topography at different taxonomic levels.
Journal of Biogeography, **29**, 1159–66.

Kessler, M. (2002b). Range size and its ecological correlates among the pteridophytes of
Carrasco National Park, Bolivia. *Global Ecology and Biogeography*, **11**, 89–102.

Kessler M. and Bach K. (1999). Using indicator families for vegetation classification in
species-rich Neotropical forests. *Phytocoenologia*, **29**, 485–502.

Kessler, M. and Siorak, Y. (2007). Desiccation and rehydration experiments on leaves of
43 pteridophyte species. *American Fern Journal*, **97**, 175–85.

Kessler, M. and Smith, A. R. (2006). Five new species of *Asplenium* (Aspleniaceae) from
Bolivia. *Candollea*, **61**, 305–13.

Kessler, M. and Smith, A. R. (2007). Ten new species and other nomenclatural changes
for ferns from Bolivia. *Brittonia*, **59**, 186–97.

Kessler, M., Parris, B. S. and Kessler, E. (2001). A comparison of the tropical montane
pteridophyte communities of Mount Kinabalu, Borneo, and Parque Nacional Carrasco,
Bolivia. *Journal of Biogeography*, **28**, 611–22.

Kessler, M., Siorak, Y., Wunderlich, M. and Wegner, C. (2007). Patterns of morphological
leaf traits among pteridophytes along humidity and temperature gradients in the
Bolivian Andes. *Functional Plant Biology*, **34**, 963–71.

Kluge, J. and Kessler, M. (2006). Fern endemism and its correlates: contribution from an
elevational transect in Costa Rica. *Diversity and Distributions*, **12**, 535–45.

Kluge, J. and Kessler, M. (2007). Morphological characteristics of fern assemblages along
an elevational gradient, patterns and causes. *Ecotropica*, **13**, 27–43.

Kluge, J., Kessler, M. and Dunn, R. (2006). What drives elevational patterns of diversity?
A test of geometric constraints, climate, and species pool effects for pteridophytes
on an elevational gradient in Costa Rica. *Global Ecology and Biogeography*,
15, 358–71.

Kluge, J., Bach, K. and Kessler, M. (2008). Elevational distribution and zonation of tropical
pteridophyte assemblages in Costa Rica. *Basic and Applied Ecology*, **9**, 35–43.

Korall, P., Pryer, K. M., Metzgar, J. S., Schneider, H. and Conant, D. S. (2006). Tree ferns:
monophyletic groups and their relationships as revealed by four protein-coding plastid
loci. *Molecular Phylogenetics and Evolution*, **39**, 830–45.

Kornaś, J. (1993). The significance of historical factors and ecological preference in the
distribution of African pteridophytes. *Journal of Biogeography*, **20**, 281–6.

Körner, C. (2000). Why are there global gradients in species richness? Mountains might
hold the answer. *Trends in Ecology and Evolution*, **15**, 513–14.

Kramer, K. U. (1993). Distribution patterns in major pteridophyte taxa relative to those of
angiosperms. *Journal of Biogeography*, **20**, 287–91.

Kramer, K. U., Schneller, J. J. and Wollenweber, E. (1995). *Farne und Farnverwandte*.
Stuttgart, Germany: Georg Thieme Verlag.

Kreft, H. and Jetz, W. (2007). Global patterns and determinants of vascular plant diversity.
Proceedings of the National Academy of Sciences, USA, **104**, 5925–30.

Kreier, H.-P. and Schneider, H. (2006). Phylogeny and biogeography of the staghorn fern
genus *Platycerium* (Polypodiaceae, Polypodiidae). *American Journal of Botany*, **93**,
217–25.

Krömer, T., Kessler, M., Gradstein, S. R. and Acebey, A. (2005). Diversity patterns of
vascular epiphytes along an elevational gradient in the Andes. *Journal of
Biogeography*, **32**, 1799–810.

Kruckeberg, A. R. and Rabinowitz, D. (1985). Biological aspects of endemism in higher
plants. *Annual Review of Ecology and Systematics*, **16**, 447–79.

Lawton, R. O., Nair, U. S., Pielke, R. A., Sr. and Welch, R. M. (2001). Climatic impact of tropical lowland deforestation on nearby montane cloud forests. *Science*, **294**, 584–7.

Lehmann, A., Leathwick, J. R. and Overton, J. McC. (2002). Assessing New Zealand fern diversity from spatial predictions of species assemblages. *Biodiversity and Conservation*, **11**, 2217–38.

Lehnert, M. (2006). The Cyatheaceae and Dicksoniaceae (Pteridophyta) of Bolivia. *Brittonia*, **58**, 229–44.

Leibold, M. A., Holyoak, M., Mouquet, N., *et al.* (2004). The metacommunity concept, a framework for multi-scale community ecology. *Ecology Letters*, **7**, 601–13.

Lester, S. E., Ruttenberg, B. I., Gaines, S. D. and Kinlan, B. P. (2007). The relationship between dispersal ability and geographic range size. *Ecology Letters*, **10**, 745–58.

Little, D. P. and Barrington, D. S. (2003). Major evolutionary events in the origin and diversification of the fern genus *Polystichum* (Dryopteridaceae). *American Journal of Botany*, **90**, 508–14.

Lloyd, R. M. and Klekowski, E. J. (1970). Spore germination and viability in Pteridophyta: evolutionary significance of chlorophyllous spores. *Biotropica*, **2**, 129–37.

Lomolino, M. V. (2001). Elevational gradients of species-density, historical and prospective views. *Global Ecology and Biogeography*, **10**, 3–13.

Lomolino, M. V., Riddle, B. R. and Brown, J. H. (2006). *Biogeography*. Sunderland, MA, USA: Sinauer Associates.

Lovis, J. D. (1959). The geographical affinities of the New Zealand pteridophyte flora. *British Fern Gazette*, **10**, 1–7.

Luebke, N. T. and Budke, J. M. (2003). *Isoëtes tennesseensis* (Isoëtaceae), an octoploid quillwort from Tennessee. *American Fern Journal*, **93**, 184–90.

Lwanga, J. S., Balmford, A. and Badaza, R. (1998). Assessing fern diversity, relative species richness and its environmental correlates in Uganda. *Biodiversity and Conservation*, **7**, 1387–98.

MacArthur, R. H. and Wilson, E. O. (1967). *The Theory of Island Biogeography*. Monographs in Population Biology, 1. Princeton, NJ, USA: Princeton University Press.

Magurran, A. E. (2004). *Measuring Biological Diversity*. Malden, MA: Blackwell Publishing.

McCain, C. M. (2005). Elevational gradients in diversity of small mammals. *Ecology*, **86**, 366–72.

Mehltreter, K. V. (1995). Species richness and geographical distribution of montane pteridophytes of Costa Rica, Central America. *Feddes Repertorium*, **106**, 563–84.

Mittelbach, G. G., Schemske, D. W., Cornell, H. V., *et al.* (2007). Evolution and the latitudinal diversity gradient, speciation, extinction and biogeography. *Ecology Letters*, **10**, 315–31.

Mönkkönen, M., Forsman J. T. and Bokma, F. (2006). Energy availability, abundance, energy-use and species richness in forest bird communities: a test of the species-energy theory. *Global Ecology and Biogeography*, **15**, 290–302.

Moolman, H. J. and Cowling, R. M. (1994). The impact of elephant and goat grazing on the endemic flora of South African succulent thicket. *Biological Conservation*, **68**, 53–61.

Moran, R. C. (1995). The importance of mountains to pteridophytes, with emphasis on neotropical montane forests. In *Biodiversity and Conservation of Neotropical Montane Forests*, ed. S. P. Churchill, H. Balslev, E. Forero and J. L. Luteyn. Bronx, NY, USA: The New York Botanical Garden, pp. 359–63.

Moran, R. C. (1996). The importance of the Andes as a barrier to migration, as illustrated by the pteridophytes of the Chocó phytogeographic region. In *Pteridology in Perspective*, ed. J. M. Camus, M. Gibby and R. J. Johns. Kew, UK: Royal Botanic Gardens, p. 75.

Moran, R. C. and Smith, A. R. (2001). Phytogeographic relationships between neotropical and African-Madagascan pteridophytes. *Brittonia*, **53**, 304–51.

Nathan, R. (2006). Long-distance dispersal in plants. *Science*, **313**, 786–8.

Ohlemüller, R. and Wilson, J. B. (2000). Vascular plant species richness along latitudinal and altitudinal gradients, a contribution from New Zealand temperate rainforests. *Ecology Letters*, **3**, 262–6.

Paciencia, M. L. B. and Prado, J. (2005). Effects of forest fragmentation on pteridophyte diversity in a tropical rain forest in Brazil. *Plant Ecology*, **180**, 87–104.

Page, C. N. (2002). Ecological strategies in fern evolution, a neopteridological overview. *Review of Palaeobotany and Palynology*, **119**, 1–33.

Palmer, D. D. (2003). *Hawai'i's Ferns and Fern Allies*. Honolulu, HI, USA: University of Hawaii Press.

Parris, B. S. (2001). Circum-Antarctic continental distribution patterns in pteridophyte species. *Brittonia*, **53**, 270–83.

Parris, B. S., Beaman, R. S. and Beaman, J. H. (1992). Ferns and Fern Allies. Vol I of *The Plants of Mount Kinabalu*. Kew, UK: Royal Botanic Gardens.

Pausas, J. G. and Sáez, L. (2000). Pteridophyte richness in the NE Iberian Peninsula, biogeographic patterns. *Plant Ecology*, **148**, 197–207.

Pautasso, M. and Gaston, K. J. (2005). Resources and global avian assemblage structure in forests. *Ecology Letters*, **8**, 282–9.

Pautasso, M. and Gaston, K. J. (2006). A test of the mechanisms behind avian generalized individuals–area relationships. *Global Ecology and Biogeography*, **15**, 303–17.

Peck, J. H., Peck, C. J. and Farrar, D. R. (1990). Influences of life history events on formation of local and distant fern populations. *American Fern Journal*, **80**, 126–42.

Perrie, L. and Brownsey, P. (2007). Molecular evidence for long-distance dispersal in the New Zealand pteridophyte flora. *Journal of Biogeography*, **34**, 2028–38.

Pickett, F. (1931). Notes on xerophytic ferns. *American Fern Journal*, **21**, 49–57.

Pole, M. (1994). The New Zealand flora – entirely long-distance dispersal? *Journal of Biogeography*, **21**, 625–35.

Polunin, N. (1951). Seeking airborne botanical particles about the North Pole. *Svensk Botanisk Tidskrift*, **45**, 320–54.

Potts, M. D., Ashton, P. S., Kaufmann, L. S. and Plotkin, J. B. (2002). Habitat patterns in tropical rain forests: a comparison of 105 plots in northwest Borneo. *Ecology*, **83**, 2782–97.

Poulsen, A. D., Tuomisto, H. and Balslev, H. (2006). Edaphic and floristic variation within 1-ha plot of lowland Amazonian rain forest. *Biotropica*, **38**, 468–78.

Pounds, J. A., Fogden, M. P. L. and Campbell, J. H. (1999). Biological response to climate change on a tropical mountain. *Nature*, **398**, 611–14.

Price, J. P. and Clague, D. A. (2002). How old is the Hawai'ian biota? Geology and phylogeny suggest recent divergence. *Proceedings of the Royal Society of London, Series B*, **269**, 2429–35.

Pryer K. M., Smith, A. R., Hunt, J. and Dubuisson, J.-Y. (2001). *rbcL* data reveal two monophyletic groups of filmy ferns (Filicopsida, Hymenophyllaceae). *American Journal of Botany*, **88**, 1118–30.

Pulliam, H. R. (1988). Sources, sinks and population regulation. *American Naturalist*, **132**, 652–61.

Punetha, N. (1991). Studies on atmospheric fern spores at Pithorgarh (northwest Himalaya) with particular reference to distribution of ferns in the Himalayas. *Annual Review of Plant Science*, **13**, 146–61.

Rahbek, C. (1995). The elevational gradient of species richness: a uniform pattern? *Ecography*, **18**, 200–5.

Rahbek, C. (1997). The relationship among area, elevation, and regional species richness in neotropical birds. *American Naturalist*, **149**, 875–902.

Ranker, T. A., Floyd, S. K. and Trapp, P. G. (1994). Multiple colonizations of *Asplenium adiantum-nigrum* onto the Hawai'ian archipelago. *Evolution*, **48**, 1364–70.

Ranker, T. A., Geiger, J. M. O., Kennedy, S. C., *et al.* (2003). Molecular phylogenetics and evolution of the endemic Hawai'ian genus *Adenophorus* (Grammitidaceae). *Molecular Phylogenetics and Evolution*, **26**, 337–47.

Ranker, T. A., Smith, A. R., Parris, B. S., *et al.* (2004). Phylogeny and evolution of grammitid ferns (Grammitidaceae): a case of rampant morphological homoplasy. *Taxon*, **53**, 415–428.

Renner, S. (2005). Relaxed molecular clocks for dating historical plant dispersal events. *Trends in Plant Science*, **10**, 550–8.

Richard, M., Bernhardt, T. and Bell, G. (2000). Environmental heterogeneity and the spatial structure of fern species diversity in one hectare of old-growth forest. *Ecography*, **23**, 231–45.

Ricklefs, R. E. (2005). Phylogenetic perspectives on patterns of regional and local richness. In *Tropical Rainforest, Past, Present, and Future*, ed. E. Bermingham, C. W. Dick and C. Moritz. Chicago, IL, USA: University of Chicago Press, pp. 16–40.

Ricklefs, R. E. (2007). Estimating diversification rates from phylogenetic information. *Trends in Ecology and Evolution*, **22**, 601–10.

Roos, M. (1996). Mapping the world's pteridophyte diversity – systematics and floras. In *Pteridology in Perspective*, ed. J. M. Camus, M. Gibby and R. J. Johns. Kew, UK: Royal Botanic Gardens, pp. 29–42.

Roos, M., Keßler, P. J. A., Gradstein, S. R. and Baas, P. (2004). Species diversity and endemism of five major Malesian islands, diversity-area relationships. *Journal of Biogeography*, **31**, 1893–1908.

Rosenzweig, M. L. and Abramsky, Z. (1993). How are diversity and productivity related? In *Species Diversity in Ecological Communities: Historical and Geographical Perspectives*, ed. R. E. Ricklefs and D. Schluter. Chicago, IL, USA: University of Chicago Press, pp. 52–65.

Rosenzweig, M. L. and Ziv, Y. (1999). The echo pattern of species diversity, pattern and process. *Ecography*, **22**, 614–28.

Rouhan, G., Dubuisson, J.-Y., Rakotondrainibe, F., *et al.* (2004). Molecular phylogeny of the fern genus *Elaphoglossum* (Elaphoglossaceae) based on chloroplast non-coding DNA sequences: contributions of species from the Indian Ocean area. *Molecular Phylogenetics and Evolution*, **33**, 745–763.

Roy, K. and Goldberg, E. E. (2007). Origination, extinction, and dispersal, integrative models for understanding present-day diversity gradients. *American Naturalist*, **170**, S71–S85.

Ruokolainen, K., Linna, A. and Tuomisto, H. (1997). Use of Melostomataceae and pteridophytes for revealing phytogeographical patterns in Amazonian rain forests. *Journal of Tropical Ecology*, **13**, 243–56.

Ruokolainen, K., Tuomisto, H., Macía, M. J., Higgins, M. A. and Yli-Halla, M. (2007). Are floristic and edaphic patterns in Amazonian rain forests congruent for trees, pteridophytes and Melastomataceae? *Journal of Tropical Ecology*, **23**, 13–25.

Salovaara, K. J., Cárdenas, G. G. and Tuomisto, H. (2004). Forest classification in an Amazonian rainforest landscape using pteridophytes as indicator species. *Ecography*, **27**, 689–700.

Samways, M. J. (1994). *Insect Conservation Biology*. London: Chapman and Hall.

Sato, T. and Sakai, A. (1980). Freezing resistance of gametophytes of the temperate fern, *Polystichum retroso-paleaceum*. *Canadian Journal of Botany*, **58**, 1144–8.

Sato, T. and Sakai, A. (1981). Cold tolerance of gametophytes and sporophytes of some cool temperature ferns native to Hokkaido. *Canadian Journal of Botany*, **59**, 604–8.

Schneider, H., Schuettpelz, E., Pryer, K. M., *et al.* (2004a). Ferns diversified in the shadow of angiosperms. *Nature*, **428**, 553–7.

Schneider, H., Russell, S. J., Cox, C. J., *et al.* (2004b). Chloroplast phylogeny of asplenioid ferns based on *rbcL* and *trnL-F* spacer sequences (Polypodiidae, Aspleniaceae) and its implications for biogeography. *Systematic Botany*, **29**, 260–74.

Schneider, H., Ranker, T. A., Russell, S. J., *et al.* (2005). Origin of the endemic fern genus *Diellia* coincides with the renewal of Hawai'ian terrestrial life in the Miocene. *Proceedings of the Royal Society of London, Series B*, **272**, 455–60.

Schneider-Pötsch, H. A. W., Kolukisaoglu, Ü., Clapham, D. H., Hughes, J. and Lamparter, T. (1998). Non-angiosperm phytochromes and the evolution of vascular plants. *Physiologia Plantarum*, **102**, 612–22.

Schneller, J. J. and Liebst, B. (2007). Patterns of variation of a common fern (*Athyrium filix-femina*; Woodsiaceae): population structure along and between altitudinal gradients. *American Journal of Botany*, **94**, 965–71.

Schuettpelz, E. and Pryer, K. M. (2006). Reconciling extreme branch length differences, decoupling time and rate through the evolutionary history of filmy ferns. *Systematic Botany*, **55**, 485–502.

Schuettpelz, E., Schneider, H., Huiet, L., Windham, M. D. and Pryer, K. M. (2007). A molecular phylogeny of the fern family Pteridaceae, assessing overall relationships and the affinities of previously unsampled genera. *Molecular Phylogenetics and Evolution*, **44**, 1172–85.

Sheffield, E. (1996). From pteridophyte spore to sporophyte in the natural environment. In *Pteridology in Perspective*, ed. M. Gibby and R. J. Johns. Kew, UK: Royal Botanic Gardens, pp. 541–9.

Shepherd, L. D., Perrie, L. R. and Brownsey, P. J. (2007). Fire and ice: volcanic and glacial impacts on the phylogeography of the New Zealand forest fern *Asplenium hookerianum*. *Molecular Ecology*, **16**, 4536–49.

Shmida, A. and Wilson, M. W. (1985). Biological determinants of species diversity. *Journal of Biogeography*, **12**, 1–20.

Smith, A. R. (1972). Comparison of fern and flowering plant distributions with some evolutionary interpretations for ferns. *Biotropica*, **4**, 4–9.

Smith, A. R. (1993). Phytogeographic principles and their use in understanding fern relationships. *Journal of Biogeography*, **20**, 255–64.

Smith, A. R. (2006). Floristics in the 21st century: balancing user-needs and phylogenetic information. *Fern Gazette*, **17**, 105–37.

Smith, A. R., Pryer, K. M., Schuettpelz, E., *et al.* (2006). A classification for extant ferns. *Taxon*, **55**, 705–31.

Soria-Auza, R. W. and Kessler, M. (2008). The influence of sampling intensity on the perception of the spatial distribution of tropical diversity and endemism, a case study of ferns from Bolivia. *Diversity and Distributions*, **14**, 123–30.

Stevens, G. C. (1989). The latitudinal gradient in geographical range: how so many species coexist in the tropics. *American Naturalist*, **133**, 240–56.

Still, C. J., Foster, P. N. and Schneider, S. H. (1999). Simulating the effects of climate change on tropical montane cloud forests. *Nature*, **398**, 608–10.

Svenning, J.-C., Kinner, D. A., Stallard, R. F., Engelbrecht, B. M. J. and Wright, S. J. (2004). Ecological determinism in plant community structure across a tropical forest landscape. *Ecology*, **85**, 2526–38.

Trewick, S. A., Morgan-Richards, M., Russell, S. J., *et al.* (2002). Polyploidy, phylogeography and Pleistocene refugia of the rockfern *Asplenium ceterach*: evidence from chloroplast DNA. *Molecular Ecology*, **11**, 2003–12.

Tryon, A. F. (1957). A revision of the fern genus *Pellaea* section *Pellaea*. *Annals of the Missouri Botanical Garden*, **44**, 125–93.

Tryon, A. F. and Lugardon, B. (1990). *Spores of the Pteridophyta*. Berlin: Springer-Verlag.

Tryon, R. M. (1970). Development and evolution of fern floras of oceanic islands. *Biotropica*, **2**, 76–84.

Tryon, R. M. (1972). Endemic areas and geographic speciation in tropical American ferns. *Biotropica*, **4**, 121–31.

Tryon, R. M. (1976). The biogeography of species, with special reference to ferns. *Botanical Review*, **52**, 116–56.

Tryon, R. M. (1985). Fern speciation and biogeography. *Proceedings of the Royal Society of Edinburgh*, **86B**, 353–60.

Tryon, R. M. (1986). The biogeography of species, with special reference to ferns. *Botanical Review*, **52**, 118–56.

Tuomisto, H. (1994). *Ecological Variation in the Rain Forests of Peruvian Amazonia, Integrating Fern Distribution Patterns with Satellite Imagery*. Reports from the Department of Biology, 45. Turku, Finland: University of Turku.

Tuomisto, H. (1998). What satellite imagery and large-scale field studies can tell about biodiversity patterns in Amazonian forests. *Annals of the Missouri Botanical Garden*, **85**, 48–62.

Tuomisto, H. (2006). Edaphic niche differentiation among *Polybotrya* ferns in Western Amazonia, implications for coexistence and speciation. *Ecography*, **29**, 273–84.

Tuomisto, H. and Poulsen, A. D. (1996). Influence of edaphic specialization of pteridophyte distribution in neotropical rain forests. *Journal of Biogeography*, **23**, 283–93.

Tuomisto, H. and Poulsen, A. D. (2000). Pteridophyte diversity and species composition in four Amazonian rain forests. *Journal of Vegetation Science*, **11**, 383–96.

Tuomisto, H. and Ruokolainen, K. (1994). Distribution of *Pteridophyta* and *Melastomataceae* along an edaphic gradient in an Amazonian rain forest. *Journal of Vegetation Science*, **5**, 25–34.

Tuomisto, H., Ruokolainen, K., Kalliola, R., *et al.* (1995). Dissecting Amazonian biodiversity. *Science*, **269**, 63–6.

Tuomisto, H., Poulsen, A. D. and Moran, R. C. (1998). Edaphic distribution of some species of the fern genus *Adiantum* in western Amazonia. *Biotropica*, **30**, 392–9.

Tuomisto, H., Ruokolainen, K., Poulsen, A. D., *et al.* (2002). Distribution and diversity of pteridophytes and Melastomataceae along edaphic gradients in Yasuni National Park, Ecuadorian Amazonia. *Biotropica*, **34**, 516–33.

Tuomisto, H., Poulsen, A. D., Ruokolainen, K., *et al.* (2003a). Linking floristic patterns with soil heterogeneity and satellite imagery in Ecuadorian Amazonia. *Ecological Applications*, **13**, 352–71.

Tuomisto, H., Ruokolainen, K., Aguilar, M. and Sarmientos, A. (2003b). Floristic patterns along a 43-km long transect in an Amazonian rain forest. *Journal of Ecology*, **91**, 743–56.

Tuomisto, H., Ruokolainen, K. and Yli-Halla, M. (2003c). Dispersal, environment, and floristic variation of Western Amazonian forests. *Science*, **299**, 241–4.

van Zanten, B. O. and Gradstein, S. R. (1988). Experimental dispersal geography of neotropical liverworts. *Beihefte zur Nova Hedwigia*, **90**, 41–94.

Vogel, J. C., Barrett, J. A., Rumsey, F. J. and Gibby, M. (1999a). Identifying multiple origins in polyploid homosporous pteridophytes. In *Molecular Systematics and Plant Evolution*, ed. P. M. Hollingsworth, R. M. Bateman and R. J. Gornall. London: Taylor & Francis, pp. 101–17.

Vogel, J. C., Rumsey, F. J., Schneller, J. J., Barrett, J. A. and Gibby, M. (1999b). Where are the glacial refugia in Europe? Evidence from pteridophytes. *Botanical Journal of the Linnean Society*, **66**, 23–37.

Vormisto, J., Phillips, O. L., Ruokolainen, K., Tuomisto, H. and Vásquez, R. (2000). A comparison of fine-scale distribution patterns of four plant groups in an Amazonian rainforest. *Ecography*, **23**, 349–59.

Wagner, W. H. (1995). Evolution of Hawai'ian ferns and fern allies in relation to their conservation status. *Pacific Science*, **49**, 31–41.

Wagner, W. H., Herbst, D. R. and Sohmer, S. H. (1990). *Manual of the Flowering Plants of Hawai'i, Volume 1*. Special Publication 83. Honolulu, HI, USA: University of Hawai'i Press and Bishop Museum Press.

Waide, R. B., Willig, M. R., Steiner, C. F., *et al.* (1999). The relationship between productivity and species richness. *Annual Review of Ecology and Systematics*, **30**, 257–300.

Watkins, J. E., Jr., Cardelús, C., Colwell, R. K. and Moran, R. C. (2006). Species richness and distribution of ferns along an elevational gradient in Costa Rica. *American Journal of Botany*, **93**, 73–83.

Watkins, J. E., Jr., Mack, M. K. and Mulkey, S. S. (2007). Gametophyte ecology and demography of epiphytic and terrestrial tropical ferns. *American Journal of Botany*, **94**, 701–8.

Wiens, J. J. and Donoghue, M. J. (2004). Historical biogeography, ecology and species richness. *Trends in Ecology and Evolution*, **19**, 639–44.

Wikström, N. and Kenrick, P. (1997). Phylogeny of Lycopodiaceae (Lycopsida) and the relationship of *Phylloglossum drummondii* Kunze based on *rbcL* sequence data. *International Journal of Plant Sciences*, **160**, 862–71.

Wikström, N. and Kenrick, P. (2000). Phylogeny of epiphytic *Huperzia* (Lycopodiaceae), paleotropical and neotropical clades corroborated by *rbcL* sequences. *Nordic Journal of Botany*, **20**, 165–71.

Wikström, N. and Kenrick, P. (2001). Evolution of Lycopodiaceae (Lycopsida), estimating divergence times from *rbcL* gene sequences by use of nonparametric rate smoothing. *Molecular Phylogenetics and Evolution*, **19**, 177–86.

Wikström, N., Kenrick, P. and Chase, M. (1999). Epiphytism and terrestrialization in tropical *Huperzia* (Lycopodiaceae). *Plant Systematics and Evolution*, **218**, 221–43.

Wild, M. and Gagnon, D. (2005). Does lack of suitable habitat explain the patchy distribution of rare calcicole fern species? *Ecography*, **28**, 191–6.

Willig, M. R., Kaufman, D. M. and Stevens, R. D. (2003). Latitudinal gradients of biodiversity, pattern, process, scale, and synthesis. *Annual Reviews of Ecology, Evolution and Systematics*, **34**, 273–309.

Wilson, D. S. (1992). Complex interactions in metacommunities, with implications for biodiversity and higher levels of selection. *Ecology*, **73**, 1984–2000.

Wolf, P. G., Sheffield, E. and Haufler, C. H. (1991). Estimates of gene flow, genetic substructure and population heterogeneity in bracken (*Pteridium aquilinum*). *Biological Journal of the Linnean Society*, **42**, 407–23.

Wolf, P. G., Schneider, H. and Ranker, T. A. (2001). Geographic distributions of
    homosporous ferns: does dispersal obscure evidence of vicariance? *Journal of
    Biogeography*, **28**, 263–70.
Wright, D. H. (1983). Species-energy theory, an extension of species-area theory. *Oikos*,
    41, 496–506.
Yatabe, Y., Masuyama, S., Darnaedi, D. and Murakami, N. (2001). Molecular systematics of
    the *Asplenium nidus* complex from Mt. Halimun National Park, Indonesia: evidence
    for reproductive isolation among three sympatric *rbc*L sequence types. *American
    Journal of Botany*, **88**, 1517–22.
Young, K. R. and León, B. (1989). Pteridophyte species diversity in the Central Peruvian
    Amazon, importance of edaphic specialization. *Brittonia*, **41**, 388–95.
Zotz, G. (2005). Vascular epiphytes in the temperate zones – a review. *Plant Ecology*,
    176, 173–83.

3

Ecological insights from fern population dynamics

JOANNE M. SHARPE AND KLAUS MEHLTRETER

Key points

1. Any comprehensive population study of ferns is based on the demography of the three major stages of the life cycle of ferns (spore, gametophyte and sporophyte) and recognizes asexual alternatives to the main sexual life cycle such as apogamy and vegetative reproduction. Knowledge of spore and gametophyte viability and development in natural habitats is critical to our understanding of the life cycle of ferns and their ecology, but most studies have focused on the larger sporophyte.
2. Classifying sequential life history stages for ferns facilitates the assessment of growth and reproductive responses to environmental stimuli. Recognition of a life history stage in ferns is based on leaf and plant morphology because age estimates for fern individuals are complicated by several problems such as gradually decomposing older tissue in rhizomes of understory ferns and variable growth rates throughout the life of an individual tree fern.
3. Basic phenological variables (e.g., leaf count, leaf and spore production rates, and leaf life span) are monitored for a better understanding of seasonal patterns, population structure and biomass turnover. Factors that influence productivity of ferns have been addressed only recently for tropical species.
4. Population studies have compared different fern species within the same habitat and the ferns have shown surprising sensitivity to microhabitat characteristics. Recent investigations of fern population responses to altitudinal and latitudinal gradients and to environmental changes over time have been based on innovative monitoring methods and transition matrices of life history stages.
5. Future research in fern population dynamics should focus on the connection between critical demographic patterns in ferns and their relevance to comprehensive community and ecosystem studies by using consistent methodologies, expanding into larger geographic ranges (especially in tropical regions) and increasing the focus on long-term monitoring.

3.1 Introduction

While life histories of many groups of seed plants (e.g., trees) are well known to ecologists and essential to interpreting ecosystem function, such background

Fern Ecology, ed. Klaus Mehltreter, Lawrence R. Walker and Joanne M. Sharpe. Published by Cambridge University Press. © Cambridge University Press 2010.

information for ferns is either lacking or found scattered throughout the literature in sources often untapped by ecologists. The individual growth patterns and reproductive processes of ferns are as varied as they are for seed plants. The general fern life cycle consists of three basic stages: spore, gametophyte and sporophyte. The fern sporophyte is a perennial plant with simple, modular architecture that has some similarities to sporophytes of seed plants such as palms or terrestrial orchids. As an individual fern grows, the stem (rhizome) elongates; and roots and leaves are produced throughout its life span. In ferns, the sporophyte produces spores in structures called sporangia that are found mostly on the lower leaf surface or on the leaf margins. When dispersed to suitable habitats, a spore may germinate into a gametophyte on which sexual reproduction takes place. Vegetative propagation can result from formation of buds on roots, stems and leaves and from rhizome branching. These offshoots subsequently become independent plants. Variations in these simple components of the fern life cycle lead to a large diversity of demographic patterns that ensure an important ecological role for ferns in many different habitats.

The specific morphological and temporal characteristics of each fern species can be used to distinguish several life history stages that allow for classification of individuals into demographic categories. Groups of individuals form populations that then can be compared based on the proportions of the various life history stages that are present (Silvertown, 1982). When standard characteristics of gametophytes and sporophytes are used to describe each life history stage, insightful comparisons can be made among fern populations found in different habitats or at different life stages, among fern species and across fern studies conducted by multiple researchers.

For a perennial plant such as a fern sporophyte, the sequence of growth from initiation through reproduction to senescence may seem predictable, yet internal triggers and environmental influences result in many variations that affect the duration of each life history stage (e.g., the life span of the individual and the frequency of vegetative reproduction).

Our main objective is to document the current status of research on fern population dynamics (including phenology) and to emphasize features of the fern life cycle that can prove useful in ecological studies. Because ferns are unfamiliar to many ecologists, we will provide several examples of the methods that have been used to describe life history stages for individual fern plants and how they have been applied to describe and compare entire populations. Leaf parameters such as growth, production rates, and life span and their effects on population structure provide a link to such ecological processes as nutrient cycling (see Chapter 4) and succession (see Chapter 6) and must be known before generalizations about the role of ferns in any ecosystem can be fully developed (Harper, 1982). Through summaries and case studies of both short- and long-term research from around the world, we also show that the results of recent studies of fern populations provide the tools for addressing problem ferns such as invasives (see Chapter 8) and the conservation of threatened species (see Chapter 9).

3.2 Life cycle of ferns and lycophytes

Life-cycle diagrams that emphasize ploidy levels and the sequence of different stages of growth document the fascinating steps of fern and lycophyte life histories (see Chapter 1, Figs. 1.2, 1.4; Jones, 1987; Sheffield, 2008). We illustrate the life cycle of one specific rain forest fern, *Danaea wendlandii* (Box 3.1, Fig. 3.1) in order to: (1) introduce the major stages of an individual fern's life history, (2) highlight examples of fern interactions with habitat, and (3) indicate patterns of growth and reproduction of the relatively long lived sporophyte that can form the basis for demographic analysis of populations of any fern.

The lives of all ferns and lycophytes follow the basic life cycle pattern shown for *D. wendlandii* in Fig. 3.1a, but the details of their life histories vary considerably. A more generalized life cycle diagram (Fig. 3.2) graphically presents the basic life history stages and highlights the ontogenetic transitions of a single individual as well as the production of new plants by spores and vegetative propagation. Growth patterns and reproduction for any individual fern species can be described and compared using variations of this diagram (e.g., Peck *et al.*, 1990). For example, some species reproduce vegetatively in more than one way and others may return to an earlier life history stage (rejuvenation), perhaps during senescence or after disturbance (Fig. 3.2). In the following sections, we describe the roles of the major elements of the fern life cycle: spore, gametophyte and sporophyte.

3.3 Fern spore banks

Spores, the haploid propagules dispersed by fern sporophytes, are well known for drifting long distances through air currents (see Chapter 2), and may germinate immediately after landing in a favorable microhabitat. They can also build up soil spore banks that are important sources of recruitment of new plants, because they can remain viable for years or decades (Dyer and Lindsay, 1992). Spores have an incredible variety of shapes, sizes and surface structures (see Chapter 1, Fig. 1.3) that facilitate identification (Tryon and Lugardon, 1990), and although these morpho-logical features currently have no known ecological function (Tryon, 1990), aero-dynamic experiments might provide some insights into limits on dispersal. The few studies of fern spore banks in soil (see Ranal, 2003 for a review) have addressed three main issues: (1) composition and stratification of the soil spore bank, (2) efficacy of laboratory methods versus field methods to document the viability of spores and (3)

Box 3.1 Life cycle of *Danaea wendlandii*: a case study

Danaea wendlandii is a small, rosette-forming fern that grows in dense patches in lowland rain forests in Central and South America (Plate 7A, Fig. 3.1). Details about the life cycle of this fern illustrate its interactions within this humid, low-light environment as we follow a population of the fern (Fig. 3.1a) from spores (A) through formation of a multi-age patch on the forest floor (J). Spores of *D. wendlandii* (A) germinate on bare soil of pits formed when large trees fall, an event that happens approximately every 100 years on its study site in the Costa Rican forest (Sanford *et al.*, 1986). Spores also germinate in small gaps in the litter on the soil surface (Watkins *et al.*, 2007) and on dirt road cuts (Sharpe, 1988).

Spores of *D. wendlandii* germinate into flat, two-dimensional, elongated gametophytes (B) that are about 2 cm long (Fig. 3.1b). Fertilization occurs on the undersurface of a bisexual or female gametophyte. The sporophyte (C) first develops a root, then a simple, rounded leaf. A remnant of the gametophyte may remain attached to the sporophyte for as long as 15 months (Sharpe, 1988). As the juvenile sporophyte develops (D), each successively emerging leaf is larger and the leaf shape becomes more complex, from simple to lobed to pinnate with increasing numbers of leaflet pairs and a terminal leaflet. Leaves emerge from the stout, decumbent rhizome at short distances, forming a compact rosette at the rhizome apex.

It takes a juvenile sporophyte approximately 7–8 years from fertilization to the development of the first sterile leaves of adult morphology (E), characterized by buds at the leaf tip rather than a juvenile terminal leaflet. Sterile leaf length stabilizes at about 11 years and thereafter the sporophyte maintains a crown size of four living leaves. A newly developing leaf exhibits circinate vernation and emerges from the shoot apex as a tightly coiled crozier. A sterile leaf of a mature plant of *D. wendlandii* expands vertically in about eight weeks and can emerge at any time of the year, but most commonly from July to August (Sharpe and Jernstedt, 1990a). Throughout its approximately three-year life span, a sterile leaf becomes increasingly horizontal in orientation and the leaves of multiple plants form a very low canopy that may be an important source of shelter and camouflage for animals (e.g., Plate 7A, bushmaster snakes). As sterile leaves age, vegetative buds at the tips begin to develop into juvenile sporophytes (ramets). As the ramet at the leaf tip develops more leaves and roots it abscises from the parent leaf (H) and becomes an independent juvenile sporophyte, developing new leaves faster (1.6 leaves year^{-1}) than juvenile sporophytes from sexual reproduction (1.1 leaves year^{-1}). Leaf-tip ramets also develop 70% fewer simple leaves before starting to produce more complex leaves (Sharpe, 1988). As the sporophyte matures and after more leaves have been produced, the distal end of the rhizome begins to deteriorate and the rhizome maintains a more or less constant net length throughout the remainder of its life.

As the sporophyte becomes reproductive, fertile leaves emerge (F). *Danaea wendlandii* is dimorphic and an expanding fertile leaf initially exhibits circinate vernation, but in contrast to the sterile leaf, it unrolls horizontally and at about 25% of its final length, lays flat against the ground (to the left in F) for several weeks. The horizontal orientation results from a light-mediated positive response to gravity (Sharpe and Jernstedt, 1990b), a process not yet observed in any other fern genus. Ten weeks later, the fertile leaf rapidly expands vertically (Fig. 3.1c) to its full height (G). A fertile leaf is yellow and succulent in contrast to the dark green of the sterile leaves. Spores are released about two weeks after the fertile leaf matures, usually in the rainy season (July, August and September) and two weeks later the fertile leaf petiole base senesces and the rapidly denaturing fertile leaf abscises. On average, every tenth leaf produced by an adult sporophyte of *D. wendlandii* is fertile, thus releasing spores every 2–3 years (Sharpe and Jernstedt, 1990a). Occasionally, the growing tip of a rhizome may die, perhaps as a result of a direct hit from falling woody debris (I). Death of the shoot apex of an adult sporophyte can result in the development of buds on small leaf base stipules on the rhizome (Sharpe and Jernstedt, 1991). Juvenile sporophytes (ramets) developing from stipule buds are similar to those arising from leaf-tip buds.

A patch of *D. wendlandii* individuals (J) expands as leaf-tip ramets are regularly established (Sharpe, 1993). These patches become multi-age populations consisting of old and juvenile sporophytes and may include sporelings, ramets from leaf-tip buds and ramets from rhizomes with damaged shoot apices. Patches can reach up to 20 m in diameter unless disturbed by tree falls. In much of its range, especially in Panama where historical herbarium locations were tracked (J. M. Sharpe, personal observation), the habitat of *D. wendlandii* has been more threatened by human development than by natural events.

correspondence between the contents of the soil spore bank and the composition of the aboveground fern community. Spores percolate with rainwater downward through the soil, and can reach depths of >1 m; spore movement in other directions may result from soil mixing (Esteves and Dyer, 2003) or disturbance by animals such as earthworms (Hamilton, 1988). Esteves and Dyer (2003) classified spores as living or dead in spore banks that extended 50 cm and 80 cm deep, respectively, at two sites along a river in Scotland. Living spores from fern species growing in the immediate vicinity were found in equal proportions throughout the depth of the soil spore bank, suggesting that viable spores are capable of movement through the soil. Total numbers of spores (both living and dead) declined with depth at one site, but were relatively abundant at the lowest levels at the other site where the profile may have been disturbed by soil movement down a steep slope (Esteves and Dyer, 2003).

Spore viability has usually been tested by germinating spores in a laboratory because of potential contamination in field or greenhouse studies (Halpern *et al.*, 1999). Esteves and Dyer (2003) used spore culture to determine the proportion of live spores at different depths in the spore bank profile and estimated that some spores survive 100 to 200 years at their sites in Scotland. Interestingly, the large percentage of

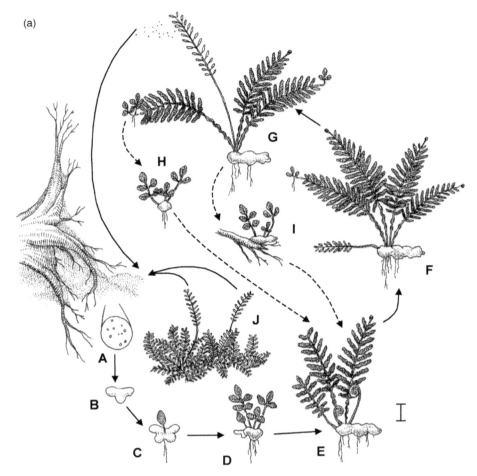

Fig. 3.1 *Danaea wendlandii* a herbaceous layer fern of neotropical rain forests. (a) Illustrated life cycle. Stages of the life cycle: A, spore dispersed to disturbed substrate; B, gametophyte; C, sporeling; D, young sporophyte; E, croziers; F, mature sporophyte with immature fertile leaf; G, mature fertile leaf; H, young sporophyte (ramet from leaf-tip bud); I, young sporophyte (ramet that develops from stipule bud on abscised petiole remnant when apex damaged); J, patch of multi-age sporophytes. Dashed lines indicate vegetative propagation. Scale bar represents 1 cm in B, C, D, H, I; 5 cm in E, F, G; 10 cm in J. (Drawn by Elizabeth Farnsworth.) (b) A population of gametophytes and sporelings. (c) A patch of sporophytes showing vertical mature sterile leaves, recumbent mature sterile leaves, and lower left, a leaf tip ramet still attached to a leaf.

Athyrium filix-femina (lady fern) spores in the total spore count was not reflected in the live spore bank, suggesting that *A. filix-femina* spores had been more abundant at the site in the past. However, Lindsay and Dyer (1996) challenged the contention that spore germination percentages in the laboratory represent a true picture of viability in

Fig. 3.1 (cont.)

the field. Their innovative field methods showed different patterns of spore germination such as seasonal differences among species. Esteves and Dyer (2003) found that the species in the live spore bank in Scotland reflected the plants growing in the immediate vicinity. In contrast, Ramirez-Trejo *et al.* (2004) showed some differences between species found in the soil spore bank and those found aboveground, because only 12 of the 23 studied fern taxa at a site in Mexico formed local soil spore banks. Such results for seed plants may have triggered studies of dispersal and allelopathy, but in ferns these processes affecting the composition of spore banks have yet to be examined. Spore bank investigation is a relatively young area of fern research. Identifying controls on fern recruitment is critical to predicting fern abundance and diversity, therefore more field studies of spore sources, dispersal, viability and germination rates are needed for a variety of fern species in different habitats.

3.4 Gametophyte growth and population structure

A fern gametophyte is a small, independent, usually haploid plant that develops when a spore germinates. For most species, the gametophyte is relatively small, ephemeral, flat and green (Fig. 3.1b, Box 1.2) and grows on aboveground substrates, although for other species the gametophyte may lack chlorophyll and germinate and develop

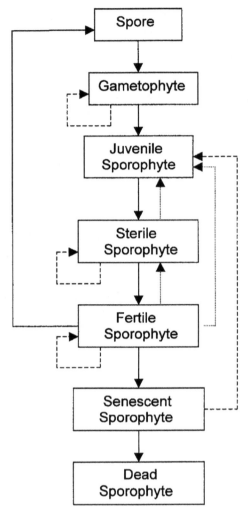

Fig. 3.2 Stages in the life history of a fern or lycophyte. Solid lines indicate direct growth. Dashed lines indicate vegetative reproduction. Dotted lines indicate rejuvenation or regression to an earlier stage.

belowground. Most photosynthetic gametophytes germinate from spores when these are exposed to light. Spores of terrestrial species may germinate after direct dispersal to open substrate or after disturbance of soil spore banks, while gametophytes of epiphytic species develop on undisturbed sites on tree trunks and canopy branches (Watkins *et al.*, 2007). Spore germination can also be induced belowground by antheridiogens (i.e., plant hormones produced by some mature female gametophytes) causing the resulting gametophytes to grow toward the surface where they develop chlorophyll (Schneller, 2008). When spores of coexisting fern species are induced to germinate in this way, but die before reaching the surface, the soil bank of viable spores is effectively reduced, suggesting the possibility of subterranean competitive inhibition in nature (Schneller, 2008). In contrast, gametophytes of some Ophioglossaceae (Johnson-Groh *et al.*, 2002; Whittier, 2003b) and Lycopodiaceae (Whittier, 2003a, 2006, Whittier and Storchova, 2007) develop subterranean gametophytes that obtain nutrition from mycorrhizal fungi (see Chapters 4 and 7). After fertilization the sporophyte develops and grows up toward the soil surface.

Homosporous ferns produce one type of spore that gives rise to potentially bisexual gametophytes. A mature gametophyte of a homosporous fern can reproduce sexually in three ways (Pryer *et al.*, 2008): (1) gametophytic selfing, when a sperm fertilizes an egg cell of the same gametophyte; (2) sporophytic selfing, when a sperm fertilizes an egg cell of a different gametophyte that developed from spores of the same sporophyte; or (3) outcrossing, when a sperm fertilizes an egg cell of another gametophyte that came from a different sporophyte. In heterosporous ferns (e.g., *Selaginella*, spike-mosses, Plate 3B; *Isoëtaceae*, quillworts, Plate 3D; *Marsilea*, water-clover; and *Salvinia*, floating ferns, Figs. 8.9, 8.10), female gametophytes germinate from megaspores and male gametophytes from microspores, which eliminates the option for gametophytic selfing. Even in homosporous ferns, gametophytic selfing seems less common, although most fern textbook illustrations of fern life cycles imply that this is the only method of sexual reproduction (but see Chapter 1, Fig. 1.2; Sheffield, 2008). Gametophytic selfing does facilitate long distance dispersal because only one spore is required to produce a new sporophyte but has the disadvantage that the offspring is completely homozygous (i.e., having identical parental genes). Movement of sperm from one gametophyte to another can result in heterozygous offspring, but requires that at least two spores have been dispersed and germinated within a distance of a few centimeters.

Gametophytes of homosporous ferns are potentially bisexual, but antheridiogens (i.e., plant hormones released by mature female gametophytes) or environmental factors (e.g., light, temperature and soil nutrients) may influence sex expression. Antheridiogens of a mature female gametophyte act at very low concentrations on gametophytes within a range of about 20 cm. Antheridiogens act on their own species, as well as other fern species with chemically similar antheridiogens, to cause immature gametophytes to develop male antheridia that produce sperm (Schneller, 2008). Once an egg cell is fertilized, its gametophyte stops antheridiogen production and neighboring male gametophytes become female or bisexual by developing archegonia (structures that produce an egg). Gametophytes that produce mature antheridia and archegonia one after the other are therefore functionally unisexual, with the supposed evolutionary advantage of decreased selfing (Schneller, 2008). The influences of environmental factors on the breeding systems of ferns are manifold and still little explored, but laboratory studies have shown that low-light conditions increase antheridiogen sensitivity of immature gametophytes. For example, gametophytes grown on light-obscuring ash and soil substrate tended to be male whereas gametophytes grown on transparent agar substrate tended to be female (Rubin *et al.*, 1985).

Although gametophytes are an important part of the fern life cycle, they are relatively short lived, with life spans generally ranging from three months to less than two years (Sato, 1982; Page, 1997; Gureyeva, 2003; Watkins *et al.*, 2007). Normally, gametophytes die after successful fertilization and initial sporophyte development. In some species, gametophytes can develop gemmae (vegetative buds) that grow and detach from the parent gametophyte and allow for the existence of large gametophyte populations independently from any nearby sporophytes (see Chapter 9; Farrar *et al.*, 2008).

For years, fern gametophytes have been considered difficult to identify and incorporate into ecological studies. However, Watkins *et al.* (2007) made detailed observations of gametophyte establishment, growth and life span in Costa Rica. Their field experiments showed that gametophyte diversity and abundance of terrestrial fern species increased with light and disturbance, whereas epiphytic fern gametophytes were negatively affected by increases in disturbance and showed no response to increased light. In studies of population structure, Gureyeva (2003) was able to classify gametophytes of six Siberian fern species (Table 3.1) into eight sequential growth stages based on (1) plant shape, (2) the presence or absence of male and female structures and (3) presence or absence of an attached young sporophyte. For example, spore germination of *Athyrium filix-femina* took place in early June; male

Table 3.1 *Plant and habitat characteristics used for comparison of different population structures by Gureyeva (2003) in Siberia*

Species	Vegetative reproduction	Spore production	Gametophyte settlement	Sporophyte prereproductive	Sporophyte reproductive
Athyrium distentifolium A. *filix-femina*	No	Regular	Regular on bare soil	In areas originally settled by gametophytes	Long lived and found only in forest gaps
Dryopteris carthusiana D. *expansa* D. *filix-mas*	Yes	Regular	Regular on bare soil or rotting wood	In gametophyte areas (from spores) and throughout (vegetative)	Long lived and found throughout habitat
Matteuccia struthiopteris	Yes	No	No	Large numbers due to vegetative reproduction	No

gametophytes (which produce sperm but cannot develop sporophytes) predominated in July; females were most common in August; and by September the majority of gametophytes had developed sporophytes. In the following spring, females were also the most commonly observed class of gametophytes in populations that had survived the winter. The observed seasonal sex expression of gametophytes in their natural habitat supplements the results of laboratory explorations reviewed in DeSoto *et al.* (2008) and Schneller (2008). Cousens *et al.* (1988) differentiated gametophytes of *Woodwardia areolata* (syn. *Lorinseria areolata*, netted chain fern) into three growth stages: first season, second season, and gametophytes attached to a sporophyte in studies that identified safe sites for establishment. The life cycle of *W. areolata* could be completed on decayed pine stumps and twigs in an area flooded only occasionally and on hummocks in swamps at the edge of a river in Florida.

It should be clear from these detailed studies of gametophyte growth and ecology that it is the gametophyte that controls the location and structure of fern populations and not the larger sporophyte. Farrar *et al.* (2008) has more thoroughly reviewed other elements of gametophyte ecology such as variations in plant longevity, effects of disturbance, water relations, desiccation tolerance and photosynthetic rates. Therefore, throughout the remainder of this chapter, we will focus on studies of demography and population dynamics of the larger and longer-lived sporophyte.

3.5 Sporophyte morphology and reproduction

A sporophyte is the plant (usually diploid) that develops from (1) a zygote after fertilization of an egg cell, (2) a single vegetative gametophyte cell in the case of apogamy or (3) vegetative reproduction by a sporophyte. Unlike many seed plants, fern sporophytes (with exception of the lycophyte *Isoëtes*, Plate 3D, and *Botrychium*, Plate 4A) are not capable of secondary growth and are therefore herbaceous perennials, although tree ferns can develop trunks that allow them to reach into the canopy. Sporophytes have stems, leaves and roots. Only the floating ferns of the Salviniaceae and Psilotaceae (Plate 4B) are rootless. Sporophyte anatomy and morphology has been studied in detail (Ogura, 1972), but growth rates and life spans have received little attention. Studies of fern roots have mainly focused on mycorrhizae (relationships with fungi), whereas less is known about root growth, therefore in the following sections we focus on the variation in the morphology of stems and leaves and identify the various ways that sporophytes reproduce vegetatively.

3.5.1 Sporophyte rhizomes

The stem of a fern is called a rhizome and usually develops horizontally, however it can also grow as a vertical trunk in tree ferns. The distance between emerging leaves determines the overall architecture of the plant. An adult sporophyte of a species with a short rhizome (whether erect, ascending or decumbent) produces a compact rosette of leaves at the vertically oriented stem apex (Fig. 3.3a, b). This rosette of leaves can serve as a collector and funnel leaf litter and rainfall toward the actively growing shoot apex, perhaps reflecting a different path for nutrient cycling than directly through the soil (see Chapter 4). In some epiphytic species including *Asplenium nidus* (bird's nest fern), the basket-shaped rosettes can support a huge variety and abundance of invertebrates such as termites (Ellwood *et al.*, 2002). The shoot apex of such rosette-forming ferns of a forest understory may also be well protected from direct blows from debris falling from the canopy, especially in plants with long, strong and vertically oriented leaves (J. M. Sharpe, personal observation). Leaves emerge from the rhizome and when they abscise often leave a distinct scar (e.g., *Cyathea arborea*, Fig. 3.4) or a remnant of the petiole called a phyllopodium on the stem (e.g., many species in *Elaphoglossum* and *Polypodium*). Large leaves of many tree ferns (e.g., *Cyathea dealbata* in New Zealand) can cause considerable damage to seedlings when they abscise and fall to the forest floor (Gillman and Ogden, 2001). In contrast, marcescent leaves are dead leaves of some tree ferns, such as found on *C. smithii* in New Zealand, that remain attached and form a skirt beneath the rosette of living leaves (Page and Brownsey, 1986). These dead leaves slowly decompose and release fine litter to the ground (Gillman and Ogden, 2001).

Few field studies have examined the morphology of belowground structures in ferns with short, compact rhizomes (but see Sharpe, 1993; Greer and McCarthy, 2000; Gureyeva, 2003). Rhizomes decay at the older end over time; their ratio of leaf bases or scars to the living leaf count is usually ≥1. Roots of *Danaea wendlandii* in Costa Rica are produced at the same rate as leaves but can function about twice as long (Sharpe, 1988). The ratio of root plus rhizome versus leaf biomass was 1.0:0.8 for a whole population of *Polystichum acrostichoides* (Christmas fern) in northeastern USA. This study also showed that larger plants allocated a higher proportion of biomass to leaves than to roots and rhizomes and that difference in ratios for individuals could not be related to environmental gradients (Greer and McCarthy, 2000). Species-specific ratios can be used to estimate whole plant biomass production in large-scale or long-term studies where destructive sampling is impossible.

The distance between individual leaves is inevitably greater in species with long, creeping rhizomes (Fig. 3.5) than in the compact, tufted rhizome of species of the rosette form. This creeping growth habit generally prevents the recognition of a single sporophyte as an individual unit (genet), especially when the rhizome grows beneath the soil surface or branches repeatedly. Instead, for creeping species a leaf, a tuft of leaves or a stand of leaves should be the basic unit used for ecological studies (e.g., *Oleandra pistillaris* in Indonesia, Takahashi and Mikami, 2006). A few species have both types of rhizome growth on the same plant. For example, the hemiepiphytic

Fig. 3.3 Adult fertile sporophytes showing leaf morphologies. (a) *Thelypteris deltoidea*, a monomorphic rosette-forming fern showing an immature leaf and fertile and sterile mature leaves. (b) *Danaea elliptica*, a dimorphic rosette-forming fern in the rain forest of Puerto Rico. Leaf ages vary from one year at the center to about four years for outer epiphyll-laden leaves. Fertile leaves are ephemeral and have abscised. (c) *Elaphoglossum* sp. showing very different sterile and fertile leaves. (d) Scrambling *Gleichenella pectinata* (Gleicheniaceae) with indeterminate leaf growth.

Fig. 3.3 (cont.)

Fig. 3.3 (cont.)

Fig. 3.4 *Cyathea arborea*, a tree fern in Puerto Rico that has distinctive leaf scars.

Fig. 3.5 *Polypodium piloselloides*, an epiphytic creeping fern with sequential leaf emergence on a tree trunk in Puerto Rico.

Lomagramma guianensis in French Guiana has a long, creeping rhizome with leaves produced at longer intervals during the initial terrestrial growth stage (Gay, 1993; Hebant-Mauri and Gay, 1993), but once the plant encounters and climbs a host tree, the leaves emerge at shorter intervals on the more compact, epiphytic stage of the rhizome (e.g., hemiepiphytic *Blechnum attenuatum*, Plate 6C).

3.5.2 Juvenile leaves

Leaves of juvenile sporophytes are often totally different from the leaves of adult plants (e.g., *Danaea wendlandii*, Fig. 3.1b, c) and this feature can provide some insight into relative ages of individuals, especially in slow-growing species. Leaf development in a young sporophyte is heteroblastic (Allsopp, 1965) in that each successive leaf is larger and often more complex than the previous one until the leaf size and shape stabilize into an adult form that is characteristic of the species. Thus, the first leaves of a fern species with compound adult leaves may be simple and unlobed while subsequent juvenile leaves are lobed and then become increasingly dissected as each new one emerges (Kato and Setoguchi, 1999). Heteroblastic leaf series have been described for very few fern species. For example, Page (1997) illustrated the sequence of juvenile leaf morphology for some ferns in Britain and Ireland based on Orth (1938) such as *Asplenium septentrionale* that has ten different juvenile leaf forms. An extreme example of heteroblastic leaf series is *Lygodium venustum* with juvenile leaves that are dichotomously divided, short (<20 cm) and have determinate growth compared with large, twining adult leaves that are 3–4 pinnate, exhibit indeterminate growth and can be >10 m long (Mehltreter, 2006). Because older plants may recover from damage by producing leaves that repeat the heteroblastic leaf sequence, leaf form and size may represent the nutritional state of the plant that is not necessarily correlated to plant age (K. Mehltreter, personal observation), but it is still a vital diagnostic tool for classifying different stages of plant growth.

3.5.3 Adult leaves

Leaves of the adult form may be of determinate or indeterminate growth. Determinate leaves stop their growth when the genetically determined leaf form is completed, whereas indeterminate leaves (Fig. 3.3d) grow until their leaf tip dies because of external factors (e.g., drought, herbivory). Indeterminate leaves are found in some species of scrambling fern genera such as *Dicranopteris*, *Gleichenia* and *Sticherus* (see Chapter 6, Table 6.3), in a neotropical high elevation (páramo) fern *Jamesonia* (see front book cover), and in climbing ferns such as *Lygodium venustum* (Mehltreter, 2006). Because leaf size and life span of indeterminate leaves are mainly the consequence of environmental factors, these species may serve as biological indicators of climatic extremes, herbivory or disturbance.

In monomorphic ferns with determinate leaves, fertile (spore-bearing) leaves may look exactly like sterile leaves except for the presence of sporangia found in various arrangements, usually on the abaxial leaf surface (Plate 5A, D, Fig. 3.3a). Leaves of the adult form can also be dimorphic (Figs. 3.1a, c, 3.3b, c) when the fertile and sterile leaves of a species differ significantly in morphology and size as well as other important characteristics. Wagner and Wagner (1977) estimate that fertile–sterile dimorphism may occur in as many as 20% of fern species (e.g., *Matteuccia struthiopteris*, Plate 5C). For example, fertile leaves of many dimorphic species may contain little chlorophyll, resulting in lower photosynthetic rates per unit area compared with the sterile leaf of the same species (e.g., *Pyrrosia lingua*, Chiou *et al.*, 2005).

The determination of whether a leaf is fertile or sterile is made while the leaf primordium is developing within the stem apex. Undeveloped leaf primordia may be present in the stem apex up to four years in advance of emergence (Crotty, 1955), but fertile leaf determination may not occur until late in leaf development (e.g., in the fall of the year before emergence as in the temperate *Osmunda cinnamomea*, cinnamon fern, Steeves, 1959). In experiments comparing clones grown in a greenhouse or outside, Siman and Sheffield (2002) showed that indoor growing conditions that included higher light levels and warmer temperature increased fertile leaf production rates for the temperate fern *Polypodium vulgare* (common polypody). In contrast, lower temperatures and shorter photoperiods not only increased fertile leaf production rates but also the proportion of fertile plants of *Matteuccia struthiopteris* (ostrich fern, Plate 5C) in populations from a latitudinal gradient from northern to southern Norway (Odland *et al.*, 2004). Even within the relatively short temperate growing season of the UK, fern spores of different species can be released at different start times and for varying time spans (Page, 1997). For example, a sea-cave fern *Cystopteris dickieana* had spore release from late May to mid August compared with the woodland oak fern *Gymnocarpium robertianum* (mid July to mid September), which suggests that their respective gametophyte populations will encounter different environmental conditions for development. In a study area at high elevation in Taiwan, Lee *et al.* (2008) found that all fertile leaves of the dimorphic temperate *Osmunda claytoniana* (interrupted fern) emerged in April whereas only 74% of the sterile leaves emerged that month, confirming that fertile and sterile leaves can have different seasonal emergence patterns. Greer and McCarthy (2000) were the first to explicitly address the timing and frequency of fertile leaf production in ferns as these factors represent tradeoffs between plant investment in growth and the costs of reproduction. For the temperate fern *Polystichum acrostichoides*, Greer and McCarthy (2000) found that fertile leaf production of an individual was sporadic from year to year and that it negatively affected leaf growth and spore production in the following year.

3.5.4 Vegetative reproduction

Young fern sporophytes are added to a population from two sources: as genets (sporelings) developing from zygotes after sexual reproduction of gametophytes or as ramets resulting from vegetative reproduction of gametophytes or sporophytes (Fig. 3.2). The presence or absence of vegetative propagules has been noted in the taxonomic fern literature, but few field studies have examined the temporal aspects of the formation, growth, independence, reproduction and death of ramets. In field surveys of sporophytes it is difficult to distinguish whether independent individuals and rhizome segments are ramets or genets, except when they are very young. However, electrophoretic analysis of leaf material has been used to explain the different ecologies in natural sympatric populations of two temperate ferns both with creeping and disintegrating rhizomes, *Deparia acrostichoides* (silvery spleen-wort) and *Diplazium pycnocarpon* (narrow-leaved glade fern; Hamilton, 1992). *Deparia acrostichoides* was considered more tolerant of its temperate forest environment because it had loosely spaced clusters of ramets, frequent spore production and a higher level of within-population genetic variation typical for a species at a late successional stage of establishment (Hamilton, 1992). In contrast, a dense mass of ramets, infrequent spore production and lower levels of genetic variation in the population of *Diplazium pycnocarpon* suggested that this species had been less successful in the Ohio forest because of its more tropical ancestry and was therefore still at an early successional stage.

Because sporophytes can originate by a combination of asexual and sexual reproduction, knowledge of their proportional recruitment from vegetative reproduction is essential in interpreting demographic patterns (Gureyeva, 2003). There are three basic means of producing sporophytes that do not involve fertilization: (1) apogamy, (2) bud production and (3) rhizome branching followed by disintegration of the connection.

Apogamy

Apogamy is the development of a sporophyte from a gametophyte cell without benefit of fertilization (Sheffield, 2008) and occurs in approximately 10% of fern species (Walker, 1985). Juvenile sporophytes that develop by apogamy tend to develop a leaf first, while in sporelings that result from fertilization the root generally appears first (Hoshizaki and Moran, 2001). Apogamous juvenile sporophytes mature faster than sporelings that develop following fertilization. This rapid development is a particularly important characteristic of those apogamous species that are used in horticulture (e.g., *Cyrtomium falcatum*, *C. fortunei*, *Dryopteris cycadina* and *Pteris cretica*, Hoshizaki and Moran, 2001). Because an apogamous gametophyte grows rapidly and does not require water for fertilization, apogamy can be advantageous in dry regions with a short growing season (see Chapter 5). For otherwise sterile ferns or hybrids, apogamy can be the only way of reproduction (e.g., *Diplazium megaphyllum*, Chiou *et al.*, 2006).

Bud production

Buds occur on sporophytes of about 5% of ferns species and can develop on roots (e.g., *Asplenium*, *Ophioglossum*, *Pecluma*) and in various locations on stems and leaves (e.g., *Asplenium*, *Diplazium*, Moran, 2004). Buds result from a meristematic tissue that develops independently of the stem apex (Harper, 1977) and can develop their own roots and leaves while still attached to the parent plant or after they detach. Sporophytes that grow from buds are juvenile in form, although they may develop more rapidly than sporelings of the same species (Sharpe, 1988). For example, the ramets of the Brazilian tree fern *Alsophila setosa* had a more rapid vertical trunk growth than the genets (Schmitt and Windisch, 2006).

Rhizome branching and disintegration

Equilateral rhizome division occurs in ferns (Johns and Edwards, 1991) when the apical cell divides isotomously (i.e., into two cells of the same size). In contrast, some species of the Lycopodiaceae exhibit anisotomous branching in which one of the two divisions is reduced in diameter and length (Øllgaard, 1979). Rhizomes of creeping ferns branch frequently while branching is rare in rosette-forming ferns. After branching, new leaves emerge from both rhizome apices at similar frequencies. Rhizome fragmentation occurs as the original connection between the branches eventually disintegrates, forming independent sporophytes that are of the same developmental stage and size as the adult parent plant. In contrast, young sporophytes that develop from apogamy and bud production start with a juvenile form. Therefore, the frequency and modes of vegetative reproduction, as well as sexual reproduction, affect the proportion of juvenile versus adult sporophytes in a population.

3.6 Classification of fern life history stages

Population profiles are created by determining the frequency of sporophytes in each sequential developmental stage of a fern's life history; the stages are defined by the characteristics of the rhizome or leaves or both. Repeated observations of sporophyte frequency over time may lead to an understanding of the dynamics of the population. Historically, demographic studies have been based on changes at specific ages (Lefkovitch, 1965), but the age of most herbaceous perennial plants, including ferns, is difficult to assess because of their modular architecture (Harper, 1977) and disintegrating rhizomes. Only the actively growing segment of a rhizome of most individual mature ferns is visible at any one time, leaving few clues as to how many years the plant had been growing, how many sterile and fertile leaves it may have produced in its past lifetime or the morphology and size of those earlier leaves. Therefore, demographic studies of ferns need to assign each fern individual to a general life history stage depending on its ontogenetic stage of development, regardless of its actual age in years. Further subdivisions of life history stages according to leaf size or leaf complexity can be used to refine this ontogenetic classification system.

3.6.1 Age determination

A direct method of assessing fern age is to count petiole bases on a harvested rhizome and then divide by the annual leaf production rate (Sharpe, 1993; Gureyeva, 2003). This method is only accurate for juvenile individuals and can lead to erroneous age estimates for mature sporophytes that exhibit necrosis of older rhizome tissue with age. For example, Gilbert (1970) harvested mature individuals of *Dryopteris villarii* in the UK and estimated their age using this method at about 30–40 years. This proved to be an underestimate as the geometry of the clone that the plants were part of suggested that it had actually been about 150 years since the establishment of the harvested sporophytes.

There are three situations for which the age of individual ferns can be estimated more reliably: (1) on a substrate of a known age, (2) for tree ferns with nondisintegrating trunks, and (3) for clones of a known history.

Substrate age

One can at least provide outside limits on the maximum age of individuals by knowing the date at which the substrate became available for colonization. Landslides, for example, open a "clean slate" with a known date for the establishment of some ferns (see Chapter 6). In the Netherlands, peat land was trenched in 1932 to reclaim it from the sea, providing a start date for evaluating population trends for the 28 fern taxa present 70 years later (Bremer, 2007). Data from fern monitoring that Bremer began in 1979 were used to estimate establishment dates retrospectively for different species. The number of taxa apparently increased between 10 and 20 years after reclamation and remained stable thereafter. The four fern species now classified as common, all established within the first ten years. *Dryopteris filix-mas* (male fern) was the first species to colonize after eight years (Bremer, 2007).

Tree fern age

A tree fern (Fig. 3.4) is a special case of a rosette-forming fern that has a vertical trunk on which leaf scars may provide evidence of leaf production history. Tree ferns are found in several genera, yet they represent only about 5.6% of the total number of fern species (Table 3.2). The trunk of a tree fern derives its support from lignified sclerenchymatic plates rather than from secondary thickening growth of vascular tissue in woody seed plants (Large and Braggins, 2004). Tree ferns have been preferred candidates for individual and leaf growth studies (Table 3.3) because of their large size, limited life spans and importance in primary succession (see Chapter 6). Two approaches have been used to estimate the age of tree ferns: (1) measurement of the increase in trunk height and (2) assessment of leaf production rates. Estimates of tree fern life spans can be very useful in spite of the shortcomings of these methods as noted by Schmitt and Windisch (2006).

In the first method, age is determined by dividing total plant height by annual trunk height growth, which is calculated from repeated measurements (Table 3.3). When Schmitt and Windisch (2006) used this method, the age estimates for *Alsophila setosa* were almost twice the known age of its substrate within a successional forest in Brazil, obviously because of differential growth during the lifetime of the fern. In contrast, age could be greatly underestimated in tree ferns such as *Cyathea muricata*, which in Guadeloupe exhibits very little increase in trunk height as a young sporophyte, but expands at a much more rapid rate as it ages (Prugnolle *et al.*, 2000).

Table 3.2 *Fern genera with trunk-forming species (i.e., tree ferns), based on Appendix B and Large and Braggins (2004)*

Family	Genus	Number of species
Blechnaceae	*Sadleria*	6
	Blechnum	7
Cibotiaceae	*Cibotium*	10
Culcitaceae	*Culcita*	2
Cyatheaceae	*Alsophila*	235
	Cyathea	185
	Gymnosphaera	30
	Sphaeropteris	100
Dicksoniaceae	*Calochlaena*	5
	Dicksonia	20
	Lophosoria	3
Lindsaeaceae	*Cystodium*	1
Onocleaceae	*Onocleopsis*	1
Osmundaceae	*Leptopteris*	6
	Osmunda	10
	Todea	2
Thyrsopteridaceae	*Thyrsopteris*	1
	Total tree fern species	624

The second method estimates tree fern age by dividing the number of leaf scars on the trunk by the number of leaves produced annually, which can be observed by tagging all new leaves during one year (Table 3.3). This method gave a more reasonable estimate of 32 years for *Alsophila setosa* in the same 36-year-old successional forest noted above. Different leaf production rates at two sites in the forest resulted in age estimates of 22 and 32 years, respectively (Schmitt and Windisch, 2006). Arens (2001) experienced the same problem with differences in annual leaf production rates among populations of *Cyathea caracasana* from three successional habitats of the Andean cloud forest in Colombia. Another difficulty of this method is that leaf scars of tree ferns may be densely covered by epiphytes (Schmitt and Windisch, 2006), or by an adventitious root mantle that, for example, can extend up to 1–4 m in height on the lower trunk of *Alsophila firma* (Mehltreter and García-Franco, 2008) or even higher on the trunk of *Dicksonia fibrosa* (G. L. Rapson, personal communication).

In conclusion, both methods must recognize that trunk height increments as well as leaf production of tree ferns may vary significantly between consecutive years and because of changing habitat conditions during a lifetime that in some species can approach 150 years (Tanner, 1983). Pinero *et al.* (1986) strongly advocated long-term monitoring in order to understand annual variation in growth rates for

Table 3.3 *Studies of tree ferns that include growth data*

Species	Leaf count (number of leaves per plant)	Annual leaf production (number of leaves per plant)	Leaf life span (months)	Annual trunk growth (cm)
Alsophila auneae[6] (syn. *Cyathea pubescens*)	6.6	8	17.5	6.6
Alsophila bryophila[5]	–	13.6	–	5.0
Alsophila erinacea[10]	4	–	–	13.6
Alsophila firma[18]	5	5	10	17.1
Alsophila polystichoides[10]	3	–	–	18.8
Alsophila salvinii[4]	6	2.5	24	5.1–6.9
Alsophila setosa[14]	–	6.92–8.7	–	6.32–14.5
Cibotium chamissoi[13]	–	4.8	11	3.0
Cibotium glaucum[9]	5–16	3–5	18–39	5–7
Cibotium splendens[2]	4.3	3.6	21	5
Cibotium taiwanense[12]	5	3	15–26	–
Cyathea arborea[3]	–	10 (12–30)	–	28.6
Cyathea caracasana[11]	–	–	–	16.8
Cyathea contaminans[1]	6–10	15	5.5–6.6	–
Cyathea delgadii[16]	7	5.75	–	4.65–81.9
Cyathea furfuracea[6]	–	–	10.9	–
Cyathea hornei[8]	3–11	3–9	13–19	–
Cyathea nigripes[10]	4	–	–	17.1
Cyathea pinnula[10]	6	–	–	10.4
Cyathea smithii[17]	–	–	–	6.7–11.1
Cyathea spinulosa[15]	–	–	7.1	8.9
Cyathea trichiata[10]	–	–	–	89.7
Cyathea woodwardioides[6]	–	–	24.3	–
Dicksonia blumei[1]	12–18	15	6.3	–
Dicksonia squarrosa[17]	–	–	–	1.86–2.4
Leptopteris wilkesiana[7]	8–18	3–9	24–30	–
Sphaeropteris cooperi[13]	–	32.4	6	15.4

–, no data available.
Sources: [1]Jaag, 1943; [2]Wick and Hashimoto, 1971; [3]Conant, 1976; [4]Seiler, 1981; [5]Tryon and Tryon, 1982; [6]Tanner, 1983; [7]Ash, 1986; [8]Ash, 1987; [9]Walker and Aplet, 1994; [10]Bittner and Breckle, 1995; [11]Arens, 2001; [12]Chiou *et al.*, 2001; [13]Durand and Goldstein, 2001; [14]Schmitt and Windisch, 2006; [15]Nagano and Suzuki, 2007; [16]Schmitt and Windisch, 2007; [17]Gaxiola *et al.*, 2008; [18]Mehltreter and García-Franco, 2008.

palms with similar architecture to tree ferns and which often occupy similar habitats. We also recommend such a long-term approach for tree ferns, and future comparative studies of the same species at different sites to understand the natural variation of growth in tree ferns.

Clone age

For some temperate species of ferns and lycophytes with dichotomously branching rhizomes, an understanding of the rate of increase in clone diameter combined with information about rhizome growth has facilitated age estimates. For example, by counting the number of living stem tips and observing the frequency of branching, the age of a clone of the lycophyte *Huperzia lucidula* (shining fir-moss, syn. *Lycopodium lucidum*), was estimated to be about 60 years (Primack, 1973). Headley and Callaghan (1990) used a similar approach with *Huperzia selago* (northern fir-moss) in open habitats in Europe and found that the clones were relatively short lived (6–16 years). Oinonen (1967b) mapped clone development in several lycophyte and fern species in forests of sandy heaths in Finland. After four years of observing branching and growth rates of different species he could reliably estimate the age of clones that were up to about 250 m in diameter, although many appeared to be older. Clones of *Pteridium aquilinum* (bracken) of that size were estimated to be about 700 years old (Oinonen, 1967a), but clones 250 m in diameter of *Lycopodium complanatum* (running-pine), which spread about 15% more slowly, were estimated to be about 800 years old (Oinonen, 1967b) and similar size clones of the common oak fern *Gymnocarpium dryopteris* (syn. *Carpogymnia dryopteris)* were estimated to be nearly 500 years old (Oinonen, 1971).

Insight into the ontogenetic history of an individual plant and the disturbance history of its site can be useful for estimating the age of some ferns and lycophytes. Even the estimates from the more successful studies of the ages of tree ferns and clones were affected by variable leaf and ramet production rates. Long-term monitoring of these rates is needed to supplement such estimates.

3.6.2 Sporophyte life history stages

Because the determination of the age of most fern sporophytes is so difficult, the developmental process for an individual sporophyte is more usefully classified through morphological characteristics of the rhizome and leaves into several sequential life history stages (Fig. 3.2, Table 3.4). Additional insights into the dynamics of a population can be gained if detailed knowledge of the morphology of individuals of a species allows subdivision of these general life history stages into additional, sequential ontogenetic categories.

Juvenile sporophyte

A juvenile sporophyte produces leaves that are smaller and may differ in shape from adult leaves (see Section 3.5.2). Characteristics of leaves and rhizomes allow the identification of several successive ontogenetic stages within the general juvenile stage. For example, the youngest juveniles may be still attached to the gametophyte

Table 3.4 *Characteristics used to classify individual rosette-forming sporophytes into general life history stages shown in Fig. 3.2*

Life history stage	Leaves	Rhizome
Juvenile	Juvenile shape, often simple, less dissected than adult leaf	Intact at both ends. Presence of gametophyte or parent tissue. Ratio of leaf bases to living leaf count ≤ 1
Sterile	Leaves of adult form but unlikely to be of maximum size or complexity. No fertile leaves present or known to have been present. Living leaf count stable	Distal tip necrotic, older leaf bases disintegrating. Net length of rhizome relatively unchanged over time. Ratio of leaf bases to living leaf count >1. Vegetative reproduction occurs
Fertile	Leaves of adult form approaching maximum size and complexity. At least one fertile leaf present or known to have been present. Living leaf count larger than in sterile adult	Distal tip necrotic, older leaf bases disintegrate. Length of rhizome relatively unchanged over time. Ratio of leaf bases to living leaf count >1. Vegetative reproduction occurs
Senescing adult	Leaves shorter or juvenile in form. Fewer living leaves. If known to be a fertile plant, ratio of fertile leaves to sterile lower than in mature fertile plant or zero. Fewer spores per leaf on fertile leaves	Shortening as more leaves die than are added. Fragmentation of branched rhizomes occurs. Vegetative reproduction probable in some species if shoot apex dies

Sources: Adapted from Sharpe (1993) and Gureyeva (2003).

or parent tissue (if it originated from vegetative propagation) and at the next stage they may have a narrower rhizome than an adult. Gureyeva (2003) used morphological characteristics to subdivide the juvenile sporophyte life history class for six species of Siberian ferns (see Table 3.1 for a list of these species) into four successive stages (embryo, sporeling, juvenile, immature). The immature juvenile class described by Gureyeva (2003) includes plants in transition with both juvenile and adult form leaves present and is often recognized in fern demographic studies (e.g., *Blechnum spicant*, Cousens, 1981; *Isoëtes lacustris*, Szmeja, 1994; some species of *Diellia*, Aguraiuja *et al.*, 2004; *Asplenium trichomanes*, Bremer, 2004; *A. scolopendrium*, Bremer and Jongejans, 2010).

Sterile adult sporophyte

The life history stage for sterile adult sporophytes (Fig. 3.2) begins when all emerging leaves are of adult form, even if they may still increase in size (Sharpe and Jernstedt, 1990a). If a species is capable of vegetative reproduction, it generally begins during this life history stage. Fern sporophytes generally pass through the sterile adult stage between the juvenile and fertile adult stage (when fertile leaves are produced). Adult sporophytes of species with fertile leaves that are either ephemeral or infrequently produced (e.g., 10% in *Thelypteris angustifolia*, Sharpe, 1997) may be erroneously classified as sterile when no fertile leaves are present during a survey (Greer and McCarthy, 2000; Sharpe, 2005). To avoid this mistake, the entire ontogenetic history of these fern individuals has to be monitored (Gureyeva, 2003).

Fertile adult sporophyte

The fertile life history stage (Fig. 3.2) begins with the production of the first fertile leaf. In strongly dimorphic tropical species, fertile leaves are often produced seasonally and are shorter lived than sterile leaves. Accurate assessment of the proportion of reproductive sporophytes will therefore be limited to the season when such ephemeral fertile leaves may be present at the study site (Moran, 1987). Rhizome architecture of fertile and sterile sporophytes is similar, but the presence of fertile leaves may lead to an increased total leaf count per plant (e.g., in *Dryopteris intermedia*, evergreen wood fern, and *Polystichum acrostichoides*, Sharpe, 2005). Fertile sporophytes can exhibit different degrees of fertility as defined by the percentage of fertile leaves in the rosette. The proportion of fertile leaves may be low in some species such as *Acrostichum danaeifolium* (giant leather fern) that annually produces 1–3 fertile and 10–13 sterile leaves per plant (Mehltreter and Palacios-Rios, 2003). In contrast, some species rarely produce any sterile leaves once the sporophyte reaches full maturity (e.g., tropical *Diplazium expansum*; K. Mehltreter, personal observation; temperate *Dryopteris marginalis*; J. M. Sharpe, personal observation). The percentage of fertile sporophytes in a population can also vary between successive years (Table 3.5).

Senescent sporophyte

Few studies have recognized that there is a unique morphology exhibited by senescing individuals, and yet this may be an important indicator of subtle changes in habitat. Bremer (2004) called this the postadult stage in studies of *Asplenium trichomanes* (maidenhair spleenwort) in The Netherlands (see Section 3.7.3) and Gureyeva (2003) describes several features of senescing sporophytes that she observed in population studies of six species in Siberia (Table 3.4). The leaves may become smaller and assume a juvenile form, but the diameter of the senescing rhizome is larger than that of a juvenile plant. Bud development along the rhizome may occur if the meristem at the shoot apex dies (Sharpe and Jernstedt, 1991). Aguraiuja *et al.* (2004) identified a senescent class of dried (possibly dormant or dead) plants of the rare, endemic genus *Diellia* in Hawaiian dry forest and used this

Table 3.5 *Percentage of fertile sporophytes producing at least one fertile leaf during each calendar year for* Dryopteris intermedia *(N = 31) and* Polystichum acrostichoides *(N = 14) in natural populations in Maine, northeastern USA*[1] *and* Matteuccia struthiopteris *(N = 34) in Manitoba, Canada*[2]

Year	*Dryopteris intermedia*	*Polystichum acrostichoides*	*Matteuccia struthiopteris*
1993	77	65	62
1994	71	71	76
1995	71	71	9
1996	48	57	53
1997	77	71	15
1998	42	79	–
1999	55	93	–
2000	64	71	–
All	64	72	

–, no data available.
Sources: [1] Sharpe, 2005.
[2] Kenkel, 1997.

evidence to classify the status of populations as dynamic, normal or regressive to guide future conservation efforts of these ferns.

3.6.3 Subdividing life history stages: leaf complexity

A general life history stage such as the fertile sporophyte (Fig. 3.2) can persist for many years because ferns are perennial plants. Consequently, these life history stages are not strongly correlated with relative age. A count of the pairs of leaflets on a leaf is one measure of leaf complexity that may document relative ages (Halleck *et al.*, 2004) but its use is limited to species with pinnate leaves and a distinct terminal leaflet or terminal bud (e.g. *Danaea wendlandii*, Fig. 3.1c). In contrast, many fern species either have simple, entire leaves, few pairs of leaflets or tapering terminal segments (e.g., *Thelypteris deltoidea*, Fig. 3.3a). Sato (1982) addressed this difficulty in demographic and phenological studies of relative age and fertility patterns in Japanese and European fern populations. Sato (1982) developed a simple index of shape complexity by counting the highest number of veins (NV) extending out from the leaf midrib on successively emerging leaves (Fig. 3.6a) and used this count as a relative indicator of developmental age, irrespective of leaf morphology or size. Sato (1985) summarized many potential applications of the NV method for assessing life history stages based on data from over 2000 leaves from several populations of *Polystichum*

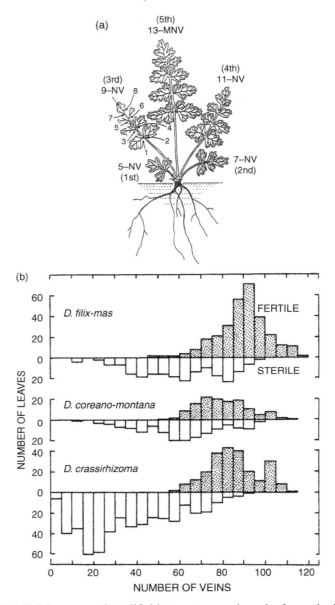

Fig. 3.6 Subdividing sporophyte life history stages using a leaf complexity index. (a) Diagram of vein counting system for assigning relative ages to ferns. Illustration of a juvenile sporophyte of *Dryopteris crassirhizoma* showing the number of veins (NV) on a leaf. The development of the sporophyte is represented by 13-NV, the maximum NV (MNV) of the sporophyte. (With permission from Sato, 1990a.) (b) Contrasting population structure of three ferns in the genus *Dryopteris* using NV. (With permission, from Sato, 1990b.)

tripteron. Sato and Tsuyuzaki (1988) tested the NV method on three species of *Dryopteris* and used the method to demonstrate differences in their population structure (Fig. 3.6b). They compared NV with a number of characteristics of leaf architecture and size including total leaf length (L), lamina length, lamina width, square of lamina length (S), petiole length and width and an index of leaf dissection $(L/2 * (3.14 * S)^{1/2})$. They found that compared with many of these measures, the NV count is not only more accurate, but is also simple to observe and can be especially useful in long-term studies as it is a nondestructive assessment of the leaf sequence. This method was then used in several studies that compared fern populations of different species in Japan and Europe according to different environmental factors (e.g., humidity levels, elevation and temperature). Each of these studies (see Sato, 1990a, 1990b for a review) produced insights into the sensitivity of various demographic parameters (e.g., leaf size, initial production of spores, and reduction in plant fertility) to habitat conditions. The demographic methodologies, graphical presentations and applications explored in Sato's work have much to offer for future studies of the life history stages of herbaceous perennials, both of ferns and seed plants, temperate and tropical.

3.7 Population structure comparisons

Population structure describes the frequency distribution in demographic categories of individuals, leaves or stands of leaves and provides a basis for simplifying the monitoring process and for comparative approaches among species. Studies of population structure can investigate how one or several species differ (1) within a single habitat, (2) among different habitats and (3) over time. Even if only two stages (sterile and fertile) are identified, insights into the ecology of fern species can be gleaned from population studies that address these three issues. For example, Odland (1998) found that fertility could be predicted in *Thelypteris limbosperma* (sweet mountain fern) based on July and January temperature, humidity and canopy cover in Norway while for *Athyrium distentifolium* (alpine lady fern) fertility levels were not correlated with any environmental variables measured. Bremer (1995) found a very low proportion of fertile sporophytes of *Polystichum setiferum* (soft shield fern) in The Netherlands in a heavily shaded population, compared with populations of open forest habitat. Year-to-year variation in the percentage of fertile sporophytes was very high (Table 3.5) for populations of the evergreen wood fern *Dryopteris intermedia* (42–77%) and *Polystichum acrostichoides* (57–93%) in temperate forest of Maine, USA (Sharpe, 2005) as well for *Matteuccia struthiopteris* (9–76%) in Manitoba, Canada (Kenkel, 1997).

Specific methods used to assess the demographic status of individuals in a population are as varied as the fern species under observation. To illustrate this, we describe three demographic studies to show some of the ways in which population structures can be compared.

3.7.1 *Siberian forest: comparing species within a habitat*

Gureyeva (2003) was able to classify six temperate, rosette-forming fern species in a forested area of Siberia into three contrasting population structures (Table 3.1) by assigning individual sporophytes to one of nine life history stages. *Athyrium distentifolium* and *A. filix-femina* exhibited little vegetative reproduction, but showed regular spore and gametophyte production. Low densities of adult sporophytes in shaded areas provided spores for continuous renewal of the population, but gametophytes and dense populations of young sporophytes were found only in gaps. In contrast, mature sporophytes of *Dryopteris carthusiana* (spinulose wood fern), *D. expansa* (northern wood fern, Plate 1C) and *D. filix-mas* were not limited to the gaps, and their regular spore production insured that both mature sporophytes and gametophytes were found throughout their forested habitat. Because of continuous vegetative production, populations of these species of *Dryopteris* also included a high proportion of juvenile plants. *Matteuccia struthiopteris* populations were characterized by the presence of long-lived adults that continuously produced juvenile sporophytes by extensive lateral branching of the rhizome. Gureyeva's comprehensive classification of life history stages, including both gametophyte stages and the impacts of vegetative propagation, resulted in contrasting population profiles that in turn could be related to environmental features.

3.7.2 Isoëtes lacustris: *comparing environments*

The lycophyte species *Isoëtes lacustris* (common quillwort) has rosettes of simple leaves and often inhabits the bottom of lakes. Using diving equipment, Vöge (1997a) collected plants in 20 Norwegian lakes along a latitudinal gradient and in southern Norway along an altitudinal gradient from sea level to 915 m elevation. She classified individuals into sterile and fertile life history stages, and determined rosette size (by summing leaf areas) and root length and counted megaspores in the sporangia of fertile leaves. Lakes were classified as optimal or suboptimal based on water characteristics and nutrient profile. There was a strong correlation between rosette size and megaspore counts for *I. lacustris* (Vöge, 1997a). Climate conditions in boreal areas (low temperatures, short summers) resulted in smaller plants and lower fertility (as indicated by spore counts), indicating that latitude limited a plant's success. At higher altitudes, lower temperatures were offset by longer summers, resulting in a less negative effect of the lower temperatures on the plants than at higher latitudes. In additional studies of 30 European lakes, she found reliable relationships between leaf length and megaspore number and habitat, as the number of leaves and spores declined in lakes with darker colored water, but did not find any correspondence to lake pH and conductivity (Vöge, 1997b). Based on these relationships, Vöge (2004) developed a nondestructive method to evaluate population vigor and fertility for *I. lacustris* in lakes where its numbers were declining. Phenological studies showed that mature spores are released from late summer to early winter (Vöge, 2006) and provided additional useful information for future demographic studies and conservation management plans for *I. lacustris*.

3.7.3 Long-term studies

Only Bremer (2004) and Bremer and Jongejans (2010) have used life history transition matrices (see Silvertown, 1982) to interpret ecological changes in fern populations over time, therefore we present in some detail the methodology and results of one of these studies. The results of long-term studies of *Asplenium trichomanes* (Bremer, 2004) in The Netherlands are graphically presented by converting Bremer's stage–transition matrix data to a loop diagram (see Gotelli, 1995) based on the five life history stages. Two populations of *A. trichomanes* were monitored for 13 years (K23) and 10 years (L72) through 2001 (Bremer, 2004) and differed dramatically in their demographic structure (Fig. 3.7). In the K23 population, most of the individuals were found in the two juvenile stages and in the adult fertile stage. Most transitions took place among those three stages, suggesting a thriving population with considerable recruitment. In contrast, in the L72 population, most of the individuals were in the adult fertile stage and some of those had become sterile and moved into a postadult stage. The few representatives of the three earlier stages (i.e., sporeling, juvenile and adult sterile) suggested that while fertile sporophytes were numerous, conditions for spore germination were suboptimal. These two different population structures for *A. trichomanes* (Fig. 3.7) reflect the history of the environmental conditions at their respective sites. The forest where K23 was located had been thinned twice during the observation period and the subsequent increase in light resulted in spore germination on eroded patches of soil, thus increasing the number of juvenile sporophytes. The L72 population had been built up in the early 1980s through considerable recruitment from a "founder plant" of *A. trichomanes*. A major storm in 1990 opened a gap in the forest and the site became overgrown with *Rubus idaeus* (red raspberry), *R. flexuosa* (bramble) and *Circaea lutetiana* (enchanter's nightshade). Although these competing plants were clipped annually to maintain the adult population of *A. trichomanes*, the site was no longer suitable for recruitment of juvenile fern plants.

Bremer and Jongejans (2010) used a similar life history matrix approach to analyze and compare structure in three populations of *A. scolopendrium* at the

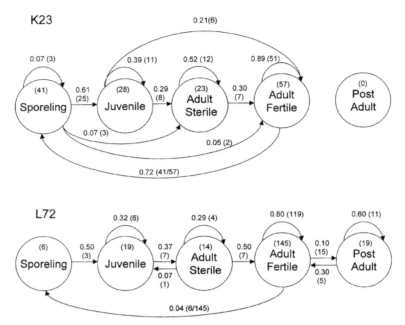

Fig. 3.7 Life cycle graph of two populations (K23, L72) of *Asplenium trichomanes* in The Netherlands. Diagram has been adapted with permission from stage-structured matrices based on 9–10 year studies by Bremer (2004) and described in Section 3.7.3. Each circle represents a life history stage defined by Bremer (2004). Sporelings are small juvenile plants that are <1.5 cm tall, while the juvenile stage includes plants that are 1.5–4.0 cm tall. Each arrow connecting the stages is labeled with the transition probability (and number of individuals).

northern edge of its ranges during three time periods (1978–1998) and found that frost had an important effect on plant performance and therefore population dynamics. Although the analyses are not as comprehensive, 15-year population studies of other species at the same location documented that higher levels of light led to higher proportions of fertile leaves (e.g., *Gymnocarpium dryopteris*, Bremer, 1994) and that growth and survival were reduced in response to severe winters (e.g., *Polystichum setiferum*, Bremer, 1995).

The work of Bremer highlights the importance of long-term (>5 years) population monitoring. There have been other recent long-term studies of temperate species and while not as comprehensive in scope, they have provided valuable ecological insights. For example, studies of repeatedly monitored individuals of *Botrychium* (grape ferns, Plate 4A) in various locations have found resistance to herbivory (*Botrychium dissectum*, 12 years; Montgomery, 1990); slow recovery from fire (*Botrychium campestre, B. gallicomontanum*, 12 years; *B. mormo*, 7 years; Johnson-Groh, 1999) and a positive relationship between light and spore production (*Botrychium australe*, 6 years, Kelly, 1994). Monitoring individuals showed annual variability in spore production, possibly in response to extremes in winter temperatures (*Dryopteris intermedia, Polystichum acrostichoides*, 8 years, Sharpe, 2005). In Japan, Sato (1990a) used repeated observations of individuals over an 8-year period to determine the age at which fertile leaves first appeared. He then noted the mean interval between the appearance of the first fertile leaf and the stage at which the maximum proportion of leaves in the crown were fertile (*Dryopteris crassirhizoma*, 7.9/8.6 years; *Polystichum braunii*, 5.8/3.5 years; *P. tripteron*, 11.0/10.9 years), and related these differences to habitat characteristics of the three species. The study site corresponded to the center of the geographic range of *D. crassirhizoma*, the southern area of *P. braunii* (the highest rate of maturation) and the northern limit for *P. tripteron* (the lowest rate of maturation). Sato (1990a) therefore suggested that maturation age of each species responds to habitat and climate conditions, although additional observations through their ranges would be needed to confirm this.

3.7.4 Summary

Few demographic studies have been performed for ferns and lycophytes although the knowledge of whether populations consist of predominantly juvenile, sterile or fertile plants is critical for relating population structure to habitat factors. For example, in a forest productivity experiment, Walker *et al.* (1996) found a higher proportion of biomass of understory ferns in a Puerto Rican rain forest after debris removal (DR) than in control (C) plots. Following up on this observation, Halleck *et al.* (2004) found a higher number of individuals of *Thelypteris deltoidea* in the DR plots, but their leaf pair counts were lower (a potential indicator of younger age), than in C plots documenting recruitment of new plants rather than growth of existing plants. Although the number of plants (C: 43, DR: 40) of another herbaceous understory fern, *Cyathea borinquena*, initially appeared unaffected by debris removal, additional observations showed that counts of fertile leaves (C: 10, DR: 1), mean leaf counts per plant (C: 4.6, DR: 2.7) and mean leaf lengths (C: 132.0 cm, DR: 95.6 cm) declined in debris removal plots compared with the control. Once the long leaf life spans of about 2.5 years (Table 3.6) and biomass estimates (dry weight of *c.* 22 g leaf^{-1}, J. M. Sharpe, unpublished data) of this species are included in the analysis, the potential effects of the debris fall on forest productivity (e.g., after a hurricane) can be more completely quantified.

Population structure reflects only the actual frequency distribution of individuals in each life history stage. For a more complete understanding of the temporal population dynamics, repeated measurements of population parameters such as leaf size, leaf production rates and biomass within each life history stage are necessary as well as monitoring of their possible environmental relationships.

Table 3.6 *Mean leaf life spans in months for common understory ferns in the Luquillo rain forest of Puerto Rico monitored for 8 years from September 1992 to January 2001[1] and in subtropical northeastern Taiwan for four years from August 1997 to August 2001[2]*

Species	Sterile leaf life span (months)	Fertile leaf life span (months)	Fertile leaf life span compared with sterile (%)	Dimorphic species
Puerto Rico				
Blechnum occidentale	12.5	12.6	101	No
Cyathea borinquena	28.4	30.4	107	No
Danaea elliptica 1	52.0	5.1	10	Yes
Danaea elliptica 2	48.9	12.0	25	Yes
Danaea nodosa	41.9	5.7	14	Yes
Thelypteris angustifolia	10.0	7.8	78	Yes
Thelypteris deltoidea	29.7	33.5	113	No
Thelypteris reticulata	21.4	14.0	65	Yes
Taiwan				
Acrorumohra hasseltii	22.4	23.7	106	No
Blechnum orientale	27.5	25.9	94	No
Cyathea podophylla	25.8	26.1	101	No
Cyathea spinulosa	8.0	6.6	83	No
Dictyocline griffithii	27.6	18.3	66	No
Dictyocline griffithii wilfordii	15.0	15.1	101	No
Diplazium dilatatum	19.8	18.3	92	No
Diplazium doederleinii	15.8	14.2	90	No
Diplazium petri	20.1	17.5	87	No
Diplazium pullingeri	16.7	16.8	101	No
Histopteris incisa	6.9	7.7	112	No
Plagiogyria adnata	30.3	4.7	16	Yes
Plagiogyria dunnii	22.1	4.4	20	Yes
Pleocnemia rufinervis	18.1	17.2	95	No
Pteris wallichiana	5.1	6.1	120	No
Sphaerostephanos taiwanensis	12.3	13.1	107	No

Nomenclature for the Puerto Rican ferns is based on Proctor (1989) who recognized two distinct variants of *Danaea elliptica* (1 and 2) that were monitored independently based on obvious differences in plant morphology.
Sources: [1] J. M. Sharpe, unpublished data.
[2] Lee *et al.*, 2009.

Destructive harvesting of a sample of individuals is needed to determine rhizome and root biomass (e.g., Greer and McCarthy, 2000), but once the ratio of leaf biomass to rhizome/root biomass is established, nondestructive observations of leaf demography can be used to estimate productivity for that population.

3.8 Leaf phenology

A fern leaf has a series of life history stages it passes through from emergence to death (Fig. 3.8). Each individual sporophyte can be regarded as a population of living leaves that has a demographic structure, determined by growth rates and life spans. This interpretation is important when considering rosette-forming sporophytes (e.g., *Polystichum setiferum*, Bremer, 1994), as well as creeping ferns where the only way to develop population structures is to classify stands of leaves (e.g., *Dennstaedtia punctilobula*, Hammen, 1993; *Gymnocarpium dryopteris*, Bremer, 1994). Environmental factors as well as internal factors (e.g., plant hormones) affect the timing of leaf emergence and death. Seasonal leaf phenological patterns are obvious in temperate deciduous ferns and tropical ferns exposed to a dry season and may be present, but more difficult to detect in apparently aseasonal climates (e.g., *Thelypteris angustifolia* in Puerto Rico, Sharpe, 1997). Leaf phenological parameters such as emergence rates, growth rates, life span and mortality have a strong influence on population structure.

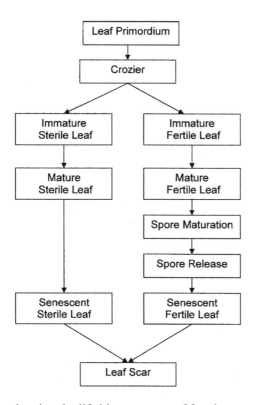

Fig. 3.8 Diagram showing the life history stages of fern leaves.

3.8.1 Leaf growth

The life history of determinate fern leaves consists of three general phases: imma-
ture, mature and senescent (Fig. 3.8). In most fern species, leaf growth is signaled by
the increase in diameter of the crozier, followed by elongation of the petiole
(Fig. 3.9a), unrolling of the rachis, and finally of the leaflets (Fig. 3.9b). This
process has been thoroughly described for *Osmunda cinnamomea* (Briggs and
Steeves, 1958). In some genera, expanded immature leaves are red (e.g.,
Blechnum, Plate 5D) or light green and of softer texture than mature leaves.
Reports of daily leaf elongation rates for three species of temperate ferns range
from 2.2 cm (*Thelypteris limbosperma*) to 8.9 cm (*Matteuccia struthiopteris*, Plate
5C: Odland, 1995). Mehltreter and Palacios-Rios (2003) found comparable rates of
4.1 cm day^{-1} for the tropical *Acrostichum danaeifolium* in the dry season and a
higher rate of 7.0 cm day^{-1} in the rainy season. Such rapid rates may confer an
adaptive advantage to rapidly expanding, soft-textured leaves that can be vulnerable
to herbivore damage in the forest understory. Many temperate fern species may
mitigate this hazard through early emergence in spring when insect abundance is
still low (see Chapter 7).

3.8.2 Leaf life span

Most temperate ferns have determinate leaves with short life spans and, using terms
defined in Lellinger (2002), leaves may be summer-green (e.g., *Dryopteris carthusiana*, 5 months), winter-green when spring and summer leaves remain alive until the
next spring (e.g., *Dryopteris filix-mas*, 12 months) or rarely evergreen with leaves
living more than a year (e.g., *Polystichum lonchitis*, 28 months; Page, 1997). Some
temperate and subtropical ferns of xeric areas have different seasonal patterns.
For example, *Cheilanthes parryi* (Parry's lip fern, syn. *Notholaena parryi*) rapidly
unfurled new leaves at the beginning of the rainy period in November after five
months of drought in the western Colorado desert (Nobel, 1978).

Growth of subtropical and tropical leaves may be indeterminate or determinate.
Long, repeatedly branching leaves often result from indeterminate growth of the leaf
rachis (Fig. 3.3d) and repetitive development and expansion of dormant leaf buds in
such scrambling ferns as *Dicranopteris linearis* in Hawaii (Russell *et al.*, 1998) and
the climbing fern *Lygodium venustum* in Mexico (Mehltreter, 2006). The leaves of
such ferns can form thickets that have profound impacts on the surrounding habitat
(see Chapters 6 and 8), but the actual leaf life spans, determined by environmental
factors, are rarely known. Although the mean leaf life span of *Lygodium venustum*
was only 5.6 months because of the high disturbance at one Mexican study site, 10%
of the leaves lived for 12–30 months (Mehltreter, 2006).

Fig. 3.9 Crozier development. (a) *Hemidictyum marginatum*, a monomorphic rosette-forming fern with sequential leaf development showing crozier as petiole expands in an intermittent stream bed in Puerto Rico. (b) *Cyathea horrida*, a tree fern rosette with sequential leaf development showing croziers as the rachis and leaflets expand.

Life spans of determinate leaves can be quite variable, even within a single habitat such as the Luquillo rain forest in Puerto Rico, where mean life spans of sterile leaves ranged from 10 months for *Thelypteris angustifolia* to 52 months for one of two very distinctive variants of *Danaea elliptica* (Table 3.6; Proctor, 1989) or a subtropical forest in northeastern Taiwan where the range was 5.1 months for *Pteris wallichiana* to 30.3 months for *Plagiogyria adnata* (Table 3.6). When the life span of a fertile leaf is shorter than that of a sterile leaf (e.g., *Danaea wendlandii*, Sharpe and Jernstedt, 1990a; *Thelypteris angustifolia*, Sharpe, 1997; *Acrostichum danaei-folium*, Mehltreter and Palacios-Rios, 2003), the differences can be dramatic, ranging from 10% of the sterile leaf life span in one variant of *Danaea elliptica* to 65% for *Thelypteris reticulata*. In most monomorphic species, fertile and sterile leaves have nearly the same life spans or fertile leaves live longer (Table 3.6) than sterile leaves (e.g., *Cibotium taiwanense*, Chiou *et al.*, 2001; *Lygodium venustum*, Mehltreter, 2006; *Alsophila firma*, Mehltreter and García-Franco, 2008). The potential underlying intrinsic or environmental factors (e.g., light, nutrients) that might explain the differences in mean leaf life spans of ferns that range from 4.4 months for fertile leaves of *Plagiogyria dunnii* (Table 3.6) to a maximum of 88 months for sterile leaves of *Danaea elliptica* (J. M. Sharpe, unpublished data) are not clearly understood. Survivorship curves for leaf cohorts can show how environmental factors may affect leaf life spans. For examples, a study of leaf cohorts of the flood-adapted tropical rheophyte *Thelypteris angustifolia* that emerged at four-month intervals over a period of two years showed that mean leaf life spans varied from 12.7 months in a humid year to 8.5 months in a year of severe drought (Fig. 3.10; Sharpe, 1997).

3.8.3 Leaf production rates

Growth and potential reproduction (i.e., fertile leaf emergence) in ferns are reflected in the number and types of leaves that are produced within a given time period (usually a year). Most temperate ferns produce one annual leaf cohort so that annual leaf production can be easily determined by counting the leaves or petioles of a sporophyte late in the season. However, it is more challenging to estimate annual leaf production in tropical ferns as rosettes or stands often consist of leaves that emerged in different years. Thus, for a tropical fern, monitoring only once a year (1) underestimates actual production if leaves are short lived (e.g., *Pteris wallichiana*, Table 3.6) and have already emerged and senesced or (2) overestimates actual production if the leaf count includes long-lived leaves (e.g., *Plagiogyria dunnii*, Table 3.6) from earlier annual cohorts. Fertile leaves of some dimorphic species are also ephemeral (e.g., *Danaea wendlandii*, Fig. 3.1c), so unless their seasonal emergence pattern is taken into consideration, the number of fertile leaves (and,

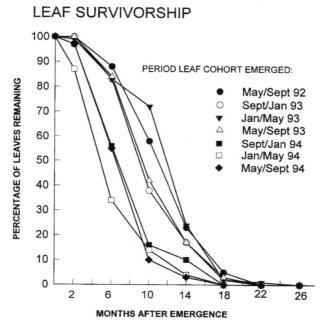

Fig. 3.10 Leaf survivorship curves for *Thelypteris angustifolia*. (With permission, from Sharpe, 1997.)

by extrapolation, the number of adult fertile sporophytes) may also be underestimated (Sharpe and Jernstedt, 1990a).

Leaves may emerge in a flush with several croziers expanding simultaneously or there may be sequential leaf emergence over discrete, although not necessarily equal, time intervals (Mehltreter, 2008). Flushes of fern leaves are more common in the temperate spring, but also occur in tropical environments with seasonal drought (Mehltreter and García-Franco, 2008). For many tropical ferns, leaves die sequentially as well, at about the same rate at which they are produced, but a severe disturbance such as a hurricane can temporarily shorten leaf life spans (e.g., *Acrostichum danaeifolium*, Sharpe, 2010). Leaf count per fern is therefore the net number of leaves gained and lost (e.g., *Acrostichum danaeifolium*, Mehltreter and Palacios-Ríos, 2003) and may be characteristic for a life history stage of a species (Sharpe, 2005). In the UK, *Dryopteris dilatata* (broad buckler fern) and *D. filix-mas* had respectively 61% and 188% higher mean leaf counts in the largest, reproductive size class (leaf length >40 cm) than in the smallest, nonreproductive size class (leaf length 1–10 cm), and smaller plants retained leaves until the following spring, while leaves of the larger plants died back in the winter (Willmot, 1989).

Leaf production rates of tropical ferns may vary considerably among species and even within a single habitat such as a lowland rain forest. Leaf production rates of rosette-forming sporophytes in a rain forest in Puerto Rico ranged from 2.2 leaves year^{-1} for the low-trunk epiphyte *Elaphoglossum crinitum*, hairy paddle fern, (J. M. Sharpe, unpublished data) to 4.7 sterile leaves year^{-1} for the rheophytic *Thelypteris angustifolia* (Sharpe, 1997). Annual leaf production rates for most species of tree ferns are considerably higher and range from 3 to 32 leaves year^{-1} (Table 3.3).

Knowledge of leaf phenology affects the interpretation of monitoring results for tropical trees (Reich *et al.*, 2004), and likewise for ferns. Leaf production may exhibit obvious seasonal patterns in temperate and in tropical habitats with distinct wet and dry seasons. Seasonal patterns in leaf production have also been detected in seemingly aseasonal climates (Mehltreter, 2008). Leaves of tropical ferns generally grow and sporulate during the rainy season, but senesce in the dry season and either abscise or remain attached to the sporophyte. In Zambia, 40% of the fern species follow this seasonal pattern (Kornaś, 1977). Another exception is the deciduous tree fern *Alsophila firma* in Mexico that sheds and replaces its leaves during the rainy season (Mehltreter and García-Franco, 2008). In the western Himalayas, Punetha (1989) found that ferns may produce leaves throughout the year although per plant leaf counts and leaf lengths were about five or six times higher from May through July than in December and January for *Cyclosorus dentatus* (syn. *Christella dentata*) and *C. canus* (syn. *Pseudocyclosorus canus*).

Experiments with defoliation to assess the horticultural potential of fern leaves (Milton, 1987a) and with fertilization to determine nutrient limitations on ferns (Walker and Aplet, 1994) have provided insights into leaf production variability. After defoliation, *Rumohra adiantiformis* maintained its leaf production rate, but the new leaves were much smaller and fewer fertile leaves emerged (Milton, 1987a). In nursery grown plants, nutrient analysis suggested that old leaves supply nutrients, particularly potassium to younger leaves, and that the rate of fertile leaf production was also related to nutrient levels. On nutrient-rich sites Walker and Aplet (1994) found increased leaf turnover rates, decreasing life span and increasing leaf productivity of the tree ferns *Cibotium glaucum* that contribute 25% of aboveground biomass on some sites in Hawaii.

3.8.4 Summary

For many ferns, especially temperate species that form rosettes, leaf count, leaf production rate and leaf life span can be estimated for an individual sporophyte if any two of these three variables are known. Although these phenological variables which describe leaf turnover have been estimated for some tree ferns (Table 3.3), they are unknown for most non-trunk-forming ferns. Leaf counts may be quickly monitored, and should remain relatively stable for adult sporophytes within a single undisturbed habitat (Sharpe, 2005). Annual leaf production rates can be determined by tagging all leaves that emerge during one year, although multiyear monitoring is desirable as the rate may vary between successive years. The calculation of leaf life span by dividing leaf counts by annual leaf production is accurate for most temperate zone ferns for which the entire year's cohort dies within the year. However, for most tropical ferns with variable leaf emergence dates and leaf life span, leaves must be monitored from emergence to senescence instead.

Data from demographic and phenological studies are often published in comprehensive formats such as graphics (e.g., Bauer *et al.*, 1991), converted to percentages to allow comparison of growth patterns in species of different sizes (e.g., Milton, 1987b) or represented by statistical metrics such as correlation coefficients (e.g., Mehltreter and Palacios-Rios, 2003). While each of these studies presents fascinating insights into some ecological aspects of ferns, original data should be made publicly available. For example, databases of long-term projects (Franklin *et al.*, 1990) would allow for cross-site comparisons and facilitate integration with future studies so as to meet the ultimate goal of identifying some general patterns (Harper, 1982) in fern ecology.

3.9 Recommendations for future fern population studies

Fern demography is still a young field of research. Disparate facts and observations are now available about various aspects of the life history, population structure and leaf growth for several fern species, but more complete demographic studies based on standard methods are needed before we can further categorize population patterns and gain further ecological insights. For example, there is almost no information about the transition times between life history stages. Future studies of fern demography and growth should be directed toward (1) developing consistent methodologies, (2) expanding geographical range with an emphasis on tropical regions and (3) encouraging long-term monitoring studies so that general patterns in essential ecosystem parameters such as leaf turnover rates or spore production can be identified and provide valuable input to models of the whole ecosystem function.

3.9.1 Consistent methodologies

Standard methods are available to develop life history matrices and to describe the structures of plant populations but these methods have rarely been applied to ferns. Future studies on fern demography should use life history stage matrices and life cycle graphs (e.g., Fig. 3.7) that can be compared among species and locations to provide insights in their commonalities and differences. Because of the simple architecture of a fern (especially that of a rosette-forming fern with a single apex), it should be possible to develop reliable and standardized methods for the field assessment of growth and biomass production based on the relationships among leaf dimensions (e.g., leaf length index for *Cibotium glaucum*, Arcand *et al.*, 2008) to make results comparable among different species and locations.

3.9.2 *Expanding geographic range*

Our ecological understanding of ferns is still based on a few species and a small number of locations worldwide. With the exception of tree ferns, demographic studies of ferns have been heavily biased towards the temperate latitudes of the northern hemisphere from Japan and Siberia to Europe, Canada and USA. Cousens *et al.* (1985) suggested that observations and experiments with each species take place throughout its geographic range in habitats where it is optimal, marginal and rare before a substantive contribution to its overall ecology and evolution can be demonstrated. For example, addressing a question such as why individuals in well-studied populations of *Polystichum setiferum* in The Netherlands were smaller than in some sampled populations in the United Kingdom would be a valuable extension of the work of Bremer (1995).

3.9.3 *Long-term monitoring*

Although several studies derive ecological insights by comparing populations along latitudinal or altitudinal gradients, or from a variety of different sites, multiyear studies of populations and even individual plants are also necessary to gain insights into the ecology of perennial ferns. Monitoring ferns for a single season does not recognize the variation among years in response to environmental parameters such as precipitation or temperature, successional changes in the habitat, or episodic disturbances such as extreme drought, flooding or storms (see Chapter 6). Unfortunately, few fern species have been studied over several years (Werth and Cousens, 1990). One historic exception is the comprehensive long-term demographic and ecological study of *Pteridium aquilinum* (bracken) in Britain since the early 1940s because of its economic importance (e.g., Watt, 1943, 1976; see Chapter 8).

Woody plants in permanent, large forest plots have been assessed at Barro Colorado Island in Panama since 1981 and similar studies are now established at various tropical and temperate sites (Condit, 1995). Long-term studies require the existence of stable site and a monitoring infrastructure that transcends the lifetime of an individual plant or researcher. Key elements of the ideal long-term project are standard sampling methods and the high priority of maintenance of long-term data sets to address different questions over time. Programs such as the LTER (Long-Term Ecological Research) network of sites in the USA (Franklin *et al.*, 1990) and the international ILTER as well as other well-established long-term research programs place a similar emphasis on maintaining data sets that will be accessible to researchers in the future. This approach should be used for monitoring the temporal aspects of growth and reproduction of ferns and other elements of the herbaceous layer. As 95% of terrestrial fern species are low-growing plants, they are easily accessible and monitoring does not require expensive equipment. Basic information about fern growth requires one day of monitoring per month during the growing season and seasonal monitoring two to four times a year for a limited number of years. Once such baseline information is recorded and available, the effects of infrequent future disturbances can be assessed as they occur with additional standardized observations.

Long-term monitoring of sample individuals in fern populations can be extremely useful, especially for conservation planning (see Chapter 9) and can be done by trained local workers. A network of such fern monitoring sites could be a natural extension of programs such as the Rapid Assessment Program for biodiversity (Rodrigues *et al.*, 2004). Knowledge of demographic attributes of fern species could provide a useful dimension to studies of biodiversity that are often based on counts of individuals without assessing the condition of the population (Noss, 1999). Conservation efforts have the best chance of success when additional information on demography and phenology of individual organisms is available.

Although in this chapter we have described many examples of fern species for which critical growth and demographic data have been collected and population structures described, only a tiny percentage of the 11 000 species have been documented. Some patterns have already been identified and the goal of developing some ecological generalizations for this important group of vascular plant species may be achievable.

Acknowledgements

Reviews by Nan Crystal Arens, Piet Bremer, Bruce and Beverly Clarkson, Gillian L. Rapson, Lawrence R. Walker, Margery Walker, Paulo Windisch and George Yatskyviech, and discussions with Elizabeth Farnsworth and Meredith Fossel greatly improved the manuscript. Joanne Sharpe was partially supported by the Luquillo Long-Term Ecological Research Program in Puerto Rico funded by the U.S. National Science Foundation. Klaus Mehltreter acknowledges support from the Instituto de Ecología, A.C., Xalapa, Veracruz, Mexico, CONACYT-SEMARNAT (2002-C01–0194) and CONACYT-SEP (2003-C02–43082).

References

Aguraiuja, R., Moora, M. and Zobel, M. (2004). Population stage structure of Hawaiian endemic fern taxa of *Diellia* (Aspleniaceae): implications for monitoring and regional dynamics. *Canadian Journal of Botany*, **82**, 1438–45.

Allsopp, A. (1965). Heteroblastic development in cormophytes. *In Handbuch der Pflanzenphysiologie*, ed. W. Ruhland. Berlin: Springer-Verlag, pp. 1172–221.

Arcand, N., Kagawa, A. K., Sack, L. and Giambelluca, T. W. (2008). Scaling of frond form in Hawaiian tree fern *Cibotium glaucum*: compliance with global trends and application for field estimation. *Biotropica*, **40**, 686–91.

Arens, N. C. (2001). Variation in performance of the tree fern *Cyathea caracasana* (Cyatheaceae) across a successional mosaic in an Andean cloud forest. *American Journal of Botany*, **88**, 545–51.

Ash, J. (1986). Demography and production of *Leptopteris wilkesiana* (Osmundaceae), a tropical tree-fern from Fiji. *Australian Journal of Botany*, **34**, 207–15.

Ash, J. (1987). Demography of *Cyathea hornei* (Cyatheaceae), a tropical tree-fern in Fiji. *Australian Journal of Botany*, **35**, 331–42.

Bauer, H. C., Gallmetzer, C. and Sato, T. (1991). Phenology and photosynthetic activity in sterile and fertile sporophytes of *Dryopteris filix-mas* (L.) Schott. *Oecologia*, **86**, 159–62.

Bittner, J. and Breckle, S. W. (1995). The growth rate and age of tree fern trunks in relation to habitats. *American Fern Journal*, **85**, 37–42.

Bremer, P. (1994). On the ecology and population dynamics of a Dutch sporophyte population of *Gymnocarpium dryopteris* (Woodsiaceae: Pteridophyta). *Fern Gazette*, **14**, 289–98.

Bremer, P. (1995). On the ecology and population dynamics of a Dutch population of *Polystichum setiferum* (Dryopteridaceae: Pteridophyta). *Fern Gazette*, **15**, 11–20.

Bremer, P. (2004). On the ecology and demography of a terrestrial population of *Asplenium trichomanes* (Aspleniaceae: Pteridophyta) in the Netherlands. *Fern Gazette*, **17**, 85–96.

Bremer, P. (2007). Ecology and colonisation of ferns in an afforested peat erosion area. In The colonisation of a former sea-floor by fern. Unpublished Ph.D. thesis, Wageningen, The Netherlands: Wageningen University, pp. 37–64.

Bremer, P. and Jongejans, E. (2010). Frost and forest stand effects on the population dynamics of *Asplenium scolopendrium* L. *Population Ecology*, **52**, 211–22.

Briggs, W. R., and Steeves, T. A. (1958). Morphogenetic studies on *Osmunda cinnamomea* L.: the expansion and maturation of vegetative fronds. *Phytomorphology*, **8**, 234–48.

Chiou, W.-L., Lin, J.-C. and Wang, J.-Y. (2001). Phenology of *Cibotium taiwanense* (Dicksoniaceae). *Taiwan Journal of Forestry Science*, **16**, 209–15.

Chiou, W.-L., Martin, C. E., Lin, T. C., *et al.* (2005). Ecophysiological differences between sterile and fertile fronds of the subtropical epiphytic fern *Pyrrosia lingua* (Polypodiaceae) in Taiwan. *American Fern Journal*, **95**, 131–40.

Chiou, W.-L., Huang, Y. M., Hsieh, T. H. and Hsu, S. Y. (2006). *Diplazium megaphyllum* (Bak.) Christ, a rare fern in Taiwan, reproduces by apogamy. *Taiwan Journal of Forestry Science*, **21**, 39–47.

Conant, D. (1976). Ecogeographic and systematic studies in American Cyatheaceae. Unpublished Ph.D. thesis, Harvard University, Cambridge, Massachusetts, USA.

Condit, R. (1995). Research in large, long-term tropical forest plots. *Trends in Ecology and Evolution*, **10**, 18–22.

Cousens, M. I. (1981). *Blechnum spicant*: habitat and vigor of optimal, marginal, and disjunct populations, and field observations of gametophytes. *Botanical Gazette*, **142**, 251–8.

Cousens, M. I., Lacey, D. G. and Kelly, E. M. (1985). Life-history studies of ferns: a consideration of perspective. *Proceedings of the Royal Society of Edinburgh*, **86**B, 371–80.

Cousens, M. I., Lacey, D. G. and Scheller, J. M. (1988). Safe sites and the ecological life history of *Lorinseria areolata*. *American Journal of Botany*, **75**, 797–807.

Crotty, W. J. (1955). Trends in the pattern of primordial development with age in the fern *Acrostichum danaeifolium. American Journal of Botany*, **42**, 627–36.

DeSoto, L., Quintanilla, L. G. and Ménez, M. (2008). Environmental sex determination: effects of nutrient availability and individual density in *Woodwardia radicans. Journal of Ecology*, **96**, 1319–27.

Durand, L. Z. and Goldstein, G. (2001). Growth, leaf characteristics, and spore production in native and invasive tree ferns in Hawaii. *American Fern Journal*, **91**, 25–35.

Dyer, A. F. and Lindsay, S. (1992). Soil spore banks of temperate ferns. *American Fern Journal*, **82**, 89–123.

Ellwood, M. D. F., Jones, D. T. and Foster, W. A. (2002). Canopy ferns in lowland dipterocarp forest support a prolific abundance of ants, termites, and other invertebrates. *Biotropica*, **34**, 575–83.

Esteves, L. M. and Dyer, A. F. (2003). The vertical distributions of live and dead fern spores in the soil of a semi-natural woodland in southeast Scotland and their implications for spore movement in the formation of soil spore banks. In *Pteridology in the New Millennium*, ed. S. Chandra and M. Srivastava. Dordrecht, The Netherlands: Kluwer Academic Publishers, pp. 261–82.

Farrar, D. R., Dassler, C., Watkins, J. E., Jr. and Skelton, C. (2008). Gametophyte ecology. In *Biology and Evolution of Ferns and Lycophytes,* ed. T. A. Ranker and C. H. Haufler. Cambridge, UK: Cambridge University Press, pp. 222–56.

Franklin, J. F., Bledsoe, C. S. and Callahan, J. T. (1990). Contributions of the Long-Term Ecological Research Program. *Bioscience*, **40**, 509–23.

Gaxiola, A., Burrows, L. E. and Coomes, D. A. (2008). Tree fern trunks facilitate seedling regeneration in a productive lowland temperate rain forest. *Oecologia*, **155**, 325–35.

Gay, H. (1993). The architecture of a dimorphic clonal fern, *Lomagramma guianensis* (Aublet) Ching (Dryopteridaceae). *Botanical Journal of the Linnean Society*, **111**, 343–58.

Gilbert, O. L. (1970). Biological flora of the British Isles, *Dryopteris villarii* (Bellardi) Woynar. *Journal of Ecology*, **58**, 301–13.

Gillman, L. N. and Ogden, J. (2001). Physical damage by litterfall to canopy tree seedlings in two temperate New Zealand forests. *Journal of Vegetation Science*, **12**, 671–6.

Gotelli, N. J. (1995). *A Primer of Ecology.* Sunderland, MA, USA: Sinauer Associates, Inc.

Greer, G. K. and McCarthy, B. C. (2000). Patterns of growth and reproduction in a natural population of the fern *Polystichum acrostichoides. American Fern Journal*, **90**, 60–76.

Gureyeva, I. I. (2003). Demographic studies of homosporous fern populations in South Siberia. In *Pteridology in the New Millennium*, ed. S. Chandra and M. Srivastava. Dordrecht, The Netherlands: Kluwer Academic Publishers, pp. 341–64.

Halleck, L. F., Sharpe, J. M. and Zou, X. (2004). Understorey fern responses to post-hurricane fertilization and debris removal in a Puerto Rican rain forest. *Journal of Tropical Ecology*, **20**, 173–81.

Halpern, C. B., Evans, S. A. and Nielson, S. (1999). Soil seed banks in young, closed-canopy forests of the Olympic Peninsula, Washington: potential contributions to understory reinitiation. *Canadian Journal of Botany*, **77**, 922–35.

Hamilton, R. G. (1988). The significance of spore banks in natural populations. *American Fern Journal*. **78**, 96–104.

Hamilton, R. G. (1992). Allozyme variation and ramet distribution in two species of athyriod ferns. *Plant Species Biology*, **7**, 69–76.

Hammen, S. C. L. (1993). Density-dependent phenotypic variation in the hay-scented fern, *Dennstaedtia punctilobula. Bulletin of the Torrey Botanical Club*, **120**, 392–6.

Harper, J. L. (1977). *Population Biology of Plants*. New York: Academic Press.

Harper, J. L. (1982). After description. In *The Plant Community as a Working Mechanism*, ed. E. I. Newman. Boston, MA, USA: Blackwell Scientific, pp. 11–25.

Headley, A. D. and Callaghan, T. V. (1990). Modular growth of *Huperzia selago* (Lycopodiaceae: Pteridophyta). *Fern Gazette*, **13**, 361–72.

Hebant-Mauri, R. and Gay, H. (1993). Morphogenesis and its relation to architecture in the dimorphic clonal fern *Lomagramma guianensis* (Aublet) Ching (Dryopteridaceae). *Botanical Journal of the Linnean Society*, **112**, 257–76.

Hoshizaki, B. J. and Moran, R. C. (2001). *Fern Grower's Manual*. Portland, OR, USA: Timber Press.

Jaag, O. (1943). Ergebnisse einer botanischen Studienreise nach Niederländisch-Indien 1937/38. I. Untersuchungen über den Rhythmus der Lauberneuerung, die Lebensdauer der Blätter and den Epiphytenbefall bei einigen Farnen in den Tropen. *Mitteilungen der Naturforschenden Gesellschaft Schaffhausen*, **18**, 205–57.

Johns, R. J. and Edwards, P. J. (1991). Vegetative reproduction in pteridophytes. *Kew Magazine*, **8**, 109–12.

Johnson-Groh, C. (1999). Population ecology of *Botrychium* (moonworts): status report on Minnesota *Botrychium* permanent plot monitoring. Report for Gustavus Adolphus College, Saint Peter, MN, USA.

Johnson-Groh, C., Riedel, C., Schoessler, L. and Skogen, K. (2002). Belowground distribution and abundance of *Botrychium* gametophytes and juvenile sporophytes. *American Fern Journal*, **92**, 80–92.

Jones, D. (1987). *Encyclopaedia of Ferns*. Portland, OR, USA: Timber Press.

Kato, M. and Setoguchi, H. (1999). An *rbcL*-based phylogeny and heteroblastic leaf morphology of Matoniaceae. *Systematic Botany*, **23**, 391–400.

Kelly, D. (1994). Demography and conservation of *Botrychium australe*, a peculiar, sparse mycorrhizal fern. *New Zealand Journal of Botany*, **32**, 393–400.

Kenkel, N. C. (1997). Demography of clonal ostrich fern (*Matteuccia struthiopteris*): a five year summary. *University of Manitoba, UFS (Delta Marsh) Annual Report*, **32**, 75–9.

Kornaś, J. (1977). Life forms and seasonal patterns in the pteridophytes of Zambia. *Acta Societatis Botanicorum Poloniae*, **46**, 669–90.

Large, M. F. and Braggins, J. E. (2004). *Tree Ferns*. Portland, OR, USA: Timber Press.

Lee, P.-H., Huang, Y.-M. and Chiou, W.-L. (2008). The phenology of *Osmunda claytoniana* L. in the Tataka area, Central Taiwan. *Taiwan Journal of Forest Science*, **23**, 71–9.

Lee, P.-H., Lin, T.-T. and Chiou, W.-L. (2009). Phenology of 16 species of ferns in a subtropical forest of northeastern Taiwan. *Journal of Plant Research*, **122**, 61–7.

Lefkovitch, L. P. (1965). The study of population growth in organisms grouped by stages. *Biometrics*, **21**, 1–18.

Lellinger, D. B. (2002). A modern multilingual glossary for taxonomic pteridology. *Pteridologia*, **3**, 1–264.

Lindsay, S. and Dyer, A. F. (1996). Investigating the phenology of gametophyte development: an experimental approach. In *Pteridology in Perspective*, ed. J. M. Camus, M. Gibby and R. J. Johns. Kew, UK: Royal Botanic Gardens, pp. 633–50.

Mehltreter, K. (2006). Leaf phenology of the climbing fern *Lygodium venustum* in a semideciduous lowland forest on the Gulf of Mexico. *American Fern Journal*, **96**, 21–30.

Mehltreter, K. (2008). Phenology and habitat specificity of tropical ferns. In *Biology and Evolution of Ferns and Lycophytes*, ed. T. A. Ranker and C. H. Haufler. Cambridge, UK: Cambridge University Press, pp. 201–21.

Mehltreter, K. and García-Franco, J. G. (2008). Leaf phenology and trunk growth of the deciduous tree fern *Alsophila firma* (Baker) D.S. Conant in a lower montane Mexican forest. *American Fern Journal*, **98**, 1–13.

Mehltreter, K. and Palacios-Rios, M. (2003). Phenological studies of *Acrostichum danaeifolium* (Pteridaceae, Pteridophyta) at a mangrove site on the Gulf of Mexico. *Journal of Tropical Ecology*, **19**, 155–62.

Milton, S. J. (1987a). Growth of seven-weeks fern (*Rumohra adiantiformis*) in Southern Cape forests: implications for management. *South African Forestry Journal*, **143**, 1–4.

Milton, S. J. (1987b). Effects of harvesting on four species of forest ferns in South Africa. *Biological Conservation*, **41**, 133–46.

Montgomery, J. (1990). Survivorship and predation changes in five populations of *Botrychium dissectum* in Eastern Pennsylvania. *American Fern Journal*, **80**, 173–82.

Moran, R. C. (1987). Sterile–fertile leaf dimorphy and evolution of soral types in *Polybotrya* (Dryopteridaceae). *Systematic Botany* **12**: 617–28.

Moran, R. C. (2004). *A Natural History of Ferns*. Portland, OR: Timber Press.

Nagano, T. and Suzuki, E. (2007). Leaf demography and growth pattern of the tree fern *Cyathea spinulosa* on Yakushima Island. *Tropics*, **10**, 47–57.

Nobel, P. S. (1978). Microhabitat, water relations, and photosynthesis of a desert fern, *Notholaena parryi*. *Oecologia*, **31**, 293–309.

Noss, R. F. (1999). Assessing and monitoring forest biodiversity: a suggested framework and indicators. *Forest Ecology and Management*, **115**, 135–46.

Odland, A. (1995). Frond development and phenology of *Thelypteris limbosperma*, *Athyrium distentifolium* and *Matteuccia struthiopteris* in Western Norway. *Nordic Journal of Botany*, **15**, 225–36.

Odland, A. (1998). Size and reproduction of *Thelypteris limbosperma* and *Athyrium distentifolium* along environmental gradients in western Norway. *Nordic Journal of Botany*, **18**, 311–21.

Odland, A., Junttila, O. and Nilsen, J. (2004). Growth responses of *Matteuccia struthiopteris* plants from northern and southern Norway exposed to different temperature and photoperiod and treatments. *Nordic Journal of Botany*, **23**, 237–46.

Ogura, Y. (1972). *Comparative Anatomy of the Vegetative Organs of the Pteridophytes*. *Handbuch der Pflanzenanatomie*. Stuttgart, Germany: Borntraeger.

Oinonen, E. (1967a). The correlation between the size of Finnish bracken (*Pteridium aquilinum* (L.) Kuhn) clones and certain periods of site history. *Acta Forestalia Fennica*, **83**(2), 1–51.

Oinonen, E. (1967b). Sporal regeneration of ground pine (*Lycopodium complanatum* L.) in southern Finland in the light of the size and age of its clones. *Acta Forestalia Fennica*, **83**(1), 1–96.

Oinonen, E. (1971). The time table of vegetative spreading in oak fern (*Carpogymnia dryopteris* (L.) Love and Love) and may-lily (*Maianthemum bifolium* (L.) F. W. Schmidt) in southern Finland. *Acta Forestalia Fennica*, **118**, 1–21.

Øllgaard, B. (1979). Studies in Lycopodiaceae. II. The branching patterns and infrageneric groups of *Lycopodium sensu lato*. *American Fern Journal*, **69**, 49–61.

Orth, R. (1938). Zur Morphologie der Primärblätter einheimischer Farne. *Flora*, **33**, 1–55.

Page, C. N. (1997). *The Ferns of Britain and Ireland*. Cambridge, UK: Cambridge University Press.

Page, C. N. and Brownsey, P. J. (1986). Tree-fern skirts: a defence against climbers and large epiphytes. *Journal of Ecology*, **74**, 787–96.

Peck, J. H., Peck, C. J. and Farrar, D. R. (1990). Influences of life history attributes on formation of local and distant fern populations. *American Fern Journal*, **80**, 126–42.

Pinero, D., Martinez-Ramoz, M., Mendoza, A., Alvarez-Buylla, E. and Sarukhan, J. (1986). Demographic studies in *Astrocaryum mexicanum* and their use in understanding community dynamics. *Principes*, **30**, 108–16.

Primack, R. B. (1973). Growth patterns of five species of *Lycopodium*. *American Fern Journal*, **63**, 3–7.

Proctor, G. R. (1989). Ferns of Puerto Rico and the Virgin Islands. *Memoirs of the New York Botanical Garden*, **53**, 1–389.

Prugnolle, F., Rousteau, A. and Belin-Depoux, M. (2000). Occupation spatiale de *Cyathea muricata* Willd. (Cyatheaceae) en foret dense humide guadeloupeenne. I. A l'echelle de l'individu. *Acta Botanica Gallica*, **147**, 361–74.

Pryer, K. M., Buler, E. Y., Farrar, D., *et al.* (2008). On the importance of portraying the plant life cycle accurately: ferns as a case study. *Abstracts of Botany 2008. Botanical Society of America.* http://2008.botanyconference.org. Viewed December 2008.

Punetha, N. (1989). Leaf growth and productivity in two western Himalayan thelypteroid ferns. *Indian Fern Journal*, **6**, 68–72.

Ramirez-Trejo, M. D. R., Perez-Garcia, B. and Orozco-Segovia, A. (2004). Analysis of fern spore banks from the soil of three vegetation types in the central region of Mexico. *American Journal of Botany*, **91**, 682–8.

Ranal, M. A. (2003). Soil spore banks of ferns in a gallery forest of the ecological station of Panga, Uberlandia, MG. Brazil. *American Fern Journal*, **93**, 97–115.

Reich, P. B., Uhl, C., Walters, M. B., Prugh, L. and Ellsworth, D. S. (2004). Leaf demography and phenology in Amazonian rain forest: a census of 40,000 leaves of 23 tree species. *Ecological Monographs*, **74**, 3–23.

Rodrigues, A. S. L., Andelman, S. J., Bakarr, *et al.* (2004). Effectiveness of the global protected area network in representing species diversity. *Nature*, **428**, 640–2.

Rubin, G., Robson, D. S. and Paolillo, D. J. (1985). Effects of population density on sex expression in *Onoclea sensibilis* L. on agar and ashed soil. *Annals of Botany*, **55**, 201–15.

Russell, A. E., Raich, J. W. and Vitousek, P. M. (1998). The ecology of the climbing fern *Dicranopteris linearis* on windward Mauna Loa, Hawaii. *Journal of Ecology*, **86**, 765–79.

Sanford, R. L., Braker, H. E. and G. S. Hartshorn. (1986). Canopy openings in a primary neotropical lowland forest. *Journal of Tropical Ecology*, **2**, 277–82.

Sato, T. (1982). Phenology and wintering capacity of sporophytes and gametophytes of ferns native to northern Japan. *Oecologia*, **55**, 53–61.

Sato, T. (1985). Quantitative expression of fern leaf development and fertility in *Polystichum tripteron* (Aspidiaceae). *Plant Systematics and Evolution*, **150**, 191–200.

Sato, T. (1990a). Estimation of chronological age for sporophyte maturation in three semi-evergreen ferns in Hokkaido. *Ecological Research*, **5**, 55–62.

Sato, T. (1990b). A quantitative comparison of foliage development among allopatric ferns, *Dryopteris crassirhizoma, D. coreano-montana* and *D. filix-mas. Botanical Magazine*, **103**, 165–76.

Sato, T. and Tsuyuzaki, S. (1988). Quantitative comparison of foliage development among *Dryopteris monticola, D. tokyoensis* and a putative hybrid, *D. kominatoensis* in northern Japan. *Botanical Magazine*, **101**, 267–80.

Schmitt, J. L. and Windisch, P. G. (2006). Growth rates and age estimates of *Alsophila setosa* Kaulf. in southern Brazil. *American Fern Journal*, **96**, 103–11.

Schmitt, J. L. and Windisch, P. G. (2007). Estrutura populacional e desenvolvimento da fase esporifitica de *Cyathea delgadii* Sternb. (Cyatheaceae, Monilophyta) no sul do Brasil. *Acta Botanica Brasilica*, **21**, 731–40.

Schneller, J. J. (2008). Antheridiogens. In *Biology and Evolution of Ferns and Lycophytes*, ed. T. A. Ranker and C. H. Haufler. Cambridge, UK: Cambridge University Press, pp. 134–58.

Seiler, R. L. (1981). Leaf turnover rates and natural history of the Central American tree fern *Alsophila salvinii*. *American Fern Journal*, **71**, 75–81.

Sharpe, J. M. (1988). Growth, demography, tropic responses and apical dominance in the neotropical fern *Danaea wendlandii* Reichenb. (Marattiaceae). Unpublished Ph.D. thesis, University of Georgia, Athens, Georgia, USA.

Sharpe, J. M. (1993). Plant growth and demography of the neotropical herbaceous fern *Danaea wendlandii* (Marattiaceae) in a Costa Rican rain forest. *Biotropica*, **25**, 85–94.

Sharpe, J. M. (1997). Leaf growth and demography of the rheophytic fern *Thelypteris angustifolia* (Willdenow) Proctor in a Puerto Rican rainforest. *Plant Ecology*, **130**, 203–12.

Sharpe, J. M. (2005). Temporal variation in sporophyte fertility in *Dryopteris intermedia* and *Polystichum acrostichoides* (Dryopteridaceae: Pteridophyta). *Fern Gazette*, **17**, 223–34.

Sharpe, J. M. (in press). Responses of the mangrove fern *Acrostichum danaeifolium* Langsd. & Fisch. (Pteridaceae, Pteridophyta) to disturbances resulting from increased soil salinity and Hurricane Georges at the Jobos Bay National Estuarine Research Reserve, Puerto Rico. *Wetlands Ecology and Management*.

Sharpe, J. M. and Jernstedt, J. A. (1990a). Leaf growth and phenology of the dimorphic herbaceous layer fern *Danaea wendlandii* (Marattiaceae) in a Costa Rican rain forest. *American Journal of Botany*, **77**, 1040–9.

Sharpe, J. M. and Jernstedt, J. A. (1990b). Tropic responses controlling leaf orientation in the fern *Danaea wendlandii* (Marattiaceae). *American Journal of Botany*, **77**, 1050–9.

Sharpe, J. M. and Jernstedt, J. A. (1991). Stipular bud development in *Danaea wendlandii* (Marattiaceae). *American Fern Journal*, **81**, 119–27.

Sheffield, E. (2008). Alternation of generations. In *Biology and Evolution of Ferns and Lycophytes*, ed. T. A. Ranker and C. H. Haufler. Cambridge, UK: Cambridge University Press, pp. 49–74.

Silvertown, J. W. (1982). *Introduction to Plant Population Ecology*. London: Longman.

Siman, S. E. and Sheffield, E. (2002). *Polypodium vulgare* plants sporulate continuously in a non-seasonal glasshouse environment. *American Fern Journal*, **92**, 30–8.

Steeves, T. A. (1959). An interpretation of two forms of *Osmunda cinnamomea*. *Rhodora*, **61**, 223–30.

Szmeja, J. (1994). An individual's status in populations of isoetid species. *Aquatic Botany*, **48**, 203–24.

Takahashi, K. and Mikami, Y. (2006). Effects of canopy cover and seasonal reduction in rainfall on leaf phenology and leaf traits of the fern *Oleandra pistillaris* in a tropical montane forest, Indonesia. *Journal of Tropical Ecology*, **22**, 599–604.

Tanner, E. V. J. (1983). Leaf demography and growth of the tree-fern *Cyathea pubescens* Mett. ex Kuhn in Jamaica. *Botanical Journal of the Linnean Society*, **87**, 213–27.

Tryon, A. F. (1990). Fern spores: evolutionary levels and ecological differentiation. *Plant Systematics and Evolution, Supplement* **5**, 71–9.

Tryon, A. F. and Lugardon, B. (1990). *Spores of the Pteridophyta*. New York: Springer-Verlag.

Tryon, R. M. and Tryon, A. F. (1982). *Ferns and Allied Plants with Special Reference to Tropical America*. New York: Springer-Verlag.

Vöge, M. (1997a). Plant size and fertility of *Isoëtes lacustris* L. in 20 lakes of Scandinavia: a field study. *Archiv für Hydrobiology*, **139**, 171–85.

Vöge, M. (1997b). Number of leaves per rosette and fertility characters of the quillwort (*Isoëtes lacustris* L.) in 50 lakes in Europe: a field study. *Archiv für Hydrobiology*, **139**, 415–31.

Vöge, M. (2004). Non-destructive assessing and monitoring of populations of *Isoëtes lacustris* L. *Limnologica*, **34**, 147–53.

Vöge, M. (2006). The reproductive phenology of *Isoëtes lacustris* L.: results of field studies in Scandinavian lakes. *Limnologica*, **36**, 228–33.

Wagner, W. H. and Wagner, F. S. (1977). Fertile–sterile leaf dimorphy in ferns. *The Gardens' Bulletin, Singapore*, **30**, 251–67.

Walker, L. R. and Aplet, G. H. (1994). Growth and fertilization responses of Hawaiian tree ferns. *Biotropica*, **26**, 378–83.

Walker, L. R., Zimmerman, J. K., Lodge, D. J. and Guzman-Grajales, S. (1996). An altitudinal comparison of growth and species composition in hurricane-damaged forests in Puerto Rico. *Journal of Ecology*, **84**, 877–89.

Walker, T. G. (1985). Some aspects of agamospory in ferns – the Braithwaite system. *Proceedings of the Royal Society of Edinburgh*, **86**B, 59–66.

Watkins, J. E., Jr., Mack, M. K. and Mulkey, S. S. (2007). Gametophyte ecology and demography of epiphytic and terrestrial tropical ferns. *American Journal of Botany*, **94**, 701–8.

Watt, A. S. (1943). Contributions to the ecology of bracken (*Pteridium aquilinum*). II. The frond and the plant. *New Phytologist*, **42**, 103–26.

Watt, A. S. (1976). The ecological status of bracken. *Botanical Journal of the Linnean Society*, **73**, 217–39.

Werth, C. R. and Cousens, M. I. (1990). Summary: The contribution of population studies on ferns. *American Fern Journal*, **80**, 183–90.

Whittier, D. (2003a). The gametophyte of *Diphasiastrum sitchense*. *American Fern Journal*, **93**, 20–4.

Whittier, D. (2003b). Rapid gametophyte maturation in *Ophioglossum crotalophoroides*. *American Fern Journal*, **93**, 137–45.

Whittier, D. (2006). Gametophytes of four tropical, terrestrial *Huperzia* species (Lycopodiaceae). *American Fern Journal*, **96**, 54–61.

Whittier, D. and Storchova, H. (2007). The gametophyte of *Huperzia selago* in culture. *American Fern Journal*, **97**, 149–154.

Wick, H. L. and Hashimoto, G. T. (1971). *Leaf Development and Stem Growth of Treefern in Hawaii*. Berkeley, California: U.S. Forest Service Research Note PSW-237, Pacific Southwest Forest and Range Experiment Station.

Willmot, A. (1989). The phenology of leaf life spans in woodland populations of the ferns *Dryopteris filix-mas* (L.) Schott and *D. dilatata* (Hoffm.) A. Gray in Derbyshire. *Botanical Journal of the Linnean Society*, **99**, 387–95.

4

Nutrient ecology of ferns

SARAH J. RICHARDSON AND LAWRENCE R. WALKER

Key points

1. Ferns both respond to and impact ecosystem nutrient cycling.
2. Most ferns and lycophytes acquire nutrients through roots and generally in association with endomycorrhizal fungi. Ferns also acquire nutrients through direct absorption from decomposing litter and exceptionally through association with either ants or nitrogen-fixing cyanobacteria.
3. At a local scale, the vegetative cover and richness of ferns appears to be greatest on sites with high fertility. However, ferns probably make the greatest proportional contribution to total plant biomass on infertile soils.
4. Nutrient levels of fern leaves span a wide range of nitrogen (N) and phosphorus (P) concentrations but ferns still have low N concentrations relative to seed plants. The decomposition rates of fern litter vary widely among species and habitats.
5. Ferns have low leaf calcium (Ca) concentrations relative to seed plants at any given site. Variation among sites in the amount of fern biomass will influence soil Ca cycling.

4.1 Introduction

Soil fertility and associated soil biodiversity are critical drivers of vegetation composition and function (Bardgett, 2005), even in human-modified ecosystems (Yaalon, 2007). Adaptations to persist and compete for nutrients on sites of varying fertility drive the development of variation among species and help to determine how species are distributed across the landscape (Grime, 2002; Callaway, 2007). Contrary to the popular notion that ferns are poorly adapted to current environmental conditions, they present a bewildering array of strategies and have radiated into the same habitats as seed plants. Consequently, ferns and lycophytes are important components in most of the world's ecosystems (see Chapter 2; Page, 1979).

Too little is known to adequately address the nutrient ecology of fern gametophytes (but see Chapter 3; Watkins *et al.*, 2007a; Farrar *et al.*, 2008), so in this chapter we will

Fern Ecology, ed. Klaus Mehltreter, Lawrence R. Walker and Joanne M. Sharpe. Published by Cambridge University Press. © Cambridge University Press 2010.

examine how fern sporophytes acquire nutrients (Section 4.2), how species are distributed across soil fertility gradients (Section 4.3), how nutrients are distributed in fern leaves and litter (Section 4.4), and how ferns respond to fertilization experiments (Section 4.5). Together, this information is used to consider how ferns influence ecosystem nutrient cycling.

We limit our scope to the well-studied nutrients, N and P, and cations such as calcium (Ca), magnesium (Mg) and potassium (K) but we do not discuss uptake or hyperaccumulation of heavy metals such as aluminum (Al). The literature on nutrient dynamics of ferns is surprisingly sparse and the significance of soils for ferns and the impact of ferns on ecosystem nutrient cycling are only now being acknowledged.

4.2 How do ferns acquire nutrients?

4.2.1 Roots and rhizomes

Nutrient uptake in vascular plants occurs mostly through morphologically "true" roots and root hairs. In ferns and some other groups of vascular plants such as sedges, roots are produced along a thickened, mostly subterranean, shoot called a rhizome. Although direct nutrient uptake through rhizomes has been demonstrated in a species of sedge (Brooker *et al.*, 1999), there is no evidence for this in ferns. The only fern family without root hairs is the Ophioglossaceae (moonworts or adder's tongue ferns, Plate 4A; Berch and Kendrick, 1982; Kelly, 1994). Sometimes, rhizoidal hairs function as root hairs in families that lack roots, as in the Psilotaceae (e.g., *Tmesipteris*, Fig. 4.1; *Psilotum*, whisk-ferns or fork ferns, Plate 4B). Some rootless Hymenophyllaceae bear root hairs on both the rhizome and petioles (Kramer *et al.*, 1995). Rhizomes and roots are also implicated in clonal connections between individual ferns.

For some species, such as *Pteridium aquilinum sensu lato* (bracken) and *Dicranopteris linearis*, the dry biomass of rhizomes can be extraordinary. In Hawaii, rhizomes of *D. linearis* can reach a biomass of $1.3 \, \text{kg} \, \text{m}^{-2}$ (Russell *et al.*, 1998), whereas rhizomes of *P. aquilinum* in English heathlands can reach $1.6 \, \text{kg} \, \text{m}^{-2}$ (Marrs *et al.*, 1993), and more than $7 \, \text{kg} \, \text{m}^{-2}$ in seral situations in New Zealand (Bray, 1991). Although rhizomes have a trivial surface area relative to fine root networks (Tyson *et al.*, 1990), it would be profitable to explore whether they contribute directly to nutrient uptake, especially given the efficacy in sequestering soil nutrients of *D. linearis* (Russell *et al.*, 1998) and *P. aquilinum* (Mitchell *et al.*, 2000).

Some epiphytic ferns have water-absorbing trichomes on their leaves, which may also take up nutrients (see Chapter 5; e.g., *Pleopeltis polypodioides*, resurrection fern, Müller *et al.*, 1981).

Fig. 4.1 *Tmesipteris elongata* (a whisk-fern) growing on *Cyathea smithii* (a tree fern with persistent petioles) in lowland rain forest, northern New Zealand. The site is an alluvial terrace that greatly favors the tree fern on which the *Tmesipteris* is dependent. While the distribution of *Tmesipteris* is not directly influenced by soil fertility, its preferred microhabitats such as tree fern trunks (caudices) tend to be found on fertile sites.

Fig. 4.2 (a) Surface roots on the stem (caudex) of *Cyathea medullaris*. These roots may intercept nutrients from decaying litter trapped in the rosette of leaves above and supplement nutrient supplies from the soil. Photograph taken on an alluvial terrace in southern New Zealand. (b) The epiphytic staghorn fern (*Platycerium superbum*) may derive nutrients from decomposing litter captured in a large leaf rosette. Photograph taken on Fraser Island, eastern Australia.

4.2.2 Litter trapping

Not all nutrients are absorbed through roots and rhizomes in the soil. Tree ferns and other species with extended surface caudices that form functional stems may retain a mantle of live and dead roots. These mats of live roots probably take up nutrients directly from throughflow (i.e., stem flow) and from decomposing organic matter trapped in petiole bases and leaf base scars (Fig. 4.2a). Some ferns may also acquire nutrients from litter trapped in their leaf rosettes (Fig. 4.2b; Wardle *et al.*, 2003; Dearden and Wardle, 2007). Litter captured in this manner is intercepted before reaching the soil, and soluble nutrients that become available through decomposition can be channeled directly toward the roots growing out of the upright, aerial rhizome. Although we cannot find examples that have used radioactively labeled nutrients in litter to demonstrate that nutrients are acquired from litter and translocated to new leaves, this seems a plausible mechanism, especially for large rosette-forming epiphytic ferns such as the bird's-nest ferns (e.g., *Asplenium nidus*, Karasawa and Hijii, 2006). This would counterbalance the disadvantage of an epiphytic existence, where nutrients such as N and P must be obtained in the canopy where there is limited soil development. Such uptake of nutrients from litter has been demonstrated for epiphytic bromeliads (Reich *et al.*, 2003). The N isotopic signature of bromeliad leaves was greater in larger plants, ranging from as low as −6.6‰ in small individuals to −0.8‰ in large individuals, indicating that larger plants were acquiring less ^{14}N directly from the atmosphere and more ^{15}N from sources such as decomposing litter. Further, epiphytic plant size strongly influences the community of litter-dwelling arthropods and rates of litter decomposition (Wardle *et al.*, 2003) such that uptake of nutrients from decomposing litter should become even more profitable as plant size increases. Experimental additions or removals of litter from fern rosettes in a range of situations (i.e., situations ranging in fertility, and across a range of plant sizes as well as habitat types) would be insightful.

4.2.3 *Mutualisms*

Sometimes mutualisms are a source of nutrients for ferns. To our knowledge, ferns are involved in three types of mutualisms that supply or enhance supply of nutrients. Mycorrhizal associations are by far the most widespread in terms of the number of fern species and of the geographic distribution involved. Free-living, N-fixing bacteria are "captured" by *Azolla* (mosquito fern), a widespread aquatic fern genus, and may be important for terrestrial ferns as well. Lastly, two genera found only in the tropics have a mutualism with ants through which they obtain nutrients (see Chapter 7).

It is likely that symbiotic associations with fungi were essential for the early vascular plant biota to make the transition from water to land (Remy *et al.*, 1994). All studied families of ferns have at least some species with mycotrophy (see Chapter 7; Brundrett, 2002; Wang and Qiu, 2006). It is interesting to consider under which environmental conditions a mycorrhizal association is important. Nonmycotrophic ferns or facultatively mycotrophic ferns are likely to be the first colonists on new terrain such as the remote volcanic islands of Hawaii, perhaps facilitating later colonization of obligately mycotrophic ferns and angiosperms (Gemma *et al.*, 1992). Endomycorrhizae were present in 66 of 89 species and 23 of 26 families examined in Hawaii and were found in both terrestrial and epiphytic ferns (Tables 4.1, 4.2; Gemma *et al.*, 1992). In another example of the importance of environment, *Equisetum* (horsetail, Plate 4C) is nonmycorrhizal in wetlands and moist soil environments (Harley and Harley, 1987; Marsh *et al.*, 2000; Read *et al.*, 2000) but the sporophytes are commonly mycorrhizal in dry environments (Brundrett, 2002). Finally, it is curious that despite mycorrhizal associations being so widespread across the ferns, ectomycorrhizal associations are sparse (e.g., Cooper, 1976; Iqbal *et al.*, 1981) relative to the dominant endomycorrhizal type. Further investigation might illuminate other examples of ectomycorrhizal ferns, or identify the environmental situations under which ectomycorrhizal associations develop with ferns.

Table 4.1 *Frequency of endomycorrhizal infection
in selected families of Hawaiian ferns*

Family	No. species	No. samples	Frequency of infection of samples (%)
Adiantaceae	4	13	69
Aspleniaceae	8	14	64
Athyriaceae	5	10	90
Blechnaceae	4	11	91
Davalliaceae	3	40	70
Dennstaedtiaceae	2	4	100
Dicksoniaceae	3	3	100
Dryopteridaceae	5	11	100
Elaphoglossaceae	6	19	63
Gleicheniaceae	3	11	91
Grammitidaceae	6	14	43
Hymenophyllaceae	4	7	57
Lindsaeaceae	3	18	100
Marattiaceae	2	3	66
Ophioglossaceae	3	14	64
Polypodiaceae	5	13	0
Pteridaceae	7	15	93
Thelypteridaceae	6	6	50

Note: The number of species sampled varied per family.
Source: From Gemma *et al.* (1992).

Table 4.2 *Frequency of endomycorrhizal infection
of Hawaiian ferns on different substrates*

Substrate preference	No. species	Frequency of infection (%)
Aquatic	2	0
Arenicolous	2	50
Epilithic	7	86
Epiphytic	22	55
Geothermal	1	100
Terricolous	59	83

Source: From Gemma *et al.* (1992).

A second form of symbiosis is found in the leaves of species of *Azolla* (the aquatic mosquito fern) where the N-fixing bacteria *Anabaena azollae* fix atmospheric nitrogen as ammonium, allowing *Azolla* to acquire nitrogen directly (Calvert and Peters, 1981; Perkins and Peters, 1993). *Azolla* is the only genus of ferns known to have this relationship with N-fixing bacteria. Developing leaves of *Azolla* have hairs that "capture" free-living *Anabaena* threads and eventually encapsulate them. *Azolla* then forms another specialized type of hair that absorbs the ammonium and transfers it to the fern (Calvert and Peters, 1981; Perkins and Peters, 1993). The high N content of *Azolla* makes it desirable as a natural fertilizer crop in rice production (Moran, 2004), but also accounts for its aggressive behavior as a weed species (see Chapter 8) in large areas of the tropics (e.g., McConnachie *et al.*, 2004) and warmer temperate zones (e.g., Janes, 1998). Free-living N-fixing bacteria may also have a role in supplying N to terrestrial ferns (Maheswaran and Gunatilleke, 1990; Russell and Vitousek, 1997).

The Southeast Asian ant ferns (*Lecanopteris* spp. in the Polypodiaceae) and neotropical *Solanopteris* (a subgenus within *Microgramma*; see Appendix B; Gómez, 1977) are distinctive epiphytic ferns that acquire their nutrients through a mutualism with ants (see Chapter 7). *Lecanopteris* rhizomes are swollen and hollow, providing domatia for ant species, and the internal rhizome walls are minutely pitted, allowing them to absorb nutrients efficiently from decaying ant middens inside their rhizomes (Gay, 1993). Additionally, fine roots on the external surface of the rhizome access nutrients from ant-built tunnels (known as carton runways) along the host tree trunk (Gay, 1993). These tunnels, consisting of chewed plant material cemented together with saliva, act as a conduit for the ants to move around on the host tree and leave their host plant. Studies on understory tree saplings fed by ants suggest ants provide <1% of the host plant's total N demand (Fischer *et al.*, 2003). However, N from ants may be more significant for epiphytic plants; Treseder *et al.* (1995) determined that 29% of the N in the epiphytic *Dischidia major* (Asclepiadaceae) came from debris deposited in leaf cavities by ants.

4.3 Soil fertility and fern distribution

4.3.1 Fern biomass and cover

Ferns occur on soils that range greatly in levels of nutrient availability (i.e., fertility), with species exhibiting clear affinities to soils of a particular fertility (e.g., Tuomisto and Poulsen, 1996). Little is known about the relationships between soil fertility and fern biomass and cover. Ferns generally contribute <10% of total biomass in woody ecosystems (Tanner, 1985; Scatena *et al.*, 1993; Raich *et al.*, 1997; Crews *et al.*, 2001), but can dominate biomass in early successional communities (see Chapter 6; Bray, 1991). In terms of forest strata, ferns often contribute a high proportion of total vegetation cover (i.e., photosynthetic leaf area), especially at the ground level (Royo and Carson, 2006), and in the subcanopy of temperate and tropical rain forests where tree ferns occur. For example, Harms *et al.* (2004) reported that ferns accounted for up to 15% of the understory cover in three lowland neotropical sites, and that cover was greater at higher soil fertility (i.e., with high extractable soil P). Likewise, Shiels *et al.* (2008) reported that fern cover was greater on landslides with fertile, volcanically derived soils, these being high in P, than on landslide soils derived from more infertile granitic soils in Puerto Rico. We elaborate below about the relationship between soil fertility and fern species richness, and how fern groups are distributed on nutrient-poor and nutrient-rich soils.

4.3.2 *Fern species richness*

How does soil fertility influence fern and lycophyte species richness? Analyses of species richness are complicated first by scale (global, regional, within a biome or ecosystem, plot size) and second by the selection of explanatory variables used. At a global scale, fern richness is highest in the wet tropics under a suite of conditions that includes low soil fertility alongside long-term stability, wide altitudinal variation, low vapor pressure deficit, and a vast array of epiphytic and terrestrial microhabitats that provide ideal conditions for ferns (see Chapter 2; Barrington, 1993). Isolating the effect that soil fertility has on richness is complicated because finding comparable sites elsewhere in the world that differ only in their soil fertility is difficult. For this reason, and because we are explicitly interested in soil fertility, it is more rewarding to examine patterns of fern richness and soil fertility at smaller scales. At these smaller scales, climate may be relatively invariant, allowing the effects of soil fertility to emerge (e.g., Tuomisto *et al.*, 1995); or climate and soil fertility may covary in such a way that the effect of soil fertility can be partitioned out from the effects of climate (e.g., Lwanga *et al.*, 1998). Lastly, studying fertility effects on fern species richness is frequently complicated by the large epiphytic component present in forests (e.g., Watkins *et al.*, 2006). In addition to the observational studies discussed below, insights can be gained from experimental manipulations of soil nutrients (see Section 4.5.2).

In the wet tropics, Tuomisto *et al.* (2002) reported that there were no significant relationships between either species richness or diversity of ferns (including lycophytes) and soil cation concentrations (Ca + Mg + Na + K, over the range of 0–30 cmol(+) kg^{-1}) in a large survey (20×25 km) of lowland rain forest in Ecuador, although a number of individual species exhibited responses to the soil fertility gradient (see Section 4.3.4). The absence of a relationship contrasted with that of a large group of tropical angiosperms, the Melastomataceae, which were sampled for comparison during the fern survey. The Melastomataceae were significantly more species rich and species diverse on sites with low soil cation concentrations (Tuomisto *et al.*, 2002). Nevertheless, several studies have reported that fern richness is highest on sites with higher nutrient availability. Lwanga *et al.* (1998) reported that fern richness in Ugandan rain forests was highest on more fertile sites (measured as the C:N ratio, over the range 8–14). Similarly, on those Hawaiian Islands with moderate precipitation (*c.* 2500 mm), data from Crews *et al.* (1995) and Kitayama and Mueller-Dombois (1995) together support that fern richness was highest on islands with high concentrations of organic P, mineralizable P and total N. Further, more structured investigations would be informative to determine whether richness follows a more typical unimodal relationship with fertility (Huston, 1996). What are the mechanisms (e.g., availability of epiphytic microhabitats) that promote high species richness at high fertility? Epiphytic fern species can account for a greater proportion of total richness than terrestrial species (Watkins *et al.*, 2006), so a clearer understanding of how environment promotes richness in these two groups separately would be useful (see Chapter 2).

Emergent patterns of total species richness obviously derive from the responses of individual species and a number of regional examples exist demonstrating how individual fern species are distributed in relation to soil fertility.

4.3.3 Tree fern distributions and soil nutrients

Several observational studies have been published that discuss the relationships between soil fertility and tree fern abundance in forests. Although soil fertility is invoked as the causal mechanism in these studies, other environmental factors may vary alongside soil fertility, such as disturbance and understory light, which could influence the relationship. Jones *et al.* (2007) tested whether the coexistence of four common tree fern species in lowland neotropical rain forest could be explained by organization of the species along fertility gradients. Three of the four species (all in the Cyatheaceae) were strongly associated with sheltered topographic positions such as terraces and valleys. In spite of this shared preference, the modeled presence of all four species was strongly determined by variation in measured soil fertility, suggesting each occupied a distinct part of the soil fertility matrix. Jones *et al.* (2007) concluded that niche partitioning of soil nutrient requirements could account for coexistence of these phylogenetically similar species, and could promote high species richness in the region. In a cool-temperate, New Zealand rain forest, the two most widespread tree fern species, *Cyathea smithii* (katote or soft tree fern) and *Dicksonia squarrosa* (wheki or rough tree fern), were present on soils ranging in total P from <5 g m^{-2} to >180 g m^{-2}, but stem density and basal area of both species were greater at higher total soil P, such that on fertile alluvial terraces (total soil P >50 g m^{-2}), tree ferns reached stem densities of 610 stems ha^{-1} (Figs. 4.3, 4.4; Coomes *et al.*, 2005). In Hawaii, the cover of *Cibotium glaucum* tree ferns was greatest on young surfaces (<20 000 years), where N and P are colimiting and cover was lower on sites with either high N or P concentrations alone (Crews *et al.*, 1995; Vitousek, 2004).

Fig. 4.3 *Cyathea smithii* fills a gully meandering across a lowland terrace in Waitutu Forest, southern New Zealand. Terraces and gullies typically have high soil total P (Richardson *et al.*, 2008) and are characterized by abundant tree ferns.

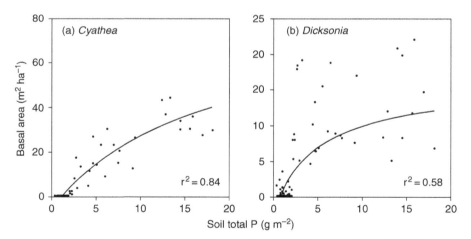

Fig. 4.4 Basal area of two tree fern species (*Cyathea smithii and Dicksonia squarrosa*) both increased with soil total P in lowland temperate rain forest, Waitutu Forest, southern New Zealand. (With permission from Coomes *et al.*, 2005.)

4.3.4 *Tropical fern distributions and soil cation concentrations*

Tuomisto *et al.* (2002) found that just over half of the common fern species in a lowland tropical rain forest in Ecuador were nonrandomly distributed with regard to soil cation concentrations. Data on uncommon species are more difficult to test statistically, and these are more likely to exhibit strong microhabitat "preferences," such as for soil nutrients, than common fern species. Of the ten most frequent fern species encountered during the survey, the occurrences of five (e.g., *Thelypteris macrophylla* and *Tectaria antioquoiana*) were biased toward soils with low cation concentrations, two toward soils with high cation concentrations, and the remaining three were indistinguishable from random (Tuomisto *et al.*, 2002). Also in western Amazonia, Tuomisto *et al.* (1998) described strong soil preferences among six co-occurring species of *Adiantum* or maidenhair ferns. The number of individuals of each species was examined along a gradient of total cation concentrations that ranged from 18 to 556 μg g^{-1}. *Adiantum humile* and *A. pulverulentum* were both strongly affiliated with cation-rich soils (>150 μg g^{-1}), while *A. tomentosum* was restricted to low-cation sites (≤75 μg g^{-1}). Finally, Costa *et al.* (2005) demonstrated that 62% of the modeled variation in fern community composition from a lowland tropical site could be explained by soil measures related to clay content, including total nitrogen content (Fig. 4.5). Together, these studies provide compelling evidence that partitioning of soil resources may promote coexistence among fern species in lowland tropical rain forests, and point to the utility of ferns as indicator species for soil fertility.

4.3.5 *Ferns on low fertility soils*

Ferns have many different life forms and ecological traits and they are present on nearly all substrate types. Based on the limited information available, we consider that terrestrial ferns reach their highest diversity on nutrient-rich soils, but make the greatest proportional contribution to biomass on nutrient-poor soils. Situations such as heathlands, early-successional soils, sand dunes and tundra are common sites for a suite of fern and lycophyte species, many of which achieve global, holarctic or pantropical distributions. Lycopodiaceae (club-mosses) are frequent in subarctic and temperate heathlands (e.g., Callaghan, 1980), alpine and open montane vegetation (Plate 3A, e.g., Flenley, 1969) and nutrient impoverished rain forests (e.g., Cullen, 1987; Aplet and Vitousek, 1994), and in many instances are reliable indicator species for low soil fertility. *Pteridium* (bracken) and three genera of scrambling ferns, *Gleichenia*, *Dicranopteris* and *Sticherus* (also known as umbrella ferns, tangle ferns or coral ferns), are often abundant on and characteristic of nutrient-impoverished sites such as heaths and oligotrophic wetlands (Pegman

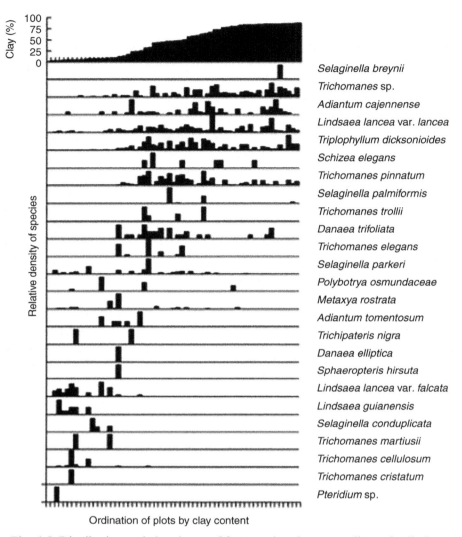

Fig. 4.5 Distribution and abundance of fern species along a gradient of soil clay content in lowland Amazonia. Soil clay content was strongly correlated with soil nutrient concentrations with low clay contents having high soil phosphate, and high clay contents having high soil nitrogen. (With permission from Costa *et al.*, 2005.)

and Ogden, 2006), landslides (see Chapter 6), highly weathered, successively retrogressive forest situations (e.g., Walker *et al.*, 1981), recent fire scars (McGlone *et al.*, 2006), coastal sand dunes, ultramafic soils, and heathlands. *Pteridium esculentum* (New Zealand bracken), *Gleichenia dicarpa* (swamp umbrella fern) and *Lycopodium diffusum* (carpet club-moss) are three of the five fern or lycophyte

Fig. 4.6 (a) *Pteridium esculentum* (a southern bracken) on ultramafic till heathland, southern New Zealand, and (b) on sand dunes on Kaitorete spit, southern New Zealand. (c) *Gleichenia dicarpa*, the swamp tangle fern, growing in nutrient-impoverished heathlands, northern New Zealand.

species that are dominant or locally common on the strongly P-limited ultramafic soils in New Zealand (Fig. 4.6; Lee, 1992).

Species such as *P. aquilinum* that colonize infertile soils often contribute to accumulation of soil organic matter through inputs of recalcitrant litter. Ganjegunte *et al.* (2005) demonstrated that *P. aquilinum* decomposed slowly relative to co-occurring species, and Johnson-Maynard *et al.* (1998) and Schrumpf *et al.* (2007) both reported that forest clearings colonized by *P. aquilinum* in northwest America and Tanzania, respectively, had high soil organic matter or soil C, and large amounts of partially decomposed litter of *P. aquilinum*. Nutrient pools of N, P and Ca are stored in this litter and soil organic matter, and substantially increase the total ecosystem pool (although not necessarily the availability) of nutrients (see Chapter 6).

4.4 Nutrient concentrations in fern leaves and litter

Do ferns have unusually high or low concentrations of nutrients in their leaves? Are ferns less effective at nutrient resorption than other vascular plants? Comparisons of leaf and litter nutrient concentrations among groups of species can be made in two ways. Recent macroecological analyses have highlighted that leaf nutrient concentrations covary with other leaf traits such as leaf mass per unit area (LMA; e.g., Reich *et al.*, 1999; Wright *et al.*, 2004). To compare leaf or litter quality between ferns and seed plants in a relative sense it is desirable to control for covariation among samples in LMA. However, in an ecosystem sense, covariation in LMA is unimportant because it is the absolute amounts of nutrients being acquired from and returned to the soil that matters. Because leaf and litter nutrient concentrations are strongly related to soil nutrient availability, we emphasize the greater insights that can be obtained from those studies that have examined nutrient concentrations while controlling for soil fertility. First, we explore some relative comparisons using data on LMA and nutrient concentrations, and then turn to differences between groups of species in absolute concentrations and the responses to soil fertility.

4.4.1 Comparisons between ferns and seed plants

Although ferns are commonly regarded as being thin leaved, they do in fact span a wide range of values for LMA and, correspondingly, of leaf N and P (Table 4.3; Vitousek *et al.*, 1995b; Wegner *et al.*, 2003; Wright *et al.*, 2004, 2005; Karst and Lechowicz, 2007; Watkins *et al.*, 2007b). That said, across all vascular plants, ferns appear to be poorly represented or absent in that part of the leaf economics spectrum (*sensu* Wright *et al.*, 2004) characterized by extremely long-lived leaves with very high LMA and very low leaf P concentrations (Table 4.3). When controlling for LMA, temperate terrestrial ferns have lower N per unit area than other vascular plants (Karst and Lechowicz, 2007).

In terms of their absolute concentrations of nutrients, a number of comparisons have been made between ferns and seed plants with contrasting insights. Using a large global dataset (>2000 vascular plant species) of leaf nutrient concentrations that contained 13 fern species, Wright *et al.* (2004) determined that nutrient concentrations in ferns (0.80–3.29% N and 0.05–0.11% P) were indistinguishable from those of seed plants (1.16–3.18% N and 0.01–0.60% P). Quested *et al.* (2003) compared leaf N concentrations in subarctic plants, and reported no differences in leaf N between five species of ferns, including lycophytes, and seed plants except N-fixers and hemiparasitic plant species. Han *et al.* (2005) reviewed >2000 leaf nutrient concentration samples from China, including 181 samples from ferns, and concluded that ferns had the lowest leaf P concentrations of any functional plant group, and similarly low N concentrations. Discrepancies among these broad-scale synthetic studies arise because of sampling biases and because comparisons are made without controlling for environment. While broad-scale reviews of published data are a logical step forward in ecology, and can provide an opportunity to consolidate valuable data, they compromise on the details that so often underpin ecologically meaningful interpretation.

An example of such detail comes from a recent study by Watkins *et al.* (2007b), who compared leaf nutrient concentrations and isotopic signatures from epiphytic and terrestrial ferns in tropical forests in Costa Rica. Epiphytic ferns had significantly lower leaf N concentrations relative to terrestrial ferns and a lower $\delta^{15}N$ signature, which suggests that epiphytic ferns acquire their nitrogen from different sources (e.g., precipitation and fog), than do terrestrial species (e.g., soil organic

Table 4.3 *Leaf traits of fern and lycophyte species*

Many published data are averages and ranges of values per species presented from within or across publications. The full range for vascular plants is presented and compared with the range of values from fern and lycophyte species in this table. Species are arranged in alphabetical order.

Species	N (% dry weight)	P (% dry weight)	LMA (g m^{-2})	Leaf life span (months)	Location
Adiantum flabellulatum	1.04	0.07			China[3]
Adiantum latifolium	3.36		31		Costa Rica[9]
Adiantum obliquum	2.84				Costa Rica[9]
Alsophila cuspidata	3.26		28		Costa Rica[9]
Anetium citrifolium	2.93		33		Costa Rica[9]
Asplenium auritum	1.57		55		Costa Rica[9]
Asplenium serra	2.58		55		Costa Rica[9]
Asplenium serratum	3.61		25		Costa Rica[9]
Athyrium wallichianum	3.14		87		Nepal[10]
Blechnum chambersii	1.25–2.34	0.16–0.42			New Zealand[5]
Blechnum discolor	0.81–1.39	0.08–0.18			New Zealand[5]
Blechnum fluviatile	1.28–1.80	0.12–0.27			New Zealand[5]
Blechnum orientale	1.41	0.04			China[3]
Blechnum sp1	2.03		26		Costa Rica[9]
Bolbitis nicotianifolia	3.58		37		Costa Rica[9]
Campyloneurum brevifolium	1.84		87		Costa Rica[9]
Cibotium chamissoi	1.4–2.2	0.11–0.17	51–120	11–12	Hawaii[2, 8]
Cibotium glaucum	1.5–1.9	0.10–0.15	60–240	13	Hawaii[1, 2, 8]
Cibotium menziesii	1.6–1.8		79–128	11–12	Hawaii[2]
Colchlidium serrulatum	1.54		112		Costa Rica[9]
Ctenitis rhodolepis	1.67	0.09			China[3]
Cyathea furfuracea	1.91	0.09	80	11	Jamaica[7]
Cyathea multiflora	2.47		37		Costa Rica[9]
Cyathea pubescens	1.47	0.08	132	17	Jamaica[7]
Cyathea smithii	1.22–2.18	0.12–0.24			New Zealand[5]
Cyathea woodwardioides	1.39	0.07	158	24	Jamaica[7]
Cyclopeltis semicordata	3.18		32		Costa Rica[9]
Danaea wendlandii	2.51		32		Costa Rica[9]
Dicksonia fibrosa	1.05–2.28	0.10–0.32			New Zealand[5]
Dicksonia squarrosa	1.10–1.81	0.10–0.23			New Zealand[5]
Dicranoglossum panamense	1.82		52		Costa Rica[9]

Table 4.3 (*cont.*)

Species	N (% dry weight)	P (% dry weight)	LMA (g m^{-2})	Leaf life span (months)	Location
Dicranopteris dichotoma	1.07	0.05			China[3]
Dicranopteris linearis	0.56–1.5	0.03–0.10	87–171	8–11	Hawaii[1, 6, 8]
Didymochlaena truncatula	2.88		32		Costa Rica[9]
Diphasiastrum complanatum	0.80				Sweden[10]
Diplazium striatastrum	3.76		32		Costa Rica[9]
Elaphoglossum herminieri	0.90		202		Costa Rica[9]
Elaphoglossum latifolium	1.35		108		Costa Rica[9]
Elaphoglossum peltatum	2.30		73		Costa Rica[9]
Equisetum arvense	2.11		101	4	Alaska[10]
Equisetum ramosissimum	1.45	0.19			China[3]
Equisetum sylvaticum	2.30		41		Sweden[10]
Equisetum variegatum	2.24		158	4	Alaska[10]
Gymnocarpium dryopteris	2.17		20–114		Canada[4], Ecuador[10]
Hemiontis palmata	2.78		34		Costa Rica[9]
Hicriopteris chinensis	1.04	0.06			China[3]
Hicriopteris glauca	0.93	0.04			China[3]
Leptopteris hymenophylloides	1.11–1.73	0.10–0.26			New Zealand[5]
Lomariopsis japurensis	3.33		34		Costa Rica[9]
Lomariopsis vestita	3.46		38		Costa Rica[9]
Lonchitis hirsuta	3.46		18		Costa Rica[9]
Lycopodiella cernua	0.80	0.06			China[3]
Lycopodium annotinum	0.86				Sweden[10]
Lycopodium clavatum	0.83	0.09			China[3]
Lycopodium complanatum	0.69	0.08			China[3]
Matteuccia struthiopteris	3.29		24–37		Canada[4], Sweden[10]
Microgramma lycopodioides	1.34		81		Costa Rica[9]

Table 4.3 (*cont.*)

Species	N (% dry weight)	P (% dry weight)	LMA (g m^{-2})	Leaf life span (months)	Location
Microgramma reptans	1.23		51		Costa Rica[9]
Nephrolepis rivularis	1.25		43		Costa Rica[9]
Oleandra articulata	1.56		42		Costa Rica[9]
Olfersia cervina	2.54		48		Costa Rica[9]
Ophioglossum reticulatum	5.98		26		Costa Rica[9]
Phlebodium pseudoaureum	2.83		22		Costa Rica[9]
Pleopeltis sp.	1.27		84		Costa Rica[9]
Polybotrya caudata	2.25		41		Costa Rica[9]
Polypodium triseriale	2.01		51		Costa Rica[9]
Polystichum wawranum	1.23–1.92	0.09–0.28			New Zealand[5]
Polytaenium ensiforme	2.30		54		Costa Rica[9]
Pteridium aquilinum	2.03–2.15	0.14	42–108		Canada[4], China[3], Spain[10]
Pteris actinopterioides	1.65	0.07			China[3]
Pteris semipinnata	1.35	0.10			China[3]
Pteris vittata	1.83	0.08			China[3]
Saccoloma inaequale	3.62				Costa Rica[9]
Salpichlaena volubilis	2.51		37		Costa Rica[9]
Sphaeropteris cooperi	2.4–2.6		35–38	6	Hawaii[2]
Struthiopteris eburnea	1.17	0.02			China[3]
Tectaria dracontifolia	3.30		31		Costa Rica[9]
Thelypteris lingulata	2.71		29		Costa Rica[9]
Thelypteris poiteana	2.85		36		Costa Rica[9]
Thelypteris sp.	5.15		45		Costa Rica[9]
Trichomanes collariatum	2.15		29		Costa Rica[9]
Trichomanes diversifrons	1.81		35		Costa Rica[9]
Trichomanes godmanii	0.21		31		Costa Rica[9]
Vittaria lineata	1.53		142		Costa Rica[9]
Vittaria stipitata	0.71		117		Costa Rica[9]

Table 4.3 (*cont.*)

Species	N (% dry weight)	P (% dry weight)	LMA (g m^{-2})	Leaf life span (months)	Location
Woodwardia japonica	1.20	0.10			China[3]
Woodwardia unigemmata	1.43	0.06			China[3]
Ferns and lycophytes	0.21–5.98	0.02–0.42	18–240	4–24	Global, this table
Vascular plants	0.25–6.35	0.008–0.60	14–1513	1–288	Global[10]

LMA, leaf mass per unit area.
Source: After[1] Baruch and Goldstein (1999); [2] Durand and Goldstein (2001); [3] Han *et al.* (2005); [4] Karst and Lechowicz (2007); [5] Richardson *et al.* (2008); [6] Russell *et al.* (1998); [7] Tanner (1983); [8] Vitousek *et al.* (1995a); [9] Watkins *et al.* (2007b), [10] Wright *et al.* (2004).

matter). At this microhabitat scale at a single site, differences among groups of fern species can be determined that could form the basis of a structured comparison with other terrestrial and epiphytic leaf nutrient concentrations. The question of whether ferns have distinctive nutrient concentrations relative to seed plants remains unanswered at a large scale and future comparisons would benefit from greater consideration of site factors (e.g., soil fertility, Richardson *et al.*, 2005), sources of nutrients (e.g., terrestrial versus epiphytic, Watkins *et al.*, 2007b), and covariation with other leaf and whole-plant traits (e.g., Wright *et al.*, 2004).

4.4.2 Tree ferns

Data from New Zealand, Hawaii, Jamaica and New Guinea together provide some evidence that tree fern leaves have higher P concentrations than other co-occurring vascular plants, and in some instances, higher N as well. In New Zealand, two tree fern species maintained consistently higher leaf P concentrations, as well as higher leaf N concentrations, than co-occurring woody angiosperms or conifers across a wide soil fertility gradient (Fig. 4.7; Richardson *et al.*, 2005). Leaves of the Hawaiian tree ferns (three species of *Cibotium*) generally contained higher P concentrations than seed plants across a chronosequence of islands ranging greatly in soil fertility (Vitousek *et al.*, 1995a). Tanner (1985) quantified the dry matter and nutrient biomass in aboveground pools from two tropical montane forests and examined whether the amount of nutrient was proportional to the amount of aboveground dry matter. The tree ferns had a disproportionately high amount of P (5.1%) compared with only 2.6% of the total leaf biomass. A similar pattern was found in

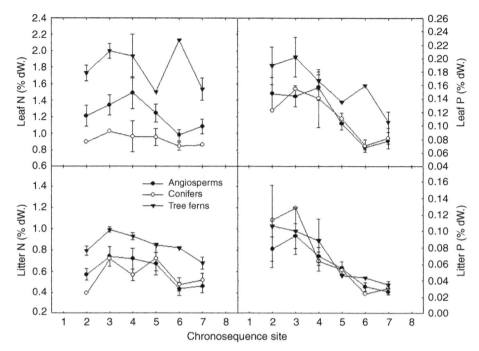

Fig. 4.7 Leaf and litter N and P concentrations in tree ferns, conifers and angiosperms along a soil chronosequence in New Zealand. Soil P and N are lower at older chronosequence stages. (With permission from Richardson *et al.*, 2005.)

subarctic heathlands for *Equisetum* (16% of P but only 5% of total biomass; Marsh *et al.*, 2000), raising the concept of a minor species (by biomass) having a disproportionately large influence on nutrient sequestration and perhaps cycling.

How do tree ferns maintain higher leaf nutrient concentrations than other plants, especially on poor soils? It is possible that tree ferns access novel sources of nutrients by channeling decomposing leaf litter from their leaf rosettes to the roots on their caudices (Dearden and Wardle, 2007). Our understanding of how plants acquire organic forms of N and P is still rather limited, but with an expanded concept of how much P may be available to plants (Turner, 2008), including forms of organic P that can be substantially more abundant than mineral forms of P, we will be in a stronger position to explore the full breadth of nutrient uptake pathways in plants, and thus understand the contribution of ferns to ecosystem functioning.

4.4.3 *Litter quality of ferns and decomposition rates*

Resorption is the process by which nutrients are mobilized during leaf senescence and either moved out of the leaf back into the plant for recycling, or mobilized into the leaf for disposal (Killingbeck, 1996). It is an important trait that contributes to plant fitness on nutrient-poor soils; the extent of resorption determines the quality of litter entering the soil and therefore has fundamental impacts on nutrient cycling and soil quality.

Comparisons of decomposition rates generally reveal that fern litter decomposes more slowly than litter of seed plants, although this is not universally the case (Shiels, 2006). Data from ferns in New Zealand forests (Wardle *et al.*, 2002), Sri Lankan forests and fernlands (Maheswaran and Gunatilleke, 1988), Hawaiian forests (Scowcroft, 1997; Allison and Vitousek, 2004a, 2004b), and subarctic heath in western Europe (Cornelissen *et al.*, 2006), all report that ferns and lycophytes decompose more slowly than seed plants. Litter decomposition rates are determined by a wide range of factors but higher concentrations of secondary defense compounds such as lignin, and lower concentrations of nutrients such as N and P, are generally related to recalcitrance and hence slower decomposition rates (Wardle *et al.*, 2002).

Nutrient concentrations in litter can be higher in ferns than in seed plants (e.g., Vitousek *et al.*, 1995a, 1995b; Richardson *et al.*, 2005) or lower (e.g., Maheswaran and Gunatilleke, 1988; Killingbeck *et al.*, 2002; Shiels, 2006), but concentrations of secondary defense compounds such as fiber, lignin or tannins are often higher (e.g., Wardle *et al.*, 2002; Ganjegunte *et al.*, 2005). However, Allison and Vitousek (2004a) demonstrated that ferns decomposed slowly relative to co-occurring angiosperms despite having higher litter N and P concentrations but lower concentrations of phenolics and similar concentrations of lignin and tannins. We suggest that slow decomposition rates in some fern species are related to high investment in fiber, lignin and tannins, irrespective of their often high nutrient concentrations. In this regard, ferns may present us with a paradox as they have high nutrient concentrations but decompose in a manner more typical of low nutrient species.

4.5 Nutrient limitation and consequences for ecosystems

4.5.1 Nutrient ratios

Insights into plant performance have been achieved by examining the ratios of nutrients in plant tissues. This approach (stoichiometry) works on the principle that plant growth requires nutrients in balance with one another, rather than having a surplus of one nutrient and insufficient amounts of others (Sterner and Elser, 2002). For example, variation in the N:P ratio among sites or species has been used as a proxy to determine whether N, P or both are limiting plant growth (e.g., Tessier and Raynal, 2003; Güsewell, 2004; Wardle *et al.*, 2004), although such proxies are only surrogates for fertilization experiments (Harrington *et al.*, 2001).

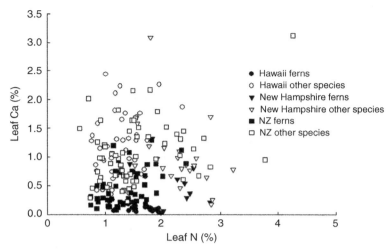

Fig. 4.8 Leaf N and Ca concentrations from rain forest species on Hawaii (Vitousek *et al.*, 1995b; Vitousek, 2004), rain forest species in New Zealand (site described in Richardson *et al.*, 2008; S. J. Richardson, unpublished data), and from understory herbs and ferns in a New Hampshire (USA) temperate hardwood forest (Siccama *et al.*, 1970). Leaf Ca concentrations are lower in ferns relative to other vascular plants, both in an absolute sense and also relative to other important nutrients, such as N.

One aspect of leaf-level and whole-plant stoichiometry that has been highlighted with ferns is low Ca concentrations relative to N and P (Vitousek *et al.*, 1995a; Amatangelo and Vitousek, 2008). Ferns appear to have low leaf Ca concentrations both in absolute terms and when expressed relative to N and P (Fig. 4.8; Amatangelo and Vitousek, 2008). Data supporting this derive from chronosequence sites in Hawaii (Vitousek *et al.*, 1995b; Vitousek, 2004, Amatangelo and Vitousek, 2008), forest plants in New Zealand (S. J. Richardson, unpublished data), and forest understory plants from a North American hardwood forest (Siccama *et al.*, 1970). A consequence of this low leaf Ca concentration is that ferns have much higher leaf N:Ca ratios (mean = 9.8) than other vascular plants (mean = 1.92; Fig. 4.8; Siccama *et al.*, 1970; Vitousek *et al.*, 1995b; Vitousek, 2004; S. J. Richardson, unpublished data). Explanations for these patterns are not obvious but Amatangelo and Vitousek (2008) recently suggested that ferns, with the exception of the Polypodiaceae, may have evolved in cation-poor environments, which led to the evolution of cell wall structures with low cation content. Regardless of the causes, these differences clearly have substantial consequences for soils and nutrient cycling, as variation in the contribution by ferns to litter inputs will impact soil Ca cycling (Amatangelo and Vitousek, 2008). Beyond the values from three forested ecosystems shown in Fig. 4.8, it would be interesting to explore a wider range of sites, including strongly cation-rich situations (such as limestone grasslands), in order to determine if such differences between ferns and seed plants are present when cations are abundant.

4.5.2 Nutrient additions

Fertilization studies provide direct evidence of fern responses to varying fertility. Such studies are particularly relevant for understanding biotic responses to the widespread but inadvertent N deposition, as from acid rain or agricultural run-off (Langan, 1999). Given the wide variation of responses of ferns to soil fertility, it is no surprise that ferns have varied responses to intentional fertilization experiments. For example, urea additions in managed forests in Washington, USA, resulted in positive growth responses by *Dryopteris austriaca* (mountain wood fern) and *Athyrium filix-femina* (lady fern) but negative growth responses by *Polystichum munitum* (sword fern), *Blechnum spicant* (deer fern) and *Pteridium aquilinum* (Thomas *et al.*, 1999).

In the N-saturated uplands of the northeastern USA, most understory plants were P- but not N-limited (Tessier and Raynal, 2003). *Huperzia lucidula* (shining fir-moss) and *Dryopteris intermedia* (evergreen wood fern) had higher N:P ratios and were likely more P-limited than other understory plants (*Acer* spp.; *Erythronium americanum*, trout lily; *Oxalis acetosella*, wood sorrel; and *Viola macloskeyi*, wild white violet). *Dryopteris intermedia* was the only species to show greater growth at higher levels of P fertilization. A comparison of different vascular plants suggested that upland vegetation is more limited by P at lower N:P ratios than is wetland vegetation (Tessier and Raynal, 2003). Whether such habitat generalizations apply to ferns is unknown.

Three studies have examined tropical fern responses to experimental fertilization. Addition of N or P increased height growth rates of *Cibotium glaucum* tree ferns in Hawaii and N addition promoted increased leaf production rates on young soils (200 years), but not on older soils (1000 years; Walker and Aplet, 1994). Fertilization had no effect on leaf longevity in either stand. This study supported the idea that N is most limiting in the early stages of primary succession (Wardle *et al.*, 2004).

A second study, also on Hawaii, examined the response by *Dicranopteris linearis* to fertilization. Annual shoot growth increased after application of N, P, or N and P together at two young (140-year-old) sites (Raich *et al.*, 1996).

A third study examined temporal changes under experimental nutrient additions. Understory ferns were measured at four and six years after initiation of a fertilization study at two elevations in Puerto Rico. Fertilization resulted in lower live fern biomass and fern cover after four years in lowland tropical rain forest, but greater live fern biomass in high elevation tropical rain forest relative to the unfertilized treatments (Walker *et al.*, 1996). The greatest negative change was seen in *Thelypteris deltoidea*, which was the most abundant terrestrial fern at the site (Halleck *et al.*, 2004). This was followed by *Blechnum occidentale* also growing terrestrially, and *Elaphoglossum rigidum* growing epiphytically. Closer examination after six years revealed that ferns in lowland fertilized plots were less dense, smaller and had lower spore production than ferns in control plots (Halleck *et al.*, 2004). Terrestrial ferns were more affected than epiphytic ferns, suggesting that the fertilization applications every three months had negative effects (Halleck *et al.*, 2004).

Fern responses to fertilization clearly depend not only on species but also on habitat factors such as successional stage, topographical status, soil chemistry and competition with other plants. Further work is needed to understand how ferns respond to alterations of nutrients.

4.6 Future research directions

We believe that ferns play a unique role in nutrient cycling, because they have different construction costs than seed plants (Amatangelo and Vitousek, 2008), and can therefore acquire and store a disproportionate amount of ecosystem nutrient capital relative to their biomass (Vitousek *et al.*, 1995a; Marsh *et al.*, 2000). However, much remains to be learned about ferns regarding their nutrient ecology. We consider that a stronger knowledge of how plants access both organic and mineral forms of nutrients will deliver significant advances in our understanding of plant–soil relationships (Turner, 2008), both for ferns and for ecosystems. Our ability to link measures of nutrient availability to vegetation properties are still impeded by the poor correspondence between those few nutrient pools that we can easily measure in laboratories versus the potentially vast range of soil and atmospheric nutrient pools that plants can draw on. We already know that species access different forms of organic N (e.g., the oak fern, *Gymnocarpium dryopteris*, preferentially stores amino acid N as glutamine in belowground structures in subarctic heathlands, Nordin and Näsholm, 1997), and similar processes have been suggested for organic P, but remain undemonstrated (Turner, 2008; Lambers *et al.*, 2008). Lastly, application of new tools for quantifying these nutrient pools and their fluxes, and assessing how ferns and other plants acquire them will be most productive along gradients of nutrient availability where ferns can be viewed across a range of situations and compared under similar conditions with other vascular plants.

Acknowledgements

We thank Peter Bellingham, Ian Dickie, Klaus Mehltreter, Jill Rapson, Joanne Sharpe and two anonymous reviewers for constructive comments and Chris Morse and Janet Wilmshurst for permission to use photographs. Sarah Richardson was supported by grant C09X0502 from the New Zealand Foundation for Research,

Science and Technology and Lawrence Walker was supported by the Luquillo Long-Term Ecological Research Program in Puerto Rico funded by the U.S. National Science Foundation.

References

Allison, S. D. and Vitousek, P. M. (2004a). Extracellular enzyme activities and carbon chemistry as drivers of tropical plant litter decomposition. *Biotropica*, **36**, 285–96.

Allison, S. D. and Vitousek, P. M. (2004b). Rapid nutrient cycling in leaf litter from invasive plants in Hawai'i. *Oecologia*, **141**, 612–9.

Amatangelo, K. L. and Vitousek, P. M. (2008). Stoichiometry of ferns in Hawai'i: implications for nutrient cycling. *Oecologia*, **157**, 619–27.

Aplet, G. H. and Vitousek, P. M. (1994). An age–altitude matrix analysis of Hawai'ian rainforest succession. *Journal of Ecology*, **82**, 137–47.

Bardgett, R. (2005). *The Biology of Soil: A Community and Ecosystem Approach*. Oxford, UK: Oxford University Press.

Barrington, D. S. (1993). Ecological and historical factors in fern biogeography. *Journal of Biogeography*, **20**, 275–9.

Baruch, Z. and Goldstein, G. (1999). Leaf construction cost, nutrient concentration, and net CO_2 assimilation of native and invasive species in Hawai'i. *Oecologia*, **121**, 183–92.

Berch, S. M. and Kendrick, B. (1982). Vesicular–arbuscular mycorrhizae of southern Ontario ferns and fern-allies. *Mycologia*, **74**, 769–76.

Bray, J. R. (1991). Growth, biomass, and productivity of a bracken (*Pteridium esculentum*) infested pasture in Marlborough Sounds, New Zealand. *New Zealand Journal of Botany*, **29**, 169–76.

Brooker, R. W., Callaghan, T. V. and Jonasson, S. (1999). Nitrogen uptake by rhizomes of the clonal sedge *Carex bigelowii*: a previously overlooked nutritional benefit of rhizomatous growth. *New Phytologist*, **142**, 35–48.

Brundrett, M. C. (2002). Coevolution of roots and mycorrhizas of land plants. *New Phytologist*, **154**, 275–304.

Callaghan, T. V. (1980). Age-related patterns of nutrient allocation in *Lycopodium annotinum* from Swedish Lapland: strategies of growth and population dynamics of tundra plants, 5. *Oikos*, **35**, 373–86.

Callaway, R. M. (2007). *Positive Interactions and Interdependence in Plant Communities*. New York: Springer-Verlag.

Calvert, H. E. and Peters, G. A. (1981). The *Azolla–Anabaena azollae* relationship. IX. Morphological analysis of leaf cavity hair populations. *New Phytologist*, **89**, 327–35.

Coomes, D. A., Allen, R. B., Canham, C. D., *et al.* (2005). The hare, the tortoise, and the crocodile: the ecology of angiosperm dominance, conifer persistence, and fern filtering. *Journal of Ecology*, **93**, 918–35.

Cooper, K. M. (1976). A field survey of mycorrhizas in New Zealand ferns. *New Zealand Journal of Botany*, **14**, 169–81.

Cornelissen, J. H. C., Quested, H. M., Logtestijn, R. S. P., *et al.* (2006). Foliar pH as a new plant trait: can it explain variation in foliar chemistry and carbon cycling processes among subarctic plant species and types? *Oecologia*, **147**, 315–26.

Costa, F. R., Magnusson, W. E. and Luizao, R. C. (2005). Mesoscale distribution patterns of Amazonian understorey herbs in relation to topography, soil and watersheds. *Journal of Ecology*, **93**, 863–78.

Crews, T. E., Kitayama, K., Fownes, J. H., *et al.* (1995). Changes in soil phosphorus fractions and ecosystem dynamics across a long chronosequence in Hawai'i. *Ecology*, **76**, 1407–24.

Crews, T. E., Kurina, L. M. and Vitousek, P. M. (2001). Organic matter and nitrogen accumulation and nitrogen fixation during early ecosystem development in Hawai'i. *Biogeochemistry*, **52**, 259–79.

Cullen, P. J. (1987). Regeneration patterns in populations of *Athrotaxis selaginoides* D. Don. from Tasmania. *Journal of Biogeography*, **14**, 39–51.

Dearden, F. M. and Wardle, D. A. (2007). The potential for forest canopy litterfall interception by a dense fern understorey, and the consequences for litter decomposition. *Oikos*, **117**, 83–92.

Durand, L. Z. and Goldstein, G. (2001). Growth, leaf characteristics, and spore production in native and invasive tree ferns in Hawai'i. *American Fern Journal*, **91**, 25–35.

Farrar, D. R., Dassler, C., Watkins, J. E., Jr. and Skelton, C. (2008). Gametophyte ecology. In *Biology and Evolution of Ferns and Lycophytes*, ed. T. A. Ranker and C. H. Haufler. Cambridge, UK: Cambridge University Press, pp. 222–56.

Fischer, R. C., Wanek, W., Richter, A. and Mayer, V. (2003). Do ants feed plants? A ^{15}N labelling study of nitrogen fluxes from ants to plants in the mutualism of *Pheidole* and *Piper*. *Journal of Ecology*, **91**, 126–34.

Flenley, J. R. (1969). The vegetation of the Wabag Region, New Guinea Highlands: a numerical study. *Journal of Ecology*, **57**, 465–90.

Ganjegunte, G. K., Condron, L. M., Clinton, P. W. and Davis, M. R. (2005). Effects of mixing radiata pine needles and understory litters on decomposition and nutrient release. *Biology and Fertility of Soils*, **41**, 310–19.

Gay, H. (1993). Animal-fed plants: an investigation into the uptake of ant-derived nutrients by the far-eastern epiphytic fern *Lecanopteris* Reinw. (Polypodiaceae). *Biological Journal of the Linnean Society*, **50**, 221–33.

Gemma, J. N., Koske, R. E. and Flynn, T. (1992). Mycorrhizae in Hawai'ian pteridophytes: occurrence and evolutionary significance. *American Journal of Botany*, **79**, 843–52.

Gómez, L. D. (1977). The *Azteca* ants of *Solanopteris brunei*. *American Fern Journal*, **67**, 31.

Grime, J. P. (2002). *Plant Strategies, Vegetation Processes and Ecosystem Properties*. Chichester, UK: Wiley.

Güsewell, S. (2004). N:P ratios in terrestrial plants: variation and functional significance. *New Phytologist*, **164**, 243–66.

Halleck, L. F., Sharpe, J. M. and Zou, X. (2004). Understorey fern responses to post-hurricane fertilization and debris removal in a Puerto Rican rain forest. *Journal of Tropical Ecology*, **20**, 173–81.

Han, W., Fang, J., Guo, D. and Zhang, Y. (2005). Leaf nitrogen and phosphorus stoichiometry across 753 terrestrial plant species in China. *New Phytologist*, **168**, 377–85.

Harley, J. L. and Harley, E. L. (1987). A check-list of mycorrhiza in the British flora. *New Phytologist*, **105**, 1–102.

Harms, K. E., Powers, J. S. and Montgomery, R. A. (2004). Variation in small sapling density, understory cover, and resource availability in four neotropical forests. *Biotropica*, **36**, 40–51.

Harrington, R. A., Fownes, J. H. and Vitousek, P. M. (2001). Patterns and components of response to nutrient limitation: comparison of long-term results in N- and P-limited tropical forest ecosystems. *Ecosystems*, **4**, 646–57.

Huston, M. A. (1996). *Biological Diversity: The Coexistence of Species on Changing Landscapes*. Cambridge, UK: Cambridge University Press.

Iqbal, S. H., Yousaf, M. and Younus, M. (1981). A field survey of mycorrhizal associations in ferns of Pakistan. *New Phytologist*, **87**, 69–79.

Janes, R. (1998). Growth and survival of *Azolla filiculoides* in Britain. I. Vegetative reproduction. *New Phytologist*, **138**, 367–75.

Johnson-Maynard, J. L., McDaniel, P. A., Ferguson, D. E. and Falen, A. L. (1998). Changes in soil solution chemistry of andisols following invasion by bracken fern. *Soil Science*, **163**, 814–21.

Jones, M. M., Olivas Rojas, P., Tuomisto, H. and Clark, D. B. (2007). Environmental and neighbourhood effects on tree fern distributions in a neotropical lowland rain forest. *Journal of Vegetation Science*, **18**, 13–24.

Karasawa, S. and Hijii, N. (2006). Determinants of litter accumulation and the abundance of litter-associated microarthropods in bird's nest ferns (*Asplenium nidus* complex) in the forest of Yambaru on Okinawa Island, southern Japan. *Journal of Forest Research*, **11**, 313–18.

Karst, A. L. and Lechowicz, M. J. (2007). Are correlations among foliar traits in ferns consistent with those in the seed plants? *New Phytologist*, **173**, 306–12.

Kelly, D. (1994). Demography and conservation of *Botrychium australe*, a peculiar, sparse mycorrhizal fern. *New Zealand Journal of Botany*, **32**, 393–400.

Killingbeck, K. T. (1996). Nutrients in senesced leaves: keys to the search for potential resorption and resorption proficiency. *Ecology*, **77**, 1716–27.

Killingbeck, K. T., Hammen-Winn, S. L., Vecchio, P. G. and Goguen, M. E. (2002). Nutrient resorption efficiency and proficiency in fronds and trophopods of a winter-deciduous fern, *Dennstaedtia punctilobula*. *International Journal of Plant Sciences*, **163**, 99–105.

Kitayama, K. and Mueller-Dombois, D. (1995). Vegetation changes along gradients of long-term soil development in the Hawai'ian montane rainforest zone. *Vegetatio*, **120**, 1–20.

Kramer, K. U., Schneller, J. J. and Wollenweber, E. (1995). *Farne und Farnverwandte*. Stuttgart, Germany: Georg Thieme Verlag.

Lambers, H., Chapin, F. S., III. and Pons, T. L. (2008). *Plant Physiological Ecology*, 2nd edn. New York: Springer-Verlag.

Langan, S. J. ed. (1999). *The Impacts of Nitrogen Deposition on Natural and Semi-Natural Ecosystems*. New York: Springer-Verlag.

Lee, W. G. (1992). New Zealand ultramafics. In *The Ecology of Areas with Serpentinized Rocks: A World View*, ed. B. A. Roberts and J. Proctor. Dordrecht, The Netherlands: Kluwer Academic Publishers, pp. 375–418.

Lwanga, J. S., Balmford, A. and Badaza, R. (1998). Assessing fern diversity: relative species richness and its environmental correlates in Uganda. *Biodiversity and Conservation*, **7**, 1387–98.

Maheswaran, J. and Gunatilleke, I. A. U. N. (1988). Litter decomposition in a lowland rain forest and a deforested area in Sri Lanka. *Biotropica*, **20**, 90–9.

Maheswaran, J. and Gunatilleke, I. A. U. N. (1990). Nitrogenase activity in soil and litter of a tropical lowland rainforest and adjacent fernland in Sri Lanka. *Journal of Tropical Ecology*, **6**, 281–9.

Marrs, R. H., Pakeman, R. J. and Lowday, J. E. (1993). Control of bracken and the restoration of heathland. V. Effects of bracken control treatments on the rhizome and its relationship with frond performance. *Journal of Applied Ecology*, **30**, 107–18.

Marsh, A. S., Arnone, J. A. A., III., Bormann, B. T. and Gordon, J. C. (2000). The role of *Equisetum* in nutrient cycling in an Alaskan shrub wetland. *Journal of Ecology*, **88**, 999–1011.

McConnachie, A. J., Hill, M. P. and Byrne, M. J. (2004). Field assessment of a frond-feeding weevil, a successful biological control agent of red waterfern, *Azolla filiculoides*, in southern Africa. *Biological Control*, **29**, 326–31.

McGlone, M. S., Dungan, R. J., Hall, G. M. J. and Allen, R. B. (2006). Winter leaf loss in the New Zealand woody flora. *New Zealand Journal of Botany*, **42**, 1–19.

Mitchell, R. J., Auld, M. H. D., Hughes, J. M. and Marrs, R. H. (2000). Estimates of nutrient removal during heathland restoration on successional sites in Dorset, southern England. *Biological Conservation*, **95**, 233–46.

Moran, R. C. (2004). *A Natural History of Ferns*. Portland, OR, USA: Timber Press.

Müller, L., Starnecker, G. and Winkler, S. (1981). Zur Ökologie epiphytischer Farne in Südbrasilien I. Saugschuppen. *Flora*, **171**, 55–63.

Nordin, A. and Näsholm, T. (1997). Nitrogen storage forms in nine boreal understorey plant species. *Oecologia*, **110**, 487–92.

Page, C. N. (1979). The diversity of ferns: an ecological perspective. In *The Experimental Biology of Ferns*, ed. A. F. Dyer. London: Academic Press, pp. 10–53.

Pegman, A. P. McK. and Ogden, J. (2006). Productivity–decomposition dynamics of *Baumea juncea* and *Gleichenia dicarpa* at Kaitoke Swamp, Great Barrier Island, New Zealand. *New Zealand Journal of Botany*, **44**, 261–71.

Perkins, S. K. and Peters, G. A. (1993). The *Azolla–Anabaena* symbiosis: Endophyte continuity in the *Azolla* life-cycle is facilitated by epidermal trichomes. I. Partitioning of the endophytic *Anabaena* into developing sporocarps. *New Phytologist*, **123**, 53–64.

Quested, H. M., Cornelissen, J. H. C., Press, M. C., *et al.* (2003). Decomposition of sub-arctic plants with differing nitrogen economies: a functional role for hemiparasites. *Ecology*, **84**, 3209–21.

Raich, J. W., Russell, A. E., Crews, T. E., Farrington, H. and Vitousek, P. M. (1996). Both nitrogen and phosphorus limit plant production on young Hawai'ian lava flows. *Biogeochemistry*, **32**, 1–14.

Raich, J. W., Russell, A. E. and Vitousek, P. M. (1997). Primary productivity and ecosystem development along an elevational gradient on Mauna Loa, Hawai'i. *Ecology*, **78**, 707–21.

Read, D. J., Duckett, J. G., Francis, R., Ligrone, R. and Russell A. (2000). Symbiotic fungal associations in 'lower' land plants. *Philosophical Transactions of the Royal Society, Series B*, **355**, 815–31.

Reich, A., Ewel, J. J., Nadkarni, N. M., Dawson, T. and Evans, R. D. (2003). Nitrogen isotope ratios shift with plant size in tropical bromeliads. *Oecologia*, **137**, 587–90.

Reich, P. B., Ellsworth, D. S., Walters, M. B., *et al.* (1999). Generality of leaf traits relationships: a test across six biomes. *Ecology*, **80**, 1955–69.

Remy, W., Taylor, T. N., Hass, H. and Kerp, H. (1994). Four hundred-million-year-old vesicular arbuscular mycorrhizae. *Proceedings of the National Academy of Sciences, USA*, **91**, 11841–3.

Richardson, S. J., Peltzer, D. A., Allen, R. B. and McGlone, M. S. (2005). Resorption proficiency along a chronosequence: responses among communities and within species. *Ecology*, **86**, 20–5.

Richardson, S. J., Allen, R. B. and Doherty, E. J. (2008). Shifts in leaf N:P ratio during resorption reflect soil P in temperate rainforest. *Functional Ecology*, **22**, 738–45.

Royo, A. and Carson, W. P. (2006). On the formation of dense understory layers in forests worldwide: consequences and implications for forest dynamics, biodiversity, and succession. *Canadian Journal of Forest Research*, **36**, 1345–62.

Russell, A. E. and Vitousek, P. M. (1997). Decomposition and potential nitrogen fixation in *Dicranopteris linearis* litter on Mauna Loa, Hawai'i, USA. *Journal of Tropical Ecology*, **13**, 579–94.

Russell, A. E., Raich, J. W. and Vitousek, P. M. (1998). The ecology of the climbing fern *Dicranopteris linearis* on windward Mauna Loa, Hawai'i. *Journal of Ecology*, **86**, 765–79.

Scatena, F. N., Silver, W., Siccama, T., Johnson, A. and Sanchez, M. J. (1993). Biomass and nutrient content of the Bisley Experimental Watersheds, Luquillo Experimental Forest, Puerto Rico, before and after Hurricane Hugo, 1989. *Biotropica*, **25**, 15–27.

Schrumpf, M., Axmacher, J. C., Zech, W., Lehmann, J. and Lyaruu, H. V. C. (2007). Long-term effects of rainforest disturbance on the nutrient composition of throughfall, organic layer percolate and soil solution at Mt. Kilimanjaro. *Science of the Total Environment*, **376**, 241–54.

Scowcroft, P. G. (1997). Mass and nutrient dynamics of decaying litter from *Passiflora mollissima* and selected native species in a Hawai'ian montane rain forest. *Journal of Tropical Ecology*, **13**, 407–26.

Shiels, A. B. (2006). Leaf litter decomposition and substrate chemistry of early successional species on landslides in Puerto Rico. *Biotropica*, **38**, 348–53.

Shiels, A. B., West, C. A., Weiss, L., Klawinski, P. D. and Walker, L. R. (2008). Soil factors predict initial plant colonization on Puerto Rican landslides. *Plant Ecology*, **195**, 165–78.

Siccama, T. G., Bormann, F. H. and Likens, G. E. (1970). The Hubbard Brook Ecosystem Study: productivity, nutrients, and phytosociology of the herbaceous layer. *Ecological Monographs*, **40**, 389–402.

Sterner, R. W. and Elser, J. J. (2002). *Ecological Stoichiometry: The Biology of Elements from Molecules to the Biosphere*. Princeton, NJ, USA: Princeton University Press.

Tanner, E. V. J. (1983). Leaf demography and growth of the tree fern *Cyathea pubescens* in Jamaica. *Botanical Journal of the Linnean Society*, **87**, 213–27.

Tanner, E. V. J. (1985). Jamaican montane forests: nutrient capital and cost of growth. *Journal of Ecology*, **73**, 553–68.

Tessier, J. T. and Raynal, D. J. (2003). Use of nitrogen to phosphorus ratios in plant tissue as an indicator of nutrient limitation and nitrogen saturation. *Journal of Applied Ecology*, **40**, 523–34.

Thomas, S. C., Halpern, C. B., Falk, D. A., Liguori, D. A. and Austin, K. A. (1999). Plant diversity in managed forests: understory responses to thinning and fertilization. *Ecological Applications*, **9**, 864–79.

Treseder, K. K., Davidson, D. W. and Ehleringer, J. R. (1995). Absorption of ant-provided carbon dioxide and nitrogen by a tropical epiphyte. *Nature*, **375**, 137–9.

Tuomisto, H. and Poulsen, A. D. (1996). Influence of edaphic specialization on pteridophyte distribution in neotropical rain forests. *Journal of Biogeography*, **23**, 283–93.

Tuomisto, H., Ruokolainen, K., Kalliola, R., *et al.* (1995). Dissecting Amazonian biodiversity. *Science*, **269**, 63–6.

Tuomisto, H., Poulsen, A. D. and Moran, R. C. (1998). Edaphic distribution of some species of the fern genus *Adiantum* in Western Amazonia. *Biotropica*, **30**, 392–9.

Tuomisto, H., Ruokolainen, K., Poulsen, A. D., *et al.* (2002). Distribution and diversity of pteridophytes and Melastomataceae along edaphic gradients in Yasuni National Park, Ecuadorian Amazonia. *Biotropica*, **34**, 516–33.

Turner, B. L. (2008). Resource partitioning for soil phosphorus: a hypothesis. *Journal of Ecology*, **96**, 698–702.

Tyson, M. J., Oughton, D. H., Callaghan, T. V., Day, J. P. and Sheffield, E. (1990). The uptake and translocation of caesium-134 and strontium-85 in bracken *Pteridium aquilinum* (Dennstaedtiaceae: Pteridophyta). *Fern Gazette*, **13**, 381–3.

Vitousek, P. M. (2004). *Nutrient Cycling and Limitation: Hawai'i as a Model System*. Princeton, NJ, USA: Princeton University Press.

Vitousek, P. M., Turner, D. R. and Kitayama, K. (1995a). Foliar nutrient during long-term soil development in Hawai'ian montane rain forest. *Ecology*, **76**, 712–20.

Vitousek, P. M., Gerrish, G., Turner, D. R., Walker, L. R. and Mueller-Dombois, D. (1995b). Litterfall and nutrient cycling in four Hawai'ian montane rain forests. *Journal of Tropical Ecology*, **11**, 189–203.

Walker, J., Thompson, C. H., Fergus, I. F. and Tunstall, B. R. (1981). Plant succession and soil development in coastal sand dunes of subtropical eastern Australia. In *Forest Succession: Concepts and Application*, ed. West, D. C., Shugart, H. H. and D. B. Botkin. New York: Springer-Verlag, pp. 107–31.

Walker, L. R. and Aplet, G. H. (1994). Growth and fertilization responses of Hawaiian tree ferns. *Biotropica*, **26**, 378–83.

Walker, L. R., Zimmerman, J. K., Lodge, D. J. and Guzmán-Grajales, S. (1996). An altitudinal comparison of growth and species composition in hurricane-damaged forests in Puerto Rico. *Journal of Ecology*, **84**, 877–89.

Wang, B. and Qiu, Y.-L. (2006). Phylogenetic distribution and evolution of mycorrhizas in land plants. *Mycorrhiza*, **16**, 299–363.

Wardle, D. A., Bonner, K. I. and Barker, G. M. (2002). Linkages between plant litter decomposition, litter quality, and vegetation responses to herbivores. *Functional Ecology*, **16**, 585–95.

Wardle, D. A., Yeates, G. W., Barker, G. M., *et al.* (2003). Island biology and ecosystem functioning in epiphytic soil communities. *Science*, **301**, 1717–20.

Wardle, D. A., Walker, L. R. and Bardgett, R. D. (2004). Ecosystem properties and forest decline in contrasting long-term chronosequences. *Science*, **305**, 509–13.

Watkins, J. E., Jr., Cardelús, C., Colwell, R. K. and Moran, R. C. (2006). Species richness and distribution of ferns along an elevational gradient in Costa Rica. *American Journal of Botany*, **93**, 73–83.

Watkins, J. E., Jr., Mack, M. C., Sinclair, T. R. and Mulkey, S. S. (2007a). Ecological and evolutionary consequences of desiccation tolerance in tropical fern gametophytes. *New Phytologist*, **176**, 708–17.

Watkins, J. E., Jr., Rundel, P. W. and Cardelús, S. L. (2007b). The influence of life form on carbon and nitrogen relationships in tropical rainforest ferns. *Oecologia*, **153**, 225–32.

Wegner, C., Wunderlich, M., Kessler, M. and Schawe, M. (2003). Foliar C:N ratio of ferns along an Andean elevational gradient. *Biotropica*, **35**, 486–90.

Wright, I. J., Reich, P. B., Westoby, M., *et al.* (2004). The world-wide leaf economics spectrum. *Nature*, **428**, 821–7.

Wright, I. J., Reich, P. B., Cornelissen, J. H. C., *et al.* (2005). Assessing the generality of global leaf trait relationships. *New Phytologist*, **166**, 485–96.

Yaalon, D. H. (2007). Human-induced ecosystem and landscape processes always involve soil change. *BioScience*, **57**, 918–19.

5

Fern adaptations to xeric environments

PETER HIETZ

Key points

1. Ferns are most prominent in shady and humid environments, but many species are also found in drought-prone habitats, either (semi) arid ecosystems or locations with discontinuous water supply within otherwise humid ecosystems. These locations include tree branches and rocks, both substrates with little water storage capacity.
2. Drought tolerance is gained through adaptations in water uptake, water loss, water storage and, in many ferns, desiccation tolerance, a feature that ferns share with other cryptogams. The little information available on the cuticle's efficiency to limit water loss suggests that it may be similar to other vascular plants. Thus many xerophytic ferns, while tolerating desiccation, normally avoid it through low cuticular and stomatal water loss and may not be considered truly poikilohydric. Exceptions are filmy ferns with very little control of water loss and whose water relations are akin to mosses rather than vascular plants.
3. Other adaptations found in xerophytic ferns include photoprotection with pigments, antioxidants, dense indument, leaf curling and drought avoidance by shedding leaves in the dry season. Crassulacean acid metabolism (CAM) is a common adaptation of xerophytic angiosperms, but is very rare in ferns. Succulence is not strongly developed in xerophytic ferns.
4. Drought adaptations of ferns are analyzed in light of their phylogenetic positions and compared with those of angiosperms. This chapter discusses the potentially underlying causes of drought tolerance in ferns and points to gaps in our understanding as well as possible future research.

5.1 Introduction

Most people correctly associate lush fern growth with humid and shady forests (Plates 1C, 2B). Indeed, the highest diversity of ferns is found in tropical rain forests, where some 65% of extant fern species are found (Page, 1979). For instance, in Southern Africa only 35 fern species are found in areas with less than 400 mm of rainfall, and many of these are restricted to humid microsites, whereas 60 species are recorded exclusively in areas with more than 1200 mm of rainfall (Jacobsen,

Fern Ecology, ed. Klaus Mehltreter, Lawrence R. Walker and Joanne M. Sharpe. Published by Cambridge University Press. © Cambridge University Press 2010.

1983). In the humid tropics, the highest fern diversity is often found in cloud forests at mid-elevation, which is basically a result of water availability (see Chapter 2; Kessler, 2001; Cardelús *et al.*, 2006).

The restriction to humid environments and the evolutionary replacement of ferns by seed plants as dominant land plants is likely due to anatomical, morphological and physiological features that distinguish ferns from seed plants. Among the reasons cited for the preference of humid environments by ferns are the inability of leaves to adapt to fluctuating environmental conditions, poor control of water loss, a less efficient water transport system and lower photosynthetic rates than in seed plants, and the fact that the independent gametophyte needs to survive (Page, 2002). Nonetheless, a number of specialized xerophytic ferns suggest that their evolutionary background does not exclude them from xeric habitats (Plate 2A). This review will look at ferns living in various xeric environments, or at least places where the ferns experience severe drought stress, and ask which adaptations enable them to survive where other ferns would die. What is assumed to prevent ferns from growing in dry locations and differences between xerophytic ferns and angiosperms will also be discussed. Unfortunately, a comparison of adaptations between ferns and seed plants is somewhat limited by the fact that few studies have been carried out on ferns, at least compared with seed plants. Still less information is available for a general comparison between drought-tolerant and drought-sensitive ferns. In contrast, the surprising drought tolerance of a number of ferns has attracted some attention, resulting in detailed case studies that provide the basis for this review.

5.2 Xerophytic ferns and their habitats

Page (1979) distinguishes five major environments in which ferns are found: tropical mesic, tropical xeric epiphytic, tropical xeric terrestrial, tropical alpine and temperate. Within each of these main environments a number of habitats are distinguished. This classification is limited in that some places are not easily assigned and many species may be found in more than one habitat. Nevertheless, Page's classification is probably the most useful for locating xerophytic ferns and Table 5.1 compiles typical genera found in various xeric habitats.

The epiphytic habitat is a place where, even in humid rain forests, plants without direct contact to soil will be exposed to recurrent drought, and xerophytic features have been found and studied in epiphytes from many distinct taxonomic groups (Lüttge, 1989). While about a third of all ferns are classified as epiphytes (Benzing, 1990), only a small proportion of these can be classified as xerophytic, and separating these from mesic species is particularly difficult. According to Tryon (1964) some 10% of all ferns are tropical xerophytic epiphytes, though this figure is

Table 5.1 *Typical fern genera found in different xeric environments*

Environment	Typical genera
Epiphytes	*Asplenium, Colysis, Crypsinus, Davallia, Dictymia, Drynaria, Humata, Lemmaphyllum, Lepisorus, Microgramma, Microsorum, Nephrolepis, Niphidium, Phymatodes, Platycerium, Pleopeltis, Polypodium, Pyrrosia, Selliguea*
Mangroves	*Acrostichum*
Ephemeral wetlands	*Blechnum, Cyclosorus, Gleichenia, Lindsaea, Lygodium, Schizoloma, Sticherus*
(Sub)tropical forest margins and disturbed, open sites	*Culcita, Dennstaedtia, Dicranopteris, Gleichenia, Histiopteris, Hypolepis, Lastreopsis, Lonchitis, Microlepia, Paesia, Pteridium, Sticherus*
Tropical dry forests and savannas	*Adiantum, Anemia, Asplenium, Blechnum, Cheilanthes, Cyclosorus, Doodia, Dryopteris, Lastreopsis, Mohria, Pellaea, Pteridium*
(Semi)deserts and dry rock outcrops	*Anemia, Argyrochosma, Astrolepis, Bommeria, Cheilanthes, Mildella, Mohria, Notholaena, Pellaea, Pentagramma*
Tropical alpine areas	*Argyrochosma, Asplenium, Cheilanthes, Cochlidium, Ctenopteris, Cystopteris, Huperzia, Jamesonia, Papuapteris, Pellaea, Pityrogramma, Pleopeltis, Polystichum, Pyrrosia*
Temperate alpine areas	*Asplenium, Athyrium, Cryptogamma, Cystopteris, Gymnocarpium, Phegopteris, Polystichum*
Temperate exposed locations	*Cheilanthes, Equisetum, Pteridium*

The list is not exhaustive and not all species in a genus are necessarily found
in xeric environments.
Source: Compiled after Page (1979); Tryon and Tryon (1982); Jacobsen (1983);
Hemp (2002); P. Hietz, personal observations.

disputable and likely to include more species than are considered xerophytic here.
In general, the drier habitats for epiphytes are the upper, sun-exposed locations in
the rain forest canopy and almost any location on trees in drier forests (Hietz and
Hietz-Seifert, 1995). Here, most epiphytic ferns will be considered xerophytic.

How xeric a location is for an epiphyte is on the one hand determined by the
climate, and on the other hand by the amount of substrate or canopy soil, which can
store water accessible for epiphyte roots. As the water-storing capacity of bare
branches is very low, epiphytes growing on these are exposed to frequent changes
in water supply (they are "pulse supplied" sensu Benzing, 1990), which distinguishes these locations from most terrestrial sites. A few epiphytic ferns in exposed
locations are large, of compact growth and either form baskets of dead leaves

Fig. 5.1 (a) *Drynaria rigidula* (Australia) can grow in the high canopy in wet and seasonally dry forests. Similar to other large epiphytic ferns, the lamina is coriaceous but not succulent; the large amount of substrate trapped by the nest leaves can store nutrients and water. (b) *Pyrrosia longifolia* (Australia), one of the few ferns with CAM. (c) *Cheilanthes kaulfussii* and *C. myriophylla* (Mexico). (d) *Pleopeltis crassinervata* (Mexico) a desiccation-tolerant epiphyte with saturated and partly wilted leaves. Necrotic parts could be the result of high light damage. (e) *Pleopeltis polypodioides* (Mexico) with desiccated and turgid leaves.

(e.g., Fig. 5.1a, *Drynaria*, oak-leaf ferns; Fig. 4.2b, *Platycerium*, staghorn ferns) or nests with leaves arranged in rosette form (e.g., *Asplenium nidus*, bird's-nest fern; *Niphidium crassifolium*). More commonly, xerophytic ferns of epiphytic habitats are midsized or small, often with a long-creeping rhizome and entire leaves (Table 5.1, Fig. 5.1).

Tropical environments that are xeric and terrestrial include wetlands that occasionally dry out, dry forests, savannas, forest margins and rocks. Mangrove swamps, although mostly waterlogged, are xeric because extracting water from brackish water with its low osmotic potential is physiologically similar to extracting water from dry soil (Schulze *et al.*, 2005), and also because leaves in this environment are exposed to very high radiation and temperature. In this habitat, *Acrostichum* (giant leather fern) is the only fern genus found in brackish water and often the only plant in the understory and open areas where herbaceous seed plants rarely establish. A number of species can be found in sometimes dry locations in tropical freshwater wetlands (Table 5.1; Page, 1979). Forest margins and disturbed open sites are characterized by higher light levels and lower relative air humidity than in the understory and are often dominated by scrambling ferns of the Gleicheniaceae (see Chapter 6).

In tropical dry forests and savannas, where fires are frequent, many ferns have subterranean rhizomes that can survive fires. Cheilanthoid ferns (Fig. 5.1c) are found in these habitats and also in rock outcrops (Plate 2A) and extremely arid environments, and thus are often classified as desert ferns. Cheilanthoid ferns and other genera and species from North American and other (semi) deserts are mostly small to midsized and include many desiccation-tolerant species (Table 5.2).

Ferns in tropical and temperate alpine locations above the tree line often grow on or between rocks and are exposed to high levels of radiation and generally harsh conditions. The species reaching the highest altitude may be *Asplenium castaneum*, found at 5100 m in the Andes (Tryon and Tryon 1982); several other species in tropical America, Africa and Asia grow above the tree line (Table 5.1). The Andes probably have the highest number of alpine fern taxa, with *Jamesonia* being represented by 19 and closely related *Eriosorus* by 18 xerophytic species (Tryon and Tryon, 1982). Among the ferns reaching the alpine zone in temperate areas, several grow on or between rocks and are found also on dry rocks at low elevations.

Fewer ferns grow in exposed lowland areas of temperate zones. *Pteridium aquilinum* (Fig. 8.2, bracken) is typically found in clear-cuts, forest margins, disturbed sites and dry grasslands. Some *Equisetum* (horsetails, Plate 4C) species grow also in the open, but rely more on water in deeper soil layers.

Table 5.2 *Desiccation-tolerant ferns and lycophytes*

Family	Genus (no. DT spp./total no. spp.)	Species	Comments	References
Anemiaceae	*Anemia* (2/120)	*tomentosa*		Gaff, 1987
		villosa		Ribeiro *et al.*, 2007
	Mohria (1/7)	*caffrorum*		Proctor and Tuba, 2002
Aspleniaceae	*Asplenium* (18/720)	*aethiopicum*	field	Hemp, 2001
		ceterach		Proctor and Tuba, 2002
		friesiorum	field	Hemp, 2001
		mannii	field	Hemp, 2001
		megalura	field	Hemp, 2001
		praegracile	field	Hemp, 2001
		sandersonii	field	Hemp, 2001
		strangeanum	field	Hemp, 2001
		theciferum	field	Jacobsen, 1983
		uhligii	field	Hemp, 2001
	Pleurosorus (1/4)	*rutifolius*		Proctor and Tuba, 2002
Cyatheaceae	*Hymenophyllopsis** (?/8)			no reference but most likely also DT
Dryopteridaceae	*Elaphoglossum* (?/700)	*acrostichoides*	field	Hemp, 2001
		petiolatum		Hietz and Briones, 2001
		spathulatum	field	Hemp, 2001
Pteridaceae	*Actiniopteris** (2/5)	*radiata*	field	Hemp, 2001
		semiflabellata		Proctor and Tuba, 2002
	Adiantum (2/200)	*hispidulum*		Hemp, 2001
		incisum		Proctor and Tuba, 2002
	Cheilanthes* (27/150)			Proctor and Tuba, 2002
	Doryopteris (3/50)			Proctor and Tuba, 2002
	Hemionitis (?/7)			Porembski and Barthlott, 2000
	Notholaena* (3/34)			Proctor and Tuba, 2002
	Pellaea* (13/25)			Proctor and Tuba, 2002
	Vittaria (?/66)	*volkensii*	field	Hemp, 2001
		isoëtifolia	field	Hemp, 2001
Hymenophyllaceae	*Hymenophyllum** (7/250)	*capillare*	field	Hemp, 2001

Table 5.2 (*cont.*)

Family	Genus (no. DT spp./total no. spp.)	Species	Comments	References
		kuhnii	field	Hemp, 2001
		splendidum	field	Hemp, 2001
		tunbrigense		Proctor, 2003
		wilsonii		Proctor, 2003
	*Trichomanes** (10/50)	*borbonicum*	field	Hemp, 2001
		bucinatum		Hietz and Briones, 2001
		chevalieri	field	Hemp, 2001
		eorsum	field	Hemp, 2001
		melanotrichum	field	Beckett, 1997; Hemp, 2001
		pyxidiferum	field	Jacobsen, 1983
		radicans	field	Hemp, 2001
		ramitrichum	field	Hemp, 2001
		rigidum	field	Hemp, 2001
Isoëtaceae	*Isoëtes* (1/150)	*australis*	only corms are DT	Proctor and Tuba, 2002
Polypodiaceae	*Ctenopteris* (1/35)	*heterophylla*		Proctor and Tuba, 2002
	Loxogramme (2/34)	*abyssinica*	field	Hemp, 2001
		lanceolata	field	Jacobsen, 1983
	Melpomene (1/30)	*flabelliformis*	field	Hemp, 2001
		peruviana	field	Lehnert, 2007
	*Pecluma** (?/40)		field	P. Hietz, personal observation
	Polypodium (2/45)	*virginianum*		Gildner and Larson, 1992
		vulgare		Stuart, 1968
	*Pleopeltis** (9/90)	*angusta*		Starnecker and Winkler, 1982
		crassinervata		P. Hietz, unpublished data
		furfuracea	field	P. Hietz, personal observation
		hirsutissima	field	Müller *et al.*, 1981
		macrocarpa	field	Hemp, 2001
		mexicana		Hietz and Briones, 2001
		plebeia		Hietz and Briones, 1998
		polypodioides		Stuart, 1968
		squalida	field	Müller *et al.*, 1981

Table 5.2 (*cont.*)

Family	Genus (no. DT spp./total no. spp.)	Species	Comments	References
	Pyrrosia (5/65)			Proctor and Tuba, 2002
	Platycerium (1/15)			Griffiths, 1989, Hovenkamp 1986
Schizaeaceae	*Schizaea* (1/20)	*pusilla*		Proctor and Tuba, 2002
Selaginellaceae	*Selaginella* (13/750)			Proctor and Tuba, 2002
Tectariaceae	*Arthropteris* (1/15)			Proctor and Tuba, 2002
Woodsiaceae	*Woodsia* (1/40)	*ilvensis*		Proctor and Tuba, 2002

DT, desiccation tolerant; *, genera with relatively high proportions of DT species; field, field observations of desiccated leaves that resaturated after rainfall, but without closer characterization to verify desiccation tolerance.

5.3 Adaptations to drought

5.3.1 Water transport in ferns and lycophytes

Life is only possible with watery solutions in cells and because cells inevitably lose some water, all organisms need to secure a balance between water uptake and loss. Plant water relations define the mechanisms and adaptations that regulate the water status of the living plant and its organs. Essential components are water transport, loss, uptake and storage, which will subsequently be discussed in relation to drought tolerance in ferns.

The water-transport system of angiosperms is based on vessels, long and relatively wide capillaries of stacked cells with more or less dissolved end walls connecting the individual cells. The water transport through vessels meets less resistance and is thus more efficient than in gymnosperms, where water flows through tracheids. Tracheids are single, elongated cells of smaller diameter than the vessels, where water needs to pass from one cell to the next through pit membranes in the cell walls. The vascular system of ferns shows numerous transitions that have been described as vessels or tracheids (Carlquist and Schneider, 2001, 2007). If the cell wall between adjacent cells is mostly dissolved, with only narrow bars remaining (Fig. 5.2a) it is a true vessel. Although fern tracheids can be quite large with diameters >100 μm, their cell walls are never completely

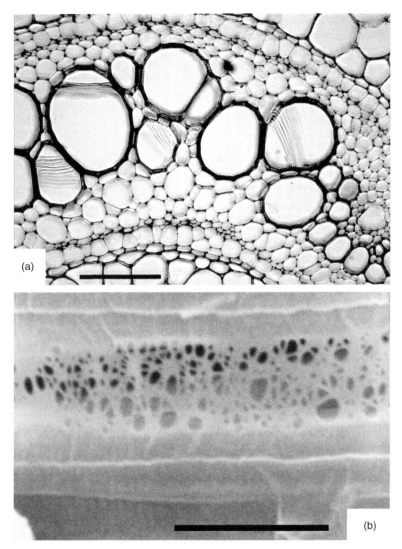

Fig. 5.2 Xylem anatomy of ferns. (a) Cross section of *Pteridium aquilinum* rhizome showing wide vessels and scalariform perforation plates between vessel elements, scale bar 100 μm. (b) Intact pit membrane of *Woodsia obtusa* tracheids, scale bar 2 μm. (After Carlquist and Schneider, 2007, reproduced with permission from *American Fern Journal*.)

dissolved because a fine pit membrane remains (Fig. 5.2b). Undisputed vessels are found at least in *Astrolepis*, *Marsilea* (water-clover), *Pteridium* and *Woodsia*, all genera found in xeric habitats (Table 5.1; Carlquist and Schneider, 2007). The presence of true vessels is likely advantageous for ferns such as *Pteridium* that grow in exposed locations where an efficient water-transport system resupplies water as fast as the leaves transpire.

The hydraulic conductance of most ferns is relatively low and typically two to three orders of magnitude lower than in angiosperms (Woodhouse and Nobel, 1982). Low conductance means that with a given water flow, the water potential strongly decreases along the water path in ferns. As a consequence, either the water potentials in leaves drop to very low levels or transpiration has to be restricted. Indeed, in seven fern species the water potential along the petiole dropped by 0.7 to 2.4 megapascals (MPa; Woodhouse and Nobel, 1982), the higher values typical for the water potential gradient along the entire axis of mesic trees. Two fern species had the lowest whole-plant hydraulic conductances: *Cheilanthes parryi* (syn. *Notholaena parryi*, Parry's lip fern), a desert fern, and *Marsilea vestita*, an aquatic fern that survives the dry season without water. Whereas low hydraulic conductance may be an advantage to an aquatic fern when exposed to drought (Woodhouse and Nobel, 1982), it is unclear why this should be an advantage for desiccation-tolerant ferns such as *Cheilanthes* (lip ferns). Clearly, more comparative and experimental data on hydraulic conductance in ferns are needed to understand its adaptive significance.

In many xerophytic ferns, drought certainly results in a strongly negative water potential, carrying the danger of xylem cavitation, which impairs water transport by air emboli (Tyree and Sperry, 1989). Some cavitation occurs also when stomata close with decreasing water potential to avoid an excessive xylem dysfunction. At least in seed plants, the cavitation threshold, i.e., the water potential at which conductance decreases to 50%, tends to be lower in the xylem of species from drier locations (Maherali *et al.*, 2004). However, one study in a tropical dry forest in Costa Rica found that the cavitation threshold did not differ between mostly xerophytic ferns and angiosperms (Brodribb and Holbrook, 2004). The water-transport system of these ferns appears to be as resistant as that of angiosperms to the low water potential resulting from drought.

5.3.2 *Control of water loss*

Interestingly, in the Costa Rican study, ferns operated with a higher safety margin; that is, they tended to close stomata at a higher water potential than the angiosperms from the same location (Brodribb and Holbrook, 2004). This response should avoid or delay cavitation but, by closing stomata earlier, ferns will forgo potential carbon uptake. Leaf hydraulic conductance of three species (the ferns *Blechnum occidentale*, hammock fern, and *Sticherus bifidus*, and the lycophyte *Selaginella pallescens*, spike-moss) growing in relatively exposed sites in the same dry forest in Costa Rica were in the lower conductance range of angiosperm trees. Given the close correlation between leaf hydraulic conductance, stomatal conductance and photosynthetic capacity, maximum rates of photosynthesis in these ferns were also lower (Brodribb *et al.*, 2005). Thus, at least in part, the hydraulic architecture appears to limit the productivity of many ferns and perhaps confine them to less productive, shady and/or humid environments.

Maximum stomatal conductance of dry forest ferns was similar to understory angiosperms (Brodribb and Holbrook, 2004), which tend to have lower stomatal conductance than canopy plants because photosynthesis is mainly limited by the available light. To understand the reason for the more conservative transpiration control in ferns and lycophytes, it is important to know the minimum conductance (i.e., uncontrolable water loss of the leaves) when stomata are closed. Perhaps ferns lose more water even with their stomata closed and thus need to limit water loss earlier than angiosperms. This appears to be assumed by Page (1979), who attributes the narrow range of transpiration control in ferns to a "poorly controlled evaporative potential", but cites only *Adiantum* (maidenhair fern) as an example. Uncontrolled water loss of detached leaves of epiphytic ferns was in the range of angiosperms and tended to be lower in ferns from exposed canopy locations (Hietz and Briones, 1998). In xerophytic ferns, the control of water loss through the cuticle and closed stomata appears no less efficient than in angiosperm leaves growing under similar conditions. Therefore, unless ferns have not mastered the biochemistry of an efficient cuticle, there is no reason to expect a systematic difference between ferns and angiosperms in avoiding water loss through leaves.

5.3.3 *Water uptake*

Osmotic potential of xerophytic ferns

Plants cannot take up water from the soil once the soil water potential becomes less than the leaf water potential, whose lower limit is the osmotic potential of its living cells. Therefore, plants with a high concentration of solutes and thus a low (i.e., more negative) osmotic potential can extract more water from a drying substrate than plants with a high osmotic potential (Larcher, 2002).

Terrestrial xerophytic *Cheilanthes parryi* and *Asplenium ceterach* (syn. *Ceterach officinarum*) had osmotic potentials between −0.98 and −1.83 MPa and values between −1.3 and −2.4 MPa were reported for uncurled leaves in other desiccation-tolerant ferns (Nobel, 1978 and references therein). These values are low compared with epiphytic ferns, and as more water is lost and leaves desiccate, the osmotic potential drops further. This enabled *C. parryi* to extract soil water during the larger part of the year in a Californian desert (Nobel, 1978). Leaves remain green and uncurled during several months, but at some point water uptake becomes impaired because the water transport through the xylem becomes blocked by cavitation.

This blockage also hinders the water uptake once the supply improves and the ability to absorb water after drought can be related to the degree of drought stress. When *C. parryi* was irrigated one day after curling, the leaves required only two hours to uncurl. However, after two weeks in the curled state water uptake, as indicated by leaf uncurling, took 12 hours and after four weeks it took 24 hours (Nobel, 1978). Similarly, curled leaves in drought-stressed *Pleopeltis polypodioides* (syn. *Polypodium polypodioides*, resurrection fern) from the southern USA could not be resaturated by water supplied to the rhizome, which would have to be transported via the petiole, but only by exposure to liquid water or very high humidity (Stuart, 1968).

In epiphytic ferns, osmotic potentials of saturated leaves were in the range of −0.7 to −1.6 MPa (Hietz and Briones, 1998; Martin *et al.*, 2004) and about −0.5 MPa in slightly succulent *Pyrrosia* (felt fern; Sinclair, 1983). These values are higher than in terrestrial xerophytic ferns. Epiphyte leaves tend to have a low concentration of solutes and consequently a high (i.e., less negative) osmotic potential, perhaps because it is difficult for them to accumulate high concentrations of organic or inorganic solutes (Martin *et al.*, 2004). Also, when branches are wet after rain, substrate water potential is high and water is easily available (Martin *et al.*, 2004). When branches are dry, even a very low leaf water potential of epiphytic ferns may not be sufficient to extract water and may therefore not be of much advantage.

The cuticle, designed to prevent water loss, invariably also hinders water uptake. Interestingly, water absorption by *Pleopeltis* leaves was delayed by an anoxic environment (Stuart, 1968). This suggests that water uptake is not only a passive process driven by the gradient in water potential, but that it can be regulated by plants, probably involving energy-consuming processes. Unfortunately, these early results from *Pleopeltis* have not been followed by more detailed studies. In contrast to other ferns, filmy ferns (Plates 6A, B), lacking an efficient cuticle, dehydrate within hours and may rehydrate within minutes, a feature discussed below in the context of their poikilohydric nature.

For epiphytes, water often becomes available in intensive but short pulses and its rapid uptake is advantageous. The roots of epiphytic orchids are covered by a velamen radicum, a mantle of dead cells with large pores in the cell walls that can rapidly absorb liquid water from its surface and conduct it to the cells of the root cortex. A strikingly similar outer root cortex composed of cells with spiral wall thickening and round pits was described in the epiphytic fern genus *Pyrrosia*, where it most likely functions similarly to the orchid velamen (Pandé, 1935).

Water-absorbing trichomes

Many xerophytic ferns have a dense cover of trichomes, either hairs or scales, particularly on the abaxial side of the lamina. Plant trichomes are very diverse in form and function (Fig. 5.3; Esau, 1965) and various aspects of the trichomes

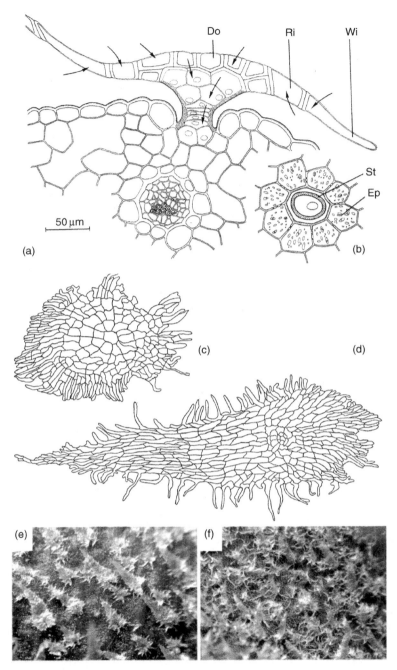

Fig. 5.3 Scale form and function in ferns. (a, b) Absorptive scale of *Pleopeltis hirsutissima* showing the likely pathway for water uptake. Do, dome cell; Ri, ring cell; Wi, wing cell; St, stalk cell; Ep, epidermal cell. (After Müller *et al.*, 1981, reproduced with permission from Elsevier.) (c, d) Peltate scales of *Pyrrosia adnascens*. (After Pandé, 1935.) (e) Abaxial and (f) adaxial leaf surface of *Elaphoglossum paleaceum*. (After Watkins *et al.*, 2006, reproduced with permission from *American Fern Journal*.)

of xerophytic ferns have been studied. Dead and air-filled trichomes that create a dense indument reflect light, reduce leaf temperature and thus reduce transpiration (Nobel, 1991). Depending on shape and density, trichomes can also increase transpiration by increasing air turbulence on the leaf surface and consequently gas exchange between the leaf and the atmosphere (Schreuder *et al.*, 2001), though this seems unlikely in xerophytic ferns with dense trichomes, scales or farina. In epiphytic bromeliads with scarce roots, shield-like trichomes are essential for the uptake of liquid water by leaves and the trichome cover is particularly dense in xerophytic species (Benzing, 1990). Many ferns also have shield-like trichomes and a drop of water placed on a trichome-covered leaf of *Pleopeltis polypodioides* spreads quickly over a wider area by capillary forces between the shield and the epidermal cells (Stuart, 1968). Trichomes on *Pleopeltis angusta*, *P. hirsutissima* and *P. squalida* from Brazil consist of living foot cells that are inserted into the epidermis, living stalk cells and a shield of dead cells (Figs. 5.3a, b; Müller *et al.*, 1981). From the foot to the distant stalk and shield cells the cuticle and cell walls become thinner. The central part of the shield, the dome, is in direct contact with the upper living stalk cells and surrounded by concentric ring cells and distal wing cells. Cell walls of shield cells are at least partially lignified, which would impede water transport, but the cell walls between dome and stalk cells have abundant pits and silver ions were taken up by the shield cells within five minutes and detected in mesophyll cells adjacent to the stalk after 15 minutes (Müller *et al.*, 1981).

Many other desiccation-tolerant ferns have water-absorbing trichomes and thus become resaturated soon after desiccated plants in the field are wetted by rain or dew, thereby avoiding the need to transport water from the roots via the petiole to the leaf lamina. In contrast, leaves of some species of *Microgramma* do not curl but remain turgescent during the dry season, although they do not absorb water through their small number of trichomes (Müller *et al.*, 1981). Apart from the species studied by Müller *et al.* (1981), a dense trichome cover is common in *Asplenium ceterach*, cheilanthoids, *Mohria*, *Pleopeltis*, xerophytic *Polypodium* species and several other groups, most of which also appear to be desiccation tolerant. Because these plants may experience frequent inactive and desiccated phases, it is of critical importance for their survival to rapidly take up water and restore an active state when favorable conditions arise.

The variety of fern trichomes exceeds that of bromeliads, and those trichomes that are more hair-like may have little importance in water uptake. Many *Elaphoglossum* species have a dense cover of shield-like trichomes (Figs. 5.3e, f), yet most grow in very humid climates and trichomes are mostly denser on the lower lamina that is less exposed to rain. There is little question that fern trichomes help in the water absorption process in some cases, but how they compare with bromeliad trichomes, how efficiency differs among types of trichomes and how much they contribute to total water uptake is unknown. Given the high diversity of fern trichomes, it is likely that their functions vary among different species, including water uptake, reducing and perhaps increasing transpiration and light protection. Comparative studies on the effects of different types and densities of trichome cover on water relations of fern leaves could be done with simple experiments and would advance our understanding of the ecology and physiology of trichomes of ferns and plants in general.

5.3.4 Water storage

The internal storage of water is common in xerophytic plants, and in succulents the morphology of leaves, stems or roots has been substantially modified to provide storage volume. In *Cyclophorus* and *Pyrrosia* (syn. *Niphobolus*) a water-storing tissue (hydrenchyma) has been described which consists of large parenchymatous cells (Pandé, 1935; Hungerbühler, 1957). In *Pyrrosia*, during periods of drought, these cells shrink with the cell wall folding like the bellows of a camera, as a consequence of which the thickness of the lamina is reduced by about 50% (Fig. 5.4). This shrinkage ensures that while water is lost first from the hydrenchyma, the photosynthetic tissue remains hydrated and active for a longer period during drought. In other, mostly midsized ferns (*Aglaomorpha, Phlebodium,*

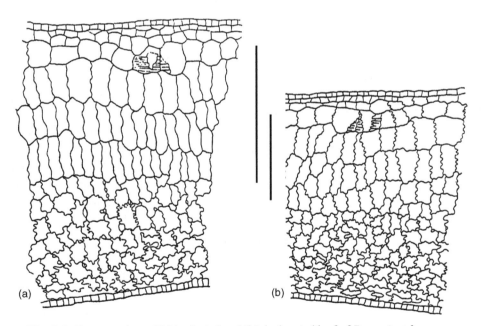

(a)   (b)

Fig. 5.4 Cross section of (a) hydrated and (b) desiccated leaf of *Pyrrosia adnascens*. The vertical bars indicate the extent of the water-storing tissue. (After Pandé, 1935.)

Phymatodes, Pleopeltis), the thick and long-creeping rhizome can also store water (Hietz and Briones, 1998). Several xerophytic species of *Ophioglossum* (adder's tongue) in the Sahel zone of Mali and seasonally dry grasslands in Africa have a thick tuberous rhizome for water storage and deciduous leaves (Jacobsen, 1983; Kramer *et al.*, 1995). Finally, large epiphytic ferns (Fig. 5.1a) trap considerable amounts of organic debris that is decomposed, becomes canopy soil and can store water externally.

5.4 Desiccation tolerance

5.4.1 Desiccation tolerance of sporophytes

Desiccation tolerance refers to the ability to survive an almost complete loss of tissue water: about 90% of the water content of a saturated leaf. At this water content, plant tissue can be at equilibrium with the ambient air (depending on humidity and temperature), so that no more water is lost unless the air becomes drier or warmer. In equilibrium with a relative humidity of 83% (and probably 25 °C) *Pleopeltis polypodioides* leaves retained about 10% of saturation water (10% relative water content or RWC), and at 0% relative humidity leaves retained 3% of saturation water (Stuart, 1968). Such a massive reduction in water content results in changes in cell shape, concentration of solutes and the hydration states of membranes and macro-molecules; few plants are able to survive these changes, leading to substantial stress at the cellular level. Plants that do survive this extreme water loss are desiccation tolerant, which is therefore an important survival strategy for many xerophytic ferns.

Desiccation tolerance is relatively common in bryophytes and lichens and very rare in seed plants (Oliver *et al.*, 2000). Ferns occupy an intermediate position and a fair number of species, mostly from xeric environments, are desiccation tolerant (Table 5.2). In contrast to the few desiccation-tolerant but much better studied angiosperms, a large proportion of desiccation-tolerant ferns have probably gone unrecorded. Even without detailed physiological studies for each species, it might be assumed that most species with desiccated, curled leaves after periods of drought and vigorous leaves after rain belong in this group. In a study on Mt. Kilimanjaro, not an arid region, 36 out of 140 species (26%) are listed as poikilohydric (Hemp, 2001). Unquestionably, desiccation tolerance is more common than previous surveys suggest and may well comprise 5–10% of all fern and lycophyte species.

Desiccation tolerance versus poikilohydry

Desiccation tolerance is sometimes used almost as a synonym of poikilohydry (e.g., Nobel, 1978). Poikilohydric organisms do relatively little to maintain their water content close to saturation; as a consequence their water content fluctuates widely and is mostly in equilibrium with the environment. In the case of plants, this generally means little control of transpiration due to the lack of or an ineffective cuticle. While both features often go hand in hand, desiccation-tolerant plants are not always poikilohydric and poikilohydric plants are not always desiccation tolerant. For instance, many aquatic plants have no barriers to limit water loss and are poikilohydric, but do not survive severe desiccation. By contrast, poikilohydric land plants also need to be desiccation tolerant in all but the most humid locations. Filmy ferns are unquestionably poikilohydric (and often desiccation tolerant) and lose water within minutes when exposed to dry air (Fig. 5.5b). For many other

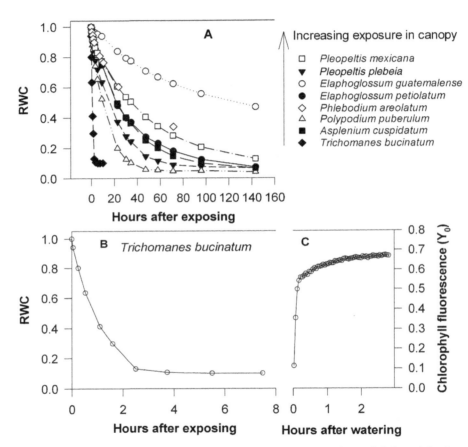

Fig. 5.5 Speed of water loss and revival in ferns. (a) Water loss of eight epiphytic ferns on a laboratory bench. The legend lists the species from the most exposed to the most protected locations on trees in a Mexican montane forest. *Pleopeltis mexicana, Pleopeltis plebeia, Elaphoglossum petiolatum* and *Trichomanes bucinatum* are desiccation tolerant. (b) Water loss of the filmy fern *Trichomanes bucinatum*. (c) Recovery of *T. bucinatum* photosynthesis after leaves were watered. RWC, relative water content. (Data from Hietz and Briones, 1998, 2001.)

desiccation-tolerant ferns that may have a coriaceous lamina and do not lose water easily, the term poikilohydric is not really appropriate.

In filmy ferns (Plates 6A, B), the leaf lamina is composed of a single cell layer, as in most mosses, and gas exchange consequently is not regulated by stomata. The largest group of filmy ferns is the family Hymenophyllaceae, but *Hymenophyllopsis* (Cyatheaceae), *Leptopteris* (Osmundaceae) as well as several species of *Asplenium* (Aspleniaceae), *Cystopteris* (Woodsiaceae) and some *Danaea* (Marattiaceae) are also filmy ferns (Kramer *et al.*, 1995). Because the cuticle is also an effective barrier for CO_2, the leaves of filmy ferns must not be covered by a waterproof cuticle, which would in effect starve the plants. As a consequence, plants with a one-cell thick lamina must be poikilohydric, and if they are to survive under moderately dry conditions on land, also need to be desiccation tolerant. While having no barrier to water loss results in rapid drying, this absence of a barrier also permits rapid rehydration and the quick return to an active state soon after plants are wetted. Most filmy ferns hardly qualify as xerophytes and desiccation tolerance here appears to be a by-product of their leaf structure. Moreover, filmy fern species differ in their tolerance to water loss. In a Jamaican forest, filmy fern species growing on exposed branches were more desiccation tolerant than those restricted to the forest floor (Shreve, 1911). Several other small, desiccation-tolerant ferns and lycophytes of rocky and more xeric habitats (e.g., *Asplenium ruta-muraria*, wall rue; *A. trichomanes*, maidenhair spleenwort; and *Selaginella* cf. *underwoodii*) also dry out quickly and show few adaptations to conserve water (Proctor and Tuba, 2002). These ferns fall between the water economy of filmy ferns and sturdier, desiccation-tolerant plants.

In many plant species, stomata close at a water content near the turgor loss point, beyond which leaves start to wilt (Brodribb *et al.*, 2003). In several epiphytic ferns, stomata were closed when as little as 2–5% of saturation water was lost, with turgor loss observed at 4–9% water loss (Hietz and Briones, 1998); that is, they operate with a safety margin. Exceptions were two desiccation-tolerant ferns from a Mexican montane forest (*Pleopeltis mexicana* and *P. plebeia*), which close their stomata after the turgor loss point (Hietz and Briones, 1998). This is risky for plants damaged by water loss, but could enable plants that are not damaged by desiccation to continue carbon uptake for longer periods. Once stomata were closed, however, their uncontrolled water loss was not higher than in other ferns and comparable to many, even xerophytic angiosperms (Hietz and Briones, 1998).

In *Pleopeltis polypodioides* photosynthesis starts to decline at a RWC of 90% and was measurable down to about 30% (Stuart, 1968). Interestingly, the relationship between RWC and photosynthesis did not differ between intact leaves and when leaflets were cut with a razor blade. Because the cuts should permit CO_2 to reach the mesophyll irrespective of stomatal opening, it appears that at least in this desiccation-tolerant fern the reduction in photosynthesis is not primarily controlled by stomata. When leaves were resaturated rapidly through cuts in the leaflets, photosynthesis and water content increased within a few minutes and were close to maximum after 40 minutes.

Responses of desiccation-tolerant ferns to drought

The physiological and biochemical mechanisms of desiccation tolerance have been studied in detail in several model plants, including the moss *Tortula* and the angiosperm *Craterostigma* (blue gem) in the Scrophulariaceae (see Oliver *et al.*, 2000 and Proctor and Tuba, 2002 for recent reviews). While species differ in some of the details, the basic mechanisms include the synthesis of proteins and carbohydrates to protect membranes and macromolecules from oxidation and mechanical damage as cells shrink. A few case studies offer insight into the biochemistry of desiccation tolerance in ferns. Abscisic acid (ABA) is a phyto-hormone that regulates various responses to drought stress. Excised leaves of desiccation-tolerant *Polypodium virginianum* (rock polypody), a fern growing on exposed rocks of the Niagara Escarpment in North America, survived slow drying but were killed by rapid and severe drying over silica (Reynolds and Bewley, 1993a). However, when leaves were immersed in ABA prior to drying they survived even very fast drying. Under drought stress, a number of small molecules are synthesized that protect essential components of the cell from the effect of reduced water content. In *P. virginianum*, a sucrose concentration of about 25% of dry weight in desiccated leaves, the synthesis of a small protein during desiccation and the synthesis of several polypeptides during rehydration probably all contribute to protecting cellular constituents during water loss and rehydration (Reynolds and Bewley, 1993a, 1993b). Because the synthesis of all these substances takes time, very rapid water loss was fatal, but such extreme rates of water loss may occur only in filmy ferns (Fig. 5.5a).

While some desiccation-tolerant plants dismantle their photosynthetic apparatus during drought, it remains intact in ferns, because their leaves remain green and actinomycin D, a fungal antibiotic that prevents protein synthesis, has no effect on recovery of *Pleopeltis polypodioides* (Stuart, 1968). In addition, photosynthesis in filmy ferns returns to predrought values within a few minutes or as soon as water content approaches saturation (Fig. 5.5c; Hietz and Briones, 2001; Proctor, 2003), which is too fast for a restoration by means of protein synthesis.

One adaptation seen in many desiccation-tolerant ferns but rare in angiosperms is leaf curling, which reduces surface area, thereby reducing water loss and light damage. The way that leaves curl is predetermined and species specific. Leaves of *Elaphoglossum calaguala* and *E. mathewsii* curl lengthwise while those of *E. hirtum* become spirally curled (Tryon and Tryon, 1982). Pinnae of *Asplenium praemorsum* fold accordion-like and those of *Pleopeltis furfuracea* bend up and inwards (Mehltreter, 2008). In the lycophyte *Selaginella lepidophylla* (resurrection spike-moss) stems curl along the axis and form a compact ball (Lebkuecher and Eickmeier, 1993; see also Plate 3B).

Desiccated ferns may absorb water primarily through their leaves because water transport through the petiole is slow under well-watered conditions and the xylem may even be largely blocked in desiccated leaves (see above). In filmy ferns (Plates 6A, B) with little or no cuticle, water absorption occurs within a few minutes, as seen by the rapid recovery of chlorophyll fluorescence (Fig. 5.5c; Proctor and Tuba, 2002). In desiccation-tolerant ferns with an efficient cuticle, water absorption is likely to be much slower but may be aided by trichomes as detailed above (Fig. 5.3). In *Pleopeltis polypodioides*, water content and photosynthesis returned in parallel to nearly predrought levels 30 minutes after leaves were immersed in liquid water, but only if leaflets were cut to aid water uptake (Stuart, 1968). Uncut leaves took about six hours to recover, which suggests that water uptake by leaves is possible but mostly impeded by the cuticle.

Other factors affecting the survival of desiccated leaves are the speed of water loss, the conditions experienced during desiccation and the duration and frequency of desiccation. For example, in the classic, desiccation-tolerant *Cheilanthes parryi* from North American deserts, leaves rapidly took up water a few days after initial curling, but became brown, brittle and necrotic after nine weeks of curling (Nobel, 1978). In *Pleopeltis*, the recovery of photosynthesis was negatively related to the time the leaves had remained desiccated (Stuart, 1968). However, some species of *Selaginella, Pellaea* and *Cheilanthes* can survive for several years in an air-dry state (Alpert, 2000).

5.4.2 Desiccation tolerance of gametophytes

One reason why most ferns and lycophytes are thought to be restricted to humid environments is their gametophyte stage. Fern gametophytes (like those of bryophytes) desiccate very rapidly, are poikilohydric and otherwise moss-like in their water relations (Watkins *et al.*, 2007a). Because fern sporophytes can only establish where their gametophytes can survive, this stage could present a bottleneck for colonizing xeric places. Very early studies on the desiccation tolerance of xerophytic ferns found that gametophytes of several species survived weeks over a strong desiccant and several months on a laboratory bench (Pickett, 1931). Species with desiccation-tolerant sporophytes (e.g., *Asplenium ceterach*, *Polypodium vulgare*) may also have desiccation-tolerant gametophytes, while those of desiccation sensitive ferns (e.g., *Athyrium filix-femina*, lady fern; *Blechnum spicant*, deer fern; *Pteridium aquilinum*) are less desiccation tolerant (Kappen, 1965). Recent studies found desiccation tolerance to be common in several tropical fern gametophytes, none of which is considered to have desiccation-tolerant sporophytes (Ong and Ng, 1998; Watkins *et al.*, 2007a). Apparently, the physiology and morphology of fern gametophytes is more akin to bryophyte gametophytes than to the sporophyte of the same fern species, but the degree of desiccation tolerance of the gametophyte is linked to the preferred species habitat (Watkins *et al.*, 2007a).

5.5 Photoinhibition and photoprotection

Most plants in a xeric environment are also exposed to high light levels. Although light drives photosynthesis and increasing light levels often promote higher photosynthetic rates, too much light damages plants primarily through a process called photoinhibition. Photoinhibition refers to a decline in photosynthesis when more light is absorbed by the chloroplast membranes than the following steps in the photosynthetic pathway can cope with (Demmig-Adams and Adams, 1992; Long *et al.*, 1994). As a result, electrons are passed from chlorophyll to molecules other than those in nonstressed photosynthesis, including O_2 (i.e., oxygen), which results in highly aggressive and potentially damaging reactive O_2 species. The result can range from a short reduction of photosynthesis to irreversible damage and death of leaves. Mechanisms to protect plants from photoinhibition are generally more prominent in species growing in high-light conditions, and are up-regulated when leaves are exposed. When stomata close under drought, CO_2 uptake is prevented, and because CO_2 is the ultimate destination of the electrons, drought often results in photoinhibition (Powles, 1984; Demmig-Adams and Adams, 1992).

Some desiccation-tolerant monocotyledonous angiosperms dismantle their photosynthetic apparatus and degrade chlorophyll upon drying (Oliver *et al.*, 2000). While this incurs energetic costs, an advantage is that these plants avoid the problems that result when too much light energy is absorbed by chlorophyll. Desiccation-tolerant ferns stay green and retain the chlorophyll, so they must have other adaptations to avoid photodamage.

The main biochemical defenses against photoinhibition are pigments and antioxidants (Fig. 5.6). In the xanthophyll cycle, the pigment violaxanthin is converted to zeaxanthin via antheraxanthin by redox reactions, in the course of which the potentially damaging oxidative energy is dissipated in a nondamaging way. Other carotenoids help reduce the excited forms of chlorophyll and O_2, and also protect thylakoid membranes. Finally, tocopherol is a lipophilic antioxidant that can reduce compounds that have been oxidized, again particularly protecting the thylakoid membrane.

A study of the desiccation-tolerant *Selaginella lepidophylla* from North American dry areas (Eickmeier *et al.*, 1993) showed that the photochemical

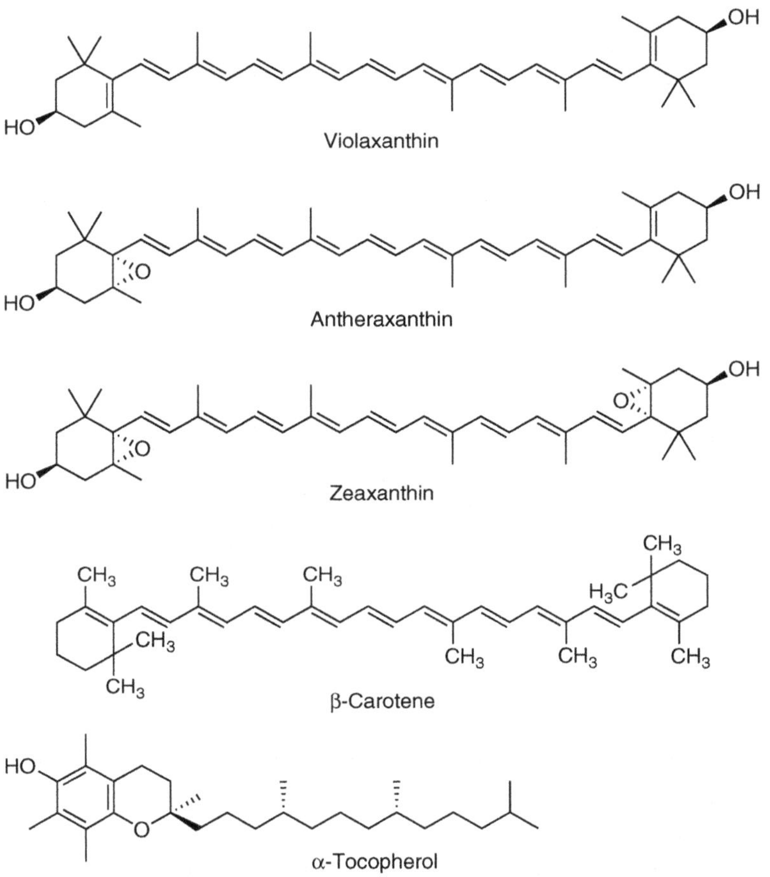

Fig. 5.6 Pigments and antioxidants involved in photoprotection.

efficiency (F_v/F_m), a measure of the proportion of light used in photosynthesis, was reduced after exposure to high light levels (2000 μmol m^{-2} s^{-1}). The reduction was stronger in plants that had been exposed for a longer time, but even after two hours of light stress, plants almost completely recovered after two hours in low light. The reduction was also more severe in plants treated with an inhibitor of zeaxanthin synthesis, showing that the xanthophyll cycle plays an important role in photoprotection. Photosynthetic rates remained high as water content decreased to 40%, but then more or less linearly declined to zero as all water was lost. The reduction in photochemical efficiency was more severe in plants exposed to high-light conditions and desiccation at the same time, but irrespective of the degree of water loss or light stress, all *S. lepidophylla* plants fully recovered after 24 hours (Eickmeier *et al.*, 1993). Drought stress resulted in an increase in the zeaxanthin concentration within two hours (Casper *et al.*, 1993) in low- as well as high-light conditions. When most xanthophylls are present as zeaxanthin, this indicates a strong protection against photoinhibition (Demmig-Adams and Adams, 1994). In the case of *S. lepidophylla*, zeaxanthin concentrations were higher in desiccating plants under low-light conditions than in well-watered plants under high-light conditions (Casper *et al.*, 1993). This shows that desiccation carries the danger of photodamage even under low-light conditions, and that desiccation-tolerant plants need photoprotective mechanisms if they do not break down their chlorophyll.

Another study measured the concentration of photoprotective pigments (xanthophylls as well as carotenoids not involved in the xanthophyll cycle) and tocopherol in epiphytic ferns from a Mexican montane forest (Tausz *et al.*, 2001). Upon exposure to full sunlight (>800 μmol m^{-2} s^{-1}) xanthophylls were largely de-epoxidized by absorption of electrons from the photosystem, but the state of oxidation did not change as plants dried. The amounts of β-carotene and α-tocopherol increased within a few hours as plants dried while exposed to high light levels. Such a short-term increase in these protective compounds had not been previously observed in other plants, but may be important for epiphytic ferns that frequently experience a rapidly changing water status.

The curling seen in most desiccation-tolerant ferns provides additional protection from high light levels. When curling was mechanically prevented in *Selaginella lepidophylla*, photoinhibition increased and all parameters of photosynthesis as well as chlorophyll content were dramatically reduced relative to plants that were left to curl (Lebkuecher and Eickmeier, 1991, 1993). *Pleopeltis polypodioides* also curls to avoid photoinhibition, but in contrast to *Selaginella*, when curling was prevented, moderately to strongly desiccated leaves exposed to high light levels recovered very inefficiently or not at all (Muslin and Homann, 1992). Scales covering the epidermis reduce the light energy absorbed by chloroplasts and thus also help to avoid photoinhibition (Figs. 5.3e, f; Watkins *et al.*, 2006).

A number of xerophytic ferns have the lower surface of their lamina covered with flavonoids secreted through specialized glands, which are thought to provide ultraviolet light protection, but may also reduce transpiration or defend against herbivores and microbes (Wollenweber, 1978; Wollenweber *et al.*, 1987; Kramer *et al.*, 1995). These include most species of *Argyrochosma*, *Pityrogramma* (gold- and silver-back ferns) and *Pentagramma*, many in *Notholaena*, and some in *Cheilanthes sensu lato*, *Pterozonium* (with sparse cover) and *Pellaea* (among sporangia) as well as the Australian endemic *Platyzoma microphyllum* (braid fern).

5.6 Crassulacean acid metabolism

Crassulacean acid metabolism (CAM) is a photosynthetic pathway found in many succulents adapted to xeric environments. In contrast to normal (C3) photosynthesis, carbon dioxide is first bound via the enzyme phospho-enol-pyruvate carboxylase (PEP-C) rather than via ribulose-1,5-bisphosphate carboxylase/oxygenase (rubisco); the resulting acid, typically malate, is stored in vacuoles (Kluge and Ting, 1978). Because CO_2 uptake and initial fixation in CAM plants do not require light, CO_2 uptake can take place in the dark, when temperatures are lower, humidity is higher and water loss through open stomata is consequently much reduced. As a consequence, the water use efficiency (WUE), which is the amount of carbon fixed per H_2O lost, is substantially higher in CAM plants than in C3 plants. This efficiency however, usually comes at the cost of rather low rates of photosynthesis and growth. Plants in full CAM mode absorb most CO_2 during the night (Fig. 5.7a), close the stomata in the morning and may fix CO_2 via rubisco in the later part of the day under favorable water supply. In weak CAM mode, substantially more CO_2 is absorbed during the day than during the night (Fig. 5.7b) and in the weakest form, called CAM cycling, only respiratory CO_2 is refixed via PEP-C during the night. In CAM cycling, stomata remain closed during the night and there is no net uptake (Fig. 5.7c), but refixation of respiratory CO_2 also helps to conserve water as some CO_2 is provided for photosynthesis without the need to open stomata, though the contribution of PEP-C to CO_2 fixation in CAM cycling is low.

The idea that CAM was primarily an adaptation of succulents to arid regions needed modification when CAM was found to be rather common in epiphytes, particularly bromeliads, orchids and cacti, but also found in a few species of *Peperomia* and gesneriads (Griffiths, 1989). The prevalence of CAM in epiphytes increases in drier forests (Hietz *et al.*, 1999), but CAM epiphytes are also found in true rain forests, supporting the view that the rain forest canopy can be a xeric environment for ferns and other epiphytes growing in exposed locations and with no access to soil water.

The first fern identified with CAM photosynthesis was *Pyrrosia piloselloides* (syn. *Drymoglossum piloselloides*) from Malaysia (Hew and Wong, 1974). Subsequent studies have found CAM in *P. confluens*, *P. dielsii*, *P. lanceolata* (syn. *P. adnans*) and *P. longifolia* (listed in Holtum and Winter, 1999). Because these ferns are phylogenetically very distant from CAM seed plants, their photosynthesis has been studied in detail in the laboratory (Ong *et al.*, 1986) as well as in the field (Kluge *et al.*, 1989b). *Pyrrosia piloselloides* and *P. longifolia* are obligate CAM plants that absorb CO_2 predominantly during the night and via PEP-C. The carbon exchange pattern (Fig. 5.7a) is typical for full CAM metabolism with uptake predominantly during the night and some uptake via rubisco in the afternoon. The finding that more carbon is stored in malate than absorbed by the leaf suggests that part of the carbon metabolized in full CAM plants is recycled from respiration. The proportion of recycled carbon in *Pyrrosia*

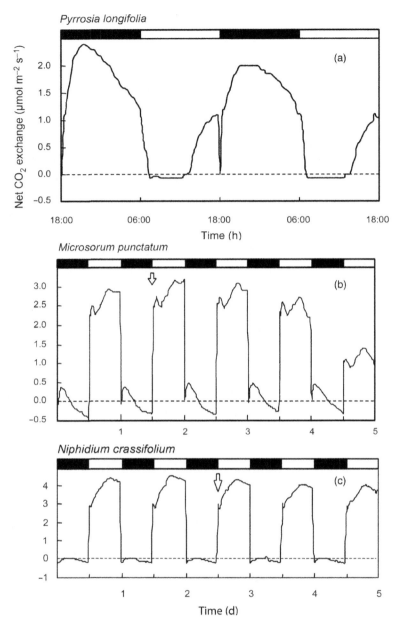

Fig. 5.7 Net CO_2 uptake (photosynthesis and respiration) of (a) a fern expressing full crassulacean acid metabolism (*Pyrrosia longifolia*), (b) a fern where weak CAM, mostly CAM cycling, is clearly seen by night-time CO_2 uptake (*Microsorum punctatum*) and (c) a fern where some CAM cycling is inferred from the nearly neutral CO_2 exchange during the night and changes in leaf acidity (*Niphidium crassifolium*). Arrows in (b) and (c) indicate when water was withheld, white and black bars above figures show the duration of light/dark hours. (After Holtum and Winter, 1999, reproduced with permission from CSIRO Publishing.)

measured in several studies ranged between 6% and 81%, and was higher in drought-stressed than in well-watered plants, while the effect of light was ambiguous (Griffiths *et al.*, 1986; Ong *et al.*, 1986; Kluge *et al.*, 1989a). Water use efficiency in *P. piloselloides* and *P. longifolia* was between 2.6 and 7.7 mmole CO_2 per mole H_2O, substantially higher than for the C3 epiphytic ferns *Asplenium nidus* (0.9) and *Nephrolepis acutifolia* (0.7) (Kluge *et al.*, 1989a). High WUE is a main advantage of CAM, though the authors note that WUE in epiphytic orchids and bromeliads was up to sixfold higher than in *Pyrrosia*. In *Pyrrosia*, gains in WUE are modest, possibly because stomatal conductance changes little in response to drought stress (Ong *et al.*, 1986; Kluge *et al.*, 1989a).

Because PEP-C discriminates less against the heavier carbon isotope (C-13) than rubisco, a CAM plant fixing a large proportion of its carbon via PEP-C has a less negative $\delta^{13}C$ signature than C3 plants. Values of $\delta^{13}C$ above about −18‰ generally indicate CAM or C4 photosynthesis, the latter never reported either for ferns or lycophytes. Values of $\delta^{13}C$ below about −23‰ are typical for C3 plants. However, there is no clear threshold, and $\delta^{13}C$ signals between −18‰ and −23‰ or also below −23‰ may result when only a portion of the carbon is first fixed via PEP-C (Pierce *et al.*, 2002). Plants that perform CAM cycling but take up most of their carbon during the day are consequently difficult to identify by their carbon isotope signature.

The obligate CAM of *P. longifolia* and *P. piloselloides* is furthermore shown by the evidence that these species exhibit CAM even under humid conditions and in the shade (Winter *et al.*, 1986; Kluge *et al.*, 1989a) as well as by $\delta^{13}C$ values between −13‰ and −16‰ (Winter *et al.*, 1983; Winter *et al.*, 1986). In contrast, $\delta^{13}C$ values in *P. confluens* ranged from −19.2‰ to −25.3‰ and in *P. dielsii* between −17.3‰ and −20.1‰ (Winter *et al.*, 1983), the lower (more negative) values indicating predominant daytime CO_2 uptake and showing that the proportion of C3 and CAM photosynthesis can vary even within one species. No evidence of CAM was found in *P. lingua* (Martin *et al.*, 2005) and *P. rupestris* ($\delta^{13}C \leq$ −23.9‰, Winter *et al.*, 1983), but given that *Pyrrosia* is a genus of some 100 species, there are likely more CAM species in the genus to discover.

The CAM species among *Pyrrosia* have fleshy leaves that are about 1 mm to 2 mm thick (Hovenkamp, 1986; Griffiths, 1989) and no more succulent than those of some C3 ferns with water-storing tissue. Interestingly, gametophytes and very small (*c.* 5 mm) sporophytes of *P. longifolia* show no evidence of CAM (Martin *et al.*, 1995), suggesting a shift in early ontogeny that is also found in cacti (Altesor *et al.*, 1992). This shift may be a result of the thin leaves of small plants, which lack the capacity to store sufficient acids for CAM metabolism. If the transition between C3 and CAM is related to size, age or stress, remains unknown.

Surprisingly, CAM was also found in the aquatic lycophyte family Isoëtaceae (quillworts, Plate 3D) and a number of other aquatic vascular plants (Keeley, 1981). Here its advantage is not related to drought but enables the uptake of CO_2 during the night, when CO_2 concentrations in the water are far higher than during the day when other plants compete for CO_2. Whether an adaptation to drought or not, *Isoëtes* shows that CAM has evolved independently in ferns and lycophytes and several times more in seed plants (Kluge and Ting, 1978). To date, $\delta^{13}C$ screening has not detected CAM in any other fern species (Winter *et al.*, 1983: 20 species screened in 13 genera; Earnshaw *et al.*, 1987: 6 species in 5 genera; Zotz and Ziegler, 1997: 21 species in 9 genera; Hietz *et al.*, 1999: 17 species in 8 genera; Watkins *et al.*, 2007b: 40 species in 31 genera), with the possible exception of *Microsorum cromwellii*, where a $\delta^{13}C$ value of −21.3‰ was found (Earnshaw *et al.*, 1987).

However, more recently CAM was also reported for *Vittaria lineata* (shoestring fern) from a Costa Rican rain forest based on nocturnal acid accumulation (Carter and Martin, 1994). *Vittaria* is an epiphytic fern with very narrow and somewhat fleshy leaves so the presence of CAM does not appear implausible. Species of *Vittaria* are generally from shady and humid canopy strata. Later and more detailed measurements could not confirm the first report, but significant increases in nocturnal acidity, albeit without carbon uptake and with C3-like $\delta^{13}C$ values, were found in *Anetium citrifolium* and *Vittaria flexuosa* (Polypodiaceae, formerly placed in Vittariaceae; Martin *et al.*, 2005).

Evidence of CAM cycling based on the change in acid concentration was also found in several more Polypodiaceae: *Dictymia brownii* (Griffiths, 1989), *Microsorum punctatum*, *Niphidium crassifolium*, *Platycerium veitchii* (Holtum and Winter, 1999) and *Platycerium bifurcatum* (Rut *et al.*, 2008). In the heterophyllous (with two types of green leaves) *P. bifurcatum*, CAM was only found in humus-accumulating cover leaves, but not in the large, lobed, spreading, sometimes spore-producing leaves. A change in acid concentration alone may not prove the existence of CAM as acids are produced by other biochemical pathways and leaves harvested at dusk and dawn may not have been of the same age, developmental stage or location, all of which can affect acid concentrations. In *Microsorum punctatum*, some nocturnal CO_2 uptake was measured (Fig. 5.7B), though the contribution to net uptake was only a fraction of daytime uptake and much less than in full CAM plants such as *Pyrrosia piloselloides*. *Niphidium crassifolium* and *Platycerium veitchii* show either a reduction in nocturnal CO_2 loss or a reduced uptake during the early light hours or both, providing additional evidence of weak CAM (Fig. 5.7C, Holtum and Winter, 1999). Detection of such small differences requires very accurate measurements of gas exchange, which have not been made on many ferns, including the purported CAM species in the genera *Anetium* and *Vittaria*.

There is little doubt that full CAM, as seen in *Pyrrosia*, is an adaptation to and a substantial advantage for the xeric conditions experienced by many species in this large epiphytic genus, in a manner similar to other epiphytic and terrestrial CAM plants. In *Niphidium crassifolium* and *Microsorum punctatum*, the contribution of carbon absorbed during the night to total carbon fixation increased with higher drought stress from 2.8% and 10% to 63.5% and 49.3%, respectively (Holtum and Winter, 1999). This carbon would be unavailable without PEP-C activity and an increase of 50% in carbon certainly is an advantage in the field. In addition, recycling of respiratory carbon in *Pyrrosia* reduces photoinhibition (Griffiths *et al.*, 1989), which is likely to also hold true for weak CAM ferns. However, the ecological advantage of water conservation or photoprotection remains unclear for the understory ferns of very humid rain forests (e.g., *Anetium, Vittaria*, Martin *et al.*, 2005).

Full CAM in ferns has only been found in *Pyrrosia* and the phylogenetically very distant lycophyte family Isoëtaceae, and mostly CAM cycling in a few other Polypodiaceae. The fact that weak CAM is difficult to identify complicates an analysis of the importance of CAM in the evolution of xerophytic ferns and lycophytes. Candidate species to look for CAM are xerophytic ferns that have some water-storing tissue. One open question is how common weak CAM is in other drought-adapted ferns, and indeed other epiphytes. If CAM cycling is relatively more common in ferns, why did ferns not more frequently achieve full CAM, such as epiphytic bromeliads and orchids did under similar ecological conditions? Given that a fair number of ferns grow in xeric habitats where CAM would be advantageous, are there barriers in the biochemical pathway to full CAM that are not easy to overcome for ferns? Water storing tissue is generally poorly developed in ferns and a precondition for CAM, but it does not seem to be an advantage for ferns.

5.7 Phenology

Temperate ferns can be either evergreen or deciduous but little information is available about fern phenology in tropical areas, where seasonality of growth appears to be common (see Chapter 3; Mehltreter, 2008). A number of ferns are deciduous. Among these are the epiphytes *Davallia* (Page, 1979), *Drynaria fortunei, Platycerium grande* (Mehltreter, 2008), *Phlebodium areolatum, Polypodium rhodopleuron* and *P. puberulum* (the latter two with a soft lamina and not xerophytic; P. Hietz, personal observation). Some terrestrial dry forest species in Costa Rica (*Adiantum lunulatum, Blechnum occidentale, Sticherus bifidus*) lose their leaves in the dry season (Brodribb and Holbrook, 2004), as do ferns in African dry forests and savannas (Jacobsen, 1983). In a detailed survey along the slopes of Mt. Kilimanjaro, about 21% of 140 fern species dropped their leaves in the dry season (50% were classified as evergreen and 26% as poikilohydric; Hemp, 2002). Deciduous fern species are common in the more arid submontane and lower montane forests below 1500 m on Mt. Kilimanjaro, are uncommon in the most humid forest zone and become the most diverse fern group in the subalpine forests and shrubland above 3200 m (Hemp, 2002). The phenology of temperate and tropical dry forest trees is directed not only by rainfall and drought but also by the length of the photoperiod (Rivera *et al.*, 2002). Whether deciduous ferns shed their leaves only in response to desiccation or are programmed to respond to the light period or other climatic triggers is unclear (see Chapter 3).

An important feature of deciduousness, particularly under an unpredictable water supply, is the ability to respond rapidly to improved conditions. The desert fern *Cheilanthes parryi*, though desiccation tolerant, sheds its leaves in the drier and hotter part of the year. Upon rewetting, young leaves unfold after three to four days and attain nearly full size after two weeks (Nobel, 1978).

Whereas passing the dry season in a dormant diaspore stage is common in seed plants from xeric ecosystems, there are hardly any annual plants among ferns, possibly because the life cycle including the gametophytic stage takes too long. A few species such as *Anogramma leptophylla* and *Nephrolepis pumicola* come close to being therophytes insofar as their sporophytic stage develops within a few weeks of favorable conditions, but the gametophytes are persistent (Kramer *et al.*, 1995). Deciduous ferns and species with the ability to recover from complete leaf loss are preadapted to survive in dry areas with recurrent fires (see Chapters 6 and 9). A deeply buried rhizome generally survives a fire and genera such as *Pteridium* that sprout or germinate fast after fires may even gain a competitive advantage by fire and as a consequence cannot be controlled by burning (see Chapters 6 and 8; Marrs and Watt, 2006). In Hawaii, *Dicranopteris linearis*, which can grow in dense thickets, sometimes several meters high, can become a substantial fire hazard when the dead leaves become dry (Stone and Pratt, 1995). A mantle of petiole bases or trichomes may also provide limited protection for the meristem, and among the club-mosses, which do not form proper rhizomes, *Lycopodiella caroliniana* (slender bog club-moss) in central Africa survives bush fires in stem tubers (Kramer *et al.*, 1995).

Some Marsileaceae (water-clovers) are semi-aquatic and need to survive dry seasons. They rely on their very resistant sporocarps, modified and bean-shaped pinnae with a hard outer layer that protects the spores. After heavy rainfall events, their life cycle continues with spore germination. If drought occurs every year, some species of *Marsilea* can effectively become annual plants, such as *Marsilea mollis*, which can grow in desert washes in North America.

5.8 Competitive ferns

A number of ferns are common in exposed locations such as forest margins and grasslands. These may not be particularly xeric habitats, but ferns growing here are more exposed to recurrent drought spells and high light exposure than on shaded sites so they are likely to require adaptations to cope with drought. The archetypal fern in this group is the cosmopolitan *Pteridium*, but *Gleichenia*, *Dicranopteris* and *Sticherus* also belong to this category. Ferns in this group apparently are not limited by angiosperms to less favorable conditions such as deep shade and are able to compete with grasses and herbs instead. Because of its wide distribution in tropical and temperate areas and its weediness, *Pteridium aquilinum sensu lato* is probably the most studied fern (see Chapter 8, Gordon *et al.*, 1999a, 1999b; Marrs and Watt, 2006). Though young leaves are damaged by drought, the rhizome is well protected against drought, frost and fire, and the mature leaves have a well-developed and efficient cuticle and rapid stomatal reactions. Indeed, fire may favor *Pteridium* by removing competitors and providing a suitable substrate for spore germination, so that after a fire *Pteridium* cover increased at some sites on Mt. Kilimanjaro from <1% to over 80% (Hemp, 2002). Maximum photosynthesis in *Pteridium* (9–$11\,\mu\mathrm{mol}\,\mathrm{m}^{-2}\,\mathrm{s}^{-1}$ and $>15\,\mu\mathrm{mol}$ for fertilized individuals) is also in the range of many herbs and substantially higher than in most other ferns (Marrs and Watt, 2006). The habitat preference and competitiveness of *Pteridium* may be a consequence of its efficient water transport through true vessels. Given the strong relationship between hydraulic supply of water to leaves and maximum photosynthesis (Brodribb and Feild, 2000), it is plausible that an evolutionarily advanced xylem structure was the first requisite to produce high-performance ferns before developing any other adaptations such as high rates of photosynthesis, stomatal conductance and low water loss.

5.9 Other extreme environments

As outlined above, drought tolerance is often accompanied by tolerance of high light levels, simply because xeric sites are often sunny and because the restriction of CO_2 uptake is associated with drought stress, which tends to result in photoinhibition. There is little information on fern heat tolerance. *Sphaeropteris cooperi*, a tree fern of the Australian rain forest, started to develop chlorosis at temperatures around $41\,^\circ\mathrm{C}$, though leaves were not damaged permanently. Desert ferns may tolerate somewhat higher temperatures in a water-saturated state, but the temperature limit of ferns in an active state is unlikely to be higher than the limit of about 45–$60\,^\circ\mathrm{C}$ found in other vascular plants (Larcher, 2002). Under drought stress, the temperature optimum for photosynthesis of the desert fern *Cheilanthes parryi* decreased from around 27 °C to 17 °C (Nobel, 1978). In *Polypodium vulgare* (common polypody), the maximum temperature tolerated increased from about 48 °C in the turgescent state to about 55 °C when desiccated (Kappen, 1966). *Selaginella lepidophylla* may experience temperatures above 60 °C in the field, but plants are negatively affected by temperatures as low as 45 °C even in a desiccated state (Eickmeier, 1986). Also, desiccation tolerance decreased with increasing temperature in *S. lepidophylla*. Thus desiccation-tolerant ferns and lycophytes may have a somewhat higher heat tolerance under field conditions, but do not appear to be substantially more tolerant than other vascular plants.

On the other end of temperature extremes, leaf desiccation generally results in a higher tolerance to low temperatures, in part because the extremely concentrated cell sap prevents intracellular ice formation that would kill affected cells. Desiccation also increases cold tolerance of fern sporophytes (Kappen, 1966) and desiccated gametophytes can even be stored in liquid nitrogen (−196 °C; Pence, 2000). As noted at the beginning of this chapter, high-mountain ferns that must endure low temperatures often grow at sites where they are likely to experience drought. Vascular epiphytes are largely restricted to tropical and subtropical areas, probably because they are limited by a combination of frost and insufficient water supply. However, fern epiphytes are less restricted geographically than other epiphytes, as they are frequently found in temperate regions (Zotz, 2005), perhaps because they are more tolerant of cold and drought than are other vascular epiphytes.

5.10 Conclusions

A substantial number of ferns and lycophytes from different systematic groups colonize xeric habitats. The intermediate anatomy of ferns, which includes features of bryophytes and seed plants, and their other adaptations range from those of mosses, exemplified in the poikilohydric filmy ferns, to those of hardy herbs, as in *Pteridium*. Desiccation tolerance is much more common in ferns than in angiosperms. Whereas water storage, efficient uptake, a resistant transport system and low water loss are the most important drought adaptations in xerophytic angiosperms, most xerophytic ferns rely on a high degree of desiccation tolerance and perhaps on transpiration control and water uptake through leaves, and less on water storage. The discovery of nocturnal carbon fixation in a number of unrelated species shows that ferns and lycophytes can master the biochemistry of CAM, but the rarity of full CAM suggests that other preconditions that make CAM ecologically advantageous have not been met. One precondition could simply be the limited amount of water-storing tissue in ferns, which again reflects their water economy.

The ultimate causes for the different strategies of drought adaptations between ferns and seed plants and also within the ferns and lycophytes are certainly genetic and reflect their evolutionary background. There could be one or several fundamental factors determining the scope of a species' drought adaptations. Possible traits that distinguish ferns are their anatomy (e.g., xylem structure), physiology (e.g., stomatal control), or biochemistry (e.g., the cuticle or the chemistry of desiccation tolerance). Which of these characteristics reflect the genetic limits and which are secondary adaptations remain largely unknown. For instance, if the leaf is composed of a single cell layer as in filmy ferns, it is inevitable that there can be no waterproof cuticle. As a consequence, these plants are poikilohydric and need a high degree of desiccation tolerance to survive in their habitats. A less efficient water transport system will result in other adaptations such as control of water loss, water absorption through leaves, or possibly desiccation tolerance. But why was the capacity to transport water not increased by producing more vessels or tracheids and why should limitations in water transport prevent the formation of water-storing tissue? Given the range of strategies to cope with drought that are found in ferns, comparative studies of ferns may provide insights into the relationship between different components of drought tolerance.

Acknowledgements

I would like to acknowledge the many useful comments of reviewers and editors to earlier versions of the manuscript. Research on Mexican ferns was partly funded by the Austrian Science Fund (FWF grants P12241 and P14775).

References

Alpert, P. (2000). The discovery, scope, and puzzle of desiccation tolerance in plants. *Plant Ecology*, **151**, 5–17.

Altesor, A., Ezcurra, E. and Silva, C. (1992). Changes in the photosynthetic metabolism during the early ontogeny of four cactus species. *Acta Oecologica*, **13**, 777–85.

Beckett, R. P. (1997). Pressure–volume analysis of a range of poikilohydric plants implies the existence of negative turgor in vegetative cells. *Annals of Botany*, **79**, 145–52.

Benzing, D. H. (1990). *Vascular Epiphytes: General Biology and Related Biota*. Cambridge, UK: Cambridge University Press.

Brodribb, T. J. and Feild, T. S. (2000). Stem hydraulic supply is linked to leaf photosynthetic capacity: evidence from New Caledonian and Tasmanian rainforests. *Plant, Cell and Environment*, **23**, 1381–8.

Brodribb, T. J. and Holbrook, N. M. (2004). Stomatal protection against hydraulic failure: a comparison of coexisting ferns and angiosperms. *New Phytologist*, **162**, 663–70.

Brodribb, T. J., Holbrook, N. M., Edwards, E. J. and Gutiérrez, M. V. (2003). Relations between stomatal closure, leaf turgor and xylem vulnerability in eight tropical dry forest trees. *Plant, Cell and Environment*, **26**, 443–50.

Brodribb, T. J., Holbrook, N. M., Zwieniecki, M. A. and Palma, B. (2005). Leaf hydraulic capacity in ferns, conifers and angiosperms: impacts on photosynthetic maxima. *New Phytologist*, **165**, 839–46.

Cardelús, C. L., Colwell, R. K. and Watkins, J. E., Jr. (2006). Vascular epiphyte distribution patterns: explaining the mid-elevation richness peak. *Journal of Ecology*, **94**, 144–56.

Carlquist, S. and Schneider, E. L. (2001). Vessels in ferns: structural, ecological, and evolutionary significance. *American Journal of Botany*, **88**, 1–13.

Carlquist, S. and Schneider, E. L. (2007). Tracheary elements in ferns: new techniques, observations, and concepts. *American Fern Journal*, **97**, 199–211.

Carter, J. P. and Martin, C. E. (1994). The occurrence of crassulacean acid metabolism among epiphytes in a high rainfall region of Costa Rica. *Selbyana*, **15**, 104–6.

Casper, C., Eickmeier, W. and Osmond, C. (1993). Changes of fluorescence and xanthophyll pigments during dehydration in the resurrection plant *Selaginella lepidophylla* in low and medium light intensities. *Oecologia*, **94**, 528–33.

Demmig-Adams, B. and Adams, W. W., III. (1992). Photoprotection and other responses of plants to high light stress. *Annual Review of Plant Physiology*, **43**, 599–626.

Demmig-Adams, B. and Adams, W. W., III. (1994). Light stress and photoprotection related to the xanthophyll cycle. In *Causes of Photooxidative Stress and Amelioration of Defense Systems in Plants*, ed. C. H. Foyer and P. M. Mullineaux. Boca Raton, FL, USA: CRC Press, pp. 105–26.

Earnshaw, M. J., Winter, K., Ziegler, H., *et al.* (1987). Altitudinal changes in the incidence of crassulacean acid metabolism in vascular epiphytes and related life forms in Papua New Guinea. *Oecologia*, **73**, 566–72.

Eickmeier, W. G. (1986). The correlation between high-temperature and desiccation tolerances in a poikilohydric desert plant. *Canadian Journal of Botany*, **64**, 611–17.

Eickmeier, W. G., Casper, C. and Osmond, B. (1993). Chlorophyll fluorescence in the resurrection plant *Selaginella lepidophylla* (Hook. & Grev.) Spring during high-light and desiccation stress, and evidence for zeaxanthin-associated photoprotection. *Planta*, **189**, 30–8.

Esau, K. (1965). *Plant Anatomy*, 2nd edn. New York: John Wiley.

Gaff, D. F. (1987). Desiccation tolerant plants in South America. *Oecologia*, **74**, 133–6.

Gildner, B. S. and Larson, D. W. (1992). Photosynthetic response to sunflecks in the desiccation-tolerant fern *Polypodium virginianum*. *Oecologia*, **89**, 390–6.

Gordon, C., Woodin, S. J., Alexander, I. J. and Mullins, C. E. (1999a). Effects of increased temperature, drought and nitrogen supply on two upland perennials of contrasting functional type: *Calluna vulgaris* and *Pteridium aquilinum*. *New Phytologist*, **142**, 243–58.

Gordon, C., Woodin, S. J., Mullins, C. E. and Alexander, I. J. (1999b). Effects of environmental change, including drought, on water use by competing *Calluna vulgaris* (heather) and *Pteridium aquilinum* (bracken). *Functional Ecology*, **13**, 96–106.

Griffiths, H. (1989). Carbon dioxide concentrating mechanisms and the evolution of CAM in vascular epiphytes. In *Vascular Plants as Epiphytes: Evolution and Ecophysiology*, ed. U. Lüttge. Heidelberg, Germany: Springer-Verlag, pp. 42–86.

Griffiths, H., Lüttge, U., Stimmel, K. H., *et al.* (1986). Comparative ecophysiology of CAM and C3 bromeliads. III. Environmental influences on CO_2 assimilation and transpiration. *Plant, Cell and Environment*, **9**, 385–93.

Griffiths, H., Ong, B. L., Avadhani, P. N. and Goh, C. J. (1989). Recycling of respiratory CO_2 during crassulacean acid metabolism: alleviation of photoinhibition in *Pyrrosia piloselloides*. *Planta*, **179**, 115–22.

Hemp, A. (2001). Ecology of the pteridophytes on the southern slopes of Mt. Kilimanjaro. II. Habitat selection. *Plant Biology*, **3**, 493–523.

Hemp, A. (2002). Ecology of the pteridophytes on the southern slopes of Mt. Kilimanjaro. I. Altitudinal distribution. *Plant Ecology*, **159**, 211–39.

Hew, C. S. and Wong, Y. S. (1974). Photosynthesis and respiration of ferns in relation to their habitats. *American Fern Journal*, **64**, 40–8.

Hietz, P. and Briones, O. (1998). Correlation between water relations and within-canopy distribution of epiphytic ferns in a Mexican cloud forest. *Oecologia*, **114**, 305–16.

Hietz, P. and Briones, O. (2001). Photosynthesis, chlorophyll fluorescence and within-canopy distribution of epiphytic ferns in a Mexican cloud forest. *Plant Biology*, **3**, 279–87.

Hietz, P. and Hietz-Seifert, U. (1995). Composition and ecology of vascular epiphyte communities along an altitudinal gradient in central Veracruz, Mexico. *Journal of Vegetation Science*, **6**, 487–98.

Hietz, P., Wanek, W. and Popp, M. (1999). Stable isotopic composition of carbon and nitrogen and nitrogen content in vascular epiphytes along an altitudinal transect. *Plant, Cell and Environment*, **22**, 1435–43.

Holtum, J. A. M. and Winter, K. (1999). Degrees of crassulacean acid metabolism in tropical epiphytic and lithophytic ferns. *Australian Journal of Plant Physiology*, **26**, 749–57.

Hovenkamp, P. (1986). A monograph of the fern genus *Pyrrosia*. *Leiden Botanical Series*, **9**, 1–80.

Hungerbühler, R. (1957). Die Xeromorphosen der Farne mit besonderer Berücksichtigung der Blattanatomie. Unpublished Ph.D. thesis, University of Zürich.

Jacobsen, W. B. G. (1983). *The Ferns and Fern Allies of Southern Africa*. Durban/Pretoria, Republic of South Africa: Butterworths.

Kappen, L. (1965). Untersuchungen über die Widerstandsfähigkeit der Gametophyten einheimischer Polypodiaceae gegenüber Frost, Hitze und Trockenheit. *Flora*, **156**, 101–15.

Kappen, L. (1966). Der Einfluss des Wassergehaltes auf die Widerstandsfähigkeit von Pflanzen gegenüber hohen und tiefen Temperaturen, untersucht an Blättern einiger Farne und von *Ramonda myconi*. *Flora*, **156**, 427–45.

Keeley, J. E. (1981). *Isoëtes howellii*: a submerged aquatic CAM plant? *American Journal of Botany*, **68**, 420.

Kessler, M. (2001). Pteridophyte species richness in Andean forests in Bolivia. *Biodiversity and Conservation*, **10**, 1473–95.

Kluge, M. and Ting, J. P. (1978). *Crassulacean Acid Metabolism: Analysis of an Ecological Adaptation*. Ecological Studies 30. Heidelberg, Germany: Springer-Verlag.

Kluge, M., Avadhani, P. N. and Goh, C. J. (1989a). Gas exchange and water relations in epiphytic tropical ferns. In *Vascular Plants as Epiphytes*, ed. U. Lüttge. Heidelberg, Germany: Springer-Verlag, pp. 87–108.

Kluge, M., Friemert, V., Ong, B. L., Brulfert, J. and Goh, C. J. (1989b). *In situ* studies of crassulacean acid metobolism in *Drymoglossum piloselloides*, an epiphytic fern of the humid tropics. *Journal of Experimental Botany*, **40**, 441–52.

Kramer, K. U., Schneller, J. J. and Wollenweber, E. (1995). *Farne und Farnverwandte*. Stuttgart, Germany: Georg Thieme.

Larcher, W. (2002). *Physiological Plant Ecology*. Berlin, Germany: Springer-Verlag.

Lebkuecher, J. G., and Eickmeier, W. G. (1991). Reduced photoinhibition with stem curling in the resurrection plant *Selaginella lepidophylla*. *Oecologia*, **88**, 597–604.

Lebkuecher, J. G. and Eickmeier, W. G. (1993). Physiological benefits of stem curling for resurrection plants in the field. *Ecology*, **74**, 1073–80.

Lehnert, M. (2007). Diversity and evolution of pteridophytes, with emphasis on the neotropics. Unpublished Ph.D. thesis, University of Göttingen.

Long, S. P., Humphries, S. and Falkowski, P. G. (1994). Photoinhibition and photosynthesis in nature. *Annual Review of Plant Physiology*, **45**, 633–62.

Lüttge, U. (1989). *Vascular Plants as Epiphytes: Evolution and Ecophysiology*. Heidelberg, Germany: Springer-Verlag.

Maherali, H., Pockman, W. T. and Jackson, R. B. (2004). Adaptive variation in the vulnerability of woody plants to xylem cavitation. *Ecology*, **85**, 2184–99.

Marrs, R. H., and Watt, A. S. (2006). Biological flora of the British Isles: *Pteridium aquilinum* (L.) Kuhn. *Journal of Ecology*, **94**, 1272–321.

Martin, C. E., Allen, M. T. and Haufler, C. H. (1995). C3 photosynthesis in the gametophyte of the epiphytic CAM fern *Pyrrosia longifolia* (Polypodiaceae). *American Journal of Botany*, **82**, 441–4.

Martin, C. E., Lin, K. C., Hsu, C. C. and Chiou, W. L. (2004). Causes and consequences of high osmotic potentials in epiphytic higher plants. *Journal of Plant Physiology*, **161**, 1119–24.

Martin, S. L., Davis, R., Protti, P., *et al.* (2005). The occurrence of crassulacean acid metabolism in epiphytic ferns, with an emphasis on the Vittariaceae. *International Journal of Plant Sciences*, **166**, 623–30.

Mehltreter, K. (2008). Phenology and habitat specificity of tropical ferns. In *Biology and Evolution of Ferns and Lycophytes*, ed. T. A. Ranker and C. H. Haufler. Cambridge, UK: Cambridge University Press, pp. 201–21.

Müller, L., Starnecker, G. and Winkler, S. (1981). Zur Ökologie epiphytischer Farne in Südbrasilien. I. Saugschuppen. *Flora*, **171**, 55–63.

Muslin, E. H., and Homann, P. H. (1992). Light as a hazard for the desiccation-resistant 'resurrection' fern *Polypodium polypodioides* L. *Plant, Cell and Environment*, **15**, 81–9.

Nobel, P. S. (1978). Microhabitat, water relations, and photosynthesis of a desert fern, *Notholaena parryi*. *Oecologia*, **31**, 293–309.

Nobel, P. S. (1991). *Physicochemical and Environmental Plant Physiology*. San Diego, CA, USA: Academic Press.

Oliver, M. J., Tuba, Z. and Mishler, B. D. (2000). The evolution of vegetative desiccation tolerance in land plants. *Plant Ecology*, **151**, 85–100.

Ong, B. L. and Ng, L. (1998). Regeneration of drought-stressed gametophytes of the epiphytic fern, *Pyrrosia piloselloides* (L.) Price. *Plant Cell Reports*, **18**, 225–8.

Ong, B. L., Kluge, M. and Friemert, V. (1986). Crassulacean acid metabolism in the epiphytic ferns *Drymoglossum piloselloides* and *Pyrrosia longifolia*: studies on responses to environmental signals. *Plant, Cell and Environment*, **9**, 547–57.

Page, C. N. (1979). The diversity of ferns. An ecological perspective. In *The Experimental Biology of Ferns*, ed. A. F. Dyer. London: Academic Press, pp. 9–56.

Page, C. N. (2002). Ecological strategies in fern evolution: a neopteridological overview. *Review of Palaeobotany and Palynology*, **119**, 1–33.

Pandé, S. K. (1935). Notes on the anatomy of a xerophytic fern *Niphobolus adnascens* from the Malay peninsula. *Proceedings of the Indian Academy of Sciences, Section B*, **1**, 556–64.

Pence, V. C. (2000). Cryopreservation of in vitro grown fern gametophytes. *American Fern Journal*, **90**, 16–23.

Pickett, F. L. (1931). Notes on xerophytic ferns. *American Fern Journal*, **21**, 49–57.

Pierce, S., Winter, K. and Griffiths, H. (2002). Carbon isotope ratio and the extent of daily CAM use by Bromeliaceae. *New Phytologist*, **156**, 75–83.

Porembski, S. and Barthlott, W. (2000). Granitic and gneissic outcrops (inselbergs) as centers of diversity for desiccation-tolerant vascular plants. *Plant Ecology*, **151**, 19–28.

Powles, S. B. (1984). Photoinhibition of photosynthesis induced by visible light. *Annual Review of Plant Physiology*, **35**, 15–44.

Proctor, M. C. F. (2003). Comparative ecophysiological measurements on the light responses, water relations and desiccation tolerance of the filmy ferns *Hymenophyllum wilsonii* Hook, and *H. tunbrigense* (L.) Smith. *Annals of Botany*, **91**, 717–27.

Proctor, M. C. F. and Tuba, Z. (2002). Poikilohydry and homoihydry: antithesis or spectrum of possibilities? *New Phytologist*, **156**, 327–49.

Reynolds, T. L. and Bewley, J. D. (1993a). Abscisic acid enhances the ability of the desiccation-tolerant fern *Polypodium virginianum* to withstand drying. *Journal of Experimental Botany*, **44**, 1771–9.

Reynolds, T. L. and Bewley, J. D. (1993b). Characterization of protein synthetic changes in a desiccation-tolerant fern, *Polypodium virginianum*: comparison of the effects of drying, rehydration and abscisic acid. *Journal of Experimental Botany*, **44**, 921–8.

Ribeiro, M. L. R. C., Santos, M. G. and Moraes, M. G. (2007). Leaf anatomy of two *Anemia* Sw. species (Schizaeaceae: Pteridophyte) from a rocky outcrop in Niterói, Rio de Janeiro, Brazil. *Revista Brasileira de Botânica*, **30**, 695–702.

Rivera, G., Elliott, S., Caldas, L., *et al.* (2002). Increasing day-length induces spring flushing of tropical dry forest trees in the absence of rain. *Trees – Structure and Function*, **16**, 445–56.

Rut, G., Krupa, J., Miszalski, Z., Rzepka, A. and Ślesak, I. (2008). Crassulacean acid metabolism in the epiphytic fern *Platycerium bifurcatum*. *Photosynthetica*, **46**, 156–60.

Schreuder, M. D. J., Brewer, C. A. and Heine, C. (2001). Modeled influences of non-exchanging trichomes on leaf boundary layers and gas exchange. *Journal of Theoretical Biology*, **210**, 23–32.

Schulze, E.-D., Beck, E. and Müller-Hohenstein, K. (2005). *Plant Ecology*. Berlin, Germany: Springer-Verlag.

Shreve, F. (1911). Studies on Jamaican Hymenophyllaceae. *Botanical Gazette*, **51**, 184–209.

Sinclair, R. (1983). Water relations of tropical epiphytes. I. Relationships between stomatal resistance, relative water content and the components of water potential. *Journal of Experimental Botany*, **34**, 1652–63.

Starnecker, G. and Winkler, S. (1982). Zur Ökologie epiphytischer Farne in Südbrasilien. II. Anatomische und physiologische Anpassungen. *Flora*, **172**, 57–68.

Stone, C. and Pratt, L. W. (1995). *Hawai'i's Plants and Animals: Biological Sketches of Hawai'i Volcanoes National Park*. Honolulu, HI, USA: University of Hawai'i Press.

Stuart, T. S. (1968). Revival of respiration and photosynthesis in dried leaves of *Polypodium polypodioides*. *Planta*, **83**, 185–206.

Tausz, M., Hietz, P. and Briones, O. (2001). The significance of carotenoids and tocopherols in photoprotection of seven epiphytic fern species of a Mexican cloud forest. *Australian Journal of Plant Physiology*, **28**, 775–83.

Tryon, R. M. (1964). Evolution in the leaf of living ferns. *Bulletin of the Torrey Botanical Club*, **21**, 73–85.

Tryon, R. M., and Tryon, A. F. (1982). *Ferns and Allied Plants with Special Reference to Tropical America*. New York: Springer-Verlag.

Tyree, M. T. and Sperry, J. S. (1989). Vulnerability of xylem to cavitation and embolism. *Annual Review of Plant Physiology and Plant Molecular Biology*, **40**, 19–38.

Watkins, J. E., Jr., Kawahara, A. Y., Leicht, S. A., *et al.* (2006). Fern laminar scales protect against photoinhibition from excess light. *American Fern Journal*, **96**, 83–92.

Watkins, J. E., Jr., Mack, M. C., Sinclair, T. R. and Mulkey, S. S. (2007a). Ecological and evolutionary consequences of desiccation tolerance in tropical fern gametophytes. *New Phytologist*, **176**, 708–17.

Watkins, J. E., Jr., Rundel, P. and Cardelús, C. (2007b). The influence of life form on carbon and nitrogen relationships in tropical rainforest ferns. *Oecologia*, **153**, 225–32.

Winter, K., Wallace, B. J., Stocker, G. C. and Roksandic, Z. (1983). Crassulacean acid metabolism in Australian vascular epiphytes and some related species. *Oecologia*, **57**, 129–41.

Winter, K., Osmond, C. B. and Hubick, K. T. (1986). Crassulacean acid metabolism in the shade: studies on an epiphytic fern, *Pyrrosia longifolia*, and other rain forest species from Australia. *Oecologia*, **68**, 224–30.

Wollenweber, E. (1978). The distribution and chemical constituents of the farinose exudates in gymnogrammoid ferns. *American Fern Journal*, **68**, 13–28.

Wollenweber, E., Scheele, C. and Tryon, A. F. (1987). Flavonoids and spores of *Platyzoma microphyllum*, an endemic fern of Australia. *American Fern Journal*, **77**, 28–32.

Woodhouse, R. M. and Nobel, P. S. (1982). Stipe anatomy, water potentials, and xylem conductances in seven species of ferns (Filicopsida). *American Journal of Botany*, **69**, 135–40.

Zotz, G. (2005). Vascular epiphytes in the temperate zones: a review. *Plant Ecology*, **176**, 173–83.

Zotz, G. and Ziegler, H. (1997). The occurrence of crassulacean acid metabolism among vascular epiphytes from central Panama. *New Phytologist*, **137**, 223–9.

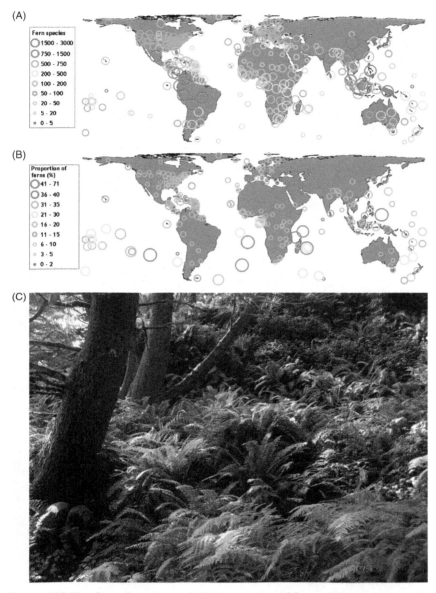

Plate 1 (A) Number of species and (B) proportion of ferns of the entire vascular plant flora in 322 floras. (After H. Kreft, personal communication). The circles for Brazil and China are drawn in the center of each country, although their highest richness is achieved in the southeastern part. These maps show the high richness of ferns in tropical regions especially where mountains are present, the relatively low diversity of Africa, and the relatively high contribution of ferns to island floras. Oceanic islands with high proportions of ferns include Easter Island, Palau, Saint Helena, and Tristan da Cunha (see Chapter 2). (C) Fern dominated understory in Olympic Peninsula rain forest, USA, with abundant *Polystichum munitum* and *Dryopteris expansa* (see Chapter 1).

Plate 2 (A) Xeric fern habitats such as rocky outcrops in San Luis Potosi, Mexico. Cheilanthoid fern genera such as *Argyrochosma*, *Cheilanthes*, *Notholaena* and *Pellaea* have evolved highly adapted species to survive under these dry conditions (see Chapter 5). (B) Cloud forest habitat along a small creek in Veracruz, Mexico. Under these favorable humid conditions at mid-elevations, fern richness reaches its maximum (see Chapter 2).

Plate 3 Lycophytes with small, single-veined microphylls. (A) *Lycopodium thyoides* with fertile strobili (i.e., cone of spore-bearing leaves) at the stem tips in Veracruz, Mexico. (B) Rosette-forming stems of desiccation-tolerant *Selaginella ribae* from Tamaulipas, Mexico. During dry periods, stems are rolled into a ball-like structure. Leaves are only a few millimeters long and cover the stems in four rows. (C) *Huperzia crassifolia*, a terrestrial species in the páramo of Costa Rica of this mostly epiphytic genus. (D) *Isoëtes pallida* from Oaxaca, Mexico with grass-like leaves. Sporangia are embedded at the inner side of the leaf bases.

(C)

(D)

Plate 3 (cont.)

Plate 4 (A) Seasonal *Botrychium pinnatum* in summer in British Colombia, Canada. Leaves are divided into a spore-bearing sporophore and the sterile trophophore, and live only for several weeks during the summer (see Chapter 3). (B) The rootless *Psilotum nudum* as an epiphyte on Reunion Island. Yellow, global structures are synangia, product of the fusion of three sporangia.

(C)

(D)

Plate 4 (cont.) (C) Dense population of *Equisetum arvense* in British Colombia, Canada. (D). *Angiopteris evecta*, cultivated at Wilson's Botanical Garden, Costa Rica. Native to Asia, this species has naturalized in Costa Rica and becomes invasive (see Chapter 8).

Plate 5 Reproduction in ferns. (A) Vegetative buds on leaves of *Asplenium daucifolium* on Reunion Island. (B) Round groups of sporangia (i.e., sori) on the lower leaf surface of *Polypodium triseriale*. (C). Leaf dimorphism in *Matteuccia struthiopteris*. The inner leaf whorl consists of young fertile leaves (i.e., sporophylls) with very narrow lamina, whereas the outer rosette-forming leaves are sterile (i.e., trophophylls). (D) Monomorphic leaves of *Blechnum appendiculatum* (syn. *B. glandulosum*) a widely distributed, stoloniferous species, native to the neotropics and invasive to Hawaii (see Chapter 8). Young leaves are white or red, supposedly an adaptation against herbivory or fungal attack (see Chapter 7).

(C)

(D)

Plate 5 (cont.)

Plate 6 Some growth forms of ferns. (A) Hemiepiphytic *Trichomanes tuerckheimii* in a Costa Rican rain forest. The rhizome is creeping up the tree stem and the leaves are attached to the stem giving this species a moss-like appearance. (B) Terrestrial *Trichomanes elegans* in a Costa Rican rain forest. The iridescent metallic-green leaves are adapted to photosynthesize in the deep shade of the understory. (C) Hemiepiphytic *Blechnum attenuatum* in Reunion Island. The rhizome is creeping straight up the tree trunk in the search for light. (D) Epiphytic *Scoliosorus ensiformis* in a cloud forest of Veracruz. This species is an indicator of undisturbed primary forest, because it quickly disappears after ecosystem disturbances (see Chapter 9).

Plate 7 Interactions between ferns and animals. (A) *Danaea wendlandii* in a Costa Rican rain forest showing a patch of sporophytes sheltering a bushmaster snake (see Chapter 3). (B) The Salvinia weevil (*Cyrtobagous salviniae*), a biological control organism on *Salvinia molesta*. Note the whisk-shaped hairs that give buoyancy to this aquatic fern (see Chapter 8). (C) Leaf galls of *Pleopeltis polypodioides* caused by an unidentified cecidomyiid in the lowlands of Mexico (see Chapter 7). (D) A spore-feeding caterpillar with soral mimicry in Mexico. The cocoon is mimicking the sorus of *Pteris orizabae* (see Chapter 7).

Plate 8 (A) *Diplopterygium bancroftii* (Gleicheniaceae) with its typical indeterminate leaf growth, often colonizing road cuts (see Chapter 6). (B) *Pteris vittata* on a wall in Guatemala. This species hyperaccumulates arsenic and is used for phytoremediation (see Chapter 8).

Plate 8 (cont.) (C) The climbing leaves of *Lygodium japonicum*, an invasive alien in the USA. (D) A severe infestation of *Lygodium microphyllum* on cypress trees in southern Florida (see Chapter 8).

6

Ferns, disturbance and succession

LAWRENCE R. WALKER AND JOANNE M. SHARPE

Key points

1. Ferns often colonize habitats disturbed by tectonic activity, wind, water, fire and humans.
2. Fern dispersal into disturbed habitats can result from long distance movement of spores but is usually by short distance spore dispersal or vegetative expansion of nearby plants.
3. Rapid establishment and dense growth can make ferns competitive with other vascular plants through light reduction and nutrient uptake or immobilization. Fern thickets can delay successional transitions but ferns also provide regeneration sites for other species and stabilize slopes. Fern influences may vary across environmental resource (e.g., light, water, nutrients) and topographic gradients.
4. Ferns can have important roles in the restoration of disturbed ecosystems.

6.1 Introduction

A typical image of fern habitat is a wet, shady forest untouched by disturbance. In fact, many ferns colonize recently disturbed and exposed areas such as scoured riverbanks (Reudink *et al.*, 2005) or the uprooted pits, mounds and trunks of fallen trees (Peterson *et al.*, 1990; Nadkarni and Wheelwright, 2000). Ferns tolerate a wide range of environmental conditions (Hemp, 2001), and some species can colonize such highly disturbed habitats as lava flows, dunes, landslides or floodplains as well as areas of forests that have been damaged by burning, ice storms, hurricanes or logging (Walker *et al.*, 1996b; Barson, 1997; Russell *et al.*, 1998; Arens and Sánchez Baracaldo, 1998, 2000; Woods, 2002). One reason some ferns readily colonize recent disturbances is that they have widely dispersed spores that reach even the most remote islands (see Chapter 2; Carlquist, 1980). Ferns also can colonize disturbed areas vegetatively from adjacent areas. However, disturbance-adapted ferns must have characteristics that allow them to establish and grow under conditions of low nutrients (see Chapter 4), drought or high light conditions (see Chapter 5). Ferns are not just members of stable communities in mesic habitats but are also an important part of plant communities impacted by both natural (Table 6.1) and anthropogenic (Table 6.2) disturbances.

Fern Ecology, ed. Klaus Mehltreter, Lawrence R. Walker and Joanne M. Sharpe. Published by Cambridge University Press. © Cambridge University Press 2010.

Table 6.1 *Characteristic ferns and lycophytes of habitats created by natural disturbances (volcanoes, glacial moraines, floodplains, hurricanes) with sample references*
Disturbances that can be of either natural or anthropogenic origin are in Table 6.2. Habitats are listed in decreasing order of severity.

Habitat	Location	Species	References
Volcanic surfaces	Hawaii	*Cibotium chamissoi*	Aplet and Vitousek, 1994
		C. glaucum	Crews *et al.*, 1995 Mueller-Dombois *et al.*, 1983 Vitousek *et al.*, 1995 Walker and Aplet, 1994
	Hawaii	*Dicranopteris linearis*	Russell *et al.*, 1998
	Hawaii	Various	Clarkson, 1997; Table 6.5
	Hawaii	Various	Kitayama *et al.*, 1995; Fig. 6.4
	Java	*Polypodium feei*	Holttum, 1938
Glacial moraines	New Zealand	*Asplenium bulbiferum* *Blechnum* spp. *Cyathea smithii* *Dicksonia squarrosa*	Wardle, 1980
Floodplains and shorelines	Alaska	*Equisetum* spp.	Atkinson, 2004
	Australia	*Marsilea drummondii*	Capon, 2005
	Borneo	*Asplenium subaquatile* *Thelypteris hispidifolia* *T. oligodictyon*	Van Steenis, 1981
	Brazil	*Salvinia* spp.	Dos Santos and Thomaz, 2007
	Central and South America	*Thelypteris francoana* *T. skinneri*	Van Steenis, 1981
	Cuba	*Thelypteris wrightii*	Van Steenis, 1981
	Haiti	*Thelypteris hottensis*	Van Steenis, 1981
	Indo-Malesia	*Microsorium pteropus*	Van Steenis, 1981

Table 6.1 (*cont.*)

Habitat	Location	Species	References
	Mexico	*Asplenium sessilifolium*	Mehltreter, 2008
	New Guinea	*Thelypteris riparia*	Van Steenis, 1981
	Norway	*Equisetum fluviatile*	Odland and del Moral, 2002
	Ohio	*Dryopteris carthusiana*	Holmes *et al.*, 2005
	Poland	*Isoëtes lacustris*	Szmeja, 1994
	Puerto Rico	*Thelypteris angustifolia*	Sharpe, 1997
	Quebec	*Onoclea sensibilis*	Désilets and Houle, 2005
	Sumatra	*Austrogramme asplenioides*	Van Steenis, 1981
	Utah	*Adiantum capillus-veneris* *Cystopteris fragilis*	Malanson, 1993
Gaps from hurricanes and tree falls	El Salvador	*Alsophila salvinii*	Seiler, 1981
	Massachusetts	*Dennstaedtia punctilobula*	Cooper-Ellis *et al.*, 1999
	New York	*Dennstaedtia punctilobula* *Dryopteris cristata* *D. intermedia* *D. marginalis* *Osmunda claytoniana* *Thelypteris noveboracensis* *Onoclea sensibilis* *Pteridium aquilinum* *Polystichum acrostichoides*	Woods, 2002
	Norway	*Athyrium filix-femina* *Gymnocarpium dryopteris* *Phegopteris connectilis*	Rydgren *et al.*, 1998
	Puerto Rico	*Acrostichum danaeifolium*	Sharpe, 2010
		Cyathea arborea	Walker *et al.*, 1996b Walker, 2000
		Danaea nodosa	Chinea, 1999

Table 6.2 *Characteristic ferns of habitats created by anthropogenic disturbances with sample references*

Disturbances that can be of either natural or anthropogenic origin (landslides, walls/cliffs, fire) are also included. Habitats are listed in decreasing order of severity.

Habitat	Location	Species	References
Mine wastes	China	*Hippochaete ramosissimum*	Wang *et al.*, 2004
	Hawaii	*Dicranopteris linearis*	Mueller-Dombois, 2000
	New Zealand	*Pteridium esculentum*	Atkinson, 2004
Landslides and road cuts	Bolivia	*Diplopterygium bancroftii*	Kessler, 1999
		Histiopteris incisa	
		Hypolepis nigrescens	
		H. parallelogramma	
		Pteridium arachnoideum	
		Sticherus spp.	
	Columbia	Cyatheaceae	Arens and Sánchez Baracaldo, 2000
	Hawaii	*Cibotium glaucum*	Restrepo and Vitousek, 2001
		Dicranopteris linearis	Scott, 1969
		Nephrolepis multiflora	
		Odontosoria chinensis	
		Sadleria pallida	
	New Zealand	*Asplenium flaccidum*	Stewart, 1986
		Blechnum discolor	
		B. minus	
		Grammitis billardieri	
		Polystichum vestitum	
	Puerto Rico	*Cyathea arborea*	Conant, 1976
		Gleichenella pectinata	Guariguata, 1990
			Shiels and Walker, 2003
		Odontosoria aculeata	Shiels *et al.*, 2008
			Walker, 1994
		Sticherus bifidus	Walker *et al.*, 1996a
	St. Lucia	*Cyathea* spp.	Wardlaw, 1931
		Sticherus spp.	
		Lycopodiella cernua	
		Lygodium spp.	

Table 6.2 (*cont.*)

Habitat	Location	Species	References
Walls and cliffs	Tanzania England	*Pteridium aquilinum* *Asplenium* 　*adiantum-nigrum* *A. ruta-muraria* *A. trichomanes* *Cystopteris fragilis* *C. montana* *Dryopteris filix-mas* *Phyllitis* 　*scolopendrium* *Polypodium vulgare* *Polystichum* 　*lonchitis* *Pteridium aquilinum*	Lundgren, 1978 Rishbeth, 1948 Rodwell, 1992
	Netherlands	*Asplenium* 　*adiantum-nigrum* *A. ceterach* *A. fontanum* *A. foreziense* *A. ruta-muraria* *A. scolopendrium* *A. septentrionale* *A. trichomanes* *A. viride* *Cyrtomium falcatum* *Cystopteris fragilis* *Gymnocarpium* 　*robertianum*	Bremer, 2003
	Quebec and Ontario	*Asplenium* 　*platyneuron* *A. ruta-muraria* *A. trichomanes* *A. viride* *Cryptogramma* 　*stelleri* *Cystopteris* 　*bulbifera* *C. fragilis* *Pellaea* 　*atropurpurea* *P. glabella* *Phyllitis* 　*scolopendrium* *Polypodium* 　*virginianum*	Brunton and Lafontaine, 1974 Larsen *et al.*, 2000
Fire-damaged 　habitats	Bolivia	*Pteridium* 　*arachnoideum*	Kessler, 1999

Table 6.2 (*cont.*)

Habitat	Location	Species	References
	Costa Rica	*Pteridium aquilinum*	Gliessman, 1978
	French Guiana	*Acrostichum aureum* *Pityrogramma calomelanos*	de Foresta, 1984
	New Zealand	*Pteridium esculentum*	Atkinson, 2004
	Norway	*Equisetum palustre*	Arnesen, 1999
	Puerto Rico	*Dicranopteris linearis* *Sticherus bifidus*	Walker and Boneta, 1995
	Sri Lanka	*Dicranopteris linearis*	Maheswaran and Gunatilleke, 1988
	Scotland	*Pteridium aquilinum*	Fletcher and Kirkwood, 1979
	UK	*Pteridium aquilinum*	Marrs *et al.*, 2000
Logged habitats	Dominican Republic	*Gleichenella pectinata*	Slocum *et al.*, 2004
	Michigan	*Pteridium aquilinum*	Roberts and Gilliam, 1995
	North Carolina	*Equisetum arvense*	Elliott *et al.*, 1998
	Papua New Guinea	Various	Van Valkenburg and Ketner, 1994, see Table 6.4
Abandoned fields	Columbia	*Cyathea caracasana*	Arens and Sánchez Baracaldo, 1998
	Dominican Republic	*Gleichenella pectinata*	Slocum *et al.*, 2004, 2006
	Indiana	*Cystopteris protrusa* *Deparia acrostichoides* *Polystichum acrostichoides*	Jenkins and Parker, 2000
	Singapore	*Dicranopteris linearis* *Gleichenia truncata*	Turner *et al.*, 1996
	Sri Lanka	*Dicranopteris linearis*	Cohen *et al.*, 1995

Ferns that colonize disturbed and open habitats in the tropics and subtropics have a variety of life forms that can be classified according to leaf and stem morphology (Table 6.3). Species can have indeterminate, determinate or rarely both forms of growth of the leaf rachis and creeping or erect stem morphology. Leaf length is also an approximate way to distinguish among ferns (Table 6.3). Creeping forms with leaf tips of indeterminate growth or dormant leaf buds often have branching rhizomes just at or above the soil surface and the leaves can either be mostly recumbent, spreading across the surface and over other vegetation (scrambling ferns; e.g.,

Table 6.3 *Life forms of common disturbance-adapted ferns*

Leaf growth	Leaf length	Creeping rhizome	Erect	Erect tree fern
Indeterminate (by long-lasting, growing leaf tip or growth of accessory leaf buds)	Variable (0.1–30 m)	*Dicranopteris*[1] *Diplopterygium*[1] *Eriosorus*[2] *Gleichenella*[1] *Gleichenia*[1] *Hypolepis*[1] *Lygodium*[2] *Odontosoria*[1] *Paesia*[2] *Sticherus*[1]		
Determinate	Long (>1 m)	*Dennstaedtia*[1] *Eriosorus* *Histiopteris*[1] *Hypolepis* *Pteridium*	*Nephrolepis* *Osmunda*	*Cibotium* *Cyathea* *Dicksonia* *Sadleria*
	Short (<1 m)	*Gymnocarpium* *Lycopodium* *Nephrolepis* *Salpichlaena*[2,3]	*Cystopteris* *Nephrolepis* *Pityrogramma*	

[1] Scramblers, [2,3] climbers, [2] climbing leaves, [3] climbing rhizomes.
Some genera fit several categories. See text and Tryon and Tryon (1982) for further details.

Dicranopteris, Gleichenella, Gleichenia, Hypolepis, Sticherus) or ascend any available vertical support such as trees or liana (e.g., *Lygodium*, climbing ferns, Plate 8C, D). Scrambling ferns often form dense thickets from branching rhizomes and interlaced leaf rachises (Fig. 6.1) that can inhibit woody plant colonization on lava flows (Russell *et al.*, 1998), landslides (Lundgren, 1978; Walker, 1994; Kessler 1999), tree fall gaps (Lyon and Sharpe, 1996) and burned habitats (Walker and Boneta, 1995). Climbing ferns frequent forest margins and tree fall gaps where they often climb into the canopy (Mehltreter, 2006) and typically form narrow, tall thickets resembling curtains. Ferns with determinate leaf growth can have creeping or erect stems. Tree ferns are the tallest of the erect ferns (Fig. 6.2).

Once established, ferns compete with other colonizing vascular plants for light and nutrients, potentially altering successional trajectories. Tree ferns can grow above and shade out competitors (Page, 1979) but also capture and retain a large proportion of limiting nutrients (N, P, K, Ca, Mg) in early successional communities (Mueller-Dombois *et al.*, 1983). Ferns also may compete chemically. The most

Fig. 6.1 Scrambling ferns *Gleichenella pectinata* and *Sticherus bifidus* on a landslide adjacent to a Puerto Rican roadside.

infamous fern in this regard is the widely distributed *Pteridium aquilinum* (bracken) that may have allelopathic impacts (Gliessman and Muller, 1978) but at least has toxic chemicals in its tissues that deter herbivory (see Chapter 8; Marrs *et al.*, 2000). When understory ferns form extensive patches they can have major influences on nutrient cycling and biodiversity (Coomes *et al.*, 2005; Gilliam, 2007). Persistence through extensive spore production, spreading of rhizomes or branching can extend the period of influence of ferns in succession (Penrod, 2000).

In this chapter, we examine the types of disturbed surfaces that ferns colonize. Then we explore the ecological role of ferns in succession, from how they disperse into new

Fig. 6.2 Tree ferns of *Cyathea arborea* on a Puerto Rican landslide.

habitats, to how they establish and grow, to how they interact with other species and alter ecosystem properties. We also discuss several benefits of studying ferns in a successional context and note the importance of ferns in ecosystem restoration.

6.2 Disturbance

A disturbance can be defined as a relatively discrete event in time and space that results in a loss of structure (as biomass) or resources (Walker and Willig, 1999). Disturbance differs from stress, which is an existing environmental condition that limits growth and productivity (Grime, 1979). Ferns have developed stress-tolerant adaptations to survive in harsh environments such as becoming poikilohydric or deciduous in dry conditions (see Chapter 5; Hemp, 2001). Disturbed habitats, in contrast, are caused by a specific event that abruptly makes conditions suboptimal for the original inhabitants of the area. Ferns that are adapted to colonize and survive on newly disturbed surfaces exhibit phenotypic plasticity, rapid growth and have a high investment in reproduction (Grime, 1979).

Disturbances often result in patchy resource distribution and can originate from inside or outside the system of interest (Glenn-Lewin and van der Maarel, 1992).

Three common attributes of a disturbance include frequency (event per time unit), extent (of the area disturbed) and magnitude. Magnitude is a measure that integrates both the intensity of the physical force (e.g., wind speed) and the severity of impact on the biota (e.g., damage caused). The net effect of all disturbances and their interactions at a given place is called the disturbance regime (Walker, 1999). The disturbance regime is often central to regeneration of fern populations (Page, 2002).

Disturbance is an omnipresent force in the lives of all organisms, including ferns. Natural disturbances in the past have influenced the evolution of ferns (Large and Braggins, 2004). Following the massive extinctions in the late Cretaceous (*c.* 65 mya), presumably from consequences of the impact of a large asteroid, ferns dominated the recovering vegetation (Vajda *et al.*, 2001). Periods of intense volcanic activity dating back to the Miocene (*c.* 20 mya) have left layers of volcanic ash in the region of Los Tuxtlas, Mexico, that to this day are still only colonized by *Pityrogramma calomelanos* (silverback fern; Riba and Reyes Jaramillo, 1990). Disturbances, both natural and anthropogenic, continue to affect ferns. Distribution, growth and survival of ferns can be impacted by a disturbance directly, as when a tree branch falls and damages existing understory ferns or epiphytic ferns on the branch itself. The disturbance effect can also be indirect, as when that same branch fall alters the physical attributes of the forest floor environment (e.g., by exposing mineral soil), or impacts canopy species (e.g., tree mortality) that in turn can have either negative (e.g., loss of necessary shade) or positive (e.g., release from root competition) effects on understory species (Fig. 6.3). Our focus in this section will be on how ferns respond in the aftermath of disturbances, usually through colonization into newly damaged areas or survival, but sometimes through increased growth and subsequent regeneration *in situ*. After reviewing the types of disturbed surfaces that ferns can colonize, we will expand on the role of ferns in the successional communities that follow disturbance.

6.2.1 Volcanoes and landslides

Ferns often colonize newly created surfaces following severe tectonic disturbances such as volcanoes, earthquakes and landslides. Ferns made up 15–44% of plant colonizers on the Krakatau islands in Indonesia in the years following the major eruption of Krakatau Volcano in 1883 (Fig. 6.4; Whittaker *et al.*, 1989) and are conspicuous on many other volcanoes in such locations as Cameroon (Benl, 1976), Reunion Island (Derroire *et al.*, 2007) and New Zealand (Crookes, 1960). Both the scrambling fern *Dicranopteris linearis* and the tree fern *Cibotium glaucum* are common colonists on lava flows and tephra deposits in Hawaii (Atkinson, 1970;

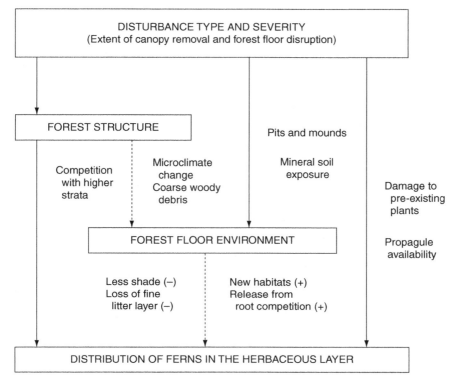

Fig. 6.3 Examples of potential disturbance effects on the distribution of species of ferns in the herbaceous layer of a forest. Solid lines indicate direct effects and dashed lines indicate indirect effects. (–) indicates a negative effect on existing ferns; (+) indicates a positive effect on new propagules or existing ferns. (Modified from Roberts and Gilliam, 2003.)

Walker and Aplet, 1994; Clarkson, 1997; Russell *et al.*, 1998). Earthquake- or rain-triggered landslides are commonly invaded by ferns in New Zealand (N. Bystriakova, personal communication), Hawaii (Scott, 1969; Restrepo and Vitousek, 2001), St. Lucia (Wardlaw, 1931), Bolivia (Kessler, 1999) and Puerto Rico (Guariguata, 1990; Walker, 1994). Several factors favor fern establishment on such barren terrain. Low-nutrient soils and dispersal barriers limit competition from seed plants and high light conditions favor sporophyte growth of those ferns that are adapted to such open habitats. The erect, perennial form of tree ferns also aids establishment and preempts the use of resources.

6.2.2 Hurricanes

Strong hurricane winds can cause extensive forest canopy destruction, which can damage epiphytic and understory ferns but also increase nutrient and light availability

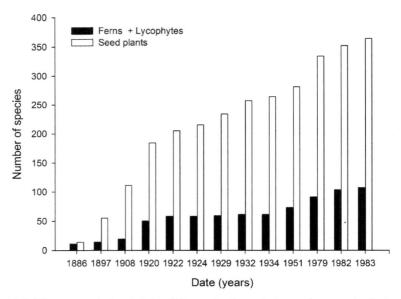

Fig. 6.4 Mean cumulative totals of ferns plus lycophytes and spermatophytes on the volcanic Krakatau islands after main eruption in 1883. (Modified from Whittaker *et al.*, 1989.)

that promotes growth of some understory ferns (Walker *et al.*, 1996b). Smaller tree fall gaps can provide partially shaded habitats for tree fern colonization (Seiler, 1981; Walker, 2000). Fern responses to wind disturbances are species specific. For example, 6 of 21 understory fern species disappeared from a study plot in Puerto Rico due to Hurricane Hugo in 1989, while 5 fern species were recruited following the hurricane despite little change in diversity of herbs and vines (Chinea, 1999).

The large deposition of litter during a hurricane can promote understory fern growth, although differential responses among fern species can occur as demonstrated by experimental litter removal in Puerto Rico (Halleck *et al.*, 2004). For the large mangrove fern, *Acrostichum danaeifolium*, biomass production declined for the first eight months after Hurricane Georges in Puerto Rico in 1998, but within two years was threefold greater than pre-hurricane levels. Possible causes of the added biomass could be increased light from canopy damage, increased nutrient availability from litter additions or higher than normal rainfall levels that increased water availability (Sharpe, 2010). In addition, *A. danaeifolium* responded to the hurricane with a 20-fold increase in spore production. In contrast, a catastrophic ice storm in northeastern USA in 1998 that resulted in hurricane-like damage to the forest canopy may have been responsible for reduced spore production of the understory fern *Dryopteris intermedia* (evergreen wood fern, see Chapter 3; Sharpe, 2005).

An experiment to directly test the effects of hurricane disturbance was conducted by pulling down trees in Harvard Forest in northeastern USA in 1990 (Cooper-Ellis *et al.*, 1999). Cover of the hay-scented fern, *Dennstaedtia puncti-lobula*, increased almost 100% in four years. Clearly, storms can have a variety of direct and indirect, positive and negative effects on fern populations.

6.2.3 Flooding

Water movement directly impacts ferns growing on exposed rock surfaces (lithophytes) or along river channels (rheophytes). Extreme flooding can damage ferns in these habitats but also can provide openings for colonization (Holttum, 1938). Ferns such as *Adiantum capillus-veneris* (maidenhair) and *Cystopteris fragilis* (brittle fern) dominated small, isolated and frequently flooded rock walls in Zion National Park in Utah (USA) but were outcompeted by herbs and trees in less frequently flooded sites (Malanson, 1993). A similar pattern occurred on a Quebec floodplain, where the dominant understory fern *Onoclea sensibilis* (sensitive fern) gave way to dominance by several other herbaceous species as flooding became less frequent (Désilets and Houle, 2005). Even on an arid floodplain in central Australia, where vegetation cover was positively rather than negatively correlated with flooding frequency, the perennial (and edible) water fern *Marsilea drummondii* (nardoo) and many annual forbs dominated frequently flooded areas (Capon, 2005). However, not all ferns are flood adapted. On a floodplain in Ohio, cover of all ferns, in particular, *Dryopteris carthusiana* (spinulose wood fern) was much higher in the uplands than on the more disturbed floodplain (Holmes *et al.*, 2005).

Rheophytic ferns often have narrow leaves or leaflets (e.g., *Thelypteris angusti-folia*) to reduce the damaging effects of the forces generated by water flowing past leaves and extensive root systems in sand, gravel bars or rock clefts (van Steenis, 1981). However, cyclic disturbances such as drought and extreme flooding do impact ferns adapted to this habitat. Long-term monitoring of the rheophytic fern, *T. angustifolia*, along a Puerto Rican riverbank (Fig. 6.5), indicated that fern leaf life spans were shorter and spore production higher when the ferns were subjected to drought (Sharpe, 1997). Severe flooding, in contrast, led to large increases in leaf loss and mortality in the same populations (J. Sharpe, unpublished data). One of the effects of severe flooding was the rearrangement of boulders in the flood zone where *T. angustifolia* individuals showed dichotomous branching around new obstacles, allowing persistence in the same space through vegetative propagation (J. Sharpe, unpublished data).

Aquatic ferns are also subject to a variety of disturbances, from running water as in the earlier *Marsilea* example, to waves, temperature changes, ice, siltation, drawdowns and nutrient loading from pollution. More wave action in shallow versus deep

Fig. 6.5 *Thelypteris angustifolia*, a rheophytic fern, on a Puerto Rican streambank following an extreme flooding event that resulted in boulder rearrangement and woody debris deposition on the fern.

water of European lakes reduced growth, surface area, root length, biomass and spore production of *Isoëtes lacustris* (lake quillwort), an aquatic lycophyte (Szmeja, 1994). Human-caused drawdowns in a Brazilian floodplain reduced the abundance of some species of the native floating *Salvinia* that are good competitors under stable water conditions (Dos Santos and Thomaz, 2007). Finally, the floating fern, *Azolla pinnata* (mosquito fern), grows very well in polluted water (Taheruzzaman and Kushari, 1995). In a nutrient enrichment experiment on an urban lake, *A. pinnata* grew so densely on the surface of enriched macrocosms that it killed the population of the submerged angiosperm *Vallisneria americana* (Morris *et al.*, 2003). Some species of ferns are clearly adapted to disturbances in floodplains or other aquatic environments.

6.2.4 Fire

Many ferns have adapted to fire-prone habitats. One strategy is to tolerate moderate damage by fire. For example, one fifth of all 114 fern species in Zambia had fire scars (Kornaś, 1978). *Blechnum auratum* in Costa Rica maintains its dominance by resisting fire (see Chapter 9; Horn, 1988; Mehltreter, 1997). Most fire-adapted ferns reestablish in postfire environments from rhizomes protected by burial or through spore germination. *Lycopodium complanatum* (running-pine; Oinonen, 1967), *Pteridium aquilinum* (Gliessman, 1978; Fletcher and Kirkwood, 1979; Atkinson, 2004) and *Blechnum spicant* (deer fern; Sykes and Horrill, 1981) recover quickly from rhizomes or spore germination following fire but generally decline when overtopped by forest species (Kessler, 1999; Marrs *et al.*, 2000). *Pteridium aquilinum* spores germinate best in the high pH levels following fire (Gliessman, 1978). Australian tree ferns (e.g., *Cyathea australis* and *Dicksonia antarctica*) regenerate after fire by resprouting from terminal buds protected by their fibrous trunks (Ough and Murphy, 2004). Periodic, light-intensity fires maintain *Blechnum spicant* in the understory of *Sequoia sempervirens* (redwood) forests in California, USA (Lenihan, 1990). When *Gleichenella pectinata* (syn. *Dicranopteris pectinata*) ferns are periodically burned to maintain roadside visibility in Puerto Rico, they reestablish within months from surviving rhizomes (Walker and Boneta, 1995). Other ferns may be short-term colonists of burned areas and their duration linked to burn severity. For example, the colonizing *Pityrogramma calomelanos* survived longer in heavily burned areas of cut forest in French Guiana than in more lightly burned areas where shrubs and low trees recovered more quickly, thereby shading out the ferns (de Foresta, 1984). Similarly, the dominance by the invasive *Pteridium arachnoideum* was most prevalent in postfire disturbances in the Bolivian Andes but some native species of *Hypolepis* dominated when fire was absent (Kessler, 1999). The response of *Pteridium aquilinum* to fire can be very rapid as leaves in one neotropical study emerged from underground rhizomes 45 days after a severe fire (Alonso-Amelot and Rodulfo-Baechler, 1996). Ferns are therefore adapted to fire in many ways.

6.2.5 Humans

Humans have created a number of different types of disturbances (del Moral and Walker, 2007), some of which are analogous to natural disturbances (e.g., pavement resembles lava, mine tailings resemble tephra, walls resemble cliffs) and some are unique (e.g., toxic mine wastes, urbanization). Some ferns have adapted to anthropogenic disturbances, including vertical wall surfaces (Rishbeth, 1948), mine wastes (Mueller-Dombois, 2000; Wang *et al.*, 2004), road cuts (Wardlaw, 1931; Conant, 1976; Arens and Sánchez Baracaldo, 2000), logging (Slocum *et al.*, 2004) and abandoned agricultural fields (Cohen *et al.*, 1995; Turner *et al.*, 1996; Arens and Sánchez Baracaldo, 1998; Jenkins and Parker, 2000). Of course, some ferns do not adapt, such as *Blechnum spicant*, which is not found in forests grazed by cattle in northern Spain (Onaindia *et al.*, 2004).

Fern responses to human disturbances are highly variable and often species specific. Light and soil conditions as well as preexisting conditions help determine the responses. For example, on intentional forest clear cuts in Michigan, *Pteridium aquilinum* dominated both recently disturbed and mature dry sites but only recently disturbed mesic sites (Roberts and Gilliam, 1995). The tree *Acer saccharum* (sugar maple) and an herb, *Maianthemum canadense* (Canada mayflower), apparently outcompeted *Pteridium aquilinum* on mature mesic sites. In other clear-cuts in North Carolina, the horsetail *Equisetum arvense* (field horse-tail, Plate 4C) was abundant for the first years, but the shade-tolerant fern *Polystichum acrostichoides* (Christmas fern) increased in relative abundance for 28 years and *Dennstaedtia punctilobula* only appeared in the later stages (Elliott *et al.*, 1998). Finally, only ferns tolerant of fire or postfire habitats are most abundant (e.g., *Pteridium aquilinum* or *Dicranopteris linearis*) in tropical slash and burn clearing in the tropics. Kessler (2001) noted that species richness of ferns (as well as other species groups) declined from mature forest to disturbed forest, secondary forest, scrub and pasture. In general, there are fewer fern species as severity of the disturbance increases. Habitat loss driven by human activities is the main cause of decreases in fern diversity (see Chapter 9).

6.3 Succession

Succession is the process of change in species composition in a given habitat following a disturbance and usually encompasses temporal scales several times the life span of the organisms involved (Walker and del Moral, 2003). Change is ubiquitous, so studies of succession provide a widely useful approach to examine temporal dynamics. Some ferns are good colonists of new disturbances while other ferns are more shade tolerant and characterize later stages of succession. Therefore, ferns may be represented in several successional stages, although they are usually less conspicuous than other vascular plants. Nevertheless, ferns are often important components of both the progressive buildup of plant biomass and soil fertility and the retrogressive phase of succession characterized by a loss of biomass and fertility (Walker *et al.*, 2001; Wardle *et al.*, 2004).

In this section, we first illustrate how ferns can be important throughout many stages of plant succession. Then we examine how ferns disperse to and establish, grow and survive in disturbed habitats. Next, we explore how ferns impact succes-sional transitions through their influences on the critical resources of light, nutrients and water. We conclude with a discussion of fern impacts on other plants in successional communities.

6.3.1 Fern succession

In some successional sequences, different species of ferns replace each other over time. For example, 16 fern species dominated a forest succession in Papua New Guinea (Van Valkenburg and Ketner, 1994) following disturbances related to mining activities (road and powerline construction, logging, firewood collection and subsistence agriculture). Several of these species were found only in 3–5 year

Table 6.4 *Ferns characteristic of five stages of forest succession following disturbance from activities related to gold mining in Papua New Guinea*

Species	Years 3–5	5–10	10–12	15–20	>100
Histiopteris incisa	X				
Nephrolepis spp.	X				
Pteridium aquilinum	X				
Pteris spp.	X				
Microlepia spp.	X	X			
Dryoathyrium spp.		X			
Thelypteris spp.		X			
Gleichenia spp.			X		
Cyathea spp.			X	X	
Dryopteris spp.			X		X
Coryphopteris spp.			X	X	X
Hymenophyllum spp.			X	X	X
Calymnodon spp.				X	
Asplenium spp.					X
Crypsinus spp.					X
Plagiogyria spp.					X

Source: Adapted from Van Valkenburg and Ketner (1994).

old disturbances, others after 5–20 years and others only in forest undisturbed for >100 years (Table 6.4).

Other good examples of fern replacement are illustrated on landslides in Bolivia and Hawaii. Some species of Lycopodiaceae (club-mosses) and Gleicheniaceae (scrambling ferns) colonized Bolivian landslides in early succession and were followed by the tree ferns *Alsophila incana* and *Dicksonia sellowiana* in mature forests (Kessler, 1999). However, as the forest canopy opened up and senescent trees fell, scrambling ferns (especially *Hypolepis nigrescens* and *H. parallelo-gramma* in the Dennstaedtiaceae, but also *Diplopterygium bancroftii* in the Gleicheniaceae, Plate 8A) colonized the light gaps and formed thickets that reduced tree regeneration. Coincidentally, species richness of epiphytic ferns and bromeliads increased over time as richness of terrestrial species declined (Kessler, 1999). On Hawaiian landslides, some species are age specific while others are not. On landslides less than 17 years old, *Odontosoria chinensis* was common while *Nephrolepis multiflora* was more abundant on landslides 18–42 years old (Restrepo and Vitousek, 2001). However, *Dicranopteris linearis* and the tree ferns *Sadleria pallida* and *Cibotium glaucum* were found on landslides of all ages, even on sites >130 years old. Only the tree ferns were also common in undisturbed forest (Restrepo and Vitousek, 2001).

Table 6.5 *Cover (%) of ferns and lycopods on a chronosequence of aa (5, 47 and 137 years old) and pahoehoe lavas (108 and 134 years old) in Hawaii*

	Aa lava			Pahoehoe lava	
	5 y	47 y	137 y	108 y	134 y
Lycopodiella cernua	0	16.0	1.3	7.3	7.3
Dicranopteris linearis	0	7.3	21.3	18.0	14.7
Nephrolepis exaltata	0	2.0	+	0	0
Polypodium pellucidum	0	+	+	+	+
Cibotium glaucum	0	1.3	1.3	+	+
Sadleria cyatheoides	0	+	10.7	+	+

+ signifies <1% cover.
Source: Clarkson (1997).

   Volcanic surfaces also provide good examples of fern replacement through succession (Whittaker *et al.*, 1989). Shifts in fern composition characterized succession over a 137-year chronosequence on Hawaiian lava flows (Table 6.5; Clarkson, 1997). When longer Hawaiian volcanic chronosequences were considered, sequential dominance by various fern species was even more evident. The scrambling fern *Dicranopteris linearis* and an erect fern, *Nephrolepis exaltata* (wild Boston fern), were early dominants (along with the club-moss *Lycopodiella cernua*) on open lava flows in Hawaii, but were replaced by *Cibotium glaucum* tree ferns after about 300 years (Figs. 6.6, 6.7; Kitayama *et al.*, 1995). The endemic *Sadleria pallida*, a genus native to Hawaii, was common in the transition between *Dicranopteris linearis* and *Cibotium glaucum* and was also found on older flows while *Athyrium sandwichianum* became important only after several thousand years of succession. Each fern was clearly adapted to a different part of this long-term sequence of habitats.

6.3.2 Dispersal

Ferns readily colonize recent disturbances, even remote ones, because of the ubiquity of their spores and the ability of spores to drift through the air (Perrie and Brownsey, 2007). Carlquist (1980) estimated that 135 of the current 168 species of ferns on the very remote Hawaiian Islands arrived via long distance dispersal, without human assistance. At increasingly smaller scales, the fern *Marsilea quadrifolia* colonized riverbanks in the Netherlands 190 km from the nearest source (Bremer, 2003), perhaps distributed by waterfowl (Johnson, 1986), the mangrove fern *Acrostichum aureum* colonized a burned forest in French Guiana from a parent about 20 km away (de Foresta, 1984) and spores of several ferns from adjacent

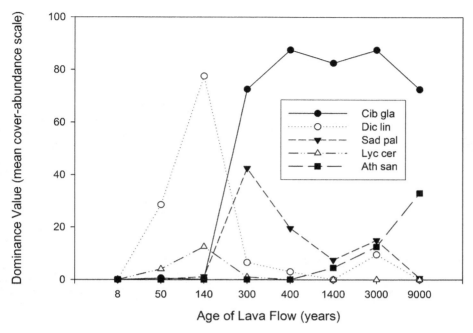

Fig. 6.6 Fern replacements along a 9000 year chronosequence on aa lava in Hawaii (from Kitayama *et al.*, 1995). The scrambling fern *Dicranopteris linearis* is the dominant species during the first two centuries, along with the club-moss *Lycopodiella cernua*. The tree fern *Cibotium glaucum* then forms dense thickets that persist for centuries, first as a canopy and then as an understory beneath the dominant tree, *Metrosideros polymorpha*. Endemic *Sadleria pallida* ferns form an important component of the vegetation just as *Cibotium glaucum* dominance begins, while other ferns such as *Athyrium sandwichianum* become important only in very old forests. Data are mean midpoints of cover-abundance scales. Fern species are abbreviated by first letters of the genus and species.

mountain vegetation colonized a glacial moraine in Alaska from a distance of 2.5 km (Matthews, 1992).

Gap colonization by ferns is also aided by the production of many spores. In temperate forests, a particularly fertile understory fern with large leaves (about 1 m long) such as *Dryopteris goldiana* (Goldie's wood fern) can produce an estimated 325 million spores per plant per year, while a species with small leaves (<40 cm long) such as *Polypodium virginianum* (rock polypody) produces an estimated 350 000 spores per plant per year (Peck *et al.*, 1990). In contrast, one tree fern leaf of *Cyathea arborea* of about 3–4 m length produced about 3–5 billion spores, which were dispersed throughout the year, enabling colonization of gaps following any disturbance (Conant, 1976). Similarly, one large understory tree fern (*Alsophila bryophila*) produced about 320 million spores (Conant, 1976). *Cyathea*

Fig. 6.7 Photos of Hawaii fern succession. (a) *Nephrolepis exaltata* on six-month-old pahoehoe lava. (b) *Nephrolepis exaltata* on six-month-old aa lava. (c) *Sadleria cyatheoides* on <50-year-old lava. Following Page: (d) *Cibotium glaucum* on 200-year-old ash (Thurston). (e) *Cibotium glaucum* on *c.* 2000-year-old ash (Ola 'a). (f) *Sadleria pallida* on 4.1-million-year-old surface (Alakai Swamp). Photos (a–e) from the Island of Hawaii; (f) from Kauai.

arborea sporophytes can produce spores when they reach about 1 m in height or 2 years of age (Conant, 1976) and spore production of *Cyathea caracasana* is generally higher in sunny than in shady habitats (Arens and Sánchez Baracaldo, 2000; Arens, 2001). Spore production can reach 8% of annual net production in *Leptopteris*

Fig. 6.7 (cont.)

(Ash, 1986). Although most spores (often still retained in sporangia) of an 8 m tall *Cyathea arborea* in Puerto Rico fell within a perimeter of 7.5 m, some spores were recorded at 30 m distance from the trunk base (Conant, 1978). In fact, the leptokurtic dispersal curve showed no diminution of spores from 7.5 to 30 m distance. Conant (1978) estimated that a single tree fern produces a spore density of 0.4 to 5.6 spores $m^{-3}\,day^{-1}$. Coupled with gametophytic selfing, *C. arborea* is well adapted to colonize forest gaps and landslides in the Puerto Rican rain forest (Conant, 1976, 1978).

Vegetative expansion is another common mode of local dispersal for ferns via rhizomes that are usually at or just below the soil surface (Russell *et al.*, 1999). A strong degree of physical and physiological integration between modules of a clonal plant (e.g., *Alsophila firma*, Mehltreter and García-Franco, 2008) allows for survival, accelerated growth and recovery following a disturbance (e.g., *Osmunda claytoniana*, interrupted fern, Hutchings and Bradbury, 1986; George and Bazzaz, 1999a). However, one study of the club-moss *Lycopodium digitatum* (syn. *Diphasiastrum digitatum*), found poor integration among clonal modules (Railing and McCarthy, 2000). Apogamy is found among about 10% of all ferns (Walker, 1985) and budding among about 5% of ferns (Moran, 2004) both of which can lead to vegetative expansion on a limited scale. When all vegetation was removed from experimental 1 m^2 plots in a forest in New York, the fern *Dryopteris intermedia* and club-moss *Huperzia lucidula* recolonized the gaps via vegetative expansion from adjacent, undisturbed plants (Tessier, 2007). Although *Dryopteris intermedia* produces abundant spores, no sporophytes developed in the artificial gaps from spores.

Sometimes vegetative expansion provides a successful way to colonize unstable or otherwise inhospitable environments. For example, *Asplenium sessilifolium* reproduces vegetatively by leaf buds on rocky stream banks in Mexico (Mehltreter, 2008). Often the parent plant remains outside the disturbed area. Recurrent erosion on Puerto Rican landslides has led to populations of *Gleichenella pectinata* that often survive undisturbed at the edges of new landslides. These ferns frequently invade newly exposed landslide surfaces at rates of 30–70 cm year^{-1} (L. R. Walker, unpublished data). Rhizomes torn from upslope can also reestablish to spread the ferns to new, downslope positions on landslides. However, vegetative expansion does not always suffice. Colonization of Hawaiian lava flows by *Dicranopteris linearis* rhizomes was limited by lack of soil and water and high surface temperatures (Russell *et al.*, 1998).

Dispersal is a big filter for the establishment of early successional communities (Walker and del Moral, 2003). With both widely dispersed spores and frequent vegetative expansion, ferns clearly have an advantage over many other species. However, successful fern establishment and growth can be more problematic.

6.3.3 Establishment, growth and longevity

Once a spore arrives at a newly exposed site which is or becomes moist and shady and is relatively free of competition, spore germination and gametophyte growth can begin (see Chapter 3; Conant, 1976). On very exposed surfaces the needed protection may come from shaded microsites or the gelatinous cover of cyanobacteria (Holttum, 1938). Gametophyte tolerance for desiccation varies greatly; the tropical epiphyte *Microgramma reptans* is very tolerant while the tropical terrestrial fern *Diplazium subsylvaticum* is not (Watkins *et al.*, 2007; Farrar *et al.*, 2008). Successful sporophyte establishment can also be dependent on microsite humidity. Sporophytes of understory ferns were more likely to colonize pits created by uprooted trees in a humid area than a drier area of a Norwegian boreal forest (Rydgren *et al.*, 1998), despite a rich spore bank at both sites. However, sporophyte growth of most disturbance-adapted ferns requires bright sunlight.

Pteridium aquilinum, for example, dominates well-lit disturbances and does poorly in a shaded understory (Watt, 1976). The paradox of shade and moisture requirements of gametophytes combined with high light requirements of sporophytes limits sexual establishment of ferns to areas with at least ephemeral shade and moisture. We focus on establishment, growth and longevity of two growth forms of terrestrial ferns: creeping ferns and tree ferns.

Creeping ferns with indeterminate leaf growth (including scrambling and climbing ferns; see Table 6.3) are very responsive to environmental conditions. When growing in shady understory conditions, climbing ferns can act like vines and grow 30 m up tall tree trunks or lianas to the canopy before spreading out (MacCaughey, 1918; Holttum, 1957). Single leaves, supported by other vegetation, can reach 20 m in length (Tryon and Tryon, 1982; Mehltreter, 2006). In high light conditions such as a lava flow, forest gap, abandoned road or landslide, scrambling ferns often grow in thickets that are 3–5 m tall (Fig. 6.8; Russell *et al.*, 1998).

Tree ferns also can grow quite densely (several hundred stems ha^{-1}; Coomes *et al.*, 2005) but have the added competitive advantage of height. Tree fern trunks can elongate between 1 and 90 cm year^{-1} although more typical growth rates are 5–50 cm year^{-1} (Ash, 1987; Bittner and Breckle, 1995; Arens and Sánchez Baracaldo, 2000; Gaxiola *et al.*, 2008) and growth rates can be relatively constant (Conant, 1976) or variable (Tanner, 1983; Arens, 2001; Mehltreter and García-Franco, 2008), depending on the species (N. Bystriakova, personal communication). Growth in tree ferns, like spore production, can be higher in more open habitats and reduced at high elevations (Ortega, 1984; Arens, 2001), but again, species differ considerably. Under some conditions, the upright part of a tree fern trunk is only a part of a longer, horizontal stem that has fallen over but continues to grow (Seiler, 1981; Walker and Aplet, 1994; Schmitt and Windisch, 2006).

Leaf longevity among all tree ferns varies from <3 months (for *Cyathea arborea* in Puerto Rico; L. R. Walker, personal observation) to 39 months (for *Cibotium glaucum* in Hawaii), giving an overall mean leaf longevity of about 18 months, production of 3–15 leaves per year and a continuous maintenance of about 10 living leaves per tree fern (Walker and Aplet, 1994). The estimated life of many tree ferns exceeds 100 years (Maxon, 1912; Tanner, 1983; Ash, 1987; Walker and Aplet, 1994; Large and Braggins, 2004), so tree ferns can have a long-term impact on litter deposition, nutrient cycling and canopy dynamics.

Ferns have adapted to disturbances in many ways. Creeping ferns and tree ferns share an ability to establish on disturbed sites quickly, and then persist, often for decades, due to rapid growth and the formation of dense stands or thickets that monopolize key resources such as light. These growth patterns have major successional implications.

Fig. 6.8 A 2-m-tall man (Aaron Shiels) in a *Gleichenella pectinata* thicket on a five-year-old Puerto Rican landslide.

6.3.4 Fern impacts on ecosystem dynamics

Light

The variety of fern life forms affects light conditions in various ways. Small, short or widely scattered ferns have minimal impact. *Pteridium aquilinum*, for example, reduced light less than other dominant understory species in jack pine (*Pinus banksiana*) plantations in Ontario (see Chapter 8; Shropshire *et al.*, 2001). Ferns that form dense thickets or grow several meters tall have much larger impacts. The clonal ferns *Dennstaedtia punctilobula* and *Osmunda claytoniana* reduced light in the understory of a Massachusetts forest from 3.4% to 1.1% of full sun (George and Bazzaz, 1999a). In Sweden, *Pteridium aquilinum* can grow quite densely, causing mortality of *Pinus sylvestris* but not *Picea abies* seedlings (Dolling, 1996). The climbing fern *Lygodium microphyllum* shades Florida forests (Plate 8D; see Chapter 8; Pemberton *et al.*, 2002). Scrambling ferns have substantial ecosystem impacts when they establish dense thickets on recent lava flows or landslides (Fig. 6.9). Reductions of incident radiation of 12- to 100-fold are typical under *Gleichenella pectinata* and *Sticherus bifidus* (syn. *Gleichenia bifida*) thickets in Puerto Rico, resulting in very dark conditions at the soil surface (Walker, 1994; Shiels and Walker, 2003). Dead rachises and leaflets often remain on the live portion of the leaf, adding to the shade effect. Dense stands of forest understory ferns such as *Blechnum discolor* can also reduce light levels (61% reduction beyond shade cast by canopy trees in New Zealand; Coomes *et al.*, 2005). Scrambling ferns and other understory ferns also can have a more open habit, as in New Zealand where *Gleichenia dicarpa*, *Hymenophyllum multifidum*, *Grammitis billardieri* and *Schizaea fistulosa* growing on a nutrient-poor terrace only intercepted 20% of incident radiation (Coomes *et al.*, 2005).

Fig. 6.9 (a) *Gleichenella pectinata* thicket on a landslide in Puerto Rico, with *Lycopodiella cernua* and *Miconia racemosa* saplings coming into a 7-year-old fern-exclusion patch in the foreground. (b) Close-up of rhizomes of *Sticherus bifidus* along a road cut in Puerto Rico.

Tree ferns can also greatly impact their light environments because of their height and long leaves. Frequent colonists of open areas such as the tree ferns *Dicksonia squarrosa* and *Cyathea smithii* can grow densely, reducing incident radiation by 44–51% in southern New Zealand (Coomes *et al.*, 2005). The shading effects of ferns can determine the outcome of competitive interactions among tree species, as in New Zealand where fern suppression of conifer seedlings allowed angiosperms to dominate (Coomes *et al.*, 2005). Tree ferns also inhibit other species in Hawaii (e.g., *Metrosideros polymorpha* trees, Burton and Mueller-Dombois, 1984). In many mesic forests, tree ferns are gap colonists that form a lower canopy under larger trees (Seiler, 1981), as long as the overstory does not become so dense that shade inhibits growth (Vitousek, 2004; N. Bystriakova, personal communication).

Nutrients

Although most ferns respond to changes in nutrient levels during succession (den Ouden and Alaback, 1996), here we focus on how ferns can alter nutrient availability in their environs through nutrient acquisition, retention and release via decomposition. The horsetails *Equisetum fluviatile*, *E. pratense* and *E. palustre* serve as nutrient pumps in Alaskan boreal wetlands, bringing P and other minerals to the surface via their deep roots and rhizomes (Marsh *et al.*, 2000). High leaf concentrations of N, P and K and low concentrations of Ca in both creeping fern and tree fern leaves relative to many angiosperm leaves suggest interesting patterns of nutrient acquisition with consequences for nutrient availability to other species and species replacements (see Chapter 4; Walker and Aplet, 1994; Russell *et al.*, 1998).

Patterns of nutrient acquisition by *Cibotium glaucum* tree ferns in Hawaii impact long-term processes. These ferns form a dominant mid-canopy in 200- to >2000-year-old forests on volcanic ash (Fig. 6.10), and can constitute >50% of the forest canopy cover for tens of thousands of years, before finally declining in numbers (Fig. 6.6; Crews *et al.*, 1995). *Cibotium glaucum* constituted only 25% of the aboveground biomass in forests, which were dominated by *Metrosideros polymorpha* trees, but contained 70% of the N and 48% of the P (Mueller-Dombois *et al.*, 1983; Aplet and Vitousek, 1994; Vitousek *et al.*, 1995).

Slow decomposition and accumulation of organic matter can conserve nutrients but also negatively impact other species. Delayed decomposition of *Dicranopteris linearis* litter immobilized N and P, perhaps extending the life span of fern thickets by reducing tree colonization after burning of tropical forests in Sri Lanka (Maheswaran and Gunatilleke, 1988). Decomposition of fern rhizomes can also be very slow. Scrambling ferns on Puerto Rican landslides develop rhizome mats that can be >30 cm deep (Slocum *et al.*, 2004). In such cases, the mineral soil surface is effectively not available for seed germination

Fig. 6.10 A cross section of a Hawaiian rainforest with a subcanopy of *Cibotium glaucum* under an overstory of *Metrosideros polymorpha*. The understory is dominated by the invasive Kahili ginger (*Hedychium gardnerianum*).

by other plants (or for soil sampling by ecologists!) and nutrient cycling is controlled by the slow decomposition of the rhizomes and suspended dead rachises and leaves. However, such slow release can favor gradual accumulation of soil carbon (Russell *et al.*, 1998; Walker and Shiels, 2008).

Leaf decomposition of the tree fern *Cyathea arborea* was faster than of the dominant woody *Cecropia peltata* on Puerto Rican landslides (Shiels, 2006) but most studies report slower decomposition of fern leaves compared with other vascular plants (see Chapter 4; Vitousek, 2004; Cornelissen *et al.*, 2006). Ecological implications are therefore likely to be very specific to species and site. For example, on some Puerto Rican landslides, *Cyathea arborea* forms nearly monospecific thickets. It may dominate due to rapid leaf turnover (L. R. Walker, unpublished data), rapid leaf decomposition (Shiels, 2006) and rapid uptake of limiting nutrients. Similar dominance of N, P and K cycling was found for *Cibotium glaucum* in Hawaii, where tree fern contributions to nutrient cycling were disproportionately high compared with their relative mass or litterfall (Vitousek *et al.*, 1995). Ultimately, widespread fern dominance on acidic or otherwise infertile soils (Tryon and Tryon, 1982) may be best explained by nutrient dynamics.

Invasive ferns can directly and indirectly alter ecosystem nutrients (see Chapter 8). The spread of the Australian tree fern *Sphaeropteris cooperi* in Hawaii is displacing three species of native *Cibotium* tree ferns (mostly *C. glaucum* but also *C. chamissoi* and *C. menziesii*) because it grows faster, produces and retains more leaves and is more fertile (Durand and Goldstein, 2001a). Displacement of the native tree fern species may result in different feeding behaviors by invasive pigs that consume native tree fern trunks, less seasonality in spore production and more rapid nitrogen cycling (Durand and Goldstein, 2001a, 2001b).

Water

Ferns also have an impact on soil water content. For example, soil water availability was higher under fern thickets than on bare mineral soil on Puerto Rican landslides, presumably due to the shade of the thicket (Walker, 1994). Fern rhizome mats (Walker and Shiels, 2008), like rhizome mats of other plants (Cammeraat *et al.*, 2007), can provide physical protection from soil erosion. Ferns may also have a role in the accumulation of soil water. On windthrow mounds in Alaska (den Ouden and Alaback, 1996), increased soil water was linked to successional changes in ferns from *Gymnocarpium dryopteris* (oak fern) mostly on young mounds, to *Dryopteris expansa* (northern wood fern, Plate 1C) on older mounds, and to *Athyrium filix-femina* (lady fern) on the forest floor. The many impacts that ferns have on their environment, from shading, to nutrient dynamics, to erosion control suggest an important role for ferns in the temporal dynamics of vegetation following disturbances.

6.3.5 Interactions among ferns and other plants

Ferns can compete with other ferns or plants for resources and they can also facilitate establishment or growth of other plants. Such interactions often have successional implications. On lava flows on Reunion Island, for example, *Blechnum tabulare* and *Nephrolepis abrupta* ferns facilitated a diverse assemblage of epiphytic mosses (Ah-Peng *et al.*, 2007) by providing favorable habitat on their rachises. In contrast, the understory fern *Dennstaedtia punctilobula* inhibited tree seedling establishment in hardwood forests of northeastern USA (Gilliam, 2007). Sometimes these negative and positive impacts occur simultaneously, and it is the net effect that is ecologically important to describe (Walker and del Moral, 2003). Competitive interactions can become more important drivers of species change when the environment is favorable, while positive interactions among species more often drive the change in species composition in unfavorable environments (Callaway and Walker, 1997).

Thicket-forming ferns

Ferns can be good competitors because of their ability to colonize disturbances quickly and then monopolize resources (light, nutrients, water), sometimes even persisting for decades. One way to monopolize resources is through the formation of thickets. Scrambling ferns are the most common ferns to form monospecific thickets but climbing ferns and some erect ferns can also form thickets. Fern thickets can slow successional transitions by inhibiting establishment and growth of woody species for several decades (Kochummen and Ng, 1977; Fletcher and Kirkwood, 1979; Guariguata, 1990; Humphrey and Swaine, 1997). Horsley (1993) described inhibition of *Prunus serotina* (black cherry) seedlings in the northeastern USA by shading from the understory fern *Dennstaedtia punctilobula*. Lyon and Sharpe (2003) suggested that *D. punctilobula* ferns compete for nutrients with red oak (*Quercus rubra*) seedlings. George and Bazzaz (1999a, 1999b) found negative impacts of several dominant understory ferns on the seedlings of all dominant canopy trees at Harvard Forest in Massachusetts, USA. Such understory ferns can impact not only woody plant colonization but the long-term nutrient dynamics of the ecosystem (see Chapter 4; Schrumpf *et al.*, 2007).

For those wanting to conserve heathlands, woody plant colonization is undesirable, as is the invasion of *Pteridium aquilinum*, which not only outcompetes heath plants but also reduces their seed bank (Pakeman and Hay, 1996; Mitchell *et al.*, 1998). *Pteridium aquilinum* had negative effects on tree seed germination and seedling growth but not on growth of more established conifers in Ontario (Bell *et al.*, 2000). The longevity, size and density of fern thickets relative to similar characteristics of species of later successional stages determine how the ferns will impact succession (Walker *et al.*, 1996a).

As noted in earlier sections, ferns colonize disturbed habitats throughout the tropics such as lava flows, landslides, burns or areas of repeated agriculture (Russell *et al.*, 1998; Slocum *et al.*, 2006). The first impact that thickets of ferns have on succession is to limit the dispersal of seeds. One study on six Puerto Rican land-slides suggested that more seeds of forest species were found on bare or grass-covered landslides than on those with fern thickets (Shiels and Walker, 2003). These results could reflect less local seed production, less direct dispersal through thickets or less indirect dispersal of seeds by birds to landslides with fewer preferred seed sources (Slocum *et al.*, 2006). Once seeds fall into a fern thicket they are unlikely to reach soil and germinate as they may be caught in a layer of living fronds, a layer of dead fronds and stems or a dense rhizome mat (Russell *et al.*, 1998; Slocum *et al.*, 2004).

On five Puerto Rican landslides, seeds of tropical trees had higher germination rates but seedlings had lower growth rates under the fern thickets than on open slopes (Walker, 1994). These results were likely due not only to higher soil water under the thickets but also to more soil nitrogen and cooler temperatures under the thickets that favored germination. Growth was likely inhibited by the reduction in light levels. Typically, fern thickets inhibit other species through competition for light or nutrients (Kochummen and Ng, 1977; Mueller-Dombois *et al.*, 1983; Stewart, 1986; Walker and Aplet, 1994; Vitousek *et al.*, 1995). This inhibition of later successional species can delay succession to forest by at least several decades (Walker, 1994; Walker *et al.*, 1996a) until the fern thickets are replaced by trees of higher shade tolerance, as experimentally confirmed by fern removals in the Dominican Republic (Slocum *et al.* 2004, 2006).

The scrambling fern *Dicranopteris linearis* is a common invader of Hawaiian lava flows. Experimental removals indicated that thickets of this fern inhibited growth of the native tree *Metrosideros polymorpha* (Russell *et al.*, 1998). However, three years after removals, no other species had colonized 40% of the exclusion area, suggesting that *D. linearis* had few competitors in early succession. This fern also had a disproportionate role in nutrient cycling, accounting for a majority of the productivity and nutrient uptake in the system (Russell *et al.*, 1998). The leaf litter of *D. linearis* was very slow to decompose (Russell and Vitousek, 1997), leading to long-term soil organic matter accumulation, impeded drainage and long-term inhibition of tree growth (Russell *et al.*, 1998).

Repeated disturbances such as fire often maintain thickets of ferns such as *D. linearis* in Sri Lanka (Maheswaran and Gunatilleke, 1988), *Gleichenella pecti-nata* and *Sticherus bifidus* in Puerto Rico (Walker and Boneta, 1995), or *Pteridium aquilinum* in Costa Rica (Gliessman, 1978). These ferns grow back quickly from buried rhizomes or through reinvasion from unburned pockets if the often shallow rhizomes are destroyed (Holttum, 1938). These ferns then outcompete grasses and herbs and inhibit establishment or growth of taller woody species (Slocum *et al.*, 2004, 2006). In Puerto Rico, *G. pectinata* and *S. bifidus* dominated the vegetation on one landslide before an intentional fire was lit by road crews. The fire promoted erosion and dominance by grasses, herbs and other ferns (including the less densely growing *Odontosoria aculeata*) during the first two years following the fire but these species all rapidly succumbed when scrambling ferns reinvaded vegetatively from unburned edges to form thickets in the third year (Walker and Boneta, 1995). The fire did not increase soil nutrients (except Mn and Ca), so the thicket-forming ferns maintained their competitive superiority.

Another potential mechanism by which ferns can inhibit growth of other plants that does not require the formation of dense thickets is the production of allelopathic exudates (see Chapter 8) as proposed for *P. aquilinum* (Gliessman and Muller, 1978; Fletcher and Kirkwood, 1979; Dolling *et al.*, 1994) or *D. linearis* (Aragon, 1975). Timing of release of toxins for *P. aquilinum* appeared to be seasonal in California and continuous in Costa Rica, presumably to match periods of establishment and growth of potentially competing plants (Gliessman, 1976). However, Walker (1994) found no evidence of allelopathy from *G. pectinata* exudates on *Tabebuia hetero- phylla* seedlings in Puerto Rico, and no iron deficiency or manganese toxicity for the seedlings as proposed by Aragon (1975).

Ferns that form thickets can facilitate as well as inhibit changes in species compo- sition by increasing soil stability and soil organic matter (Gliessman and Muller, 1978; Shiels *et al.*, 2008) or by providing safe sites for germination as noted above (Walker, 1994). On Puerto Rican landslides, the dense cover of fern thickets effec- tively reduced erosion and permitted the buildup of soil organic matter, soil nitrogen and soil moisture over several decades (Shiels *et al.*, 2008; Walker and Shiels, 2008). These changes likely benefit the trees that eventually overtop and shade out the ferns (Walker, 1994). Early colonists may therefore have an initial negative influence on other species and slow the rate of succession while ultimately having a positive influence during later stages of succession (Callaway and Walker, 1997).

Tree ferns

Colonizing species of tree ferns can also have both negative and positive influences on the species around them. They can form dense and sometimes self-perpetuating populations (Seiler, 1981) that often inhibit colonization by other species (Wick and Hashimoto, 1971; Page, 1979; Burton and Mueller-Dombois, 1984; Mueller- Dombois, 1985). Tree fern thickets persisted in Hawaii for >2000 years on tephra (Vitousek *et al.*, 1995), regenerating through both spores and rooting of trunk buds (Becker, 1976). Such persistence may be due to the ability to sequester limiting nutrients, thus competing successfully with the dominant *Metrosideros polymorpha* tree (Burton and Mueller-Dombois, 1984).

Tree ferns usually outcompete creeping ferns on fertile patches of soil as demonstrated for landslides in the Dominican Republic (Slocum *et al.*, 2006), Tanzania (Lundgren, 1978), Bolivia (Kessler, 1999) and New Zealand (Stewart, 1986). Physical damage from falling leaves of tree ferns may kill as many as 15% of tree seedlings, according to an experimental study in Hawaii that used artificial seedlings (Drake and Pratt, 2001). The high mortality estimates were attributed to rapid leaf turnover and slow tree seedling growth. Species of *Cibotium* tree ferns can reach densities of 2500 individuals ha^{-1} in Hawaiian montane rain forests (Drake and Mueller-Dombois, 1993). With rapid vertical growth for 10–15 years they compete successfully with woody plants. However, they usually reach a maximum trunk height of 15–20 m (Large and Braggins, 2004), so they eventually become vulnerable to competition from taller woody plants. When overtopped by woody plants, the tree fern *Cyathea caracasana* in Colombia grew elongated petioles to forestall eventual overtopping by fast-growing canopy trees (Arens and Sánchez Baracaldo, 2000). As indicated previously, some tree ferns (e.g., *Cibotium* in Hawaii) may also contain a disproportionate concentration of nutrients (e.g., higher N and P and lower Ca ; see Chapter 4) than angiosperms, which may help explain the eventual decline of the dominant *M. polymorpha* tree on infertile volcanic soils (Mueller-Dombois *et al.*, 1983; Burton and Mueller-Dombois, 1984). Positive influences of tree ferns include the support their trunks provide to epiphytic communities (Medeiros *et al.*, 1993), as well as serving as nurse logs for both herbaceous and woody species while still erect (Derroire *et al.*, 2007; Gaxiola *et al.*, 2008) or after they topple (Buck, 1982; Clarkson, 1997; Coomes *et al.*, 2005).

6.4 Conclusions and future directions

Ferns can be critical drivers of succession so there are several reasons why their role in future studies should not be overlooked. First, such studies will elucidate principles of dispersal, competition, facilitation and community assembly. Ferns are important participants in succession (Fig. 6.11). The ferns that dominate early succession on many severely disturbed habitats do so because of widespread spore dispersal and tolerance of low nutrient conditions. As we have demonstrated, their numerous interactions with other species often have important implications for rates and trajectories of community assembly, particularly through the frequent formation of fern thickets. The diversity of their interactions is not surprising when one considers that such ferns fit all three of Grime's (1979) plant strategies (stress tolerant, ruderals adapted to disturbance, and good competitors). Experimental manipulations of ferns help explore their role in community assembly (Russell *et al.*, 1998; George and Bazzaz, 1999a, 1999b; Runk *et al.*, 2004).

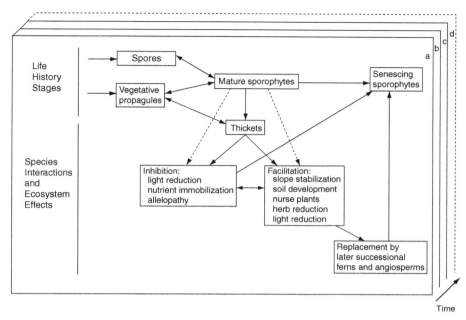

Fig. 6.11 Fern effects on succession. Dispersal of spores or vegetative propagules to a site results in development of mature sporophytes that eventually senesce. Sporophytes can either inhibit or facilitate transitions to sequential successional stages (planes b, c, d, etc.) as individuals (dotted arrows) or through a variety of changes that thickets of sporophytes create in the local environment.

Second, the important role of ferns in ecosystem dynamics is clarified when they are studied in an ecological context, including how ferns impact nutrient cycling and soil formation over time. Where ferns form dense, slowly decomposing thickets that contain a large proportion of the nutrients in an ecosystem, they have long-term impacts on succession and soil development (Maheswaran and Gunatilleke, 1988; Russell *et al*., 1998). Even when leaf turnover and decomposition is faster, ferns often have a disproportionately large influence on nutrient dynamics (Tryon and Tryon 1982; Vitousek *et al*. 1995; Marsh *et al*. 2000). Further research can address the implications of this important role.

Third, in a successional context, one can examine temporal responses of ferns across gradients of disturbance frequency, severity and nutrient status. Although such comparative studies are still uncommon (Wardle *et al*., 2004), there is much potential for exploring questions that contrast the role of ferns in successional sequences with the influences of various environmental drivers. For example, can one generalize about the roles of ferns across different environmental gradients or are those roles limited to species-specific examples?

Finally, the variable importance of ferns across physical gradients suggests questions about the spatially explicit role that ferns have on landscape processes. What are the consequences of fern-induced reductions of slope erosion on stream sediment loads, regional carbon pools or landslide reoccurrences (Walker and Shiels, 2008)? Do fern communities, like other plant communities, differ from ridge top to swale (Silver *et al.*, 1994)? If so, what are the consequences for landscape-level processes (Coomes *et al.*, 2005; Heartsill Scalley, 2005)?

By examining the ecological roles of ferns in successional systems, one provides the necessary temporal perspective for studying the influences of species interactions, nutrient dynamics, and spatial gradients. One additional research area of promise is to examine the role of ferns in ecosystem restoration. Direct use can be made of ferns in phytoremediation because some ferns have the ability to hyperaccumulate harmful elements such as arsenic (Plate 8B; see Chapter 10; Nishizono *et al.*, 1987; Tu and Ma, 2005; Mehltreter, 2008). Indirectly, ferns can improve soil stability, sequester nutrients and facilitate establishment of desired forest species. Restoration sometimes focuses on the removal of such ferns as *Pteridium aquilinum* that both invades heathlands and inhibits tree establishment (Marrs *et al.*, 2000). Cautionary lessons for restoration come from damage by overfertilization of ferns (see Chapter 4; Railing and McCarthy, 2000; Halleck *et al.*, 2004), although some ferns, particularly aquatic ones, respond well to fertilization (Taheruzzaman and Kushari, 1995). Ferns fill many important ecological roles in disturbed ecosystems and should be actively considered as components of restoration planning.

Acknowledgements

Reviews by David Coomes, Michael Kessler and Klaus Mehltreter greatly improved the manuscript. Both authors acknowledge support from the Luquillo Long-Term Ecological Research Program in Puerto Rico funded by the U.S. National Science Foundation.

References

Ah-Peng, C., Chuah-Petiot, M., Descamps-Julien, B., *et al.* (2007). Bryophyte diversity and distribution along an altitudinal gradient on a lava flow in La Réunion. *Diversity and Distributions*, **13**, 654–62.

Alonso-Amelot, M. E. and Rodulfo-Baechler, S. (1996). Comparative spatial distribution, size, biomass and growth rate of two varieties of bracken fern (*Pteridium aquilinum* (L.) Kuhn) in a neotropical montane habitat. *Vegetatio*, **125**, 137–47.

Aplet, G. H. and Vitousek, P. M. (1994). An age–altitude matrix analysis of Hawaiian rain-forest succession. *Journal of Ecology*, **82**, 137–47.

Aragon, E. L. (1975). Inhibitory effects of substances from residues and extracts of staghorn fern (*Dicranopteris linearis*). M.Sc. thesis, University of Hawai'i, Manoa, HI, USA.

Arens, N. C. (2001). Variation in performance of the tree fern *Cyathea caracasana* (Cyatheaceae) across a successional mosaic in an Andean cloud forest. *American Journal of Botany*, **88**, 545–51.

Arens, N. C. and Sánchez Baracaldo, P. (1998). Distribution of tree ferns (Cyatheaceae) across the successional mosaic in an Andean cloud forest, Nariño, Colombia. *American Fern Journal*, **88**, 60–71.

Arens, N. C. and Sánchez Baracaldo, P. (2000). Variation in tree fern stipe length with canopy height: tracking preferred habitat through morphological change. *American Fern Journal*, **90**, 1–15.

Arnesen, T. (1999). Succession on bonfire sites following burning of management waste in Solendet Nature Reserve, central Norway. *Gunneria*, **76**, 1–64.

Ash, J. (1986). Demography and production of *Leptopteris wilkesiana* (Osmundaceae), a tropical tree-fern from Fiji. *Australian Journal of Botany*, **34**, 207–15.

Ash, J. (1987). Demography of *Cyathea hornei* (Cyatheaceae), a tropical tree-fern from Fiji. *Australian Journal of Botany*, **35**, 331–42.

Atkinson, I. A. E. (1970). Successional trends in the coastal and lowland forest of Mauna Loa and Kilauea volcanoes. *Pacific Science*, **24**, 387–400.

Atkinson, I. A. E. (2004). Successional processes induced by fires on the northern offshore islands of New Zealand. *New Zealand Journal of Ecology*, **28**, 181–93.

Barson, M. M. (1997). Dune zonation on Fraser Island, Queensland. In *Dry Coastal Ecosystems: General Aspects. Ecosystems of the World: 2C*, ed. E. van der Maarel. Amsterdam, The Netherlands: Elsevier, pp. 497–504.

Becker, R. E. (1976). The phytosociological position of tree ferns (*Cibotium* spp.) in the montane rainforests on the Island of Hawaii. Ph.D. dissertation, University of Hawai'i, Manoa, HI, USA.

Bell, F. W., Ter-Mikaelian, M. T. and Wagner, R. G. (2000). Relative competitiveness of nine early-successional boreal forest species associated with planted jack pine and black spruce seedlings. *Canadian Journal of Forest Research*, **30**, 790–800.

Benl, G. (1976). Studying ferns in the Cameroons. I. The lava ferns and their occurrence on Cameroon Mountain. *Fern Gazette*, **11**, 207–15.

Bittner, J. and Breckle, S.-W. (1995). The growth rate and age of tree fern trunks in relation to habitats. *American Fern Journal*, **85**, 37–42.

Bremer, P. (2003). Some aspects of the fern flora (Filicopsida) of the Netherlands. In *Pteridology in the New Millenium*, ed. S. Chandra and M. Srivastava. Dordrecht, The Netherlands: Kluwer Academic Publishers, pp. 327–40.

Brunton, D. F. and Lafontaine, J. D. (1974). An unusual escarpment flora in western Quebec. *Canadian Field Naturalist*, **88**, 337–44.

Buck, M. G. (1982). Hawaiian treefern harvesting affects forest regeneration and plant succession. Research note PWS-355. Pacific Southwest Forest and Range Experiment Station, USDA Forest Service, Berkeley, CA, USA.

Burton, P. J. and Mueller-Dombois, D. (1984). Response of *Metrosideros polymorpha* seedlings to experimental canopy opening. *Ecology*, **65**, 779–91.

Callaway, R. M. and Walker, L. R. (1997). Competition and facilitation: a synthetic approach to interactions in plant communities. *Ecology*, **78**, 1958–65.

Cammeraat, E., van Beek, R. and Kooijman, A. (2007). Vegetation succession and its consequences for slope stability in southeastern Spain. In *Eco- and Ground Bio-Engineering: The Use of Vegetation to Improve Slope Stability. Proceedings of the*

First International Conference on Eco-Engineering, ed. A. Stokes, I. Spanos, J. E. Norris and E. Cammeraat. Berlin: Springer-Verlag, pp. 273–85.

Capon, S. J. (2005). Flood variability and spatial variation in plant community composition and structure on a large arid floodplain. *Journal of Arid Environments*, **60**, 283–302.

Carlquist, S. (1980). *Hawai'i: A Natural History*. Kauai, HI, USA: Pacific Tropical Botanical Garden.

Chinea, J. D. (1999). Changes in the herbaceous and vine communities at the Bisley experimental watersheds, Puerto Rico, following Hurricane Hugo. *Canadian Journal of Forest Research*, **29**, 1433–7.

Clarkson, B. C. (1997). Vegetation succession (1967–1989) on five recent montane lava flows, Mauna Loa, Hawaii. *New Zealand Journal of Ecology*, **22**, 1–9.

Cohen, A. L., Signakumara, B. M. P. and Aston, P. M. S. (1995). Releasing rain forest succession: a case study in the *Dicranopteris linearis* fernlands of Sri Lanka. *Restoration Ecology*, **3**, 261–70.

Conant, D. S. (1976). Ecogeographic and systematic studies in American Cyatheaceae. Ph.D. dissertation, Harvard University, Boston, MA, USA.

Conant, D. S. (1978). A radioisotope technique to measure spore dispersal of the tree fern *Cyathea arborea* (L.) Sm. *Pollen et Spores*, **20**, 583–93.

Coomes, D. A., Allen, R. B., Bentley, W. A., *et al.* (2005). The hare, the tortoise and the crocodile: the ecology of angiosperm dominance, conifer persistence and fern filtering. *Journal of Ecology*, **93**, 918–35.

Cooper-Ellis, S., Foster, D. R., Carlton, G. and Lezberg, A. (1999). Forest response to catastrophic wind: results from an experimental hurricane. *Ecology*, **80**, 2683–96.

Cornelissen, J. H. C., Quested, H. M. and Logtestijn, R. S. P. (2006). Foliar pH as a new plant trait: can it explain variation in foliar chemistry and carbon cycling processes among subarctic plant species and types? *Oecologia*, **147**, 315–26.

Crews, T., Kitayama, K., Fownes, J., *et al.* (1995). Changes in soil phosphorus fractions and ecosystem dynamics across a long chronosequence in Hawaii. *Ecology*, **76**, 1407–24.

Crookes, M. (1960). On the lava fields of Rangitoto. *American Fern Journal*, **50**, 257–63.

de Foresta, H. (1984). Heterogeneity in early tropical rain-forest regeneration after cutting and burning: ARBOCEL, French Guiana. In *Tropical Rain Forest: The Leeds Symposium*, ed. A. C. Chadwick and S. L. Sutton. Leeds, UK: Central Museum, pp. 243–53.

del Moral, R. and Walker, L. R. (2007). *Environmental Disasters, Natural Recovery and Human Responses*. Cambridge, UK: Cambridge University Press.

den Ouden, J. and Alaback, P. B. (1996). Successional trends and biomass of mosses on windthrow mounds in the temperate rainforests of southeast Alaska. *Vegetatio*, **124**, 115–28.

Derroire, G., Schmitt, L., Rivière, J.-N., Sarrailh, J.-M. and Tassin, J. (2007). The essential role of tree-fern trunks in the regeneration of *Weinmannia tinctoria* in rain forest on Réunion, Mascarene Achipelago. *Journal of Tropical Ecology*, **23**, 487–92.

Désilets, P. and G. Houle. (2005). Effects of resource availability and heterogeneity on the slope of the species-area curve along a floodplain–upland gradient. *Journal of Vegetation Science*, **16**, 487–96.

Dolling, A. H. U. (1996). Interference of bracken (*Pteridium aquilinum* (L.) Kuhn) with Scots pine (*Pinus sylvestris* L.) and Norway spruce (*Picea abies* L. Karst.) seedling establishment. *Forest Ecology and Management*, **88**, 227–35.

Dolling, A., Zackrisson, O. and Nilsson, M.-C. (1994). Seasonal variation in phytotoxicity of bracken (*Pteridium aquilinum* (L.) Kuhn). *Journal of Chemical Ecology*, **20**, 3163–72.

Dos Santos, A. M. and Thomaz, S. M. (2007). Aquatic macrophyte diversity in lagoons of a tropical floodplain: the role of connectivity and water level. *Austral Ecology*, **32**, 177–90.

Drake, D. R., and Mueller-Dombois, D. (1993). Population development of rain forest trees on a chronosequence of Hawaiian lava flows. *Ecology*, **74**, 1012–19.

Drake, D. R. and Pratt, L. W. (2001). Seedling mortality in Hawaiian rain forests: the role of small-scale physical disturbance. *Biotropica*, **33**, 319–23.

Durand, L. Z. and Goldstein, G. (2001a). Growth, leaf characteristics, and spore production in native and invasive tree ferns in Hawaii. *American Fern Journal*, **91**, 25–35.

Durand, L. Z., and Goldstein, G. (2001b). Photosynthesis, photoinhibition, and nitrogen use efficiency in native and invasive tree ferns in Hawaii. *Oecologia*, **126**, 345–54.

Elliott, K. J., Boring, L. R. and Swank, W. T. (1998). Changes in vegetation structure and diversity after grass-to-forest succession in a southern Appalachian watershed. *American Midland Naturalist*, **140**, 219–32.

Farrar, D. R., Dassler, C., Watkins, J. E., Jr. and Skelton, C. (2008). Gametophyte ecology. In *The Biology and Evolution of Ferns and Lycophytes*, ed. T. A. Ranker and C. H. Haufler. Cambridge, UK: Cambridge University Press, pp. 222–56.

Fletcher, W. W. and Kirkwood, R. C. (1979). The bracken fern (*Pteridium aquilinum* (L.) Kuhn); its biology and control. In *The Experimental Biology of Ferns*, ed. A. F. Dyer. London: Academic Press, pp. 591–635.

Gaxiola, A., Burrows, L. E. and Coomes, D. A. (2008). Tree fern trunks facilitate seedling regeneration in a productive lowland temperate rain forest. *Oecologia*, **155**, 325–35.

George, L. O. and Bazzaz, F. A. (1999a). The fern understory as an ecological filter: emergence and establishment of canopy-tree seedlings. *Ecology*, **80**, 833–45.

George, L. O. and Bazzaz, F. A. (1999b). The fern understory as an ecological filter: growth and survival of canopy-tree seedlings. *Ecology*, **80**, 846–56.

Gilliam, F. S. (2007). The ecological significance of the herbaceous layer in temperate forest ecosystems. *BioScience*, **57**, 845–58.

Gliessman, S. R. (1976). Allelopathy in a broad spectrum of environments as illustrated by bracken. *Botanical Journal of the Linnean Society*, **73**, 95–104.

Gliessman, S. R. (1978). The establishment of bracken following fire in tropical habitats. *American Fern Journal*, **68**, 41–4.

Gliessman, S. R. and Muller, C. H. (1978). The allelopathic mechanisms of dominance in bracken (*Pteridium aquilinum*) in southern California. *Journal of Chemical Ecology*, **4**, 337–62.

Glenn-Lewin, D. C. and van der Maarel, E. (1992). Pattern and process of vegetation dynamics. In *Plant Succession: Theory and Prediction*, ed. D. C. Glenn-Lewin, R. K. Peet and T. T. Veblen. London: Chapman and Hall, pp. 11–59.

Grime, J. P. (1979). *Plant Strategies and Vegetation Processes*. New York: Wiley.

Guariguata, M. R. (1990). Landslide disturbance and forest regeneration in the upper Luquillo Mountains of Puerto Rico. *Journal of Ecology*, **78**, 814–32.

Halleck, L. F., Sharpe, J. M. and Zou, X. (2004). Understorey fern responses to post-hurricane fertilization and debris removal in a Puerto Rican rain forest. *Journal of Tropical Ecology*, **20**, 173–81.

Heartsill Scalley, T. (2005). Characterization of riparian zone vegetation and litter-fall production in tropical, montane rainforest headwater streams along an environmental gradient. Dissertation, Utah State University, Logan, UT, USA.

Hemp, A. (2001). Ecology of the pteridophytes on the southern slopes of Mt. Kilimanjaro. II. Habitat selection. *Plant Biology*, **3**, 493–523.

Holmes, K. L., Goebel, P. C., Hix, D. M., Dygert, C. E. and Semko-Duncan, M. E. (2005). Ground-flora composition and structure of floodplain and upland landforms of an old-growth headwater forest in north-central Ohio. *Journal of the Torrey Botanical Society*, **132**, 62–71.

Holttum, R. E. (1938). The ecology of tropical pteridophytes. In *Manual of Pteridology*, ed. F. Verdoorn. The Hague: Martinus Nijhoff, pp. 420–50.

Holttum, R. E. (1957). Morphology, growth-habit, and classification in the family Gleicheniaceae. *Phytomorphology*, **7**, 168–84.

Horn, S. (1988). Effect of burning on a montane mire in the Cordillera de Talamanca, Costa Rica. *Brenesia*, **30**, 81–92.

Horsley, S. B. (1993). Mechanisms of interference between hay-scented fern and black cherry. *Canadian Journal of Forest Research*, **23**, 2059–69.

Humphrey, J. W. and Swaine, M. D. (1997). Factors affecting the natural regeneration of *Quercus* in Scottish oakwoods. I. Competition from *Pteridium aquilinum*. *Journal of Applied Ecology*, **34**, 577–84.

Hutchings, M. J. and Bradbury, I. K. (1986). Ecological perspectives on clonal perennial herbs. *BioScience*, **36**, 178–82.

Jenkins, M. A. and Parker, G. R. (2000). The response of herbaceous-layer vegetation to anthropogenic disturbance in intermittent stream bottomland forests of southern Indiana, USA. *Plant Ecology*, **151**, 223–37.

Johnson, D. M. (1986). Systematics of the New World species of *Marsilea* (Marsileaceae). *Systematic Botany Monographs*, **11**, 1–87.

Kessler, M. (1999). Plant species richness and endemism during natural landslide succession in a perhumid montane forest in the Bolivian Andes. *Ecotropica*, **4**, 123–36.

Kessler, M. (2001). Maximum plant-community endemism at intermediate intensities of anthropogenic disturbance in Bolivian montane forests. *Conservation Biology*, **15**, 634–41.

Kitayama, K., Mueller-Dombois, D. and Vitousek, P. M. (1995). Primary succession of Hawaiian montane rain forest on a chronosequence of eight lava flows. *Journal of Vegetation Science*, **6**, 211–22.

Kochummen, K. M. and Ng, F. S. P. (1977). Natural plant succession after farming in Kepon. *Malaysian Forester*, **40**, 61–78.

Kornaś, J. (1978). Fire-resistance in the pteridophytes of Zambia. *Fern Gazette*, **11**, 373–84.

Large, M. F. and Braggins, J. E. (2004). *Tree Ferns*. Portland, OR, USA: Timber Press.

Larsen, D. W., Matthes, U. and Kelly, P. E. (2000). *Cliff Ecology: Pattern and Process in Cliff Ecosystems*. Cambridge, UK: Cambridge University Press.

Lenihan, J. M. (1990). Forest assessment of Little Lost Man Creek, Humboldt County, California: Reference level in the hierarchical structure of old-growth coastal redwood vegetation. *Madroño*, **37**, 69–87.

Lundgren, L. (1978). Studies of soil and vegetation on fresh landslide scars in the Mgeta Valley, Western Ulugura Mountains, Tanzania. *Geografiske Annaler*, **60**, 91–127.

Lyon, J. and Sharpe, W. E. (1996). Hay-scented fern (*Dennstaedtia punctilobula* (Michx.) Moore) interference with growth of northern red oak (*Quercus rubra* L.) seedlings. *Tree Physiology*, **16**, 923–32.

Lyon, J. and Sharpe, W. E. (2003). Impacts of hay-scented fern on nutrition of northern red oak seedlings. *Journal of Plant Nutrition*, **26**, 487–502.

MacCaughey, V. (1918). The genus *Gleichenia* (*Dicranopteris*) in the Hawaiian Islands. *Torreya*, **18**, 41–52.

Maheswaran, J. and Gunatilleke, I. A. U. N. (1988). Litter decomposition in a lowland rain forest and a deforested area in Sri Lanka. *Biotropica*, **20**, 90–9.

Malanson, G. P. (1993). *Riparian Landscapes*. Cambridge, UK: Cambridge University Press.

Marrs, R. H., Le Duc, M. G., Mitchell, R. J., *et al.* (2000). The ecology of bracken: its role in succession and implications for control. *Annals of Botany*, **85** (Suppl. B), 3–15.

Marsh, A. S., Arnone, J. A., III, Bormann, B. T. and Gordon, J. C. (2000). The role of *Equisetum* in nutrient cycling in an Alaskan shrub wetland. *Journal of Ecology*, **88**, 999–1011.

Matthews, J. A. (1992). *The Ecology of Recently Deglaciated Terrain: A Geoecological Approach to Glacier Forelands and Primary Succession*. Cambridge, UK: Cambridge University Press.

Maxon, O. (1912). Tree Ferns of North America. Annual Report, 1911. *Smithsonian Institution, Publication*, **2120**, 463–91.

Medeiros, A. C., Loope, L. L. and Anderson, S. J. (1993). Differential colonization by epiphytes on native (*Cibotium* spp.) and alien (*Cyathea cooperi*) tree ferns in a Hawaiian rain forest. *Selbyana*, **14**, 71–4.

Mehltreter, K. (1997). Farne der neotropischen Hochgebirge. III. Die Gattung *Blechnum* Linné. *Palmengarten*, **61**, 58–61.

Mehltreter, K. (2006). Leaf phenology of the climbing fern *Lygodium venustum* in a semideciduous lowland forest on the Gulf of Mexico. *American Fern Journal*, **96**, 21–30.

Mehltreter, K. (2008). Phenology and habitat specificity of tropical ferns. In *The Biology and Evolution of Ferns and Lycophytes*, ed. T. A. Ranker and C. H. Haufler. Cambridge, UK: Cambridge University Press, pp. 201–21.

Mehltreter, K. and García-Franco, J. G. (2008). Leaf phenology and trunk growth of the deciduous tree fern *Alsophila firma* (Baker) D. S. Conant in a lower montane Mexican forest. *American Fern Journal*, **98**, 1–13.

Mitchell, R. J., Marrs, R. H. and Auld, M. H. D. (1998). A comparative study of the seedbanks of heathland and successional habitats in Dorset, southern England. *Journal of Ecology*, **86**, 588–96.

Moran, R. (2004). *A Natural History of Ferns*. Portland, OR, USA: Timber Press.

Morris, K., Bailey, P. C., Boon, P. I. and Hughes, L. (2003). Alternative stable states in the aquatic vegetation of shallow urban lakes. II. Catastrophic loss of aquatic plants consequent to nutrient enrichment. *Marine and Freshwater Research*, **54**, 201–15.

Mueller-Dombois, D. (1985). Ohi'a dieback in Hawaii: 1984 synthesis and evaluation. *Pacific Science*, **39**, 150–70.

Mueller-Dombois, D. (2000). Rain forest establishment and succession in the Hawaiian Islands. *Landscape and Urban Planning*, **51**, 147–57.

Mueller-Dombois, D., Vitousek, P. M. and Bridges, K. W. (1983). Canopy dieback and dynamic processes in Pacific ecosystems. *Hawaii Botanical Report 44*. University of Hawaii, Manoa, HI, USA.

Nadkarni, N. and Wheelwright, N. T. (eds.) (2000). *Monteverde: Ecology and Conservation of a Tropical Cloud Forest*. Oxford, UK: Oxford University Press.

Nishizono, H., Suzuki, S. and Ishii, F. (1987). Accumulation of heavy metals in the metal-tolerant fern *Athyrium yokoscense*, growing on various environments. *Plant and Soil*, **102**, 65–70.

Odland, A. and del Moral, R. (2002). Thirteen years of wetland vegetation succession following a permanent drawdown, Myrkdalen Lake, Norway. *Plant Ecology*, **162**, 185–98.

Oinonen, E. (1967). Sporal regeneration of ground pine (*Lycopodium complanatum* L.) in southern Finland in the light of size and age of its clones. *Acta Forestalia Fennica*, **83**, 1–85.

Onaindia, M., Dominguez, I., Albizu, I., Garbisu, C. and Amezaga, I. (2004). Vegetation diversity and vertical structure as indicators of forest disturbance. *Forest Ecology and Management*, **195**, 341–54.

Ortega, F. (1984). Notas sobre la autecología de *Sphaeropteris senilis* (Kl.) Tryon (Cyatheaceae) en el Parque Nacional El Avila, Venezuela. *Pittieria*, **12**, 31–53.

Ough, K. and Murphy, A. (2004). Decline in tree-fern abundance after clearfell harvesting. *Forest Ecology and Management*, **199**, 153–63.

Page, C. N. (1979). Experimental aspects of fern ecology. In *The Experimental Biology of Ferns*, ed. A. F. Dyer. London: Academic Press, pp. 551–89.

Page, C. N. (2002). The role of natural disturbance regimes in pteridophyte conservation management. *Fern Gazette*, **16**, 284–9.

Pakeman, R. J. and Hay, E. (1996). Heathland seedbanks under bracken (*Pteridium aquilinum* (L.) Kuhn) and their importance for re-vegetation after bracken control. *Journal of Environmental Management*, **47**, 329–39.

Peck, J. H., Peck, C. J. and Farrar, D. R. (1990). Influences of life history attributes on formation of local and distant fern populations. *American Fern Journal*, **80**, 126–42.

Pemberton, R. W., Goolsby, J. A. and Wright, T. (2002). Old World climbing fern. In *Biological Control of Invasive Plants in the Eastern United States*, ed. R. Van Driesche, B. Blossey, M. Hoddle, S. Lyon and R. Reardon. Publication FHTET-2002–04. Morgantown, WV, USA: USDA Forest Service, pp. 139–147.

Penrod, K. A. (2000). Ecology of hay-scented fern: spore production, viability and germination. Dissertation, Pennsylvania State University, University Park, PA, USA.

Perrie, L. and Brownsey, P. (2007). Molecular evidence for long-distance dispersal in the New Zealand pteridophyte flora. *Journal of Biogeography*, **34**, 2028–38.

Peterson, C. J., Carson, W. P., McCarthy, B. C. and Pickett, S. T. A. (1990). Microsite variation and soil dynamics within newly created treefall pits and mounds. *Oikos*, **58**, 39–46.

Railing, C. A. and McCarthy, B. C. (2000). The effects of rhizome severing and nutrient addition on growth and biomass allocation in *Diphasiastrum digitatum*. *American Fern Journal*, **90**, 77–86.

Restrepo, C. and Vitousek, P. (2001). Landslides, alien species, and the diversity of a Hawaiian montane mesic ecosystem. *Biotropica*, **33**, 409–20.

Reudink, M. W., Snyder, J. P., Xu, B., Cunkelman, A. and Balsamo, R. A. (2005). A comparison of physiological and morphological properties of deciduous and wintergreen ferns in southeastern Pennsylvania. *American Fern Journal*, **95**, 45–56.

Riba, R. and Reyes Jaramillo, I. (1990). *Pityrogramma calomelanos* (L.) Link (Adiantaceae) on layers of volcanic ash in Los Tuxtlas, State of Veracruz, Mexico. *Annals of the Missouri Botanical Garden*, **77**, 287–9.

Rishbeth, J. (1948). The flora of Cambridge walls. *Journal of Ecology*, **36**, 136–48.

Roberts, M. R. and Gilliam, F. S. (1995). Disturbance effects on herbaceous layer vegetation and soil nutrients in *Populus* forests of northern lower Michigan. *Journal of Vegetation Science*, **6**, 903–12.

Roberts, M. R. and Gilliam, F. S. (2003). Response of the herbaceous layer in eastern forests. In *The Herbaceous Layer in Forests of Eastern North America*, ed. F. S. Gilliam and M. R. Roberts. Oxford, UK: Oxford University Press, pp. 302–20.

Rodwell, J. S. (1992). *British Plant Communities*. Cambridge, UK: Cambridge University Press.

Runk, K., Moora, M. and Zobel, M. (2004). Do different competitive abilities of three fern species explain their different regional abundances? *Journal of Vegetation Science*, **15**, 351–6.

Russell, A. E., Raich, J. W. and Vitousek, P. M. (1998). The ecology of the climbing fern *Dicranopteris linearis* on windward Mauna Loa, Hawaii. *Journal of Ecology*, **86**, 765–79.

Russell, A. E., Ranker, T. A., Gemmill, C. E. C. and Farrar, D. R. (1999). Patterns of clonal diversity in *Dicranopteris linearis* on Mauna Loa, Hawaii. *Biotropica*, **31**, 449–59.

Russell, A. E. and Vitousek, P. M. (1997). Decomposition and potential nitrogen fixation in *Dicranopteris linearis* litter on Mauna Loa, Hawaii. *Journal of Tropical Ecology*, **13**, 579–94.

Rydgren, K., Hestmark, G. and Økland, R. H. (1998). Revegetation following experimental disturbance in a boreal old-growth *Picea abies* forest. *Journal of Vegetation Science*, **9**, 763–76.

Schmitt, J. L. and Windisch, P. G. (2006). Growth rates and age estimates of *Alsophila setosa* Kaulf. in southern Brazil. *American Fern Journal*, **96**, 103–11.

Schrumpf, M., Axmacher, J. C., Zech, W., Lehmann, J. and Lyaruu, H. V. C. (2007). Long-term effects of rainforest disturbance on the nutrient composition of throughfall, organic layer percolate and soil solution at Mt. Kilimanjaro. *Science of the Total Environment*, **376**, 241–54.

Scott, G. A. J. (1969). Relationships between vegetation and soil avalanching in the high rainfall areas of Oahu, Hawaii. M.A. thesis, University of Hawaii, Manoa, HI, USA.

Seiler, R. L. (1981). Leaf turnover rates and natural history of the Central American tree fern *Alsophila salvinii*. *American Fern Journal*, **71**, 75–81.

Sharpe, J. M. (1997). Leaf growth and demography of the rheophytic fern *Thelypteris angustifolia* (Willdenow) Proctor in a Puerto Rican rainforest. *Plant Ecology*, **130**, 203–12.

Sharpe, J. M. (2005). Temporal variation in sporophyte fertility in *Dryopteris intermedia* and *Polystichum acrostichoides* (Dryopteridaceae: Pteridophyta). *Fern Gazette*, **17**, 223–34.

Sharpe, J. M. (2010). Responses of the mangrove fern *Acrostichum danaeifolium* Langsd. & Fisch. (Pteridaceae, Pteridophyta) to disturbances resulting from increased soil salinity and Hurricane Georges at the Jobos Bay National Estuarine Research Reserve, Puerto Rico. *Wetlands Ecology and Management*, **18**, 57–68.

Shiels, A. B. (2006). Leaf litter decomposition and substrate chemistry of early successional species on landslides in Puerto Rico. *Biotropica*, **38**: 348–53.

Shiels, A. B. and Walker, L. R. (2003). Bird perches increase forest seeds on Puerto Rican landslides. *Restoration Ecology*, **11**, 457–65.

Shiels, A. B., West, C. A., Weiss, L., Klawinski, P. D. and Walker, L. R. (2008). Soil factors predict initial plant colonization on Puerto Rican landslides. *Plant Ecology*, **195**, 168–78.

Shropshire, C., Wagner, R. G., Bell, F. W. and Swanton, C. J. (2001). Light attenuation by early successional plants of the boreal forest. *Canadian Journal of Forest Research*, **31**, 812–23.

Silver, W. L., Scatena, F. N., Johnson, A. H., Siccama, T. G. and Sánchez, M. J. (1994). Nutrient availability in a montane wet tropical forest: spatial patterns and methodological considerations. *Plant and Soil*, **164**, 129–45.

Slocum, M. G., Aide, T. M., Zimmerman, J. K. and Navarro, L. (2004). Natural regeneration of subtropical montane forest after clearing fern thickets in the Dominican Republic. *Journal of Tropical Ecology*, **20**, 483–6.

Slocum, M. G., Aide, T. M., Zimmerman, J. K. and Navarro, L. (2006). A strategy for restoration of montane forest in anthropogenic fern thickets in the Dominican Republic. *Restoration Ecology*, **14**, 526–36.

Stewart, G. H. (1986). Forest dynamics and disturbance in a beech/hardwood forest, Fiordland, New Zealand. *Vegetatio*, **68**, 115–26.

Sykes, J. M. and Horrill, A. D. (1981). Recovery of vegetation in a Caledonian pinewood after fire. *Transactions of the Botanical Society of Edinburgh*, **43**, 317–25.

Szmeja, J. (1994). Effect of disturbance and interspecific competition in isoetid populations. *Aquatic Botany*, **48**, 225–38.

Taheruzzaman, Q. and Kushari, D. P. (1995). Biomass and concentrations of macronutrients and mercury in *Azolla pinnata* R. Br. in Indian ponds enriched by anthropogenic effluents. *Netherlands Journal of Aquatic Ecology*, **29**, 157–60.

Tanner, E. V. J. (1983). Leaf demography and growth of the tree-fern *Cyathea pubescens* Mett. ex Kuhn in Jamaica. *Botanical Journal of the Linnean Society*, **87**, 213–27.

Tessier, J. T. (2007). Re-establishment of three dominant herbaceous understory species following fine-scale disturbance in a Catskill northern hardwood forest. *Journal of the Torrey Botanical Society*, **134**, 34–44.

Tryon, R. M. and Tryon, A. F. (1982). *Ferns and Allied Plants with Special Reference to Tropical America*. New York: Springer-Verlag.

Tu, C. and Ma, L. Q. (2005). Effects of arsenic on concentration and distribution of nutrients in the fronds of the arsenic hyperaccumulator *Pteris vittata* L. *Environmental Pollution*, **135**, 333–40.

Turner, I. M., Wong, Y. K., Chew, P. T. and Ibrahim, A. B. (1996). Rapid assessment of tropical rain forest successional status using aerial photographs. *Biological Conservation*, **77**, 177–83.

Vajda, V., Raine, J. I. and Hollis, C. J. (2001). Indication of global deforestation at the Cretaceous-Tertiary boundary by New Zealand fern spike. *Science*, **294**, 1700–2.

van Steenis, C. G. C. J. (1981). *Rheophytes of the World*. Alphen aan den Rijn, The Netherlands: Sijthoff and Noordhoff.

Van Valkenburg, J. L. C. H. and Ketner, P. (1994). Vegetation changes following human disturbance of mid-montane forest in the Wau area, Papua New Guinea. *Journal of Tropical Ecology*, **10**, 41–54.

Vitousek, P. M. (2004). *Nutrient cycling and limitation: Hawai'i as a Model System*. Princeton, NJ, USA: Princeton University Press.

Vitousek, P. M., Gerrish, G., Turner, D. R., Walker, L. R. and Mueller-Dombois, D. (1995). Litterfall and nutrient cycling in four Hawaiian montane rainforests. *Journal of Tropical Ecology*, **11**, 189–203.

Walker, J., Thompson, C. H., Reddell, P. and Rapport, D. J. (2001). The importance of landscape age in influencing landscape health. *Ecosystem Health*, **7**, 7–14.

Walker, L. R. (1994). Effects of fern thickets on woodland development on landslides in Puerto Rico. *Journal of Vegetation Science*, **5**, 525–32.

Walker, L. R. (ed.) (1999). *Ecosystems of Disturbed Ground*. Ecosystems of the World, 16. Amsterdam, The Netherlands: Elsevier.

Walker, L. R. (2000). Seedling and sapling dynamics of treefall pits in Puerto Rico. *Biotropica*, **32**, 262–75.

Walker, L. R. and Aplet, G. H. (1994). Growth and fertilization of Hawaiian tree ferns. *Biotropica*, **26**, 378–83.

Walker, L. R. and Boneta, W. (1995). Plant and soil responses to fire on a fern-covered landslide in Puerto Rico. *Journal of Tropical Ecology*, **11**, 473–9.

Walker, L. R. and del Moral, R. (2003). *Primary Succession and Ecosystem Rehabilitation*. Cambridge, UK: Cambridge University Press.

Walker, L. R. and Shiels, A. B. (2008). Post-disturbance erosion impacts carbon fluxes and plant succession on recent tropical landslides. *Plant and Soil*, **313**, 205–16.

Walker, L. R. and Willig, M. R. (1999). An introduction to terrestrial disturbances. In *Ecosystems of Disturbed Ground*, ed. L. R. Walker. Amsterdam, The Netherlands: Elsevier, pp. 1–5.

Walker, L. R., Zarin, D. J., Fetcher, N., Myster, R. W. and Johnson, A. H. (1996a). Ecosystem development and plant succession on landslides in the Caribbean. *Biotropica*, **28**, 566–76.

Walker, L. R., Zimmerman, J. K., Lodge, D. J. and Guzmán-Grajales, S. (1996b). An altitudinal comparison of growth and species composition in hurricane-damaged forests in Puerto Rico. *Journal of Ecology*, **84**, 877–89.

Walker, T. G. (1985). Some aspects of agamospory in ferns: the Braithwaite system. *Proceedings of the Royal Society of Edinburgh*, **86**B, 59–66.

Wang, Y.-B., Liu, D.-Y., Zhang, L., Li, Y. and Chu, L. (2004). Patterns of vegetation succession in the process of ecological restoration on the deserted land of Shizishan copper tailings in Tongling City. *Acta Botanica Sinica*, **46**, 780–7.

Wardlaw, C. W. (1931). Observations on the dominance of pteridophytes on some St. Lucia soils. *Journal of Ecology*, **19**, 60–3.

Wardle, D. A., Walker, L. R. and Bardgett, R. D. (2004). Ecosystem properties and forest decline in contrasting long-term chronosequences. *Science*, **305**, 509–13.

Wardle, P. (1980). Primary succession in Westland National Park and its vicinity, New Zealand. *New Zealand Journal of Botany*, **18**, 221–32.

Watkins., J. E., Jr., Mack, M. C., Sinclair, T. R. and Mulkey, S. S. (2007). Ecological and evolutionary consequences of desiccation tolerance in tropical fern gametophytes. *New Phytologist*, **176**, 708–17.

Watt, A. S. (1976). The ecological status of bracken. *Botanical Journal of the Linnean Society*, **73**, 217–39.

Whittaker, R. J., Bush, M. B. and Richards, K. (1989). Plant recolonization and vegetation succession on the Krakatau Islands, Indonesia. *Ecological Monographs*, **59**, 59–123.

Wick, H. L. and Hashimoto, G. T. (1971). Leaf development and stem growth of tree-fern in Hawaii. U.S. Forest Service Research Note PSW-237. Pacific Southwest Forest and Range Experimental Station, Berkeley, CA, USA.

Woods, S. S. (2002). Response of ferns to overstory disturbance: Effects of ice storm and timber harvest on four common fern species in hardwood forests of New York. Masters thesis, State University of New York, Syracuse, NY, USA.

7

Interactions of ferns with fungi and animals

KLAUS MEHLTRETER

Key points

1. Ferns and lycophytes have developed a wide spectrum of antagonistic and mutualistic relationships with fungi and animals. While some of these interactions, such as endomycorrhizae, are old and may have coexisted with their host plants for a long time, other interactions may have originated more recently, such as some herbivorous insects that have switched from seed plants to ferns.
2. More than 80% of sporophytes possess endomycorrhizae, while for a few species fern–ericoid mycorrhizae and ectendomycorrhizae have been reported. In the gametophytic stage, mycorrhizae are obligate in older fern and lycophyte lineages but are facultative or can be absent in more modern lineages.
3. Interactions with parasitic, symbiotic and neutral endophytic fungi that infect aerial parts of the ferns seem to be as common as in seed plants, while the proportion of interactions with insects seems to be 3–7 times lower than in seed plants.
4. Fern herbivores are most often members of the insect orders Coleoptera, Hemiptera and Lepidoptera and can be either generalists or insect species that have specialized on ferns. Two fern genera have strong mutualistic relationships with ants: *Microgramma* subgenus *Solanopteris* in the New World and *Lecanopteris* in the Old World tropics; a third, more facultative relationship has recently been described for *Antrophyum* in Costa Rica.
5. Most ferns have few specific biochemical defense mechanisms in comparison with seed plants, yet ferns and seed plants sustain similar levels of herbivore damage.

7.1 Introduction

Ferns and lycophytes are often regarded as old vascular plants because some of their ancestors date back to the Devonian (410–355 million years ago, Rothwell, 1996; see Appendix C for geological timescale). Consequently, it has often been assumed that they would have developed few interactions with fungi and animals or that they would interact primarily with old fungal and animal lineages

Fern Ecology, ed. Klaus Mehltreter, Lawrence R. Walker and Joanne M. Sharpe. Published by Cambridge University Press. © Cambridge University Press 2010.

(Boullard, 1957; Gerson, 1979; Auerbach and Hendrix, 1980). In contrast, one could argue exactly the opposite: ferns and lycophytes have had more time to develop interactions because of a longer coexistence with other organisms, such as mycorrhizal fungi since the Devonian, and with insects since the Carboniferous (355–290 mya). However, the oldest fern lineages had become extinct and extant ferns and lycophytes developed during two radiations, one in the Mesozoic (250–65 mya) and another during the Cretaceous (135–65 mya, Schneider *et al.*, 2004; Rothwell and Stockey, 2008). It is still debated whether (1) ferns have fewer interactions with other organisms (e.g., insects) because of the lack of such complex reproductive structures as flowers, fruits and seeds, and a less diverse biochemical array of secondary metabolites in comparison with seed plants, or (2) ferns and seed plants have developed the same number of interactions. Such counts may be biased because ferns are of less economic importance and have received less attention from field biologists than other plants (Balick *et al.*, 1978, Hendrix, 1980, Cooper-Driver, 1985a).

In this chapter, I review antagonistic, neutral and mutualistic interactions of ferns and lycophytes with fungi and animals. Competition with other plants is dealt with in more detail in Chapter 6. Whereas parasitic fungi can damage ferns, mycorrhizal fungi are beneficial and enhance water and nutrient uptake. Interactions with animals are mostly antagonistic, with a predominance of herbivorous insects caus-ing damage to fern leaves. Ferns have developed a variety of biochemical defense mechanisms to address these challenges and some species have evolved mutualistic relationships with ants. This chapter can only provide a preliminary overview because of the disparity between the limited number of studies on this subject and the almost infinite number of potential interactions of ferns with other organisms. However, some generalizations will be drawn from the available information and open questions will be addressed in order to encourage future generations to focus on this broad and fascinating field of research.

7.2 Types of interactions

The three main types of interactions that can take place between a fern and another organism are neutralism, antagonism and mutualism (Begon *et al.*, 2006). An example of a neutralism is the host–epiphyte relationship, where the epiphytic fern grows on a tree and benefits from the available space, but affects its host neither positively nor negatively. If we consider that an epiphyte increases the weight of its supporting branch and can cause its premature breakage, the interaction could be considered antagonistic. In antagonistic interactions, one organism receives some benefit at a cost of the other (e.g., parasitism and herbivory) or one is inhibiting another (e.g., through competition). In mutualistic relationships, both interacting organisms benefit from each other (e.g., the aquatic mosquito fern *Azolla* and its symbiotic nitrogen-fixing blue-green algae *Anabaena azollae*) and in its extreme form neither can survive independently (e.g., obligate mycorrhizae). Sometimes it is assumed that only antagonistic and mutualistic interactions may develop into very specific relationships but the aforementioned neutral epiphyte–host interaction can also be quite specific. For example, tree ferns that develop a root mantle with high water storage capacity (e.g., *Alsophila firma*, *Cyathea divergens*, *Dicksonia sellowiana*) are excellent hosts for epiphytes (Moran *et al.*, 2003; Mehltreter *et al.*, 2005). Although such habitat appears to be preferred by many fern epiphytes (e.g., *Elaphoglossum petiolatum*, *Trichomanes capillaceum*), some closely related species were more frequent on angiosperm trunks (e.g., *Elaphoglossum peltatum*, *Trichomanes reptans*, Mehltreter *et al.*, 2005). Fern lineages may switch from the terrestrial to the epiphytic habitat with difficulty because several ecological adaptations (e.g., anchoring roots, drought resistance) seem to be necessary for this step. In contrast, reversal from the epiphytic to the terrestrial habitat seems to be easier. For example, Wikström *et al.* (1999) found one single origin of epiphytism in the lycophyte genus *Huperzia* (fir-moss), but multiple reversals of epiphytic species to the terrestrial habitat (Plate 3C).

7.3 Interactions with fungi

Endophytic fungi are parasitic, neutral or mutualistic fungi that can live for a variable period of time within living tissues (e.g., aerial parts or roots) of their host plants. Most of these fungi become apparent at some point in their development by causing symptoms on the host. For example, fungi that parasitize aerial parts may become apparent as dead leaf spots and symbiotic ectomycorrhizal fungi that colonize roots may produce a hyphal sheath around the root. However, some detrimental or beneficial fungi show no visible signs of their presence on the host plant (Petrini, 1991). Because the ecological role of these "symptomless" fungi is unclear, they were first considered to be neutral to the host plant, but some may be parasitic or mutualistic at some stage of the life of their host plant. Because of their hidden life form, their presence and diversity within nearly every host plant is therefore surprising. For example, from three unidentified tropical fern species, Petrini (1992) isolated strains of a total of 19 different fungal genera. In a detailed study of endophytes of *Pteridium aquilinum* (bracken) in the UK, Petrini (1993) reported over 66 fungal taxa. Most individual ferns were infected with several endophytic species. The composition of the endophyte assemblage within *Pteridium aquilinum* was seasonally changing but did not differ between two study sites.

Endophytes are often tissue specific (e.g., *Cylindrocarpon destructans* in rhizomes, one species of *Stagonospora* in petioles and leaves, *Aureobasidium pullulans* in the leaf rachis and veins). They can become parasitic and affect the host at some later stage of its life (e.g., *A. pullulans* is a known pathogen of *P. aquilinum* that can be isolated from apparently healthy fern individuals). Endophytic fungi can also be neutral (e.g., saprophytic fungi often become apparent only when the host plants are dying) as well as mutualistic. Most endophytic fungi of aerial parts remain symptomless and may even form a symbiotic relationship with their host. For example, they may produce antibiotics that protect their host plants from pathogens (Carroll, 1986; Petrini, 1992) or increase the host's herbivore resistance, as has been shown for some grasses (Clay, 1989).

The study of endophytic fungi has recently received renewed interest, but ferns have received only minor attention. Further research is needed to understand the actual role of fern endophytes, the number of pathogenic or symbiotic fungi that infect ferns, their host and tissue specificity, and the chemical and ecological interactions between ferns and their symbiotic counterparts. In the following two sections I will discuss endophytic fungi that have well-known interactions with ferns: (1) saprophytic and parasitic fungi and (2) mutualistic mycorrhizae.

7.3.1 Saprophytic and parasitic fungi

Saprophytic fungi are widely distributed and feed on dead or decaying organic matter. Because they cause little or no visible damage during the lifetime of the fern, this group has received less attention than more conspicuous parasitic fungi of living ferns that cause problems in horticulture. Medel and Lorea-Hernandez (2008) reported five species of *Lachnum* (Helotiales, Ascomycota) on decaying tree fern leaves in Mexican cloud forests. *Lachnum oncospermatis* was host specific in this study (found only on leaves of *Dicksonia sellowiana*), but other species such as *L. fimbriferum* were found on all examined tree fern genera (*Alsophila, Cyathea* and *Dicksonia*). Because there are few studies of saprophytic fungi, new discoveries are common. For example, Samuels and Rogerson (1990) described from fern hosts three new species of ascomycetes that also show the transition between parasitic and saprophytic fungi. In Colombia, *Crocicreas sessilis* grows on living leaves of the tree fern *Cyathea divergens* and might be considered parasitic, whereas the saprophytic *Bioscypha pteridicola* was found on dead spots of living leaves of *Cnemidaria uleana*. The third fungal species, *Dimeriella polypodii*, is found in leaf scales of *Polypodium montigenum* and *P. madrense* in Mexico and might be considered either saprophytic or parasitic, because the scales consist of both living and dead cells. In North America and Europe, the saprophytic basidiomycete *Mycena pterigena* (Agaricales) grows on dying leaves and petioles of a wide variety of fern species but not on angiosperms (Redhead, 1984; Kramer *et al.* 1995). The study of saprophytic fungi on decaying fern leaves promises new discoveries in the future.

Parasitic fungi, in contrast to the saprophytes, infect living ferns. Although horticulturists are necessarily aware that ferns are susceptible to a range of fungal diseases, information about interactions between parasitic fungi and ferns in their natural habitat is scarce, perhaps because most of these fungi are inconspicuous (Kramer *et al.*, 1995). Furthermore, infections from parasitic fungi do not have a significant economic impact because most can be controlled with fungicides. An exception is a fungal infection in *Rumohra adiantiformis* (leatherleaf fern), the most important fern for ornamental foliage. This fern species is cultivated in Hawaii, Florida and Costa Rica, where parasitic species of *Colletotrichum* (Sordariomycetes, Ascomycota) cause severe anthracnosis. Anthracnosis is a soil-borne disease caused by several fungi on fern and seed plant hosts that is characterized by dead spots on the leaves (Leahy *et al.*, 1995). *Colletotrichum* spores are transmitted by splashing water, wind or gardening tools. These spores infect unfurling croziers. Because the mycelia kill the outermost parts of the croziers, those develop abnormally and appear scorched. The rhizomes and mature leaves do not appear to be susceptible to anthracnosis. Control with fungicide trials has not been effective, so only preventive management methods such as isolation of infected plants and sterilizing soil and tools are used (Leahy *et al.*, 1995).

Although most fungal parasites of ferns do not cause severe damage they are very common and in some cases aid in the identification of fern species. Fern taxonomists have recognized that some fern species are consistently infected by specific fungi. Because of the easy dispersal of fungal spores, they seem to colonize their host fern species throughout their geographic range. Thus, the presence of a specific fungus can be used to identify the fern host species. Fungi of the orders Uredinales (Basidiomycota) and Taphrinales (Ascomycota) seem to be the most relevant for host-specific interactions, but even lower fungi (i.e., fungi that do not belong to Ascomycota or Basidiomycota) such as *Synchytrium* (Chytridiales, Chytridiomycota) have been reported on specific ferns (e.g., *Synchytrium athyrii* on *Athyrium filix-femina*, lady fern, in Europe, Müller and Schneller, 1977).

Bennell and Henderson (1985) reported that parasitic Taphrinales infect at least ten fern genera. Some examples may illustrate their degree of host specificity. *Taphrina polystichi* occurs on *Polystichum acrostichoides* (Christmas fern) in the eastern USA, while the northwestern North American *Polystichum munitum* (sword fern) is infected by *Taphrina faulliana*. In Mexico, several species of *Terpsichore* (e.g., *T. subtilis* and *T. taxifolia*) are infected by the ascomycete genus *Acrospermum*, a black clavate fungus. This fungus has not been found in any species of the *T. asplenifolia* group or other closely related genera such as *Melpomene* (Mickel and Smith, 2004).

Fern rusts (Uredinales, Basidiomycota) are obligate, parasitic fungi that develop on two alternating hosts (heteroecy). Specific fern species (not lycophytes) can serve as the primary host and any local species of the gymnosperm genus *Abies* (fir) as the alternate host to complete the complex fungal life cycle (Bennell and Henderson, 1985). Rust fungi infect host plants through their leaf stomata and their mycelia develop haustoria, specialized nutrient-absorbing structures that penetrate the living host cells. On the primary host, they develop rust-colored pustules on the leaves that release large numbers of wind-dispersed fungal spores to infect the alternative host. The entire life cycle consists of five or six spore stages, but may be shortened if no alternative host is available. Rust fungi damage their host plants but rarely kill them. Most rust fungi infect seed plants, but Bennell and Henderson (1985) have reported 119 fern host species, mainly those in modern fern families (e.g., Blechnaceae, Dryopteridaceae, Polypodiaceae, Pteridaceae; Schuettpelz and Pryer, 2008), and new fern rust species are continuously discovered (Table 7.1). The lineage of fern rusts have been regarded as the oldest of rust fungi (Uredinales), but a phylogenetic study revealed that fern rusts may actually belong to derived lineages (Table 7.1, e.g., *Uredinopsis* and *Hyalopsora* on *Athyrium*) as well as old lineages (e.g., *Mixia* on *Osmunda*; Sjamsuridzal *et al.*, 1999). Because rust fungi infect mostly seed plants, many fern rusts may have evolved by switching from seed plant hosts to ferns. For example, *Puccinia lygodii* is the only species of its genus that occurs on the climbing fern *Lygodium* (Plates 8C, D; Bennell and Henderson, 1985). Berndt (2008) found species of three rust genera (Uredinales) on ferns in South Africa: *Milesia*, *Milesina* and *Uredinopsis*. *Milesina blechni* is an introduced rust fungus that in its natural range in the northern hemisphere required two host plant species, a fern (*Blechnum spicant*) and a gymnosperm (*Abies*) for its complete life cycle. Because neither of its host species occurs in South Africa, the rust adapted its life cycle to a single new, but related, fern host species, *Blechnum punctulatum*, and reproduces there without alternating host plants.

Some fern species are appreciated by horticulturalists because their young leaves initially develop in pink, red, purple or white colors (e.g., *Blechnum*, Plate 5D). In general, young leaves have a higher water and nutrient content than older leaves and are more prone to predation (Crawley, 1983), but they may also contain highly toxic chemical compounds that are no longer present in older leaves (Zangerl and Bazzaz, 1992). In fern species with young red leaves, anthocyanins are synthesized during leaf development before chlorophyll production initiates and causes leaves to turn green by the time they have nearly expanded. Two hypotheses have been suggested to explain red coloration of young leaves: (1) protection against herbivory by indicating to potential herbivores an unpalatable leaf (Coley, 1983), and (2) protection against fungi (Coley and Aide, 1989). Evidence has been found only for the second hypothesis. Coley and Aide (1989) tested their antifungal defense hypothesis by taking advantage of specialists in cultivation of fungi, the leaf cutter ants *Atta colombica*. These ants harvest a wide variety of leaves and bring it to their nests to

Table 7.1 *Examples of fungi that interact with ferns*
Glomeromycota and some Ascomycota are mycorrhizal fungi. All other taxa are
parasitic (i.e., feeding on living plant material), saprophytic (i.e., feeding on dead
material) or neutral.

Fungi		Fern and lycophyte	
Division/Phylum	Taxa	taxa	References
Glomeromycota	*c.* 150 mycorrhizal species	~ 80% of fern species	Boullard, 1957; Cooper, 1976; Iqbal *et al.*, 1981; Gemma *et al.*, 1992; Moteetee *et al.*, 1996, Schüßler *et al.*, 2001
Chytridiomycota	*Synchytrium athyrii*	*Athyrium filix-femina*	Müller and Schneller, 1977
Ascomycota	Unidentified mycorrhizal species	*Ceradenia, Cochlidium, Elaphoglossum, Grammitis, Hymenophyllum, Melpomene, Lellingeria, Trichomanes*	Schmid *et al.*, 1995, Lehnert *et al.*, 2007
	Acrospermum spp.	*Terpsichore* spp.	Mickel and Smith, 2004
	Bioscypha pteridicola	*Cnemidaria uleana*	Samuels and Rogerson, 1990
	Colletotrichum spp.	*Rumohra adiantiformis*	Leahy *et al.*, 1995
	Crocicreas sessilis	*Cyathea divergens*	Samuels and Rogerson, 1990
	Dimeriella polypodii	*Polypodium madrense, P. montigenum*	Samuels and Rogerson, 1990
	Lachnum spp.	Cyatheaceae, Dicksoniaceae	Medel and Lorea-Hernández, 2008
	Taphrina faulliana	*Polystichum munitum*	Bennell and Henderson, 1985
	Taphrina polystichi	*Polystichum acrostichoides*	Bennell and Henderson, 1985
Basidiomycota	*Hyalopsora cheilanthis*	*Cheilanthes pringlei*	Berndt, 2008
	Hyalopsora polypodii	*Athyrium yokoscense, Leptogramma mollissima*	Sjamsuridzal *et al.*, 1999
	Milesia silvae-knysnae	*Polystichum pungens*	Berndt, 2008
	Milesina blechni	*Blechnum punctulatum, B. spicant*	Berndt, 2008

Table 7.1 (*cont.*)

Fungi		Fern and lycophyte taxa	References
Division/Phylum	Taxa		
	Mixia osmundae	*Osmunda* sp.	Sjamsuridzal *et al.*, 1999
	Puccinia lygodii	*Lygodium* sp.	Bennell and Henderson, 1985
	Uredinopsis intermedia	*Athyrium* sp.	Sjamsuridzal *et al.*, 1999
	Uredinopsis pteridis	*Pteridium aquilinum, Stenochlaena tenuifolia*	Berndt, 2008
	Uredo vetus	Selaginellaceae	Farr and Horner, 1958; Berndt, 2008

cultivate their subterranean fungus gardens, avoiding leaves with antifungal compounds. The cultivated fungi are the main diet of leaf cutter ants. Coley and Aide (1989) set up an experiment to determine whether these ants can distinguish between leaves rich and poor in anthocyanin, and among oat flakes with different anthocyanin concentrations. In both experiments, ants preferred anthocyanin-poor material, suggesting that anthocyanins might serve as an antifungal defense; direct evidence for their negative effects on fungi has not yet been found.

7.3.2 Mycorrhizae

Mycorrhizae are symbiotic associations between fungi and plant roots. Mycorrhizae increase water and nutrient absorption, especially of nitrogen and phosphorus (Siddiqui and Pichtel, 2008), and also increase the drought tolerance of their host plants (Aboul-Nasr, 1998; Al-Karaki, 1998). Mycorrhizal fungi may compete with parasitic fungi and consequently protect plants from parasitic fungal infections (Brundrett, 2002). Most of the evidence of mycorrhizal benefits for host plants comes from experimental studies on seed plants. In ferns, it has been demonstrated that mycorrhizae promote faster growth, as shown by a comparison of mycorrhizal and uninfected plants of *Pteridium aquilinum* in New Zealand (Cooper, 1976).

At least seven types of mycorrhizae have been distinguished (Siddiqui and Pichtel, 2008), of which at least two occur in ferns, the most common endomycorrhizae and the uncommon fern–ericoid mycorrhizae (Schmid *et al.*, 1995), while the presence of ectomycorrhizae in ferns is controversial. All mycorrhizae invade the root cortex of their host plants but differ by their fungal partners and their further development. Endomycorrhizae (ENM) are formed by aseptate Glomeromycota that invaginate the cell membranes of the root cells with arbuscules and may also develop vesicles. Ectomycorrhizae (ECM) are built up by septate Ascomycota or Basidiomycota, or aseptate Zygomycota that form an internal network of hyphae that surround intact root cells (i.e., a Hartig net) and an external hyphal sheath around the root tips. Fern–ericoid mycorrhizae (FEM) are formed by septate Ascomycota and are intermediate between endomycorrhizae and ectomycorrhizae (i.e., ectendomycorrhizae).

Ectomycorrhizae occur in about 3% of seed plant families (Smith and Read, 1997), but their occurrence in ferns is controversial. Some authors have reported a shift from ENM to ECM in forests where ectomycorrhizal tree species (e.g., beech forests, pine forests, pine plantations) are dominant (Cooper, 1976, Iqbal *et al.*, 1981). Others suggest that the observed roots did not belong to ferns (Smith and Read, 1997; Brundrett, 2002). Repeated studies of ferns from forests dominated by ectomycorrhizal trees would be necessary to clarify this discussion.

Dhillion (1993) and Jumpponen and Trappe (1998) have found ascomycetous dark septate fungi (DSF) in several species of ferns and lycophytes. Jumpponen (2001) suggests that DSF comprise a diversity of fungi with some species forming ectendomycorrhizae (i.e., mycorrhizal fungi with ecto- and endomycorrhizal features). Fernández *et al.* (2008) reported DSF in all their samples of *Equisetum bogotense* and *Lycopodium paniculatum* from *Nothofagus* forests in Patagonia. Because the taxonomic identities of the DSF are still unknown, it remains unclear if they are primarily symbiotic or neutral endophytes. Clearly, the association of DSF and ferns is poorly understood.

The most common mycorrhizal fungi of ferns are endomycorrhizae (ENM). All ENM are formed by aseptate Glomeromycota, a distinct fungal phylum, separated from the Zygomycota (Schüßler *et al.*, 2001; Schüßler, 2004). Because some ENM only produce arbuscules but no vesicles, some scientists now prefer to call them arbuscular mycorrhizae (AM) rather than vesicular–arbuscular mycorrhizae (VAM). Arbuscules are branched, tree-like structures that promote nutrient exchange between fungus and host plant, while vesicles are terminal, lipid-storing swellings of hyphae. Endomycorrhizal fungi are asexual and obligate symbionts (i.e., they cannot grow independently from their host plant roots) and occur in more than 80% of extant vascular plants. The mutualistic relationship between mycorrhizal fungi and land plants developed early in Earth's history even before roots evolved (Brundrett, 2002). *Aglaophyton major*, an Early Devonian (*c.* 400 mya) land plant at Rhynie, Scotland had ENM-like arbuscules in its specialized rhizome meristem (Taylor *et al.*, 1995). The mutualistic relationships between fungi and early ancestors of ferns and lycophytes may have been crucial for the colonization of the terrestrial environment, because mycorrhizal fungi improve the host plant's access to water and inorganic nutrients such as phosphorus, nitrogen, and some metals (e.g., copper and zinc), thereby promoting increased host plant growth (Gange *et al.*, 1990; Smith and Read, 1997, Goltapeh *et al.*, 2008, Duponnois *et al.*, 2008). To test for the mycorrhizal benefits to host plants, Gange *et al.* (1990) studied the effect of reduced mycorrhizae on plant succession in the UK. They found that plots of ploughed, bare ground that had been treated with fungicide to reduce mycorrhizal infections slowed down the rate of colonization and establishment during the first year of succession. There were fewer individuals and plant species and a lower vegetation cover in the fungicide treated plots than in control areas.

Within the mycorrhizal symbiosis, the fungal partner benefits from the organic exudates of the roots. Early soil fungi may have evolved into mycorrhizal fungi to escape parasitic soil fungi, growing within plant organs while protecting their host plants (Brundrett, 2002). However, some types of mycorrhizae developed from a balanced mutualistic relationship into an exploitative antagonistic interaction where the fungus or the plant becomes a parasite of the other partner. Both fern generations, rhizoidal gametophytes and root-bearing sporophytes may host mycorrhizae. All gametophytes of the Lycopodiaceae, Psilotaceae and Ophioglossaceae have obligate mycorrhizae, while in all other families mycorrhizae are facultative or missing at the gametophytic stage. In Lycopodiaceae, gametophytes show a variety of life forms and differing interactions with their mycorrhizal fungi. For example, photosynthetic, green gametophytes of *Lycopodiella cernua* live on the soil surface, whereas gametophytes of *Lycopodium clavatum* are subterranean, slow growing and lack chlorophyll. Both species depend on mycorrhizal fungi, but the gametophyte of *L. cernua* only receives water and minerals from its fungus whereas that of *L. clavatum* also exploits its fungus as a source for carbohydrates (Schmid and Oberwinkler, 1993). Because in the latter case the achlorophyllous gametophytes do not synthesize carbohydrates, the mycorrhizal fungus must exploit another host to provide nutrients for itself and the host gametophyte (Harley and Harley, 1987). The fungus of *L. clavatum* infects the gametophyte at some distance from the growing margin through older rhizoids, develops in intracellular spaces and penetrates the cortical gametophyte cells where it then forms vesicles, but no arbuscules. Gametophyte cells of the epidermis, central tissue and sex organs are not infected. In older, intact host tissue, the fungus degenerates (Schmid and Oberwinkler, 1993). In contrast, the mycorrhizal fungus of the chlorophyllous gametophyte of *L. cernuum* forms arbuscules but no vesicles, and it infects epidermal tissue but does not grow into intercellular spaces (Duckett and Ligrone, 1992).

The two examples above demonstrate that mycorrhizal interactions with fern gametophytes are not as uniform as often stated, which becomes even more evident in the sporophyte. Mycorrhizal fungal infection of fern sporophytes occurs through the root hairs, and not through the gametophyte–sporophyte junction as one might expect (Schmid and Oberwinkler, 1995). Because the sporophyte does not directly inherit the mycorrhizae from the gametophyte, the sporophyte may interact with different fungi. However, it seems probable that both generations will interact with the same mycorrhizal fungus because of their close proximity during the early sporophyte development (Schmid and Oberwinkler, 1995).

Mycorrhizae in fern sporophytes are almost as common as in most seed plants. In a worldwide literature review of 180 fern species studied since 1960, 87% of the species always had mycorrhizae, 7% had mycorrhizae only at some of the study sites, and 6% had no mycorrhizae (Newman and Reddell, 1987). This study does not allow further generalizations or conclusions, because the reported fern species did not adequately represent all families, climatic regions (tropical versus temperate) or habitats (e.g., terrestrial, epiphytic). Subsequent regional surveys from Hawaii (Gemma *et al.*, 1992) and South Africa (Moteetee *et al.*, 1996) were more comprehensive and contributed further insights. These studies showed that over 70% of fern species have nonspecific arbuscular mycorrhizae. Ferns can often be infected by a mixture of up to five different mycorrhizal fungi and one fungus can infect more than one fern or seed plant species. In Hawaii, 85% of native angiosperms (Koske *et al.*, 1992) and 74% of ferns possess mycorrhizae (Gemma *et al.*, 1992). In contrast, Moteetee *et al.* (1996) surveyed 49 African fern species from Lesotho, and found only 37% of the species with mycorrhizal associations. The low incidence of mycorrhizae is unexpected if plants of the nutrient-poor soils such as those in this region have a greater dependence on mycorrhizae. Moteetee *et al.* (1996) suggest that other studies (e.g. Cooper, 1976; Iqbal *et al.*, 1981) may have overestimated fern mycorrhizae because they did not distinguish saprophytic–parasitic fungi of older and dead roots from mycorrhizal fungi of young and living roots. Without this differentiation among fungi, 82% of the surveyed species of Lesotho would have been counted as associated with mycorrhizae.

Some angiosperm families (e.g., Brassicaceae, Caryophyllaceae; Brundrett, 1991) and fern groups (e.g., Salviniaceae, Boullard, 1957, 1979) are nonmycor-rhizal. These groups may have lost their fungal partner because of the development of biochemical metabolites that negatively affect the mycorrhizae (e.g., Brassicaceae) or in the case of Salviniaceae, by evolutionary change into aquatic habitats (Boullard, 1957, 1979) where mycorrhizal fungi may develop poorly. Mycorrhizae have never been reported for water ferns such as Marsileaceae and Salviniaceae, but have been found in some species of Isoëtaceae (Plate 3D) that root in aquatic habitats (Beck-Nielsen and Madsen, 2001; Radhika and Rodrigues, 2007), and in *Lycopodium paniculatum* in waterlogged peat bogs (Fernández *et al.*, 2008). In Hawaii, epiphytic ferns had considerable but lower mycorrhizal infection rates (55% of tested species) than epilithic (86%) and terrestrial ferns (83%; Gemma *et al.*, 1992). There are two possible explanations for this result: either the lower abundance of fungi, especially on soilless tree branches, decreases the opportunity for epiphytic ferns to find a fungal partner, or the epiphytic ferns developed adaptations to their new habitat (e.g., drought tolerance) that negatively affected the chance of fungal infections. The second alternative is not conclusive, because mycorrhizae are supposed to increase the water and nutrient availability for their host ferns, which is especially important in drought-prone epiphytic and epilithic habitats. For example, Palmieri and Swatzell (2004) documented arbuscular mycorrhizae in *Cheilanthes lanosa*, a xerophytic fern species growing on rocky outcrops from Missouri to Illinois. They concluded that the ferns benefit from mycorrhizae by increasing drought tolerance in a similar way to that demonstrated for squash and wheat plants (Aboul-Nasr, 1998; Al-Karaki, 1998). For this reason, it seems more probable that mycorrhizal fungi are less abundant in epiphytic than in rocky habitats. This pattern might also explain why some epiphytic ferns have changed their mycorrhizal partners.

Fern–ericoid mycorrhizae (FEM) build a Hartig net as in ectomycorrhizae, but also may penetrate the root cells as in endomycorrhizae (Schmid *et al.*, 1995, Lehnert *et al.* 2007). FEM have been reported only for some epiphytic fern genera, first in Costa Rica (Table 7.1, e.g. *Elaphoglossum, Grammitis, Hymenophyllum, Lellingeria*; Schmid *et al.*, 1995), then in 27 Ecuadorian fern species of the afore-mentioned genera as well as in *Ceradenia, Cochlidium, Melpomene* and *Trichomanes* (Lehnert *et al.* 2007). Fernández *et al.* (2008) suggested that the degree of endomycorrhizal colonization depends on the substrate, because they could not detect ENM in epiphytic *Lycopodium paniculatum* in Patagonia but could in some of terrestrially growing individuals. Future studies should investigate whether the proportion of endomycorrhizal Glomeromycota and fern–ericoid Ascomycota varies among habitats and whether the fern–ericoid mycorrhizae are competitively more beneficial to epiphytic ferns.

7.4 Interactions with animals

Most studies of fern interactions with animals document direct effects because these are easier to observe and to correlate than indirect effects. One exception is a study that associated the presence of exotic earthworms (*Lumbricus rubellus*) with the extirpation of the endangered fern *Botrychium mormo* (moonwort) in Minnesota (Gundale, 2002). The earthworms did not feed on the fern but decreased the depth of the organic soil horizon on which *B. mormo* relies for nutrient uptake. In another study, an introduced slug (*Deroceras reticulatum*) in New Zealand extensively defoliated *B. australe*. This direct herbivore damage increased with the presence of an introduced grass species (*Agrostis capillaris*; Sessions and Kelly, 2002) that provided shelter for the slug. However, herbivory on nearby flowering plants can have an indirect positive effect on ferns if it results in a slower rate of succession at disturbed sites (Brown, 1984) so that colonizing fern species may gain additional time until they are displaced by other species (see Chapter 6).

The predominant type of direct fern–animal interaction is antagonistic (e.g., herbivory), while few mutualistic interactions between ferns and animals are known (Auerbach and Hendrix, 1980; Hendrix 1980; Cooper-Driver, 1985a). In the next section, I first discuss herbivory (i.e., the damage caused by herbivores and the degree of specialization of herbivores on ferns as well as the biochemical defense strategies of ferns) and I conclude with a review of fern–ant mutualisms.

7.4.1 Herbivory

Fossil evidence for fern-feeding arthropods exists from the Carboniferous (355–290 mya; Smart and Hughes, 1973) and the Triassic (250–205 mya; Ash, 1997, 1999, 2000). Feeding traces on pinnules of *Cynepteris lasiophora* (Ash, 1997), leaf excisions on *Sphenopteris arizonica* (Ash, 1999), and coprolite-bearing borings of a mite in the stem of the filicalean tree fern *Itopsidema vancleaveii* (Ash, 2000) are proof of fossil fern–herbivore interactions. Extant fern herbivores are mainly arthropods but a few examples come from vertebrates. For example, introduced feral pigs in Hawaii feed on tree fern trunks that contain starch (Diong, 1982; Arcand, 2007) and cattle can graze on the leaves of *Pteridium aquilinum* but in limited amounts due to the fern's toxicity (see Chapter 8). Both examples result from anthropogenic interference into natural biotic interactions. One unusual example of fern feeding by a native species is *Pyrrhula murina*, the Azores bullfinch that switches its diet from invertebrates, fruits and seeds during the summer and autumn to fern leaves and sporangia (Plate 5B) as well as tree seeds and flower buds during the winter and spring when food is scarce (Ramos, 1994, 1995). These birds feed on the sporangia of *Culcita macrocarpa* and *Woodwardia radicans* (European chain fern), of which they reject the indusium. They also ingest the leaves of *Pteridium aquilinum* and *Osmunda regalis* (royal fern) when food is scarce. Generally, it appears that most vertebrates avoid ferns as a food source.

Leaf damage

Leaf damage is usually reported as the percentage of the total available leaf area of a plant that has been removed or damaged by herbivores. Terrestrial and epiphytic ferns sustain similar proportions of leaf damage that normally range between 5% and 15% with a maximum of up to 36% (Balick *et al.*, 1978; Hendrix and Marquis, 1983; Mehltreter and Tolome, 2003; Mehltreter *et al.*, 2006). These levels of leaf damage are similar to those reported for angiosperms of tropical forests (Lowman, 1984, 1985; Coley and Aide, 1991; Williams-Linera and Baltazar, 2001). Fertile and sterile leaves are generally consumed at the same rates. Young leaves have considerably less damage than older leaves, suggesting that herbivores may feed on leaves at all developmental stages, so that damage levels increase with leaf age (Raupp and Denno, 1983; Mehltreter and Tolome, 2003). Consequently, it should be expected that mean leaf damage increases with longer leaf life spans of the species (Mehltreter *et al.*, 2006). In comparison to coexisting orchids and bromeliads at the same site, ferns had higher mean leaf damage. Whereas less than 32% of individual plants of orchids and 15% of bromeliads in a montane Mexican forest showed some traces of herbivore damage, 60–95% of fern individuals were affected by herbivory (Winkler *et al.* 2005). Plants with longer leaf life spans often possess more and better biochemical defenses (Coley *et al.*, 1985; Coley and Aide, 1991; Coley and Barone, 1996) and lower leaf damages should be expected. Thus, the high damage levels reported by Winkler *et al.* (2005) indicate either that these fern species do not possess an efficient defense mechanism or that the insects have become resistant to it.

Interactions with arthropods

Brues (1920) stated that for herbivorous insects "only an extremely small, almost negligible proportion subsist on ferns". Subsequent research has postulated a variety of hypotheses to explain why so few insects feed on ferns. Because ferns do not possess the wide array of structural and chemical defenses of angiosperms (Ehrlich and Raven, 1964; Soo Hoo and Fraenkel, 1964; Southwood, 1973), it was thought that fern leaves are of lower nutritional value than angiosperm foliage (Coe *et al.* 1987; Midgley *et al.*, 2002). It was further assumed that most insects would have focused and coevolved with nutrient-rich angiosperms rather than specializing on ferns (Balick *et al.*, 1978; Hendrix, 1980; Cooper-Driver, 1985b). Hummel *et al.* (2008) questioned the former hypothesis by demonstrating that leaf tissues of ferns and seed plants provide similar amounts and types of nutrients. They argued that in the Mesozoic (205–65 mya) ferns and gymnosperms could not have been of low nutritional value because they had been the dominant vegetation and food source that supported the evolution and maintenance of the largest herbivorous dinosaur species on Earth. Before the evolutionary radiation of angiosperms in the Cretaceous (135–65 mya), all herbivorous animals had to rely mainly on the leaves of ferns and gymnosperms for food. Reproductive structures that evolved in angiosperms such as nectar-producing flowers, nutrient-rich fruits and seeds provided a variety of new food sources for insects and may have resulted in an increased number of insect associations with angiosperms. These hypotheses still do not explain why the number of leaf-feeding insect species per plant species should differ between ferns and angiosperms. For example, if ferns really do possess fewer biochemical defense strategies than the flowering plants and both plant groups offer the same nutritional value, it might be expected that insects would prefer ferns as a food source rather than angiosperms.

Could the low number of known fern-feeding insects be an artifact of under-sampling? In search of biological control organisms for *Pteridium aquilinum* (see Chapter 8), scientists were surprised at the considerable diversity of its associated arthropods (Weiczorek, 1973; Lawton, 1976, Shuter and Westoby, 1992) and its wide variety of chemical compounds to deter insects (Carlisle and Ellis, 1968, Hendrix, 1977). Ferns have not received as much attention from entomologists as seed plants have and even large and conspicuous insects have been overlooked for some time. For example, the moth *Papaipema speciosissima* (Noctuidae) with a wingspan of 30–46 mm had been described in 1868, but their larvae that can be up to 55 mm long were not found until 1913 because they feed on the rhizomes of *Osmunda* (Bird, 1938). In three classic papers Balick *et al.* (1978), Gerson (1979) and Hendrix (1980) addressed the question of whether or not ferns are under-utilized as a food source by herbivorous insects in comparison with angiosperms. In their literature review, Balick *et al.* (1978) listed 420 fern-feeding insect species. During two months of field work at six study sites, they showed that 19% of 137 fern species in Veracruz, Mexico, sustained a variety of types of damage representing the impact of a diversity of insects. If 19% of all 11 000 fern and lycophyte species worldwide (see Appendix A) were to host just a single herbivorous insect species, about 2000 fern-feeding insect species would be expected. Consequently, Balick *et al.* (1978) questioned the assumption of under-utilization of ferns by insects as did Gerson (1979) by arguing that species of 12 insect orders (in a classification system of 29 orders) interact with ferns. They also noted that cultivated, ornamental ferns suffer severe damage from generalist herbivores such as aphids (Aphidoidea) and scale insects (Coccoidea), or from some specialized insects (e.g., the leatherleaf fern borer, *Undulambia polystichalis* on *Rumohra adiantiformis*). In contrast, Hendrix (1980) supported the opposite point of view after critically reviewing the list of Balick *et al.* (1978), from which he excluded detritivores and predatory species but integrated additional data. Most fern-feeding insects belong to three orders: Hemiptera, Coleoptera and Lepidoptera (Table 7.2). The ratio of Hendrix's revised list of 465 insect species to the total number of extant fern species is 25 times lower than the reported herbivorous insect–angiosperm ratio of 1.3:1, and 11 000 more insect species would have to be discovered to adjust both ratios to the same level. Hendrix (1980) agreed that ferns have been undersampled, but concluded that the difference detected between insect–fern and insect–angiosperm ratios would withstand future investigations.

Table 7.2 *Systematic distribution of fern-feeding arthropods, expressed as relative percentage of 63 species of the United Kingdom and 465 species worldwide*

Order of arthropods	Feeding habit	UK[1] (%)	World[2] (%)
Collembola (springtails)	Chewing–biting or piercing–sucking	1.6	0.6
Psocoptera (booklice and barklice)	Chewing–biting	0.0	0.4
Orthoptera (grasshoppers and locusts)	Chewing–biting	0.0	2.4
Thysanoptera (thrips)	Puncturing–sucking	0.0	2.6
Hemiptera (aphids, cicadas, leafhoppers, scale insects, shield bugs, whiteflies, etc.)	Piercing–sucking	27.0	38.7
Hymenoptera (ants, bees, wasps)	Diverse (e.g., leaf chewers, gall formers)	23.8	7.7
Coleoptera (beetles)	Chewing–biting	3.2	22.2
Diptera (flies)	Diverse (e.g., leaf miners, stem borers)	22.2	5.6
Lepidoptera (moths and butterflies)	Caterpillars, chewing–biting	22.2	19.8

Source: From [1]Ottosson and Anderson (1983); [2]Hendrix (1980).

While I support Hendrix's hypothesis of fewer herbivores on ferns than on angiosperms, I propose that the herbivorous insect–fern ratio is only 3–7 times lower (rather than 25 times lower) than the insect–angiosperm ratio because all former studies were strongly geographically and species biased and the undersampling of ferns is notorious. For example, 15% of the known fern-feeding species are from the UK and 12% from Hawaii (Hendrix, 1980). In these two regions with good biological inventories, herbivorous insects are only three times (UK) and six times (Hawaii) more frequent on angiosperms than on ferns (Table 7.2; Zimmerman, 1970; Hendrix, 1980; Ottosson and Anderson, 1983). If these ratios are extrapolated to ferns world-wide, there are 1500–3500 insect species that either are not known to feed on ferns or have not yet been discovered. In recent years, more fern-feeding species have been reported, although mostly in search of biological control organisms for problem fern species (Table 7.3). In a detailed search of gall-forming species in Costa Rica, Hanson and Gómez-Laurito (2005) discovered 17 species of Cecidomyiidae attacking ferns, more than had been previously counted for ferns worldwide.

Generalists versus specialists

Herbivores use two strategies to counteract plant toxic defenses that may deter them from feeding. The strategy of generalist insects is to ingest either low amounts of

Table 7.3 *Examples for recently described interactions between ferns and fern feeding insects, mainly discovered during searches for biological control organisms of weedy fern species (see Chapter 8)*

Insect species	Distribution	Fern host species	References
Hymenoptera, Tenthredinidae	Venezuela,	*Pteridium*	Smith, 2005; Avila-
Anegmenus merida	Ecuador	*aquilinum*	Nuñez *et al.*, 2007
Anegmenus colombia	Colombia	*Pteridium*	Smith, 2005
		aquilinum	
Lepidoptera, Crambidae	Australia,	*Lygodium*	Yen *et al.*, 2004
Austromusotima	Southeast	*microphyllum*	
camptozonale	Asia		
Lygomusotima stria	Southeast	*Lygodium*	Solis *et al.*, 2004
	Asia	*microphyllum*	
Neomusotima conspurcatalis	Australia	*Lygodium*	Solis *et al.*, 2004
		microphyllum	
Lepidoptera, Tortricidae	Brazil	*Microgramma*	Brown *et al.*, 2004
Tortrimosaico polypodivora		*squamulosa*	

each plant species to avoid accumulated effects of any of the different toxins or to consume only plant species with low concentrations of toxins. The strategy of specialist insects is to adapt to the specific defense mechanism of one or several plant species by resisting or neutralizing the toxins or even by accumulating them for their own defense. Fern herbivores have followed both strategies. Most generalists tend to be leaf chewers and sap suckers, while specialists are often leaf miners, gall formers, and spore feeders. The very active caterpillar of *Tarchon felderi* (Apatelodidae) is an extreme generalist that feeds on more than 65 plant species belonging to over 30 families of ferns and angiosperms in Costa Rica (Fig. 7.1; Miller *et al.* 2006, 2007). Individuals of *T. felderi* consumed leaf tissue of more than 15 fern species in Mexico (K. Mehltreter, unpublished data). Rowell *et al.* (1983) conducted laboratory experiments with two fern-feeding grasshoppers in Costa Rica, a generalist (*Homeomastax dentata*) and a specialist (*Hylopedetes nigrithorax*). Both species concurred on the palatability of all 13 fern species that were offered to them, but the specialist had a strong preference for two of the fern species whereas the generalist only disliked one of the fern species. The specialized grasshopper species also lived on its favorite food plant while the generalist grasshoppers moved around.

Most fern-feeding insects consume leaves, but some also ingest roots (3.3%) and spores (2.4%), while others are miners and borers (5.9%) or gall formers (4.3%; Plate 7C; Balick *et al.*, 1978). Of 465 fern-feeding insect species, 35% were

Fig. 7.1 *Tarchon felderi* (Apatelodidae), a caterpillar of a generalist species feeding on several species of ferns and seed plants in Mexico.

classified by Hendrix (1980) as specialists on ferns, 45% as generalists consuming ferns and angiosperms, and 20% as indeterminate species with unknown feeding habit. A large number of these insect species are members of a few insect families and genera, which suggests coevolution of the insect and fern taxa (Hendrix, 1980). For example, 88.9% of the genera of Anthomyiidae (Diptera) with fern specialists have evolved two or more fern-feeding species, as well as 65.2% of the genera of Tenthredinidae (Hymenoptera) and 56.7% of the genera of Aphididae (Hemiptera). The overrepresentation of more primitive hemimetabolous insects (with simple metamorphosis) on ferns in comparison to homometabolous insects (with complete metamorphosis) is also interpreted as a consequence of the longer coevolutionary history of ferns with older insect groups (Hendrix, 1980). However, insect genera with only one specialized fern-feeding species or several species that feed on a wide range of fern hosts may have originated more recently by host-plant switching from seed plants to ferns and may occasionally occur in all insect groups (Weintraub *et al.*, 1995).

The first gall-inducers on ferns that were described by Docters van Leeuwen (1938) belonged to *Physothrips* (Thysanoptera: Thripidae). Ferns are underrepresented as targets for gall-inducing insects in Costa Rica (Plate 7C). In a collection of 967 gall-forming arthropods of Costa Rica, only 17 species of gall formers (all within Cecidomyiidae) were found on 18 fern species from ten genera and five different families (Aspleniaceae, Cyatheaceae, Dryopteridaceae, Polypodiaceae, Woodsiaceae), while the remaining 950 species were registered on 693 angiosperm species (Hanson and Gómez-Laurito, 2005). The vascular flora of Costa Rica consists of about 1100 ferns and lycophytes (Mehltreter, 1995) and 9000 angiosperms (Hammel *et al.*, 2004); therefore the ratio of gall-forming insects is about seven times larger for angiosperms than for ferns.

There are about 2500 species of stick insects (Order Phasmatodea) in the tropics. Only two of these species, *Chondrostethus woodfordii* from the Solomon Islands and *Oroephoetes peruana* from South America, seem to feed on (several) ferns of the forest understory (Golding, 2007). *Oroephoetes peruana* insects produce a malodorous and irritant substance in their thoracic glands (supposedly quinoline) that is discharged as defense against predators. Eisner *et al.* (1997) suggested that individuals of *O. peruana* absorb this substance directly from their food source, *Nephrolepis exaltata* (Boston fern) on which the stick insect is reared in the laboratory, but could find no trace of the chemical compound in the fern, indicating that the stick insect synthesizes it on its own.

Barker *et al.* (2005) reported a very special case of interaction between lepidopteran larvae and ferns that they called soral crypsis. They found caterpillars (superfamily Gelechioidea) feeding on the leaves of five species of Puerto Rican ferns. To hide from possible enemies (e.g., parasitoid wasps), these caterpillars harvest sporangia from their host fern, feed on the spores and attach the remains on the outer layer of their 3–7 mm long cocoons, mimicking the appearance of the natural sori (Plate 7D). The elongate form of the cocoon does not always correspond to the shape of the sori of the host fern as three of the fern species had round instead of elongate sori. Some butterfly species lay their eggs on tree ferns. Because their eggs resemble in shape and size the globose sori of these tree ferns, they might also be interpreted as an example of soral egg mimicry (K. Mehltreter, personal observation), but further studies are needed to test this hypothesis (Fig. 7.2). In southern USA, *Herpetogramma aeglealis* (Pyralidae) builds increasingly larger and more complex feeding shelters on its host fern *Polystichum acrostichoides* (Christmas fern) corresponding to the increasing size and food demand of the caterpillar (Ruehlmann *et al.*, 1988). Shelters are built by folding and weaving together parts of the fern leaves. After creating the first ephemeral bundle shelter and the second shelter resembling a natural fern crozier, the caterpillars build three conspicuous globe shelters at the leaf tips. Ruehlmann *et al.* (1988) suggested that the successive construction of five shelters during the life history of the lepidopterous larvae not only fits its growing demand for food and space but also minimizes the risk of

Fig. 7.2 An example of egg mimicry. Butterfly eggs on a leaf of the Mexican tree fern *Alsophila firma* are of similar shape, size and form as the fern sori. Only the placement of the eggs on the midvein and in two rows rather than one single row differs from the soral arrangement in *Alsophila firma*.

predation. Predators presumably become discouraged as the number of persistent empty shelters increases during the season every time the caterpillar builds a new one.

7.4.2 Responses of ferns to herbivory

Plants respond in several ways to herbivore damage, passively by compensating for damages with new plant growth or actively by constitutive or induced resistance (Rhoades, 1979). Plants that have no biochemical resistance have economized their investment in biochemical defenses, but need to compensate for losses in biomass caused by herbivores. One way for a fern to minimize herbivore damage is to produce an overabundance of young leaves to satiate herbivore populations. The Mexican tree fern species *Alsophila firma* perhaps uses this strategy of herbivore satiation when its populations develop all leaves synchronously during two months of the rainy season when insects are most abundant (Mehltreter and García-Franco, 2008). Most plant species do resist herbivores either by continually producing

protective secondary metabolites as preventive constituents (i.e., constitutive resistance), a costly investment, or by synthesizing phytoalexins in direct response to
herbivore attack (i.e., induced resistance), which economizes the expenses against
herbivory. Phytoalexins are toxic substances that are released locally where bacterial or fungal infections occur on a plant (Cooper-Driver, 1985b) and deter herbivorous insects by indicating that the plant is no longer palatable. Phytoalexins may also
be volatile and emitted to attract predators to feed on the herbivores (e.g., tobacco,
Nicotiana attenuata; Kessler and Baldwin, 2001). In ferns, however, evidence for
phytoalexins has not yet been found, suggesting that they are usually plants with
constitutive resistance (see above).

The major metabolic pathways of photosynthesis (e.g., chlorophyll a and b,
carotenoids) and respiration (e.g., cytochromes) are similar for ferns and seed
plants, and although the array of secondary metabolites of ferns is not as diverse
as in seed plants (Cooper-Driver, 1985b), some of these chemical compounds are
produced only by ferns (Hegnauer, 1962, 1986). The most common defense
mechanisms in ferns include high fiber content and high tannin concentrations in
completely differentiated, mature tissues. High fiber content combined with lignified cell walls makes feeding on ferns more difficult for herbivores and is effective
against chewing insects. It is, however, a relatively ineffective defense against
insects with piercing–sucking mouthparts that directly exploit phloem sap, a type
of herbivore that is consequently overrepresented on ferns (Balick *et al.*, 1978;
Hendrix, 1980). Tannins are astringent and make leaf tissues unpalatable and
indigestible because they bind strongly to proteins and can even denature enzymes.

In addition to tannins, ferns and lycophytes do possess other herbivore-deterrent
substances. These substances include alkaloids such as lycopodin in lycopods and
palustrin in horsetails (but alkaloids are missing in all other ferns), carcinogenic
sesquiterpene glycosides (e.g., ptaquiloside in *Onychium contiguum* and *Pteridium
aquilinum*; Somvanshi *et al.*, 2006), cyanogenic glycosides (e.g., prunasin of
Cystopteris fragilis and *Pteridium aquilinum* and vicianin of *Davallia*), thiaminase
(e.g., *Nephrolepis exaltata* and *Pteridium aquilinum*; Hendrix, 1977), phytoecdysones (e.g., Blechnaceae, Dennstaedtiaceae, Dryopteridaceae, Gleicheniaceae,
Osmundaceae, Polypodiaceae, Pteridaceae; Kaplanis *et al.* 1967; Carlisle and
Ellis, 1968; Russell and Fenemore, 1971; Hikino *et al.*, 1973) and various defensive
proteins (Markham *et al.*, 2006). Alkaloids and carcinogenic sesquiterpene glycosides of ferns and lycophytes are supposed to be toxic to most herbivores, but
their effectiveness in deterring herbivores from feeding has not been studied.
Phytoecdysones (i.e., insect-molting hormones) are quite common in ferns. For
example, Hikino *et al.* (1973) confirmed variable insect-molting hormone activity
for 170 out of 283 studied Japanese ferns, but 51 of these species had high levels of
phytoecdysone activity. Thiaminase has been blamed for toxic effects on livestock

feeding on *Pteridium*, causing a shortage of vitamin B (see Chapter 8). Cyanogenic glycosides have been detected in 3% of studied ferns (Balick *et al.*, 1978) and become effective after leaf damage, when enzymes break down the glycoside into sugar and highly toxic, volatile hydrocyanic acid. Populations of *Pteridium aquilinum* are polymorphic and vary in their content of cyanogenic glycosides depending on light conditions. Plants of shady habitats contain higher concentrations of cyanogenic glycosides than in sunny habitats, and populations with cyanogenic glycosides are attacked less often by insects than populations without cyanogenic glycosides (Cooper-Driver *et al.*, 1977; Schreiner *et al.*, 1984). Protein extracts of *Asplenium platyneuron* (ebony spleenwort), *Athyrium pycnocarpon* (glade fern) and *Onoclea sensibilis* (sensitive fern) reduce herbivore damage in bioassays with caterpillars, although the specific active compounds have not yet been identified (Markham *et al.*, 2006).

Other biochemical compounds in ferns that need to be tested for their possible herbivore deterrent effects are flavonoids, phloroglucides, metals and silicic acid. Flavonoids are produced by glands and accumulate on the lower leaf surface (e.g., *Argyrochosma, Notholaena, Pityrogramma*) forming a white or yellow farina. Flavonoids are common in ferns of dry habitats (Plate 2A) and leaf surfaces coated with flavonoids may reduce transpiration and act as protection against high ultraviolet radiation, but they may also deter herbivores (Kramer *et al.* 1995). Phloroglucides (e.g., in Dryopteridaceae) are taeniafuges (i.e., substances for expelling tapeworms) suggesting that other deterrent properties are possible. Other chemical substances that occur in higher concentrations in plant tissues of ferns and lycophytes and that might prove to be herbivore deterrents are aluminum (e.g., in Lycopodiaceae, Gleicheniaceae, Marattiaceae), arsenic (e.g., in *Pteris vittata*, Plate 8B; see Chapter 8) or silicic acid (e.g., in Equisetaceae, Plate 4C; Lycopodiaceae, Plate 3A).

7.4.3 *Mutualistic interactions of ferns and ants*

The interactions between ferns and ants show all degrees of intimacy from opportunistic neutral interactions with no benefit for the fern to highly specialized mutualism where both partners benefit to obligate mutualism when neither partner can survive on its own. A highly opportunistic interaction was reported for ants living in the giant leather fern, *Acrostichum danaeifolium* in the Mexican mangroves (Mehltreter *et al.* 2003). The leaves of this mangrove fern can be 2–4 m in length and are attacked by microlepidopterous larvae that feed and pupate inside the petiole and rachis, which can be 1–2.5 cm in diameter. The adults emerge from an exit hole through which ants then can access the empty tunnel system that had been built by the feeding activity of the microlepidopteran. Most of the ant species that take advantage of this microhabitat are alien species that have been introduced to the neotropics such as *Tapinoma sessile* and *Wasmannia auropunctata*. These ants are not able to create holes or burrow tunnels on their own, nor do they seem to protect the fern from any natural enemy. Furthermore, the availability of this microhabitat for the ants is limited by the mean leaf life span of 7.7 months of its host, *A. danaeifolium* (Mehltreter and Palacios-Rios, 2003).

Fern–ant mutualisms become more intimate where ferns supply ants with nectaries or domatia (i.e., hollow plant organs in which ants can live). Both morphological structures occur in some fern species and can provide a direct incentive for ants.

Nectaries

Nectaries are known from some species of mostly epiphytic fern genera (e.g., (*Aglaomorpha*, *Drynaria*, *Polypodium* and *Platycerium*) but also from a few terrestrial genera (e.g., *Pteridium*). In *Polypodium*, nectaries may have been derived from simple hydathodes, structures at the vein ends near the leaf margins that exude water and diluted salts such as calcium carbonate. Species of the *Polypodium squamatum* group have pinnatifid to pinnate leaves with numerous hydathodes (de la Sota, 1966). The basal acroscopic lobe of each leaf segment bears one nectary in place of a hydathode (Koptur *et al.*, 1982). The nectary does not exhibit the white residues of salts typical for fern hydathodes, but later often becomes blackened in color as a result of fungal infections (K. Mehltreter, personal observation). In other fern species, nectaries are located on the rachis or petiole (e.g., *Drynaria rigidula*, *Polybotrya osmundacea*, *Pteridium aquilinum*; Jolivet, 1996), and their developmental origin remains unclear. Koptur *et al.* (1982) analyzed the nectar in eight fern species (five species of *Pleopeltis*, two of *Drynaria*, and one *Polybotrya*) cultivated in the California Botanical Garden, Berkeley, and found glucose, fructose, sucrose and other sugars together with a variety of amino acids. Nectar had high sugar concentrations (20–60% of fresh weight) and generally low concentrations of amino acids (<400 μmol l^{-1}). Sugars were mostly sucrose poor, as is typically reported for extrafloral nectaries in angiosperms. For ants, the fern nectar of the five species of *Pleopeltis* is a good source of energy but not of nitrogen. In contrast, *Drynaria quercifolia* had 100-fold higher concentrations of amino acids than the species of *Pleopeltis* (Koptur *et al.*, 1982) and also may serve ants as a nitrogen source. To show that fern nectaries attract ants in sufficient numbers and that these ants protect ferns against herbivores, Koptur *et al.* (1998) conducted a field experiment in Mexico by comparing the leaf damage of ferns with and without nectaries and by excluding the access of ants to the ferns. Because *P. plebeia* produces nectar only on young, still expanding leaves, ants were not attracted to adult plants where they might have facilitated spore dispersal. Sawflies and caterpillars removed 9.8 times less of the total leaf area from ant-protected plants of *P. plebeia* than from plants from which ants were excluded. In three fern species without nectaries (i.e., *Phlebodium pseudoaureum*, *Pleopeltis furfuracea* and *Polypodium plesiosorum*), leaf damage levels did not differ between plants with free and restricted ant access, indicating that these fern species do not benefit from ant protection.

Heads (1986) demonstrated some evidence for ant protection of *Pteridium* against sucking insects providing a second example of fern–ant mutualism. Three earlier field experiments could not prove that ants protect *Pteridium* from herbivore damage even when nectar production was continuous through-out the summer (Tempel, 1983, Heads and Lawton, 1984; Lawton and Heads, 1984). Finally, Rashbrook *et al.* (1992) found that it is not the nectaries that increase ant density, but the presence of honeydew-producing aphids (Homoptera), which gives *Pteridium* an ant protection against caterpillars. If these results are confirmed, the fern–ant interaction of *Pteridium* would be indirect and dependent on a third species, the aphids, that damage their fern host by sucking its phloem sap, but increase the ant protection of its host plant against other herbivores.

Domatia

The highest level of fern–ant mutualism occurs in two epiphytic genera, *Microgramma*, subgenus *Solanopteris* (Lellinger, 1977, 1984) with 5 species in the New World tropics (Wagner, 1972, Gómez, 1974, León and Beltrán, 2002), and *Lecanopteris* with 13 species in the Old World tropics (Table 7.4; Walker, 1986; Gay 1991, 1993a, 1993b). Some species of both genera develop domatia, strongly modified hollow rhizomes with several chambers where ants build their nests. At least some of these fern species benefit from the ants' waste, because their roots grow into the chambers to absorb nutrients, as shown by laboratory experiments with radioactively labeled nitrogen (Gay, 1993a). Occasional ant associations were also reported for two other fern species from Malaysia: *Lepisorus longifolius* with *Camponotus* ants and *Pyrrosia lanceolata* with mainly *Crematogaster* ants (Kaufmann and Maschwitz, 2006). Recently, Watkins *et al.* (2008) provided an example of another fern species that is associated with ants that do not build domatia. The understory epiphyte *Antrophyum lanceolatum* from Costa Rica har-bored ant nests of *Pheidole flavens* in 62% of the fern plants. Fern individuals with ant nests obtained about 54% of their nitrogen budget from the ant debris in the rhizome mats. Because this fern–ant interaction demonstrates that ferns do not require domatia to attract ants and to receive a nutritional benefit from them, Watkins *et al.* (2008) refer to it as a cryptic association.

The 13 species of *Lecanopteris* vary considerably in their rhizome morphology (Haufler *et al.*, 2003). *Lecanopteris mirabilis* does not even have hollow rhizomes.

Table 7.4 *Fern species that are often associated with ants*
Most species of Lecanopteris and Microgramma provide domatia (i.e., ant houses).
Some more frequent interactions with ant species are indicated in the text.

Species	Distribution	Reference
Antrophyum lanceolatum	Mexico-Brazil, Antilles	Watkins *et al.*, 2008
Lecanopteris balgooyi	Sulawesi	Hennipman, 1986
Lecanopteris carnosa	Malaysia, Sumatra, Borneo, Sulawesi, Philippines	Hennipman, 1986
Lecanopteris celebica	Celebes	Hennipman and Verduyn, 1987
Lecanopteris crustacea	Indonesia, Malaysia	Hennipman and Hovenkamp, 1998
Lecanopteris darnaedii	Sulawesi	Hennipman, 1986
Lecanopteris deparioides	Borneo	Hennipman and Hovenkamp, 1998
Lecanopteris holttumii	Sulawesi	Hennipman and Verduyn, 1987
Lecanopteris lomarioides	Sulawesi	Hennipman and Hovenkamp, 1998
Lecanopteris luzonensis	Philippines	Hennipman and Verduyn, 1987
Lecanopteris mirabilis	Papua New Guinea	Walker, 1986
Lecanopteris pumila	Java	Hennipman and Hovenkamp, 1998
Lecanopteris sinuosa	Borneo	Janzen, 1974
Lecanopteris spinosa	Sulawesi	Jermy and Walker, 1975
Lepisorus longifolius	Malaysia	Kaufmann and Maschwitz, 2006
Microgramma bifrons	Colombia–Peru	Wagner, 1972
Microgramma bismarckii	Peru	Rauh, 1973; León and Beltrán, 2002
Microgramma brunei	Costa Rica–Colombia	Wagner, 1972; Lellinger, 1977
Microgramma fosteri	Peru	León and Beltrán, 2002
Microgramma tuberosa	Ecuador	Maxon, 1943; Lellinger, 1984
Pyrrosia lanceolata	Malaysia	Kaufmann and Maschwitz, 2006

Instead, ants live in the external space between the host trunks and attached, disk-like curved segments of the fern rhizome that are built in a linear sequence (Walker 1986). The architecture of the rhizome thus allows ants to associate with this fern, but there is no further evidence of a stronger mutualistic relationship. However, *L. mirabilis* has another unique feature that may facilitate spore dispersal by ants (Tryon 1985). Its spores have 1–4 conspicuous, nonhygroscopic filaments which are 17–25 times the length of the spore and are similar to hygroscopic elaters of spores of *Equisetum*. Walker (1986) suggested that such entangled filaments allow for synaptospory (i.e., dispersal of the spores in groups). The main advantage of these examples of synaptospory might be that intergametophytic fertilization can more easily occur if spores are dispersed together and germinate in close proximity. Spore dispersal by ants has never been proven; therefore one might question this advantage for ferns that can produce millions of wind-dispersed spores and colonize any suitable habitats.

In *L. spinosa*, the main rhizome can form lateral lobes, which together can completely encircle its host branch, and even overgrow its own dead rhizome again, thus forming balls of up to 25 cm diameter (Jermy and Walker, 1975). The rhizome is only 2–5 cm thick, but contains water-storing tissue beneath the epidermis. Because roots also grow into the internal system of interconnected galleries, perhaps the primary function of such rhizome features is water storage. *Crematogaster* ants that inhabit the rhizomes apparently do not protect the fern, and leaves have been reported to be regularly damaged by herbivores (Jermy and Walker, 1975). The species of *Lecanopteris* may be called myrmecophilous to distinguish their less intimate relationship with ants from true mutualism of *Microgramma* with its associated *Azteca* ants. The potato fern *Microgramma brunei* develops normal primary rhizomes and secondary modified tuber-like rhizomes of 1.5–3 cm diameter, which house the ants and give the plant its common name (Wagner, 1972). The tubers are inhabited by various ant species of four different genera (Gómez, 1974). In Costa Rica, *Azteca* ants have always been present and very bellicose (Gómez, 1977), appearing to defend the plants against any intruder, but there has been no further experimental investigation, either qualitative or quantitative, of the benefit that mutualistic ferns receive from the ants.

It is surprising how few studies have been performed to gain a deeper understanding of the restricted number of fern–ant interactions. Field experiments are needed to clarify: (a) at which developmental stage of these fern species and under which environmental circumstances ant colonization occurs, (b) whether ant species favor some fern species while neglecting others and (c) to quantify the net gain of nutrient uptake and protection against herbivores of ant-inhabited ferns in comparison with ant-free fern plants.

7.5 Conclusions

Ferns and lycophytes possess a wide array of interactions with animals and fungi that is still poorly explored. Because ferns and seed plants are phylogenetic sister lineages (see Chapter 1; Schuettpelz and Pryer, 2008) one might expect that extant fern species would have evolved a similar proportion of interactions with parasitic and mycorrhizal fungi as well as herbivores, as have the extant seed plant species. Finding similar levels of mycorrhizal infection and herbivory of ferns and seed plants might support this point of view. However, seed plants have evolved additional interactions with animals and fungi such as feeding on flowers, seeds and other modified plant organs (e.g., tubers, tap roots) and tissues (e.g., floral nectaries, elaiosomes), which are lacking in ferns. Because seed plants, especially angiosperms, have developed interactions with plant pollinators and seed dispersers, it is not surprising that the proportion of interactions of ferns with animals and fungi might remain 3–7 times lower than in seed plants, even after adjusting for less research focus on ferns than on seed plants. For example, it is estimated that at least 1500–3500 interactions of ferns with herbivorous insects have not yet been documented, and that a high percentage of these interactions will lead to the discovery of new, still undescribed insect species.

Future research on fern–animal and fern–fungus interactions will require more field studies and multidisciplinary collaborative research efforts (e.g., among biochemists, mycologists, zoologists, pteridologists and ecologists), updated and comprehensive reviews of these interactions at a global scale, and comparative studies at different taxonomic levels to understand the origin of interactions (e.g., coevolutionary processes and host-plant switching). Results of studies in the last three decades have already indicated that ferns deserve more attention from field biologists as they have demonstrated some unusual and new interactions with coexisting organisms. Research efforts in this discipline include promising new biochemical and taxonomic discoveries as well as interesting ecological insights. The study of some of the opportunistic and facultative fern–insect interactions as well as other types of interactions (e.g., spore feeders) may shed light on the evolution of these interactions.

Acknowledgements

I thank Mike Barker, Paul Hanson, Peter Hietz, Rosario Medel, Joanne M. Sharpe and Lawrence R. Walker for advice and comments that considerably amended the manuscript. I also acknowledge support from the Instituto de Ecología, A. C., Xalapa, Veracruz, Mexico, CONACYT-SEMARNAT (2002-C01–0194) and CONACYT-SEP (2003-C02–43082).

References

Aboul-Nasr, A. (1998). Effects of inoculation with *Glomus intraradices* on growth, nutrient uptake and metabolic activities of squash plants under drought stress conditions. *Annals of Agricultural Science*, **1**, 119–33.

Al-Karaki, G. N. (1998). Benefit, cost and water-use efficiency of arbuscular mycorrhizal durum wheat grown under drought stress. *Mycorrhiza*, **8**, 41–5.

Arcand, N. (2007). Population structure of the Hawaiian tree fern *Cibotium chamissoi* across intact and degraded forests, Oʻahu, Hawaiʻi. Unpublished Master thesis, University of Hawaiʻi.

Ash, S. (1997). Evidence of arthropod-plant interactions in the Upper Triassic of the southwestern United States. *Lethaia*, **29**, 237–48.

Ash, S. (1999). An upper Triassic *Sphenopteris* showing evidence of insect predation from Petrified Forest National Park, Arizona. *International Journal of Plant Science*, **160**, 208–15.

Ash, S. (2000). Evidence of oribatid mite herbivory in the stem of a late Triassic tree fern from Arizona. *Journal of Paleontology*, **74**, 1065–71.

Auerbach, M. and Hendrix, S. D. (1980). Insect–fern interactions: Macrolepidopteran utilization and species–area association. *Ecological Entomology*, **5**, 99–104.

Avila-Nuñez, J. L., Otero, L. D., Silmi, S. and Calcagno-Pisarelli, M. P. (2007). Life history of *Anegmeus merida* Smith (Hymenoptera: Tenthredinidae) in the Venezuelan Andes. *Neotropical Entomology*, **36**, 22–7.

Balick, M. J., Furth, D. G. and Cooper-Driver, G. (1978). Biochemical and evolutionary aspects of arthropod predation on ferns. *Oecologia*, **35**, 55–89.

Barker, M. S., Shaw, S. W, Hickey, R. J., Rawlins, J. E. and Fetzner, J. W., Jr. (2005). Lepidopteran soral crypsis on Caribbean ferns. *Biotropica*, **37**, 314–16.

Beck-Nielsen, D. and Madsen, T. V. (2001). Occurrence of vesicular-arbuscular mycorrhiza in aquatic macrophytes from lakes and rivers. *Aquatic Botany*, **71**, 141–8.

Begon, M., Harper, J. L. and Townsend, C. R. (2006). *Ecology*, 4th edn. Malden, MA, USA: Blackwell.

Bennell, A. P. and Henderson, D. M. (1985). Rusts and other fungal parasites as aids to pteridophyte taxonomy. *Proceedings of the Royal Society of Edinburgh*, **86B**, 115–24.

Berndt, R. (2008). The rust fungi (Uredinales) on ferns in South Africa. *Mycological Progress*, **7**, 7–19.

Bird, H. (1938). The longevity of *Osmunda cinnamomea* with notes on some fern-feeding larvae. *American Fern Journal*, **28**, 151–7.

Boullard, B. (1957). La mycotrophie chez les pteridophytes. Sa fréquence, ses caractères, sa signification. *Botaniste*, **41**, 1–185.

Boullard, B. (1979). Considerations sur la symbiose fongique chez les Pteridophytes. *Syllogeus*, **19**, 1–61.

Brown, J. W., Baixeras, J., Solórzano-Filho, J. A. and Kraus, J. E. (2004). Description and life history of an unusual fern-feeding tortricid moth (Lepidoptera: Tortricidae) from Brazil. *Annals of the Entomological Society of America*, **97**, 865–71.

Brown, V. K. (1984). Secondary succession: insect–plant relationships. *BioScience*, **34**, 710–6.

Brues, C. T. (1920). The selection of food plants by insects, with special reference to lepidopterous larvae. *American Naturalist*, **54**, 313–32.

Brundrett, M. C. (1991). Mycorrhizas in natural ecosystems. In *Advances in Ecological Research*, vol. 21, ed. A. Macfayden, M. Begon and A. H. Fitter. London: Academic Press, pp. 171–313.

Brundrett, M. C. (2002). Coevolution of roots and mycorrhizas of land plants. *New Phytologist*, **154**, 275–304.

Carlisle, D. B. and Ellis, P. E. (1968). Bracken and locust ecdysones: their effects on molting in the desert locust. *Science*, **159**, 1472–4.

Carroll, G. C. (1986). The biology of endophytism in plants with particular reference to woody perennials. In *Microbiology of the Phyllosphere*, ed. N. J. Fokkema and J. Van den Heuvel. Cambridge, UK: Cambridge University Press, pp. 205–22.

Clay, K. (1989). Clavicipitaceous endophytes of grasses: their potential as biocontrol agents. *Mycological Research*, **92**, 1–12.

Coe, M. L., Dilcher, D. L., Farlow, J. O., Jarzen, D. M. and Russell, D. A. (1987). Dinosaurs and land plants. In *The Origins of Angiosperms and their Biological Consequences*, ed. E. M. Friis, W. G. Chaloner and P. R. Crane. Cambridge, UK: Cambridge University Press, pp. 225–58.

Coley, P. D. (1983). Herbivory and defense characteristics of tree species in a lowland tropical forest. *Ecological Monographs*, **53**, 209–33.

Coley, P. D. and Aide, T. M. (1989). Red coloration of tropical young leaves: a possible antifungal defence? *Journal of Tropical Ecology*, **5**, 293–300.

Coley, P. D. and Aide, T. M. (1991). Comparison of herbivory and plant defenses in temperate and tropical broad-leaved forests. In *Plant–Animal Interactions: Evolutionary Ecology in Tropical and Temperate Regions*, ed. P. W. Price, T. M. Lewinsohn, G. W. Fernandes and W. W. Benson. New York: John Wiley & Sons, pp. 25–49.

Coley, P. D. and Barone, J. A. (1996). Herbivory and plant defenses in tropical forests. *Annual Review of Ecology and Systematics*, **27**, 305–35.

Coley, P. D., Bryant, J. P. and Chapin, F. S. (1985). Resource availability and plant antiherbivore defenses. *Science*, **230**, 895–9.

Cooper, K. M. (1976). A field survey of mycorrhizas in New Zealand ferns. *New Zealand Journal of Botany*, **14**, 169–81.

Cooper-Driver, G. A. (1985a). The distribution of insects on ferns. *American Journal of Botany*, **72**, 921.

Cooper-Driver, G. A. (1985b). Anti-predation strategies in pteridophytes: a biochemical approach. *Proceedings of the Royal Society of Edinburgh*, **86B**, 397–402.

Cooper-Driver, G. A., Finch, S., Swain, T. and Bernays, E. (1977). Seasonal variation in secondary plant compounds in relation to palatability of *Pteridium aquilinum*. *Biochemical Systematics and Ecology*, **5**, 177–83.

Crawley, M. J. (1983). *Herbivory: the Dynamics of Animal–Plant Interactions*. Oxford, UK: Blackwell Scientific.

de la Sota, E. R. (1966). Revisión de las especies americanas del grupo "Polypodium squamatum" L. ("Polypodiaceae" s. str.). *Revista del Museo de La Plata, Sección Botánica*, **10**, 69–186.

Dhillion, S. S. (1993). Vesicular-arbuscular mycorrhizae of *Equisetum* species in Norway and USA: occurrence and mycotrophy. *Mycological Research*, **97**, 656–60.

Diong, C. H. (1982). Population biology and management of the feral pig (*Sus scrofa* L.) in Kipahulu Valley, Maui. Unpublished Ph.D. thesis, University of Hawai'i.

Docters van Leeuwen W. (1938). Zoocecidia. In *Manual of Pteridology*, ed. F. Verdoorn. The Hague: Nijhoff, pp. 192–5.

Duckett, J. G. and Ligrone, R. (1992). A light and electron microscope study of the fungal endophytes in the sporophyte and gametophyte of *Lycopodium cernuum* with observations on the gametophyte–sporophyte junction. *Canadian Journal of Botany*, **70**, 58–72.

Duponnois, R., Galiana, A. and Prin, Y. (2008). The mycorrhizosphere effect: a multitrophic interaction complex improves mycorrhizal symbiosis and plant growth. In *Mycorrhizae: Sustainable Agriculture and Forestry*, ed. Z. A. Siddiqui, M. S. Akhtar and K. Futai. Springer Science & Business Media B.V., pp. 227–40.

Ehrlich, P. R. and Raven, P. H. (1964). Butterflies and plants: a study in coevolution. *Evolution*, **18**, 586–608.

Eisner, T., Morgan, R. C., Attygalle, A. B., *et al.* (1997). Defensive production of quinoline by a phasmid insect (*Oreophetes peruana*). *Journal of Experimental Biology*, **200**, 2493–500.

Farr, M. L. and Horner, H. T. (1958). Fungi on *Selaginella*. *Nova Hedwigia*, **15**, 239–83.

Fernández, N., Messuti, M. I. and Fontenla, S. (2008). Arbuscular mycorrhizas and dark septate fungi in *Lycopodium paniculatum* (Lycopodiaceae) and *Equisetum bogotense* (Equisetaceae) in a Valdivian temperate forest of Patagonia, Argentina. *American Fern Journal*, **98**, 117–27.

Gange, A. C., Brown, V. K. and Farmer, L. M. (1990). A test of mycorrhizal benefit in an early successional plant community. *New Phytologist*, **115**, 85–91.

Gay, H. (1991). Ant-houses in the fern genus *Lecanopteris* Reinw. (Polypodiaceae): the rhizome morphology and architecture of *L. sarcopus* Teijsm. & Binnend. and *L. darnaedii* Hennipman. *Botanical Journal of the Linnean Society*, **106**, 199–208.

Gay, H. (1993a). Animal-fed plants: an investigation into the uptake of ant-derived nutrients by the far-eastern epiphytic fern *Lecanopteris* Reinw. (Polypodiaceae). *Biological Journal of the Linnean Society*, **50**, 221–33.

Gay, H. (1993b). Rhizome structure and evolution in the ant-associated epiphytic fern *Lecanopteris* Reinw. (Polypodiaceae). *Botanical Journal of the Linnean Society*, **113**, 135–60.

Gemma, J. N., Koske, R. E. and Flynn, T. (1992). Mycorrhizae in Hawaiian pteridophytes: occurrence and evolutionary significance. *American Journal of Botany*, **79**, 843–52.

Gerson, U. (1979). The association between pteridophytes and arthropods. *Fern Gazette* **12**, 29–45.

Golding, Y. (2007). Fern sticks. *Pteridologist*, **4**, 190.

Goltapeh, E. M., Danesh, Y. R., Prasad, R. and Varma, A. (2008). Mycorrhizal fungi: what we know and what should we know? In *Mycorrhiza*, ed. A. Varma. Berlin: Springer, pp. 3–27.

Gómez, L. D. (1974). Biology of the potato-fern *Solanopteris brunei*. *Brenesia*, **4**, 37–61.

Gómez, L. D. (1977). The Azteca ants of *Solanopteris brunei*. *American Fern Journal*, **67**, 31.

Gundale, M. J. (2002). Influence of exotic earthworms on the soil organic horizon and the rare fern *Botrychium mormo*. *Conservation Biology*, **16**, 1555–61.

Hammel, B. E., Grayum, M. H., Herrera, C. and Zamora, N. (eds.) (2004). *Manual de Plantas de Costa Rica*, vol. I. St. Louis, MO, USA: Missouri Botanical Garden Press.

Hanson, P. E. and Gómez-Laurito, J. (2005). Diversity of gall-inducing arthropods of Costa Rica. In *Biology, Ecology and Evolution of Gall-Inducing Arthropods*, vols. 1 and 2, ed. A. Raman, C. W. Schaefer and T. M. Withers. Enfield, NH, USA: Science Publishers, pp. 673–692.

Harley, J. L. and Harley E. L. (1987). A check-list of mycorrhiza in the British flora. *New Phytologist*, **105**, 1–102.

Haufler, C. H., Grammer, W. A., Hennipman, E., *et al.* (2003). Systematics of the ant-fern genus *Lecanopteris* (Polypodiaceae): testing phylogenetic hypotheses with DNA sequences. *Systematic Botany*, **28**, 217–27.

Heads, P. A. (1986). Bracken, ants and extrafloral nectaries. IV. Do wood ants (*Formica lugubris*) protect the plants against insect herbivores? *Journal of Animal Ecology*, **55**, 795–809.

Heads, P. A. and Lawton, J. H. (1984). Bracken, ants and extrafloral nectaries. II. The effect of ants on the insect herbivores of bracken. *Journal of Animal Ecology*, **53**, 1015–31.

Hegnauer, R. (1962). *Chemotaxonomie der Pflanzen, Bd. I*. Basel, Switzerland: Birkhäuser.

Hegnauer, R. (1986). *Chemotaxonomie der Pflanzen, Bd. VII*. Basel, Switzerland: Birkhäuser.

Hendrix, S. D. (1977). The resistance of *Pteridium aquilinum* (L.) Kuhn to insect attack by *Trichoplusia ni* (Hübn.). *Oecologia*, **26**, 347–61.

Hendrix, S. D. (1980). An evolutionary and ecological perspective of the insect fauna of ferns. *American Naturalist*, **115**, 171–96.

Hendrix, S. D. and Marquis, R. J. (1983). Herbivore damage to three tropical ferns. *Biotropica*, **15**, 108–11.

Hennipman, E. (1986). Notes on the ant-ferns of *Lecanopteris sensu stricto* in Sulawesi, with description of two new species. *Kew Bulletin*, **41**, 781–8.

Hennipman, E. and Hovenkamp, P. (1998). *Lecanopteris*. In *Flora Malesiana*, Series II, vol. 3, *Ferns and Fern Allies*, ed. C. Kalkman and H. Nooteboom. Leiden, The Netherlands: Rijksherbarium/Hortus Botanicus, pp. 59–76.

Hennipman, E. and Verduyn, G. P. (1987). A taxonomic revision of the genus *Lecanopteris* (Polypodiaceae) in Sulawesi, Indonesia. *Blumea*, **32**, 313–19.

Hikino, H., Okuyama, T., Jin, H. and Takemoto, T. (1973). Screening of Japanese ferns for phytoecdysones. I. *Chemical and Pharmaceutical Bulletin*, **21**, 2292–302.

Hummel, J., Gee, C. T., Südekum, K.-H., *et al.* (2008). In vitro digestibility of fern and gymnosperm foliage: implications for sauropod feeding ecology and diet selection. *Proceedings of the Royal Society, Series B*, **275**, 1015–21.

Iqbal, S. H., Yousaf, M. and Younus, M. (1981). A field survey of mycorrhizal associations in ferns of Pakistan. *New Phytologist*, **87**, 69–79.

Janzen, D. H. (1974). Epiphytic myrmecophytes in Sarawak: mutualism through the feeding of plants by ants. *Biotropica*, **6**, 237–59.

Jermy, A. C. and Walker, T. G. (1975). *Lecanopteris spinosa*: a new ant-fern from Indonesia. *Fern Gazette*, **11**, 165–76.

Jolivet, P. (1996). *Ants and Plants*. Leiden, The Netherlands: Backhuys Publishers.

Jumpponen, A. (2001). Dark septate endophytes: are they mycorrhizal? *Mycorrhiza*, **11**, 207–211.

Jumpponen, A. and Trappe, J. M. (1998). Dark septate endophytes: a review of facultative biotrophic root-colonizing fungi. *New Phytologist*, **140**, 295–310.

Kaplanis, J. N., Thompson, M. J., Robbins, W. E. and Bryce, B. M. (1967). Insect hormones: alfa ecdysone and 20-hydroxyecdysone in bracken fern. *Science*, **157**, 1436–8.

Kaufmann, E. and Maschwitz, U. (2006). Ant-gardens of tropical Asian rainforests. *Naturwissenschaften*, **93**, 216–27.

Kessler, A. and Baldwin, I. T. (2001). Defensive function of herbivore-induced plant volatile emissions in nature. *Science*, **291**, 2141–4.

Koptur, S., Smith, A. R. and Baker, I. (1982). Nectaries in some neotropical species of *Polypodium* (Polypodiaceae): preliminary observations and analyses. *Biotropica*, **14**, 108–13.

Koptur, S., Rico-Gray, V. and Palacios-Rios, M. (1998). Ant protection of the nectaried fern *Polypodium plebeium* in central Mexico. *American Journal of Botany*, **85**, 736–9.

Koske, R. E., Gemma, J. N. and Flynn, T. (1992). Mycorrhizae in Hawaiian angiosperms: a survey with implications for the origin of the native flora. *American Journal of Botany*, **79**, 853–62.

Kramer, K. U., Schneller, J. J. and Wollenweber, E. (1995). *Farne und Farnverwandte*. Stuttgart, Germany: Georg Thieme Verlag.

Lawton, J. H. (1976). The structure of the arthropod community on bracken (*Pteridium aquilinum* (L.) Kuhn). *Botanical Journal of the Linnean Society*, **73**, 187–216.

Lawton, J. H. and Heads, P. A. (1984). Bracken, ants and extrafloral nectaries. I. The components of the system. *Journal of Animal Ecology*, **53**, 995–1014.

Leahy, R., Schubert, T., Strandberg, J., Stamps, B. and Norman, D. (1995). Anthracnose of leatherleaf fern. *Florida Department of Agriculture and Customer Services, Plant Pathology Circular*, **372**, 4 pp.

Lehnert, M., Kottke, I., Setaro, S. and Kessler, M. (2007). New insights on the mycorrhizal infections in ferns: an example from southern Ecuador. In 'Diversity and evolution of Pteridophytes, with emphasis on the neotropics', ed. M. Lehnert. Unpublished Ph.D. thesis, University of Göttingen, Germany, pp. 437–49.

Lellinger, D. B. (1977). Nomenclatural notes on some ferns of Costa Rica, Panama and Colombia. *American Fern Journal*, **67**, 58–60.

Lellinger, D. B. (1984). New combinations and some new names in ferns. *American Fern Journal*, **74**, 56–60.

León, B. and Beltrán, H. (2002). A new *Microgramma* subgenus *Solanopteris* (Polypodiaceae) from Peru and a new combination in the subgenus. *Novon*, **12**, 481–5.

Lowman, M. D. (1984). An assessment of techniques for measuring herbivory: is rain forest defoliation more intense than we thought? *Biotropica*, **16**, 264–8.

Lowman, M. D. (1985). Insect herbivory in Australian rain forests: is it higher than in the Neotropics? *Proceedings of the Ecological Society of Australia*, **14**, 109–19.

Markham, K., Chalk, T. and Stewart, C. N., Jr. (2006). Evaluation of fern and moss protein-based defenses against phytophagous insects. *International Journal of Plant Sciences*, **167**, 111–17.

Maxon, W. R. (1943). New tropical American ferns. XIV. *American Fern Journal*, **33**, 133–7.

Medel, R. and Lorea-Hernández, F. (2008). Hyaloscyphaceae (Ascomyota) growing on tree ferns in Mexico. *Mycotaxon*, **106**, 209–18.

Mehltreter, K. (1995). Species richness and geographical distribution of montane pteridophytes of Costa Rica, Central America. *Feddes Repertorium*, **106**, 563–84.

Mehltreter, K. and García-Franco, J. G. (2008). Leaf phenology and trunk growth of the deciduous tree fern *Alsophila firma* (Baker) D. S. Conant in a lower montane Mexican forest. *American Fern Journal*, **98**, 1–13.

Mehltreter, K. and Palacios-Rios, M. (2003). Phenological studies of *Acrostichum danaeifolium* (Pteridaceae, Pteridophyta) at a mangrove site on the Gulf of Mexico. *Journal of Tropical Ecology*, **19**, 155–62.

Mehltreter, K. and Tolome, J. (2003). Herbivory on three tropical fern species of a Mexican cloud forest. In *Pteridology in the New Millennium*, ed. S. Chandra and M. Srivastava. Dordrecht, The Netherlands: Kluwer Academic Publishers, pp. 375–81.

Mehltreter, K., Rojas, P. and Palacios-Rios, M. (2003). Moth-larvae damaged giant leather-fern *Acrostichum danaeifolium* as host for secondary colonization by ants. *American Fern Journal*, **93**, 48–54.

Mehltreter, K., Flores-Palacios, A. and García-Franco, J. (2005). Host preferences of vascular trunk epiphytes in a cloud forest of Veracruz, Mexico. *Journal of Tropical Ecology*, **21**, 651–660.

Mehltreter, K., Hülber, K. and Hietz P. (2006). Herbivory on epiphytic ferns of a Mexican cloud forest. *Fern Gazette*, **17**, 303–9.

Mickel, J. T. and Smith, A. R. (2004). *The Pteridophytes of Mexico*. New York: The New York Botanical Garden.

Midgley, J. J., Midgley, G. and Bond, W. J. (2002). Why were dinosaurs so large? A food quality hypothesis. *Evolutionary and Ecological Research*. **4**, 1093–5.

Miller, J. C., Janzen, D. H. and Hallwachs, W. (2006). *100 Caterpillars: Portraits from the Tropical Forests of Costa Rica*. Cambridge, MA, USA: Harvard University Press.

Miller, J. C., Janzen, D. H. and Hallwachs, W. (2007). *100 Butterflies and Moths: Portraits from the Tropical Forests of Costa Rica*. Cambridge, MA, USA: Harvard University Press.

Moran, R. C., Klimas, S. and Carlsen, M. (2003). Low-trunk epiphytic ferns on tree ferns versus angiosperms in Costa Rica. *Biotropica*, **35**, 48–56.

Moteetee, A., Duckett, J. G. and Russell, A. J. (1996). Mycorrhizas in the ferns of Lesotho. In *Pteridology in Perspective*. ed. J. M. Camus, M. Gibby and R. J. Johns. Kew: Royal Botanic Gardens, pp. 621–31.

Müller, E. and Schneller, J. J. (1977). A new record of *Synchytrium athyrii* on *Athyrium filix-femina*. *Fern Gazette*, **11**, 313–4.

Newman, E. I. and Reddell, P. (1987). The distribution of mycorrhizas among families of vascular plants. *New Phytologist*, **106**, 745–51.

Ottosson, J. G. and Anderson, J. M. (1983). Number, seasonality and feeding habits of insects attacking ferns in Britain: an ecological consideration. *Journal of Animal Ecology*, **52**, 385–406.

Palmieri, M. and Swatzell, L. J. (2004). Mycorrhizal fungi associated with the fern *Cheilanthes lanosa*. *Northeastern Naturalist*, **11**, 57–66.

Petrini, O. (1991). Fungal endophytes of tree leaves. In *Microbial Ecology of Leaves*, ed. J. A. Andrews and S. S. Hirano. New York: Springer-Verlag, pp. 179–97.

Petrini, O. (1992). Ecology, metabolite production, and substrate utilization in endophytic fungi. *Natural Toxins*, **1**, 185–96.

Petrini, O. (1993). Fungal endophytes of bracken (*Pteridium aquilinum*) with some reflections on their use in biological control. *Sydowia*, **44**, 282–93.

Radhika, K. P. and Rodrigues, B. F. (2007). Arbuscular mycorrhizae in association with aquatic and marshy plant species in Goa, India. *Aquatic Botany*, **86**, 291–4.

Ramos, J. A. (1994). Fern frond feeding by the Azores bullfinch. *Journal of Avian Biology*, **25**, 344–7.

Ramos, J. A. (1995). The diet of the Azores bullfinch *Pyrrhula murina* and floristic variation within its range. *Biological Conservation*, **71**, 237–49.

Rashbrook, V. K., Compton, S. G. and Lawton, J. H. (1992). Ant–herbivore interactions: reasons for the absence of benefits to a fern with foliar nectaries. *Ecology*, **73**, 2167–74.

Rauh, W. (1973). *Solanopteris bismarckii* Rauh. *Abhandlungen der Akademie der Wissenschaften und der Literatur Mainz, Mathematisch-Naturwissenschaftliche Klasse*, **5**, 223–56.

Raupp, M. J. and Denno, R. F. (1983). Leaf age as a predictor of herbivore distribution and abundance. In *Variable Plants and Herbivores in Natural and Managed Systems*, ed. R. F. Denno and M. S. McClure. New York: Academic Press, pp. 91–124.

Redhead, S. A. (1984). Two fern-associated mushrooms, *Mycena lohwagii* and *M. pterigena* in Canada. *Le Naturaliste Canadien*, **111**, 439–42.

Rhoades, D. F. (1979). Evolution of plant chemical defences against herbivores. In *Herbivores: Their Interaction with Secondary Plant Metabolites*, ed. G. A. Rosenthal and D. H. Janzen. New York: Academic Press, pp. 3–54.

Rothwell, G. W. (1996). Phylogenetic relationships of ferns: a paleobotanic perspective. In *Pteridology in Perspective*. ed. J. M. Camus, M. Gibby and R. J. Johns. Kew, UK: Royal Botanic Gardens, pp. 395–404.

Rothwell, G. W. and Stockey, R. A. (2008). Phylogeny and evolution of ferns: a paleontological perspective. In *Biology and Evolution of Ferns and Lycophytes*, ed. T. A. Ranker and C. H. Haufler. Cambridge, UK: Cambridge University Press, pp. 332–66.

Rowell, C. H. F., Rowell-Rahier, M., Braker, H. E., Cooper-Driver, G. and Gómez, P. L. D. (1983). The palatability of ferns and the ecology of two tropical forest grasshoppers. *Biotropica*, **15**, 207–16.

Ruehlmann, T. E., Matthews, R. W. and Matthews, J. R. (1988). Roles for structural and temporal shelter-changing by fern-feeding lepidopteran larvae. *Oecologia*, **75**, 228–32.

Russell, G. B. and Fenemore, P. G. (1971). Insect moulting hormone activity in some New Zealand ferns. *New Zealand Journal of Science*, **14**, 31–5.

Samuels, G. J. and Rogerson, C. T. (1990). Some ascomycetes (fungi) occurring on tropical ferns. *Brittonia*, **42**, 105–15.

Schmid, E. and Oberwinkler, F. (1993). Mycorrhiza-like interaction between the achlorophyllous gametophyte of *Lycopodium clavatum* L. and its fungal endophyte studied by light and electron microscopy. *New Phytologist*, **124**, 69–81.

Schmid, E. and Oberwinkler, F. (1995). A light- and electron-microscopic study on a vesicular-arbuscular host–fungus interaction in gametophytes and young sporophytes of the Gleicheniaceae (Filicales). *New Phytologist*, **129**, 317–24.

Schmid, E., Oberwinkler, F. and Gómez, L. D. (1995). Light and electron microscopy of a host endophyte interaction in the roots of some epiphytic ferns from Costa Rica. *Canadian Journal of Botany*, **73**, 991–6.

Schneider, H., Schuettpelz, E., Pryer, K. M., *et al.* (2004). Ferns diversified in the shadow of angiosperms. *Nature*, **428**, 553–7.

Schreiner, I., Nafus, D. and Pimentel, D. (1984). Frequency of cyanogenesis in bracken in relation to shading and winter severity. *American Fern Journal*, **74**, 51–5.

Schüßler, A. (2004). Das fünfte Pilz-Phylum: die *Glomeromycota. Biospektrum*, **10**, 741–2.

Schüßler, A., Schwarzott, D. and Walker, C. (2001). A new phylum, the *Glomeromycota*: phylogeny and evolution. *Mycological Research*, **105**, 1413–21.

Schuettpelz, E. and Pryer, K. M. (2008). Fern phylogeny. In *Biology and Evolution of Ferns and Lycophytes*. ed. T. A. Ranker and C. H. Haufler. Cambridge, UK: Cambridge University Press, pp. 395–416.

Sessions, L. and Kelly, D. (2002). Predator-mediated apparent competition between an introduced grass, *Agrostis capillaris*, and a native fern, *Botrychium australe* (Ophioglossaceae), in New Zealand. *Oikos*, **96**, 102–9.

Shuter, E. and Westoby, A. (1992). Herbivorous arthropods on bracken *Pteridium aquilinum* (L.) Kuhn in Australia compared with elsewhere. *Australian Journal of Ecology*, **17**, 329–39.

Siddiqui, Z. A. and Pichtel, J. (2008). Mycorrhizae: an overview. In *Mycorrhizae: Sustainable Agriculture and Forestry*, ed. Z. A. Siddiqui, M. S. Akhtar and K. Futai. Springer Science & Business Media B.V., pp. 1–35.

Sjamsuridzal, W., Nishida, H., Ogawa, H., Kakishima, M. and Sugiyama, J. (1999). Phylogenetic positions of rust fungi parasitic on ferns: evidence from 18S rDNA sequence analysis. *Mycoscience*, **40**, 21–7.

Smart, J. and Hughes, N. (1973). The insect and the plant: progressive palaeoecological integration. In *Insect Plant Relationships*, no. 6, ed. H. Van Emden. London: Royal Entomological Society, pp. 143–55.

Smith, D. R. (2005). Two new fern-feeding sawflies of the genus *Anegmeus* Hartig (Hymenoptera: Tenthredinidae) from South America. *Proceedings of the Entomological Society of Washington*, **107**, 273–8.

Smith, S. E. and Read, D. J. (1997). *Mycorrhizal Symbiosis*, 2nd edn. London: Academic Press.

Solis, M. A., Yen, S.-H. and Goolsby, J. (2004). Description and life history of *Lygomusotima* new genus, and *Neomusotima conspurcatalis* (Lepidoptera: Crambidae) from Australia and Southeastern Asia feeding on *Lygodium microphyllum* (Schizaeaceae). *Annals of the Entomological Society of America*, **97**, 64–76.

Somvanshi, R., Lauren, D. R., Smith, B. L., *et al.* (2006). Estimation of the fern toxin, ptaquiloside, in certain Indian ferns other than bracken. *Current Science*, **91**, 1547–52.

Soo Hoo, C. and Fraenkel, G. (1964). The resistance of ferns to the feeding of *Prodenia eridania* larvae. *Annals of the Entomological Society of America*, **57**, 788–90.

Southwood, T. (1973). The insect plant relationship: an evolutionary perspective. In *Insect Plant Relationships*, no. 6, ed. H. van Emden. London: Royal Entomological Society, pp. 3–30.

Taylor, T. N., Remy, W., Hass, H. and Kerp, H. (1995). Fossil arbuscular mycorrhizae from the Early Devonian. *Mycologia*, **87**, 560–73.

Tempel, A. S. (1983). Bracken fern (*Pteridium aquilinum*) and nectar-feeding ants: a nonmutualistic interaction. *Ecology*, **64**, 1411–22.

Tryon, A. F. (1985). Spores of myrmecophytic ferns. *Proceedings of the Royal Society of Edinburgh*, **86B**, 105–10.

Wagner, W. H. (1972). *Solanopteris brunei*, a little-known fern epiphyte with dimorphic stems. *American Fern Journal*, **62**, 33–43.

Walker, T. G. (1986). The ant-fern *Lecanopteris mirabilis*. *Kew Bulletin*, **41**, 533–45.

Watkins, J. E., Jr., Cardelús, C. and Mack, M. (2008). Ants mediate nitrogen relations of an epiphytic fern. *New Phytologist*, **180**, 5–8.

Weiczorek, H. (1973). Zur Kenntnis der Adlerfarninsekten: Ein Beitrag zum Problem der biologischen Bekämpfung von *Pteridium aquilinum* (L.) Kuhn in Mitteleuropa. *Annalen für Angewandte Entomology*, **72**, 337–58.

Weintraub, J. D., Lawton, J. H. and Scoble, M. J. (1995). Lithinine moths on ferns: a phylogenetic study of insect-plant interactions. *Biological Journal of the Linnean Society*, **55**, 239–50.

Wikström, N., Kenrick, P. and Chase, M. (1999). Epiphytism and terrestrialization of tropical *Huperzia* (Lycopodiaceae). *Plant Systematics and Evolution*, **218**, 221–43.

Williams-Linera, G. and Baltazar, A. (2001). Herbivory on young and mature leaves of one temperate deciduous and two tropical evergreen trees in the understory and canopy of a Mexican cloud forest. *Selbyana*, **22**, 213–18.

Winkler, M., Hülber, K., Mehltreter, K., García-Franco, J. and Hietz, P. (2005). Herbivory in epiphytic bromeliads, orchids and ferns in a Mexican montane forest. *Journal of Tropical Ecology*, **21**, 147–54.

Yen, S.-H., Solis, M. A. and Goolsby, J. (2004). *Austromusotima*, a new musotimine genus (Lepidoptera: Crambidae) feeding on old world climbing fern, *Lygodium microphyllum* (Schizaeaceae). *Annals of the Entomological Society of America*, **97**, 397–410.

Zangerl, A. R. and Bazzaz, F. A. (1992). Theory and pattern in plant defense allocation. In *Plant Resistance to Herbivores and Pathogens: Ecology, Evolution, and Genetics*, ed. R. S. Fritz, and E. L. Simms. Chicago, IL, USA: The University of Chicago Press, pp. 363–91.

Zimmerman, E. C. (1970). Adaptive radiation in Hawaii with special reference to insects. *Biotropica*, **2**, 32–8.

8

Problem ferns: their impact and management

RODERICK C. ROBINSON, ELIZABETH SHEFFIELD
AND JOANNE M. SHARPE

Key points

1. Despite the popular image of ferns as decorative, innocuous plants, certain fern species can become substantial problems where human activities disturb the natural equilibrium. Making the distinction between native and alien species helps us to understand how some ferns become problematic in the first place and how such problems can be managed.
2. About 60 species of ferns create problems for ecology and conservation in terrestrial and aquatic environments. Some of these ferns have significant negative impacts on human and animal health, food production and management of both land and water.
3. Where legislative or other preemptive controls fail, problem ferns need to be managed by timely and effective combinations of physical, chemical and biological methods. Researchers continue to improve methods of managing problem fern species in order to enhance efficacy and to minimize damage to nontarget vegetation and the local environment.
4. The full human, economic and environmental costs of problem ferns have not been investigated on a global basis. Continued international development of effective legislation and chemical and biological management of problem ferns will be required in order to contain their further spread which, in some cases, may be extensive and catastrophic.

8.1 Introduction

At least 60 fern species (see Table 8.1) have the proven or potential ability to occupy areas where they may create a variety of problems. The terrestrial ferns in this group can disrupt local ecosystems, conservation efforts, wildlife management and the productivity of land (including grazing lands, certain crops and forestry). The aquatic ferns in this group can also occupy freshwater lakes, rivers and canals, interfering with navigation and transport, water flow, flood management, fishing, water quality and ecology. Some of these ferns are also deleterious to human and animal health due to their innate toxicity, carcinogenicity or their ability to harbor

Fern Ecology, ed. Klaus Mehltreter, Lawrence R. Walker and Joanne M. Sharpe. Published by Cambridge University Press. © Cambridge University Press 2010.

Table 8.1 *Ferns that are problematic because of their invasiveness, potential to threaten or displace native species in natural habitats, or otherwise negatively impact the environment, resource use or human health*

Species	Growth form	Native range	Areas of concern
Native terrestrial			
Blechnum orientale	Long creeping	Australia, East and Southeast Asia, Pacific islands	Tropics of Southeast Asia and Australia
Dicksonia squarrosa	Tree fern	New Zealand	New Zealand
Dicranopteris curranii	Scrambling	Southeast Asia	Malaysia
Dicranopteris linearis	Scrambling	Old World tropics	Malaysia, Sri Lanka
Dicranopteris flexuosa	Scrambling	New World tropics	Southeastern USA
Dryopteris filix-mas	Rosette	Temperate northern hemisphere	UK
Equisetum arvense	Long creeping and tuberous	Northern hemisphere	Northern hemisphere
Equisetum telmateia	Creeping	African, Asia (temperate), Europe, western USA	USA (OR)
Gleichenella pectinata	Scrambling	New World tropics	New World tropics
Gleichenia dicarpa	Scrambling	Eastern Australia, New Zealand	New Zealand
Gleichenia japonica	Scrambling	Japan	Japan
Histiopteris incisa	Long creeping	Southern hemisphere and Taiwan.	Temperate southern hemisphere (e.g., Queensland, Tasmania, New Zealand)
Hypolepis muelleri	Long creeping	East Australia	Western Australia
Lygodium venustum	Climbing	South and Central America, Caribbean.	Mexico
Paesia scaberula	Creeping	New Zealand	New Zealand
Pteridium spp.	Long creeping	Cosmopolitan	Europe, Azores, Turkey, New Zealand, Australia, Central and South America, Caribbean, India, Japan, China
Sticherus bifidus	Scrambling	New World tropics	New World tropics

Table 8.1 (*cont.*)

Species	Growth form	Native range	Areas of concern
Alien terrestrial			
Adiantum capillus-veneris	Short creeping	West and south Europe, Africa, North America, Central America, Hawaiian Islands	New Zealand, Mediterranean
Adiantum hispidulum	Rosette	Asia, East Africa, eastern Australia, India, New Zealand	Hawaiian Islands, southern hemisphere, USA
Adiantum raddianum	Rosette	Tropical America, West Indies	Hawaiian Islands, New Zealand
Adiantum tenerum	Short creeping	Central America, Florida, Puerto Rico, Venezuala	New Zealand
Angiopteris evecta	Rosette	Tropical Asia (e.g., Madagascar, northeastern Australia, India, eastern Himalaysa, Sikkim, Assam	Hawaiian Islands, Jamaica, Costa Rica
Blechnum appendiculatum (syn. *B. occidentale*)	Long creeping (stolons)	Tropical America	Hawaiian Islands
Cheilanthes viridis (syn. *Pellaea viridis*)	Short creeping	Southern Africa, India, Asia	Australia, New Zealand, Florida, Hawaiian Islands
Cyclosorus dentatus (syn. *Thelypteris dentata*)	Rosette	Paleotropics	Costa Rica, Hawaiian Islands, Mediterranean, southeastern USA
Cyrtomium falcatum	Rosette	Asia	Hawaiian Islands, Western Europe, Macaronesia
Deparia petersenii (syn. *Athyrium japonicum*)	Short creeping	Africa, New Zealand, Polynesia, southeast Asia	Southeastern USA, southeastern Brazil, Hawaiian Islands, Azores, Madeira, Reunion Island
Doodia caudata	Rosette	Australia, New Zealand	Madeira
Hypolepis rugosula	Long creeping	Southern hemisphere	Southeastern and southwestern Australia
Lindsaea ensifolia	Short creeping	Africa, India, Pacific (Malesia to west Melanesia)	Hawaiian Islands
Lygodium flexuosum	Climbing	Asia, Australia	USA

Table 8.1 (*cont.*)

Species	Growth form	Native range	Areas of concern
Lygodium japonicum	Climbing	China, eastern and tropical Asia	Australia, Hawaiian Islands, Philippines, Puerto Rico, Taiwan, southeastern USA (AL, FL)
Lygodium microphyllum	Climbing	Africa, Asia, Australia, Pacific (Micronesia, Fiji)	Malaysia, Vietnam, Jamaica, Guyana, southeastern USA
Macrothelypteris torresiana	Rosette	Natal, Madagascar, Japan, Indonesia, Thailand, Malaysia and Queensland	Neotropics, southeastern USA, Puerto Rico, southern Mexico, Costa Rica, Trinidad, Samoa, Cook Islands, French Polynesia, north cape of New Zealand, Hawaiian Islands, Mexico (Oaxaca, Veracruz, Chiapas)
Microsorum pustulatum	Creeping (epiphyte)	Australia and New Zealand	Ireland, UK
Microsorum scolopendria	Long creeping	Old World tropics	Tropical America
Nephrolepis brownii (= N. multiflora)	Stoloniferous	India, tropical Asia, Taiwan, Philippines	Carribbean, Florida, Hawaiian Islands, Puerto Rico
Nephrolepis cordifolia	Stoloniferous and tuberous	Australia, Himalayas	Western Australia, Ecuador, Galapagos, Jamaica, New Zealand, southeastern USA
Nephrolepis falcata	Stonoliferous	Australia	Hawaiian Islands, southeastern USA
Nephrolepis hirsutula	Stonoliferous	Australia, Asia, Pacific	Hawaiian Islands
Phlebodium aureum	Long creeping (epiphyte)	Tropical America	Hawaiian Islands, Cook Islands
Phymatosorus grossus	Long creeping	Tropical Africa, southern Asia, Australia, New Guinea, South Pacific	Hawaiian Islands
Pityrogramma austroamericana	Rosette	New World tropics	Hawaiian Islands
Pityrogramma calomelanos	Rosette	Central America, Mexico, South America, West Indies	Hawaiian Islands, southern USA, tropical Africa, Southeast Asia and Pacific

Table 8.1 (*cont.*)

Species	Growth form	Native range	Areas of concern
Platycerium bifurcatum	Rosette (epiphyte)	Australia (Queensland, New South Wales), New Caledonia, New Guinea	Florida, Hawaiian Islands
Platycerium superbum	Rosette (epiphyte)	Australia	Hawaiian Islands
Pteris cretica	Rosette	Worldwide tropics	UK
Pteris vittata	Short creeping	China	Worldwide tropics and subtropics; especially USA, Hawaiian Islands, Caribbean and Mediterranean, UK
*Sphaeropteris cooperi (*syn. *Cyathea cooperi)*	Tree fern	Eastern Australia	Western Australia, South Africa, Mauritius, Reunion Island, New Zealand, Hawaiian Islands, French Polynesia
Tectaria incisa	Rosette	Caribbean and tropical America	Florida
Alien aquatic			
Azolla filiculoides	Floating	Western USA and South America	Many countries worldwide
Azolla pinnata	Floating	Africa, Asia, Australia	Australia
Ceratopteris thalictroides	Emergent or floating, rosette	Worldwide in tropical areas	USA (Florida, California), Hawaiian Islands
Marsilea macropoda	Emergent, long creeping	Western USA	Florida, Louisiana
Marsilea minuta	Emergent, long creeping	Africa, India, Malaysia	Southeastern USA
Marsilea mutica	Emergent, long creeping	Australia, New Caledonia	Southeastern USA
Marsilea quadrifolia	Emergent, long creeping	Europe, Azores	Northeastern USA
Marsilea vestita	Emergent, long creeping	Western USA	Florida
Salvinia auriculata	Aquatic, floating	Mexico, Central and South America	USA (CA, MA, NC, OK, SC, UT)
Salvinia biloba	Aquatic, floating	Bolivia, Brazil, Paraguay	USA (CA, MA, NC, OK, SC, UT)

Table 8.1 (*cont.*)

Species	Growth form	Native range	Areas of concern
Salvinia herzogii	Floating	Bolivia, Brazil, Paraguay and Argentina	USA (CA, MA, NC, OK, SC, UT)
Salvinia minima	Floating	Southeastern United States, West Indies, Central and South America	Neotropics, Louisiana USA
Salvinia molesta	Floating	Southern Brazil, Paraguay, Uruguay	Many countries worldwide

Source: The information is drawn from authors cited in the text; Hawaiian Ecosystems at Risk Project (2008); USDA, ARS, National Genetic Resources Program (2008); USDI, US Geological Survey, (2008); and a considered assessment of minor online sources too numerous to list.

pests that carry disease. Ferns that create such problems are herein called "problem ferns". Many of these ferns tend to be invasive and share two essential characteristics. First, they are capable of rapid spread by means of spores, rhizome growth, fragmentation or various combinations of these mechanisms. Second, they tend to be sun-loving plants during their sporophyte stage, which sets them apart from many other ferns. Although these characteristics are without doubt contributory, other factors primarily determine whether a fern species becomes a problem or not.

In their natural environment, ferns that are potentially invasive are held in check by competition, herbivory or disease. In a stable environment, the spread of opportunistic, sun-loving ferns may be severely restricted by shade from woodland and forest canopies. When the natural equilibrium is disturbed (e.g., due to fire, erosion or landslides) potentially invasive fern species can exploit opportunities for colonization (see Chapter 6). However, it is when the disturbance is caused by human interference that ferns most often become a nuisance due to the problems they create. This situation arises in two ways.

First, human manipulation of the environment (e.g., drainage, tree felling or burning) can alter the habitat substantially so that light levels increase and the number of natural competitors, herbivores and disease vectors decrease. Native, shade-intolerant ferns that largely exist on woodland margins can then spread and become a problem by taking advantage of a newly disturbed, or continuously disturbed, environment (Page, 1982). After natural disturbances, these native fern species may represent a primary stage in the regeneration of woodland and forest, but it is after human disturbance that this colonizing behavior tends to create problems.

Second, humans may transport a fern deliberately or accidentally into a nonnative terrestrial or aquatic habitat where it subsequently escapes into a region that lacks the natural competition, herbivory or diseases that would otherwise impose restraint. These introduced, alien fern species do not require further human inter-ference to spread once they have escaped into their new environment and therefore can proliferate rapidly, sometimes with catastrophic results. The unintended escape of alien fern species from parks and gardens, or subsequently from garden waste, has had serious consequences in many parts of the world. Spores provide the most common mode of dispersal for such alien terrestrial ferns. Many species that pose problems in the wild are also those that, under glasshouse cultivation, can appear unexpectedly in neighboring pots of other species (C. N. Page, personal commu-nication). This suggests that these invasive species may have particularly effective mechanisms for spore dispersal, spore germination or both. Alternatively, budding and fragmentation prove a more effective means of dispersal for some of the aquatic problem ferns.

The Hawaiian Islands provide a good example of problems created by introduc-tion of alien fern species. Of some 260 species of alien ferns that have been, or are currently cultivated on the islands, mostly in botanical gardens and arboreta, over 30 species have escaped and naturalized into wild habitats (Wilson, 1996). These naturalized ferns represent less than 16% of the total fern species on the islands but over half of these alien fern species have become widespread and abundant (Palmer, 2003). Some of these alien problem ferns are displacing native species while others are hybridizing with native ferns, posing a serious threat to Hawaiian ecosystems. The large number of alien problem ferns in the Hawaiian Islands stands out as an exceptional case. In other parts of the world it is often a single fern species, native or alien, that may be capable of creating widespread environmental and economic disruption.

The distinction between native and alien problem ferns is maintained throughout this chapter based on the following definitions (Marinelli, 2004). A *native species* is a plant that naturally grows wild in a particular area, whereas an *alien species* is a plant occurring outside its natural range as a result of intentional or accidental dispersal by human activities. This distinction is necessary because it reflects not only the manner in which ferns become problems but also their management, especially when native problem ferns provide benefits for the local environment. Conversely, although an introduced alien species may have long-standing, tradi-tional uses in parts of the world where it is indigenous, a goal of complete eradica-tion is not unreasonable when it is invasive in a nonnative habitat.

Management of problem ferns at all times must be tempered with respect for the ecology of nontarget vegetation and the wider environment. Without efficient and timely implementation of effective control measures, problem ferns have the potential to become rampant with rapid, catastrophic consequences for both the environment and human welfare. Early detection of problem ferns and use of preemptive control measures rarely occur and, therefore, management usually requires various combinations of physical, chemical and biological controls. The escape of an alien fern, possibly from just a single garden, can lead to multimillion dollar annual control programs requiring very long-term commitments. In this chapter, we describe terrestrial and aquatic problem ferns in categories of native or alien, their consequences, their management and economic impact. Not all the problem ferns listed in Table 8.1 can be discussed in this chapter due to limitations of space. Instead, we focus on the worst problem ferns and some other selected species that illustrate principles of ecology and problem management.

8.2 Native problem ferns

Of all native ferns (Table 8.1), *Pteridium* causes the greatest range of problems (Table 8.2). The resulting ecological and management challenges are complex and will be discussed in detail. We will also describe some examples of other native terrestrial ferns that cause problems, usually because of human alteration of local environments. Aquatic ferns that do not appear to create problems in their native habitats but create significant problems as alien species are discussed in Section 8.4.

8.2.1 Pteridium

The genus *Pteridium* comprises several species, subspecies and varieties (as reclassified by Thomson, 2004), collectively known as bracken. Despite regional variations in *Pteridium* (Taylor, 1990), its principal taxa appear to create a similar range of potential problems and therefore these ferns will be discussed at the generic level. De Winter and Amoroso (2003) imply that *Pteridium* poses few problems in Southeast Asia but this may reflect highly efficient land use required to support a large human population rather than any differences in climate or the varieties of *Pteridium* found there. Southwestern China is certainly not exempt from *Pteridium*-related problems (Xu, 1992).

Page (1982) claimed that *Pteridium* is the world's most widespread vascular plant. Its success may be explained by its comprehensive package of survival strategies and its exceptional ability to capitalize on human disturbance of the landscape. *Pteridium* is associated with problems across many parts of the world with multiple impacts on ecology and human welfare. A large body of work, reviewed by Marrs and Watt (2006), refers to the UK where the climate is ideal for *Pteridium* and where the problems created are especially severe. Pakeman *et al.* (1996) estimated that *Pteridium* occupies an area of about 1.7 million ha,

Table 8.2 *A summary of problems caused by* Pteridium *in the UK*

Category	Habitat	Consequences
Productivity of land	Permanent grazing (rough pasture, natural grasslands, heather moorland, heathland) especially when poorly managed	Decreased areas for grazing. Decreased breeding and hunting areas for game birds. Decreased levels of health in livestock, wild birds and game. Hindrance to shepherding. Places for sick animals to escape detection.
Conservation areas	Woodland, moorland, heathland, grassland, sand dunes, coastal cliffs	Hinders access. Increases tick-related health risks. Impedes rewilding schemes.
Forestry	Forest plantations	Collapsed leaves break or disfigure seedling trees, assisted by snow. Litter accumulation increases fire risk.
Tick-borne diseases	Rough grazing Risks to farm animals, wild animals, game birds and humans	Sheep tick, vector for various diseases including louping-ill, tick-borne fever, staphylococcal pyaemia, redwater fever and Lyme disease.
Acute and chronic toxicity to livestock	Grazing or feeding areas (all farm animals)	Loss of condition or death of single animals or whole herds after eating bracken. Acute poisoning caused by toxins attacking bone marrow. Thiaminase enzyme causes vitamin B1 deficiency (anorexia, ataxis and death). Indirect effect on horses that may eat other poisonous weeds (e.g., *Senecio*) made more palatable by spraying of *Pteridium*.
Cancer risks (ptaquiloside being one of several carcinogens)	All habitats with *Pteridium*. All parts of bracken are carcinogenic. Highest levels in croziers and young leaves. Carcinogens are leached into the environment where they can be persistent.	Risks of cancer in humans or livestock after longer-term *Pteridium* consumption. Cancers develop in organs that are alkaline; e.g., bladder tumors in cattle (causing bovine enzootic haematuria, BEH). Risks associated with *Pteridium* beds in drinking water catchments. Indirect risks to humans from drinking water, cow's milk, goat's milk. Risks from inhaling *Pteridium* spores.

representing about 7% of the total UK land surface. *Pteridium* further occupies an estimated 122 000 km of linear features in the UK, such as hedgerows, roadside verges, tracks and woodland edges. These linear habitats also provide refuges from which new encroachment into the adjoining landscape can occur. A similar scale of infestation exists in New Zealand (Taylor, 1985).

Pteridium leaves develop a very thick cuticle when growing in drier conditions, enhancing their control of water loss. Sporophytes of *Pteridium* often grow on open ground in full sun, occupying open woodland only when confined there by farming, forest shade, or climates that would otherwise prove too cold or too hot (Marrs and Watt, 2006). *Pteridium* grows in a wide range of soil conditions (pH 2.8–8.6) as long as the land is free of prolonged waterlogging, drought or frost. The wide range of conditions tolerated by *Pteridium* is reflected by the height of its stands, which may reach a few centimeters on exposed mountains or exceed 4 m in sheltered places (Marrs and Watt, 2006).

Pteridium frequently produces spores in tropical zones but less regularly in cool regions (Wynn *et al.*, 2000). Spread of *Pteridium* by spores explains colonization of new pastures created in South America by large-scale deforestation and burning (Hartig and Beck, 2003), the colonization of ash fields on Mount St. Helens (Wolf *et al.*, 1988) and the genetic diversity of *Pteridium* on the isolated volcanic islands of Hawaii (Sheffield *et al.*, 1995). *Pteridium* gametophytes are able to colonize new ground after fire or other recent disturbances but are rarely observed in other circumstances. *Pteridium* is absent from Chile where mountain ranges and prevailing winds have apparently prevented colonization by spores (Alonso-Amelot *et al.*, 2001). Once established, *Pteridium* spreads by means of underground rhizomes (Fig. 8.1) that can encroach at rates of 1.8 m year^{-1} to form extensive stands that may dominate the landscape for decades (Marrs and Watt, 2006). The rhizomes are not only protected from fire but contain extensive starch reserves that permit overwintering in cool climates. When it is dominant, *Pteridium* develops a deep litter layer on the soil surface that further protects its rhizomes from frost and desiccation. Rhizomes under established stands can account for about 300 tonnes ha^{-1} fresh weight. Plants can possess about 1500 dormant leaf buds m^{-2} and survive repeated annual cutting for over 20 years (Le Duc *et al.*, 2007). These traits reflect an effective antigrazing strategy that is reinforced by numerous toxic and carcinogenic compounds that are found in every part of *Pteridium*, including its spores (Evans *et al.*, 1982; Simán *et al.*, 1999; Rasmussen *et al.*, 2003).

The combined result of these traits is that *Pteridium* can readily invade disturbed land and form dominant stands over extensive areas (Fig. 8.2) that perhaps resemble the fern plains that existed at the end of the Jurassic, some 70 million years before grasses and other flowering plants evolved (Ingrouille, 1992). Left to itself, *Pteridium* consolidates its competitive advantage by (1) shading; (2) occupying

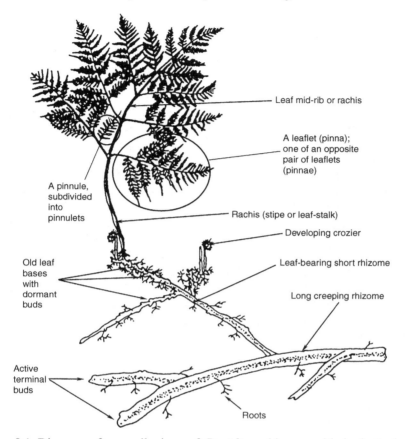

Leaf mid-rib or rachis

A leaflet (pinna); one of an opposite pair of leaflets (pinnae)

A pinnule, subdivided into pinnulets

Rachis (stipe or leaf-stalk)

Developing crozier

Old leaf bases with dormant buds

Leaf-bearing short rhizome

Long creeping rhizome

Active terminal buds

Roots

Fig. 8.1 Diagram of a small piece of *Pteridium* rhizome with leaf attached. (Adapted from Holloway, 1994.)

all the available soil space with its rhizomes; (3) forming a litter layer that inhibits the establishment of other species by burial of their seeds and harboring seed predators; and (4) allelopathic effects attributed to *Pteridium* that inhibit the growth of other vegetation (Taylor and Thomson, 1990, 1998). In some places, *Pteridium* is so successful that it can postpone indefinitely the subsequent natural succession to scrub, woodland and forest (Shaw, 1988; Den Ouden, 2000; Marrs *et al.*, 2000, Hartig and Beck, 2003).

Problems created by Pteridium

Pteridium ranks as one of the world's worst "weeds" (Taylor, 1989; Alonso-Amelot *et al.*, 2001), creating a variety of problems that are summarized in Table 8.2. Its ability to occupy and dominate land creates some of the worst problems caused by the partial or complete loss of permanent grazing and the consequent economic

Fig. 8.2 *Pteridium* creating an extensive fern plain after invasion of heather moorland, England.

effects (see Section 8.6). *Pteridium* occupies and dominates permanent pasture at the direct expense of livestock production in many parts of the world. The problems mostly relate to extensive, low-level grazing on uncultivated, "rough" pasture that is unimproved by reseeding. Such effects are severe over large areas of South America where the fern invades rough grazing after deforestation (Taylor, 1980, Hartig and Beck, 2003). However, even highly productive, intensive dairy farms are not immune from the effects of *Pteridium* (Pinto *et al.*, 2004).

The insulated, humid conditions under *Pteridium* litter are ideal for the development and hibernation of pests such as *Ixodes ricinus* (sheep tick), a blood-sucking parasite that frequents rough grazing and is responsible for several diseases in humans, livestock and game (Sheaves and Brown, 1995a). There is an increased risk that people will contract Lyme disease from tick bites where *Pteridium* occupies recreational lands, tracks and footpaths (Sheaves and Brown, 1995b).

Another problem with *Pteridium* is that all parts of the plant are toxic or carcinogenic and potentially dangerous to animals that consume it, causing acute poisoning, chronic disease or cancer (Evans *et al.*, 1982). These properties can also have consequences for human health (Wilson *et al.*, 1998). Although many herbivorous insects consume *Pteridium* without apparent harm, birds may not be immune. For example, I. A. Evans (1986) showed that *Pteridium* extracts affected the embryos of *Cotornix japonicus* (Japanese quail). *Pteridium* produces numerous harmful compounds, including at least 30 pterosides and pterosins that resemble illudanes found in poisonous fungi (W. C. Evans, 1986; Alonso-Amelot *et al.*, 2001). The best known of these compounds is ptaquiloside, a norsesquiterpene glycoside, a potent carcinogen that is held responsible for some 50% of the cancers caused by *Pteridium* consumption. Ptaquiloside is also present in less widespread ferns such as *Cheilanthes sieberi*, *Diplazium esculentum*, *Dryopteris juxtaposita*, *Histiopteris incisa*, *Onychium contiguum*, *Polystichum squarrosum* and *Pteris cretica*, and can produce symptoms similar to those caused by *Pteridium* ingestion (Smith *et al.*, 2000; Somvanshi *et al.*, 2006). Eating young *Pteridium* shoots as a delicacy is still commonplace in several countries of the world (e.g., Japan, Brazil) where unusually high levels of esophageal and gastric cancers are reported (Hirayama, 1979; Marliere *et al.*, 2000). In Venezuela, Alonso-Amelot *et al.* (1996) found high levels of ptaquiloside in milk up to 120 hours after cows had ingested *Pteridium*, strongly supporting a connection to an increase in human gastric cancer in Central and South American countries with dense *Pteridium* growth. Similar findings were reported in Turkey (Pamukcu *et al.*, 1978) and in Costa Rica (Villalobos-Salazar *et al.*, 2000). *Pteridium* can also have chronic and lethal toxic effects on sheep (Fig. 8.3), cattle, water buffalos, horses and pigs,

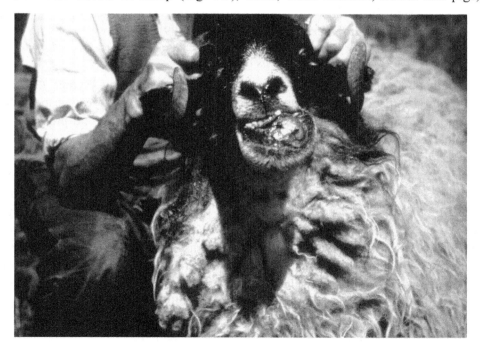

Fig. 8.3 Carcinoma of the jaw attributed to consumption of *Pteridium*, UK.

producing a range of different symptoms after ingestion (Evans *et al.*, 1982; Rasmussen, 2003). The potential loss of farm animals caused by the carcinogenic effects of *Pteridium* can be judged from the upland pastures of the Azores where, until recent control measures were taken, up to 18% of dairy cows in intensive grazing developed bladder tumors (Pinto *et al.*, 2004). Similar incidences of *Pteridium*-induced cancers in farm animals have been reported from at least 15 other countries worldwide.

Ptaquiloside is highly water soluble and can be leached from both green leaves and surface litter by rainfall. Wherever *Pteridium* occurs within water catchments, ptaquiloside is a potential contaminant of local run-off, groundwater, wells and drinking-water abstractions, especially in early summer. Rasmussen *et al.* (2005) showed that water in wells surrounded by *Pteridium* can contain ptaquiloside levels of 45 µg L^{-1} that pose risks to any person or animal that drinks it. Where extensive *Pteridium* stands infest water catchments and the periphery of surface waters, spores may also contaminate the water. Several studies show that ingested *Pteridium* spores can cause cancers in rodents, while extracts of *Pteridium* spores are known to damage DNA (Simán *et al.*, 2000). The presence of ptaquiloside itself in the spores of *Pteridium* remains under investigation. Although most *Pteridium* spores will usually fall within meters of the parent leaves, Caulton *et al.* (1995) showed that airborne spores can be found many kilometers away. Dust masks and other safety measures may be recommended to reduce exposure to spores when working in, or close to, spore-releasing *Pteridium* (Simán *et al.*, 1999).

Benefits of Pteridium

The ethnobotanical aspects of *Pteridium* were reviewed by Rymer (1976). *Pteridium* has been useful to humans probably for thousands of years, especially as an aid to survival in cooler climates. Some typical uses past and present are summarized in Table 8.3.

Pteridium may have future uses for biofuel production (Callaghan and Sheffield, 1985; Lawson *et al.*, 1986). Uncultivated, dense stands of *Pteridium* can typically produce 4–16 tonnes ha^{-1} of dry leaf material annually having an average calorific value of 21 MJ kg^{-1}, a greater energy value than timber (Watt, 1976; Donnelly, 2003, 2004). The potential of cultivating *Pteridium* in deeper, fertile soils as a perennial crop (with few effective weeds, pest or diseases) has not been investigated but there is potential for cutting by forage harvesters for transport direct to on-farm digesters to produce biogas (Callaghan *et al.*, 1981). In addition to its established use as compost (Pitman and Webber, 1998; Donnelly, 2003), *Pteridium* can also serve as an animal feed, in the form of silage (Macdonald, 1887) or as steamed hay (Williams and Evans, 1959). Heat, such as produced by fermentation, detoxifies the leaf material (Pitman and Webber, 1998).

Table 8.3 *Examples of ancient and modern uses of problem fern species*

Species	Function
Angiopteris evecta	Rhizome, petiole base: food source; flavoring; decoration; aromatic oil; medicinal (pain, scabies, headaches, food-poisoning, cancer, leprosy).
Azolla filiculoides	Prolific carbon sink. Intercropped green-manure; fertilizer; green-mulch; livestock and poultry feed; mosquito control.
Azolla pinnata	Intercropped green-manure; fertilizer; green-mulch; livestock and poultry feed; mosquito control; substrate for mushroom culture; sewage treatment; heavy metal extraction; biogas.
Ceratopteris thalictroides	Food source.
Deparia petersenii	Food source in its native areas of Southeast Asia, Polynesia, Australia and New Zealand.
Dicranopteris linearis	Stems used as ropes and leaves used for personal adornment (New Guinea).
	Crushed leaves are applied as a poultice to control fever; pain control; anti-inflammatory agent (Malaysia). Treatment for intestinal worms (Indochina). Treatment of boils, ulcers and wounds (New Guinea). Also used as a laxative.
Equisetum sp.	Food source; source of silica (abrasives); heavy metal and gold extraction; medicinal (osteoporosis, kidney infections, diuretic, antibleeding properties).
Lygodium japonicum	Rachis: used to make baskets, brooms, wok brushes.
	Spores: medicinal (anti-inflammatory, anticongestive, diuretic, urinary stones, gonorrhea).
Lygodium microphyllum	Medicinal.
Marsilea spp.	Food source; medicinal (neuroasthenia, edema, leprosy, skin diseases, fever, blood poisoning).
Phlebodium aureum	Medicinal (skin disorders, dementia, respiratory problems, autoimmune disorders).
Phymatosorus grossus	Perfuming of tapa and preparation of leis (Hawaiian Islands).
Pityrogramma calomelanos	Arsenic extraction (phytoremediation of soils).
Pteridium	Crozier: cooked vegetable. Leaf/litter smoke: insect repellent.
	Leaf petioles: thatching; fiber for cord and paper; control of toothache and mouth infections.
	Leaf/litter ash: rehydrated as crude soap; flux to lower melting temperature of glass; potash fertilizer.
	Leaves, dried: insulation of winter shelters; animal and human bedding; tinder.
	Leaves, fresh: packing material and preservative; dyeing; curing leather; green mulch; composting; treament for intestinal worms. Also potential for silage, biofuel and biogas production.

Table 8.3 (*cont.*)

Species	Function
	Plant: seasonal landscape enhancement; erosion control; people barriers.
	Rhizome: food source (starch flour); antirheumatic; treatment for intestinal worms.
Pteris cretica	Arsenic extraction (phytoremediation of soils).
Pteris vittata	Arsenic extraction (phytoremediation of soils).
Salvinia molesta	Mulch; compost; fodder supplement for livestock; potential for paper, biofuel, sewage treatment.

Note: The uses listed are far from exhaustive.
Source: The information is drawn from authors cited in the text and additionally; Veitch (1990), Ortega (1994), Valier (1995), Tan (2001), Zakaria *et al.* (2006).

Ecological interactions of Pteridium *with other organisms*

As a native species, *Pteridium* can have both positive and negative consequences for ecology, biodiversity and conservation (Marrs and Watt, 2006). A careful balance therefore needs to be struck when planning land management where *Pteridium* is present (Burn, 1988). Three situations in which organisms rely directly or indirectly on *Pteridium* are discussed below. Such interactions, mostly derived from detailed studies in the UK, may reflect those of other, less studied, native ferns.

Species that use Pteridium *as a direct food source:* At least 40 species of fungi are found on *Pteridium*, some of which may represent food sources for the arthropod species that also frequent *Pteridium*. Few of these fungi cause significant disease in *Pteridium* although "curl-tip" disease (attributed to attack by *Ascochyta pteridis* and *Phoma aquilina*) or "pinnule blight" (attributed to *Corticium anceps*) can defoliate *Pteridium* stands prematurely in wet, warm summers in the UK (Irvine *et al.*, 1987). *Ascochyta pteridis* is also said to attack *Pteridium* in Ecuador (Hartig and Beck, 2003). "Leaf curl" or "tar spot" in *Pteridium*, caused by *Cryptomycena pteridis*, appears in Europe, America and Northern Asia. Although gametophytes of *Pteridium* are very susceptible to disease, *Pteridium* sporophytes appear to resist fungal attack by lignification (McElwee and Burge, 1990).

There is one known case where other plants derive nutrition from *Pteridium*. *Orobanche* species (plants lacking chlorophyll) parasitize *Pteridium* rhizomes in the Canary Isles (De Winter and Amoroso, 2003), perhaps the only area in which this phenomenon has been observed.

Several insect groups, at different stages of growth, interact with ferns (see Chapter 7) and *Pteridium* is no exception. As many as 24 insect species were found feeding, mostly as specialists, on *Pteridium* in Papua New Guinea (Kirk, 1977). Some 40 insect species regularly feed on *Pteridium* in the UK and 27 of these feed on aerial parts of the plant (Marrs and Watt, 2006). Sawfly larvae (e.g., *Aneugmenus padi*) and some moths feed on external parts of *Pteridium*. Larvae of many species of flies, especially *Chirosia* and *Dasineura*, feed internally, creating mines and galls. Bugs that feed on *Pteridium* sap (e.g., *Ditropis pteridis* and *Monalocoris filicis*) emerge from their overwintering sites in *Pteridium* litter to coincide with the emergence of the croziers. The aphid *Macrosyphum ptericolens* feeds on *Pteridium* in both Europe and North America late in the growing season. *Popillia japonica* (Japanese beetle) attacks *Pteridium* in Vermont, USA (De Winter and Amoroso, 2003). The larvae of a number of regionally native moths, such as *Callopistria juventina* (latin moth) and *Petrophora chlorosata* (brown silverline moth), feed exclusively on European *Pteridium*, while a score of others may additionally use a range of other food plants. The larvae of several species of *Hepialis* (swift moth) can overwinter in *Pteridium* rhizomes and are minor pests of cultivated root crops.

Insect herbivores may have little effect on *Pteridium*, possibly because they either are rare in relation to the extensive areas of *Pteridium* or are themselves subject to high levels of predation. The cyanogenic properties of *Pteridium* may also play a part in controlling herbivory (Low and Thomson, 1990). *Pteridium* produces several phytoecdysteroids but at levels considered too low to significantly disrupt insect molting patterns (Alonso-Amelot *et al.*, 2001). Hendrix (1977) showed that a toxin, not an ecdysone, was responsible for interrupting the feeding of the moth *Trichloplusia ni* on *Pteridium*.

Species partially dependent on Pteridium *habitat:* After removal of natural woodland habitat, *Pteridium* sometimes provides a "substitute woodland canopy" as alternative shelter for the survival of certain herbaceous plants. A dozen or more shade-loving species in the UK, whether common (e.g., *Scilla non-scripta*, English bluebell), localized (e.g., *Trientalis europaea*, chickweed wintergreen) or scarce (e.g., *Gladiolus illyricus*, wild gladiolus), can benefit from *Pteridium* in this role (Anonymous, 1988; Pakeman and Marrs, 1992; Warren, 1993; Mabey, 1998; Backshall, 1999). It is likely that *Pteridium* has a similar role in many other parts of the world (e.g., protection of *Galanthus krasnovii* in Turkey; Davis, 1999). These seed plants are not opportunists like *Epilobium angustifolium* (fireweed) that occurs by chance alongside *Pteridium* on suitable disturbed sites. In the absence of other shelter, these low-growing species are dependent upon *Pteridium* for protection against excessive heat, cold or grazing.

Some of these *Pteridium*-dependent plants themselves support dependent fauna in the UK (e.g., *Procas granulicollis*, a rare weevil associated with *Corydalis claviculata*). *Mythimna turca* (double line moth) can show indirect dependency on *Pteridium* where the fern either provides a microclimate for the overwintering caterpillars or facilitates the establishment and growth of grass food-plants such as *Holcus mollis* (Wigglesworth *et al.*, 2004). Certain fritillary butterflies can be locally reliant on *Pteridium* habitat, especially following the decline in traditional woodland management such as rotational coppicing (Warren and Oates, 1995). Species include *Argynnis adippe* (high brown fritillary), *Boloria euphrosyne* (pearl-bordered fritillary; Fig. 8.4) and *Mellicta athalia* (heath fritillary) that may be regionally endangered. These butterflies not only depend on the larval food plants (e.g., species of *Melampyrum* or *Viola*) that are protected by the *Pteridium*, but also the thermal insulation and radiated heat provided by *Pteridium* litter during each stage of their life cycle. The butterflies may further depend on early nectar-providing plants (e.g., species of *Ajuga* and *Primula*), that also benefit from the shelter of the *Pteridium* canopy (Oates, 2004).

Pteridium in the UK can be valuable as shelter for some species of birds for feeding and nesting purposes (Tyler, 1988; Pakeman and Marrs, 1992; Backshall,

Fig. 8.4 *Boloria euphrosyne* (pearl-bordered fritilliary), a species partly dependent on *Pteridium* in the UK, basking on *Pteridium* litter.

1999). Examples include habitats such as moorland edge (e.g., *Turdus torquatus*, ring ouzel), woodland edge (e.g., *Phylloscopus sibilatrix*, wood warbler) and scrubby areas either with trees (e.g., *Anthus trivialis*, tree pipit) or without (e.g., *Saxicola rubetra*, whinchat). Regional action plans and other prescriptions exist for some of these species (e.g., *Carduelis flavirostris*, twite, Anonymous, 2006). Moth larvae such as *Philudoria potatoria* (drinker) can be plentiful in grass under *Pteridium* and are taken avidly by *Cuculus canorus* (cuckoo) which uses *Pteridium* leaves as feeding posts. *Caprimulgus europaeus* (nightjar) is a nocturnal bird that uses dense *Pteridium* beds for secure daytime roosts. Small birds, rodents and lizards in *Pteridium* also serve as prey for avian predators such as *Falco columbarius* (merlin) or *Circus cyaneus* (hen harrier). *Pteridium* litter offers an ideal home for the underground nests of field mice (*Apodemus sylvaticus*; Brown, 1986) which, along with squirrels and jays (*Garrulus glandarius*), assist the succesional replacement of *Pteridium* by woodland trees due to their burial of nuts as winter food supplies. Squirrels will use *Pteridium* litter in their nests while *Pteridium* beds also provide refuges for deer and badgers.

Other ferns occur rarely within dominant *Pteridium* stands in the UK but do occur along tracks and watercourses, under boulders or on walls that may be associated with the stands (Robinson and Page, 2000). Exceptions are *Botrychium lunaria* (moonwort) and *Ophioglossum vulgatum* (adder's tongue), which grow in grass beneath a partial *Pteridium* canopy and, occasionally, clumps of *Dryopteris filix-mas* (male fern, Fig. 8.5). All three species were probably present before *Pteridium* invaded. In Costa Rica and Mexico, two other species of Dennstaedtiaceae, *Histiopteris incisa* and *Hypolepis repens*, have been observed growing in association with *Pteridium* (K. Mehltreter, personal communication).

Species threatened by Pteridium*:* Limited stands of *Pteridium* encourage biodiversity at woodland and moorland edges, especially within a mosaic of other vegetation types. However, in the absence of appropriate control, *Pteridium* tends to become a dominant species, usually at the expense of other wildlife (Fig. 8.2). Thus, *Pteridium* usually requires some form of intermittent control on wildlife management sites in order to preserve optimum conditions for target conservation species. A fine balance is required, especially on sites where rarer species may be dependent on *Pteridium*, such as fritillary butterflies (Brereton and Warren, 1999), lizards (e.g., *Lacerta agilis*, sand lizard; *L. vivipara*, common lizard) or snakes (*Coronella austriaca*, smooth snake; *Vipera berus*, adder; Pakeman and Marrs 1992). *Pteridium* threatens the habitats of unusual plants such as *Epipactis atrorubens* (dark-red helleborine) and *Geranium sanguineum* (bloody cranesbill), or rare butterflies such as *Maculinia arion* (large blue), or birds with specialized breeding areas such as *Sylvia undata* (Dartford warbler) or *Burhinus oedicnemus* (stone

Fig. 8.5 Clumps of *Dryopteris filix-mas* covered by litter of *Pteridium*, England.

curlew). In abandoned grazing on coastal grasslands and cliffs, *Pteridium* can diminish feeding areas for birds with restricted distribution, such as *Pyrrhocorax pyrrhocorax* (chough, MacLochlainn, 2004).

8.2.2 Other native problem ferns

No other native problem ferns (Table 8.1) are as widespread as *Pteridium* but they share the trait of being invasive opportunists on disturbed land as a first stage in regeneration of woodland and forest. Their roles may represent a problem or an asset depending on circumstances, although these species are usually much less studied.

Histiopteris incisa (batswing fern) is highly adaptable and occurs widely in tropical, subtropical and temperate regions of the southern hemisphere and else-where (e.g., Taiwan). *Histiopteris incisa* is a close relative of *Pteridium*, which it resembles ecologically. *Histiopteris incisa* can be a dominant fern associated with Tasmanian rain forest "fernlands" (Anonymous, 2007) and occupies light woodland and woodland margins on mainland New Zealand but is not confined to forest areas. Its creeping rhizomes also thrive on disturbed areas and can dominate other species where trees have fallen or after logging. In Taiwan, *H. incisa* colonizes hostile environments such as sulfurous thermal areas or soils polluted with high levels of copper (Wei *et al.*, 1997). Like *Pteridium*, *H. incisa* adopts a deciduous habit in colder areas and is recorded as colonizing the Snares Islands off New Zealand (Horning, 1983) and the remote, windswept Auckland, Campbell and Antipodes islands (southeast of New Zealand). On these islands, larvae of the moth *Musotima nitidalis* feed off this fern (Patrick, 1994), just as related *Musotima* species do in South Africa (or on *Lygodium* in Australia; see Section 8.5.2). *Histiopteris incisa* can spread like *Pteridium* (Nicholls, 1999) and has been described as a nuisance, along with *Hypolepis muelleri*, in the Noosa National Park in southeastern Queensland (Anonymous, 1999). In Australia, *Histopteris incisa* is currently listed as an environmental weed (i.e., plant that represents a threat to the conservation values of a natural ecosystem), and may deserve more attention as a potential invasive species elsewhere.

Several genera of ferns include species that possess a scrambling habit made possible by appreciable extension growth and repeated branching of the leaves. However, such growth does not match the marked indeterminate growth seen in the leaves of climbing ferns (see Section 8.3.1). In the scrambling species, the extended leaf midrib first penetrates the understory vegetation followed by the expansion of lateral leaflets that anchor the subsequent growth on top of the vegetation (Fig. 8.6). In certain *Dicranopteris*, *Gleichenella*, *Gleichenia* and *Sticherus* species (see Table 8.1), this growth habit can cause problems where it permits the creation of dense, smothering thickets that can be 4 m high and which dominate the underlying vegetation (Palmer, 2003; Moran, 2004). These scrambling species are important in succession following natural disturbance (see Chapter 6) but their invasive vigor tends to be problematic where humans have caused changes in the environment.

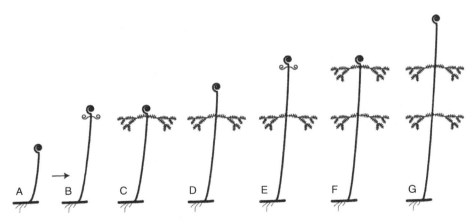

Fig. 8.6 Pattern of leaf development in *Gleichenia* and other scrambling ferns. This type of growth assists initial penetration of ground cover followed by an ascending, scrambling habit that can dominate supporting vegetation. (Adapted from Moran, 2004.)

Fig. 8.7 *Dicranopteris linearis* (uluhe), a scrambling fern, Kopiliula, Maui, Hawaii.

Because scrambling ferns are difficult to grow, they have not been transported as often as other ferns for ornamental purposes (Hoshizaki and Moran, 2001) and so they remain problematic essentially as native species.

Of the scrambling species, both *Gleichenella pectinata* (syn. *Dicranopteris pectinata*) and *Sticherus bifidus* (syn. *Gleichenia bifida*), though important to land-slide succession (see Chapter 6), require the expense of treatment by chemicals, cutting or burning when their exuberant growth creates problems as a roadside hazard in Puerto Rico (J. M. Sharpe, personal observation). In Malaysia, *Dicranopteris linearis* (resam) appears to exert a strong allelopathic effect on other weedy species (Tet-Vun and Ismail, 2006) which may explain its ability to dominate substantial areas of land. In Sri Lanka, thickets of *D. linearis* can suppress the regeneration of rain forest on land previously subject to regular burning (Cohen *et al.*, 1995). In the Hawaiian Islands, litter accumulated from the short-lived but very persistent (marcescent) leaves of *D. linearis* (uluhe; Fig. 8.7) affects natural soil processes and may reduce the suitability of sites for the subsequent recovery of native species. In Japan, abandonment of traditional management of *Pinus densiflora* forests has resulted in regeneration problems due to the proliferation of *Gleichenia japonica* (Fig. 8.8) and *D. linearis*, probably facilitated by deer grazing and the cutting and removing of dead pine trees (Kuroda *et al.*, 2006).

However, some of these native, scrambling species may also confer appreciable ecological benefits as yet little studied. Russell *et al.* (1998) suggested that

Fig. 8.8 Dense ground-cover created by *Gleichenia japonica*, a scrambling fern, Japan.

Dicranopteris linearis may help to resist invasions by alien plants in the Hawaiian Islands because of its ability to rapidly occupy land that may be unavailable to other native species in the short term. The dense canopies of this species also help to protect the burrows of threatened birds such as Hawaiian petrels (*Pterodroma sandwichensis*) and Newell's shearwaters (*Puffinus newelli*) (Day *et al.*, 2003). *Dicranopteris curranii* is capable of dominating bare, disturbed ground after logging activities in Malaysia, but this is not considered a problem as it protects disturbed areas from subsequent erosion by tropical rainfall and prepares a more suitable habitat for the reestablishment of natural tree species (Negishi *et al.*, 2006).

Equisetum arvense (field horsetail) is another, sun-loving native species that can be locally invasive and a persistent nuisance (Anonymous, 1996). *Equisetum arvense* causes problems as a weed of pastures, orchards, nurseries, hayfields, roadsides and landscaped areas, particularly on poorly drained soils. It also invades superficially dry sites such as railway ballast. With storage tubers that enhance its tenacity and creeping rhizomes that achieve depths of over 1.5 m, this species is extremely difficult to control. *Equisetum arvense* also has tough and waxy shoots with a high silica content that helps the plant to penetrate hard soil (even tarmac) and makes control difficult using herbicides (Young, 1990). Like *Pteridium* (Table 8.2),

Equisetum can cause thiamine deficiency symptoms in horses and its safety is suspect, especially in Japan where the buds of *E. arvense* are eaten as a vegetable. Horsetails, especially when present in dried hay, have long been known to be poisonous to young horses and sheep. *Equisetum telmateia* (giant horsetail) is another species from similar habitats and causes similar problems.

In New Zealand, the herbaceous *Paesia scaberula* (ring fern) and the arborescent *Dicksonia squarrosa* sometimes have a strong association after forest clearance and occupy pastures where they represent successional changes back to woodland. *Paesia scaberula* has been described as one of the worst indigenous weeds of improved pasture (i.e., exotic grassland) in New Zealand (Silvester, 1964). Its thin, superficial rhizomes spread from all sides of the circular patches that are characteristic of this fern, whereas the clumping, colonial habit of *Dicksonia squarrosa* arises from its ability to reproduce vegetatively by runners arising from subterranean adventitious buds. The spread and formation of *Dicksonia* groves is a slow process but the emergence of *Dicksonia* through either grass or *Paesia* is often observed (Silvester, 1964). Both native fern species create problems in productive grassland comprising exotic grass species (e.g., hybrids of *Lolium multiflorum* and *L. perenne* raised in New Zealand as "short rotation ryegrass", Hubbard, 1968). This situation is different from the usual circumstances where invasive alien ferns threaten native species (see Section 8.3) and illustrates the subjective decision of whether the invader is welcome or not.

8.3 Alien terrestrial problem ferns

An appreciable number of alien ferns create problems as invasive species, mostly as a result of proliferation by spores after their initial introduction as ornamental or (rarely) as food plants. The genus *Lygodium* includes about 25 species, at least two of which have proved to be invasive with dire consequences for alien habitats. *Lygodium* therefore ranks with *Pteridium* and *Salvinia* as the worst of the problem ferns. Several other alien terrestrial ferns (Table 8.1) will also be discussed but their problems are of generally lesser consequence.

8.3.1 Lygodium

Some species of *Lygodium* can climb trees and smother ground vegetation with more vigor and tenacity than better known, vine-like, flowering plants such as *Clematis*, *Convolvulus* (bindweeds, morning glory), *Polygonum* (knotweed) or *Pueraria* (kudzu). Such climbing ferns are unusual in making their ascent using a climbing leaf structure rather than a climbing stem. A *Lygodium* leaf has a terminal meristem of potentially indeterminate growth of the rachis until it becomes a very long, twining vine with a succession of branches in the form of short leaflets (Plate 8C). The lateral leaflets also end in dormant buds that can continue indeterminate leaf growth after the death of the main leaf tip (Mehltreter, 2006). These vines eventually become semirigid to create, in alien habitats, immense, 3-dimensional carpets over 1 m thick that can strangle whole forests, including the trees, their epiphytes and associated ground flora (Plate 8D; Hutchinson *et al.*, 2006). When exposed to sunlight, fertile leaflets are produced toward the top of the forest canopy so that spores can be wind dispersed above the canopy. In addition to long distance dispersal, the spores have proven longevity, while the gametophyte generation has the ability to self-fertilize so that abundant young sporophytes are readily created (Lott *et al.*, 2003). Two species of *Lygodium* cause serious problems as invasive aliens: *L. microphyllum* and *L. japonicum*.

Lygodium microphyllum is native over large areas of the Old World (Table 8.1) but is aggressively invasive in places where it has become naturalized. In Florida, *L. microphyllum* is a disastrous weed that probably originated from two garden nurseries in 1958 and its advance has been charted by Pemberton and Ferriter (1998). By 1999, this fern had spread extensively and was responsible for degrading more than 40 000 ha of unique native habitat and wildlife refuges (Volin *et al.*, 2004). Populations of *L. microphyllum* in Florida have been subjected to an aggressive control program based on several years of study of its growth and spread. The summaries of this work by Volin *et al.* (2004) and Hutchinson *et al.* (2006) are the basis for the following discussion of *L. microphyllum* unless otherwise noted.

Lygodium microphyllum grows in the southern zones of Florida where its leaves can climb to 30 m in height. It is now common in *Taxodium distichum* (bald cypress) stands, pine flatwoods, wet prairies, *Cladium jamaicense* (sawgrass sedge) marshes, mangrove communities and disturbed areas. This fern not only smothers bald cypress forests, it also blankets entire tree-islands in the Everglades; they are so completely covered by the fern that it is impossible to see trees and other vegetation beneath the mat (Plate 8D). The fern also spreads laterally across *C. jamaicense* marshes, effectively bridging areas of open water. These fern bridges carry the natural wildfires that occur within the swamp vegetation beyond the natural water barriers.

Vegetative growth and spore production of *L. microphyllum* continue year round in the tropics. About 80% of spores remain viable after 5 months and this means that spores may prove capable of reinfesting land over a period of years. Fire will kill the present fern growth but regrowth is usually rapid from underground rhizomes and from spores. *Lygodium microphyllum* spreads most successfully in damp, shady habitats subject to seasonal inundation, and an invasion can often be predicted by the presence of indicator species such as the fern *Blechnum serrulatum*. These habitat conditions exist throughout the forest-dominated wetlands of southern

Florida. It appears likely that conservation projects in the Everglades, designed to combat other invasive seed plant species (e.g., *Melaleuca quinquenervia* and *Schinus terebinthifolius*) by restoring the wetter conditions that prevailed prior to widespread drainage schemes, may actually be assisting the spread of *L. microphyllum*. Where standing water protects its rhizomes from frost, this fern also may spread into cooler areas than latitude would otherwise suggest.

Lygodium japonicum, native to India, Queensland and both temperate and tropical areas of eastern Asia, is considered a weed in Taiwan and the Philippines. It has naturalized in Puerto Rico, Hawaii and in nine states of the southeastern USA. Its escape from cultivation in the latter case was noted prior to 1906 but its presence has only become significant since the 1940s (Nelson, 2000). *Lygodium japonicum* occupies damp, disturbed areas such as yards, roadsides, woodland margins and the edges of wetlands. In cooler areas, the current year's leaflets die each winter, leaving the ascending, entwined midribs intact as a "ladder" for the new growth that arises from the rhizomes the following spring. Compared with *L. microphyllum*, a more serious problem in Florida, *L. japonicum* is predominant further north in cooler areas with a shorter growing season. In recent years, *L. japonicum* has also moved south and its range now overlaps with *L. microphyllum* in Florida.

Both *Lygodium* species threaten rare species in Florida including ferns such as *Acrostichum danaeifolium* (giant leather fern) and *Actinostachys pennula* (syn. *Schizaea germanii*, ray fern), and flowering plants such as *Aristolochia tomentosa* (Dutchman's pipe), *Cucurbita okeechobeensis* (Okeechobee gourd), *Polygonum meisnerianum* (branched tearthumb), *Sideroxylon thornei* (Thorne's buckthorn), *Tillandsia utriculata* (giant airplant) and *Vanilla mexicana* (Mexican vanilla). The climbing ferns also endanger habitats that are refuges for birds and animals dependent on the natural plant life for feeding or nesting. The current control programs (see Section 8.5) are not yet able to halt the ever-increasing area of devastation inflicted by these ferns. Without a substantial change in their effectiveness, it may be surmised from Volin *et al.* (2004) that over 1 million ha of Florida could be devastated by *Lygodium* infestations by 2014.

8.3.2 Other alien terrestrial problem ferns

We briefly summarize problems posed by certain other terrestrial alien ferns (Table 8.1), although none have created problems on such a scale in their new environments as those described for *Lygodium*. Problem species have been especially well documented for the Hawaiian Islands where the remote island flora appears especially vulnerable to alien invasion. In contrast, other oceanic islands with a harsh climate such as the Falklands have not proved vulnerable to introduced alien ferns (R. Lewis, personal communication) although they do share fern species with distant lands at similar latitudes. Surprisingly, *Pteridium* is not one of these species, although *Histiopteris incisa* is potentially invasive in a severe climate as is found on islands south of New Zealand (see Section 8.2.2).

Of the alien terrestrial ferns listed in Table 8.1, 18 have been noted by Palmer (2003) as problems in the Hawaiian Islands because they are common, widespread and abundant, with many first collected as naturalized escapes from gardens in the early 1900s. One of the earliest collections (1887) was *Cyclosorus dentatus* (syn. *Christella dentata*), an understory fern in rain forests that was also spreading in parts of the neotropics by the 1930s. *Cyclosorus dentatus* has hybridized in Hawaii with both the alien *C. parasitica* (syn. *Christella parasitica*) and the endemic *C. cyatheoides* (syn. *Christella cyatheoides*; Palmer, 2003). Another problem fern, *Lindsaea ensifolia*, has hybridized with the indigenous *Sphenomeris chinensis*. *Pityrogramma* × *mckenneyi* is yet another hybrid (Wilson, 1996). Such hybrids have unknown consequences for the native flora and all may be potentially problematic where derived from alien species.

Platycerium bifurcatum was the most recently noted alien problem fern in the Hawaiian Islands (first collected in 1991) after being reported as invasive by Mason (2004). It is one of three problem ferns in Hawaii that are epiphytic, along with *Phlebodium aureum* and *Platycerium superbum*. Control of their colonization in the wild will require a different approach from methods used for true terrestrial species. The growth habit of all but four of the alien, terrestrial problem ferns in Hawaii is compact (rosette or short, creeping rhizomes, Table 8.1) and most have fairly short leaves. Only *Phymatosorus grossus*, *Blechnum appendiculatum*, *Deparia petersenii* and *Nephrolepis brownii* fit the model of invasion seen in *Pteridium* and the scrambling ferns. *Nephrolepis brownii* (syn. *N. multiflora*) has also spread rapidly in Puerto Rico since its introduction after 1940, becoming one of the most common ferns on the island (Proctor, 1989), but in the Hawaiian Islands, its spread may be underestimated because of its similarity to native species of *Nephrolepis*. *Phymatosorus grossus*, although historically encountering similar identification problems, is now considered to be one of the most common ferns in urban areas of Hawaii (Palmer, 2003). *Blechnum appendiculatum* has been reported as a threat to the rare endemic fern species of *Diellia* in the Hawaiian Islands. More specifically, *Blechnum appendiculatum* and *Cyclosorus parasitica* competed spatially with a single remaining plant of *Diellia mannii* on the Island of Kauai (Aguraiuja and Wood, 2003).

Deparia petersenii (syn. *Athyrium japonicum*), an edible fern native to Southeast Asia, Polynesia, Australia and New Zealand, was first collected as an alien on the Island of Hawaii in 1928 (Wilson, 1996). Known only from the Kohala Mountains in 1950, it has since spread rapidly on all Hawaiian Islands. Along with other alien species, *D. petersenii* is considered a threat to the rare herbaceous flowering plant

Cyrtandra cyaneoides (Anonymous, 2000). *Deparia petersenii* has also naturalized in the southeastern USA and in southeastern Brazil, is present in the Azores, is spreading in Madeira, and is displacing a native *Cyclosorus* species in Reunion Island (P. J. Acock, personal communication).

The effects and spread of two other alien ferns in the Hawaiian Islands have been objects of special study. *Sphaeropteris cooperi* (syn. *Cyathea cooperi*) is a hardy tree fern from Australia that can become invasive in warmer climates. Although international tree fern trade is now prohibited through CITES (Hoshizaki and Moran, 2001; see Chapter 9), this species was a popular ornamental in the Hawaiian Islands and had escaped from cultivation by the 1950s (Wagner, 1995). *Sphaeropteris cooperi* is especially invasive in the warm, wet forests of Hawaii (Medeiros *et al.*, 1992) where it grows faster and produces more fertile leaves than the four endemic species of *Cibotium* tree ferns, particularly in high-light environments generated by high levels of disturbance (Durand and Goldstein, 2001). Thus, *Sphaeropteris cooperi* can rapidly displace even the most important plants in the wet forest of Hawaii (Durand and Goldstein, 2001). Based on the Hawaiian experience, reports of *S. cooperi* spreading in Madeira (P. J. Acock, personal communication) and its presence on seven of the nine islands of the Azores (Borges *et al.*, 2005) where much of the natural flora has already been lost, gives cause for concern.

The eusporangiate *Angiopteris evecta* (mule's-foot fern), with leaves that can be 6 m long and 3 m wide, resembles fossil ferns of 300 million years ago. This fern is an invasive alien in Jamaica, where it was introduced from Tahiti by Captain Bligh of the *Bounty* in 1793. In Hawaii, it was first introduced to the Lyon Arboretum on Oahu in 1927 and is now spreading rapidly in mountain valleys, threatening the indigenous flora (Wilson, 1996). Since its introduction as an ornamental in Costa Rica in the 1950s (Christenhusz and Toivonen, 2007), thousands of young sporophytes have been observed, and De Winter and Amoroso (2003) reported its spread by spores over distances of at least 80 km in an appropriate environment. While most alien ferns are suited to open ground created by disturbance, *A. evecta* is unusual because it also thrives in shaded environments such as rain forests and cloud forests. Not only does *A. evecta* produce prodigious numbers of spores, it also reproduces vegetatively from buds on stipules at the base of every leaf petiole. In contrast, within its native ranges in Australia (Anonymous, 2001) and India (Srivastava, 2008), *A. evecta* is increasingly rare and endangered. It seems likely that increasing aridity of the Australian sites, combined with the effects of fire and overcollection for a range of uses, are causing this decline.

Pteris vittata (Chinese brake fern, Plate 8B) presents an interesting alternative for an alien problem fern. *Pteris vittata* was one of the earliest recorded fern introductions to the Hawaiian Islands (1887) and is an invasive sun plant with prolific spore production. Native to China, it is widely naturalized in the USA (e.g., Florida, MacDonald *et al.*, 2008), the Caribbean and the Mediterranean in lime-rich areas. It has been described as an "urban weed" (P. J. Acock, personal communication) due to its preference for calcareous sites, as offered by concrete, masonry and pavements. *Pteris vittata* accumulates arsenic in its leaves (Ma *et al.* 2001) and offers a natural remediation process for treating soils contaminated with arsenic derived from industrial processes (Bondada and Ma, 2003). Strains have been bred in the USA with enhanced arsenic-retentive properties and increased cold tolerance (Ruder, 2004). Issues such as proper disposal of fern leaves used in phytoremediation have been discussed and the contaminant removal process has even been patented in the USA (Ma *et al.*, 2006). While other ferns hyperaccumulate arsenic, such as *Pteris cretica* (naturalized in the UK; Wilkins and Salter, 2003) and *Pityrogramma calomelanos* (Francesconi *et al.*, 2002), none are as fast growing or as well studied as *Pteris vittata*. As worldwide planting of this fern for industrial purposes proceeds, especially as a low-cost option in developing countries, measures are likely to be needed to limit further spread of this fern in areas where it has the potential to become as invasive as in Hawaii.

While these examples illustrate that some alien ferns are well documented, especially in the Hawaiian Islands, noting potential invasives before they become problems is also important. Further examples include *Cyrtomium falcatum*, recorded as an alien problem fern in several places (Table 8.1), but only recently observed displacing *Asplenium marinum* (sea spleenwort) in Macaronesia (P. J. Acock, personal communication). Similarly, the alien *Adiantum raddianum* may be displacing the native *A. capillus-veneris* in the Hawaiian Islands (Wilson, 1996) but the latter also appears as an alien colonist of limestone soils in the Mediterranean (C. N. Page, personal communication). Both species are also naturalized in New Zealand where *A. raddianum* appears less aggressive (Wilson, 1996). *Cyclosorus dentatus* is also reportedly colonizing wet, disturbed sites in the Mediterranean (C. N. Page, personal communication).

The many terrestrial aliens identified as problem ferns have a wide range of growth habits, dispersal methods and sometimes unexpected responses to a non-native habitat. Understanding their biology will help develop control methods as and when required.

8.4 Alien aquatic problem ferns

Aquatic ferns are restricted to freshwater environments but the problems they create as aliens (Table 8.1) can be as severe and far reaching as those created by terrestrial ferns.

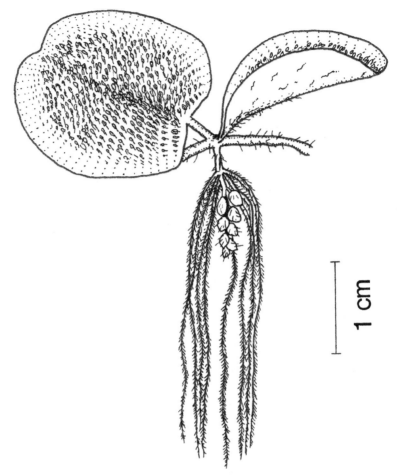

1 cm

Fig. 8.9 Features typical of *Salvinia* aquatic ferns. In *S. molesta* the sporocarps seen among the roots are sterile. (Adapted from Moran, 2004.)

8.4.1 Salvinia

Native to just a small area of southeastern Brazil, *Salvinia molesta* is a floating fern (Fig. 8.9) that has been introduced into fresh waters of many countries as an aquatic ornamental and now has a virtually global distribution where it is usually problematic (Miller and Sheffield, 2004). The first significant invasion by *S. molesta* was recognized in Sri Lanka in 1930 but its common name "Kariba weed" dates from the 1950s and the problems it created for the Kariba Dam project on the Zambezi River. The Global Invasive Species Database (Anonymous, 2005) offers a concise list of the wide variety of problems created by this fern in 17 areas throughout the world.

(a)

(b)

Fig. 8.10 Lake Moondarra, Australia (a) infested by *Salvinia molesta* prior to biocontrol. (b) After biocontrol using the Salvinia weevil (*Cyrtobagous salviniae*).

McFarland *et al.* (2004) extensively reviewed the ecology of *S. molesta* and much of the information below is derived from this report. *Salvinia molesta* grows rapidly and spreads across water surfaces, forming dense, buoyant mats (Fig. 8.10a), often 1 m deep, that cut off light to other aquatic plants, reduce oxygen content and degrade water quality for fish and other aquatic organisms. In warm, nutrient-rich water, *S. molesta* has the capacity to double the number of its leaves, its coverage of the water surface and its fresh weight, in as little as two days. In its native environment, natural herbivory appears to keep the spread of *S. molesta* in check. In the absence of such control, *S. molesta* can spread over water in tropical and subtropical regions at the staggering rate of $400 \, km^2 \, year^{-1}$ and under favorable conditions it can produce fresh biomass at a rate exceeding 200 tonnes ha^{-1} in just ten days (Julien *et al.*, 2002).

The invasive success of *S. molesta*, the sporocarps of which are effectively sterile, is achieved entirely by vegetative means, by budding and fragmentation. Natural fragmentation of the older stems occurs in nutrient-rich conditions, even in calm waters. The plant can also withstand very cold air as long as the water itself does not freeze. Once introduced, fragments of the fern are spread naturally by wind and water currents, or by hitching a ride on boats or birds. *Salvinia molesta* grows from a horizontal stem that produces each leaf in three parts. Two leaf-like green lobes, only about 1 cm across, float on the water surface while a third, submerged, lobe produces root-like growths that absorb nutrients and also serve as a drogue anchor (Fig. 8.9). The plant is unsinkable, due to air trapped between specialized, waxy trichomes that create a whisk-like cage structure (Plate 7B) covering the upper surface of the floating lobes. Buoyancy is further aided by aerenchyma (i.e., air-filled tissue) within the leaf lobes. In periods of drought, the plants die back to dormant buds that resume growth immediately when the rains return. Outside its native habitat, only freezing conditions or salinity seem to limit the potential of this plant, which means that control measures are essential to protect surface waters in a wide variety of situations.

By restricting the amount of light penetrating the water and by removing available nutrients, *Salvinia* outcompetes native aquatic plants, decreases habitat diversity and reduces food for the dependent elements of food chains, especially fish (McFarland *et al.*, 2004). The gradual decay of older parts of the mats can add large amounts of organic material to lake-bottom sediments that further reduce the depth of water for fish and diminish their breeding potential. Water under the mats usually shows a drop in both pH and dissolved O_2, together with increased CO_2 and H_2S levels. These changes can increase corrosion of machinery in hydroelectric and water-pumping installations.

The floating mats of *S. molesta* have negative economic impacts as they accumulate in filter-grids, sluices and dams and block spillways, drainage systems and irrigation canals. The reduction in water flow enhances silting of channels, increases the risks of flooding and requires extra dredging. Flooding also disperses the weed. The physical barriers created by the mats on the water surface can impede travel by boat which, in certain areas, may be the only access to food, markets, medical facilities and schools for whole communities (e.g., the floodplain of the Sepik River,

Papua New Guinea). Encroachment into many Asian and African waterways has led to declines in hunting, fishing and tourism industries. In Sri Lanka, India and Borneo, the plant is a serious pest in rice crops. To contain the risks of spread in the USA it is necessary to check contamination of boat hulls that ply waterways at risk from *S. molesta*.

Infestations of *S. molesta* also have many other ecological consequences (McFarland *et al.*, 2004). Various other plants can colonize the thick mats of *S. molesta* to create floating islands of mixed vegetation communities. Boughey (1963) documented more than 40 plant species colonizing *S. molesta* mats in Lake Kariba. The deceptive "islands" caused livestock fatalities when animals fell through the mats and drowned in deep water. In the Hawaiian Islands, the mats of *Salvinia* have threatened the habitats for endangered birds (Richerson and Jacono, 2005) such as *Fulica alai* (Hawaiian coot), *Gallinula chloropus sandvicensis* (Hawaiian gallinule) and *Himantopus mexicanus knudseni* (Hawaiian stilt).

Salvinia molesta promotes stagnant water conditions that encourage mosquitoes as vectors for encephalitis, dengue fever, malaria and elephantiasis. In Lake Kariba, the mats assisted the spread of *Biomaphalaria boissyi*, the snail that is the vector of bilharzia (schistosomiasis). Occasionally, the infestations of *S. molesta* may offer incidental benefits (Table 8.3), but the presence of this alien aquatic usually represents a serious problem. One exception occurred on Lake Kariba where decaying mats of *S. molesta* along the shoreline assisted establishment of *Panicum repens*, a grass that now remains available to herbivores at the end of the dry season when other grazing is depleted.

Salvinia minima, native to Central and South America, is another troublesome weed where it has been introduced in Texas and Louisiana. In parts of Louisiana, it has completely displaced native floating plants such as *Lemna* (duckweed) which is an important food source for waterfowl (Jacono, 2003). In the past, *S. auriculata*, also native to Central and South America, was widely confused with *S. molesta* from which it is distinguished only by small differences in the sporocarps. Although potentially invasive, it is now difficult to judge how much of a problem this species represents.

The *Salvinia* species that create problems should be contrasted with *S. natans*, which is rare and endangered in areas of Europe such as the Danube Delta in Romania where the wetlands are susceptible to drainage, pollution and other human interference (Dyatlov and Vasilieva, 2008).

8.4.2 Azolla, Ceratopteris *and* Marsilea

Additional problem ferns of aquatic habitats have been reported for species of the genera *Azolla*, *Ceratopteris* and *Marsilea* (Table 8.1). While *Azolla* floats on the surface of the water and therefore presents problems similar to *Salvinia*, *Ceratopteris* and *Marsilea* (water-clover) are mostly rooted and emergent. *Ceratopteris thalictroides* is listed as a noxious weed in California and has been a long-standing, alien, agricultural weed of taro crops in the Hawaiian Islands (Wilson, 1996). It was recorded there in 1907 and was probably introduced by Chinese rice growers and cultivated for its edible young leaves. It now has a limited distribution in the Hawaiian Islands because of reduced wetland habitat and improved methods of weed control (Palmer, 2003).

Five species of *Marsilea* are listed as problem plants in the USA (Table 8.1). *Marsilea macropoda* and *M. vestita* from western USA became naturalized in southern subtropical USA along with warm-climate species from other parts of the world (*M. minuta* and *M. mutica*). Depending on water levels, *Marsilea* can be fully aquatic, amphibious or marginal. Plants have tough, rooted rhizomes that permit rapid spread and overwintering in colder areas as far north as Oklahoma (Jacono, 2002) and thus can create large colonies. *Marsilea* produces drought-resistant and heat-resistant sporocarps that can be spread by feeding waterfowl (Nelson, 2000). When invasive, *Marsilea* tends to dominate the marginal and shallow water of wetland communities but does not smother huge areas of open water as do *Azolla* and *Salvinia*. In contrast, conservation of *M. villosa*, an endangered native of the Hawaiian Islands, has been necessary to prevent extinction where human activity has upset natural water-level cycles in its seasonal, wetland habitats (Orr, 2008). *Marsilea drummondii* (nardoo), native to Australia, has long been used as a food plant by the indigenous people (Jones and Clemesha, 1980) and poses no problems, as long as it is realized that these plants contain levels of thiaminase higher even than in *Pteridium* and can therefore be dangerous to both farm animals and humans (Moran, 2004).

The aquatic genus *Azolla* includes two species of floating ferns that can be invasive when introduced to alien waters (Table 8.1): *Azolla filiculoides* from the New World and *A. pinnata* from the Old World. Each causes problems in some part of the other species' range. The two *Azolla* species share a number of characteristics. Each free-floating leaf is just 1–3 mm wide (Fig. 8.11) which makes these some of the smallest ferns in the world. In addition to aerenchymatic tissue within the leaves themselves, buoyancy is achieved by air trapped between pairs of leaves closely folded against each other. An important feature of *Azolla* is that chambers near the leaf-bases contain filaments of *Anabaena azollae*, a symbiotic cyanobacterium (blue-green alga) that fixes atmospheric nitrogen and thus permits the ferns to flourish even in nutrient-poor waters. The ability to spread rapidly by budding and fragmentation and the huge floating rafts that result, produce problems very similar to those described for *Salvinia molesta*. Unlike *Salvinia* species that only spread vegetatively, *Azolla* species occasionally produce fertile sporocarps that lie

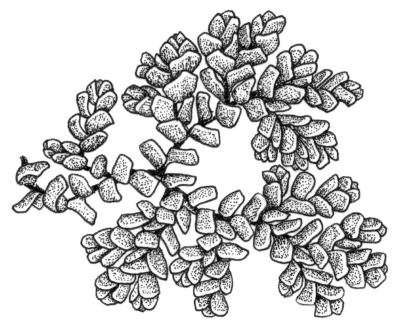

Fig. 8.11 Features typical of *Azolla* aquatic ferns: very small (1–3 mm), densely packed floating leaflets. (Adapted from Moran, 2004.)

dormant at the bottom of shallow waters until spore germination and subsequent fertilization of gametophytes results in the addition of young sporophytes to the population. *Azolla* species may also be spread by waterfowl (Hussner and Lösch, 2005). *Azolla filiculoides* can double its surface area in 7–10 days in South Africa (McConnachie *et al.*, 2004) while *A. pinnata* doubles its fresh weight in 3–5 days in Southeast Asia (De Winter and Amoroso, 2003). Dense mats of *Azolla* negatively affect biodiversity, degrade water quality, clog irrigation systems and interfere with recreational activities (De Winter and Amoroso, 2003; McConnachie *et al.* 2004). In northern Iran, encouraged by sewage and fertilizer run-off, *A. filiculoides* has even dominated invasive *Salvinia* and *Lemna* species (Filizadeh and Zolfinejad, 2006). Ambient water conditions and the management regime greatly affect the relative success of competing aquatic species. *Azolla filiculoides* thrives under high light levels, low flow rates and low levels of disturbance (Sabbatini *et al.*, 1998).

As with several other invasives, the vigorous growth of *Azolla* can also result in positive effects (Table 8.3). The nitrogen-fixing ability of *Azolla pinnata* makes it a valued source of nitrogen. Under optimum conditions, *A. pinnata* can produce 9 tonnes ha^{-1} year^{-1} of protein and it has been and still is of great importance for intercropping with rice, taro (*Colocasia esculenta*) and arrowhead (*Sagittaria sagittifolia*), as a fertilizer, a green manure and as a feed for animals, poultry and fish in many parts of Asia (De Winter and Amoroso, 2003). In such areas, farmers go to great lengths to promote growth of *A. pinnata*, to prevent relentless attack by fungi and snails, and to ensure its survival during winter in cooler areas: a plant that proves surprisingly difficult to cultivate (Moran, 2004). In parts of China, the use of the *A. filiculoides* has replaced the traditional *A. pinnata*, being more cold tolerant. In stretches of the river Erft in Germany, which are heated by geothermal activity, the appearance of *A. filiculoides* was welcome due to its enhancement of biodiversity without significant affect on other vegetation (Hussner and Lösch, 2005). Thick, continuous mats of *Azolla* have been used to reduce levels of mosquitoes by reducing their access to water to lay eggs or preventing the larvae from reaching the surface to breathe. A study of arctic climatology (Brinkhuis *et al.*, 2006) even suggests that some species of *Azolla* may have had a significant role in reversing a greenhouse effect that caused the region around the North Pole to turn into a hot tropical environment 55 mya. Huge, dense patches of *Azolla* growing in freshwater lakes formed by the climate change may eventually have consumed enough CO_2 for the greenhouse effect to reverse.

8.5 Management of problem ferns

A single species of problem fern such as *Pteridium* may require management for many different reasons (Robinson, 2008). Management is usually an extended, costly process because of the scale of infestation, often exacerbated by difficulties of access. Where invasive problem ferns occupy thousands or even millions of hectares (e.g., *Lygodium*, *Pteridium* and *Salvinia*), control measures may some-times take on the aspect of warfare. The finance, labor and resources for such undertakings are substantial and regional planning in terms of decades may be necessary. In some cases, the control of encroachment by invasive ferns may be required in perpetuity.

A salient requirement of managing problem ferns is to "know your enemy." In all control programs, the correct species identification and the distinction between native and alien fern species is fundamental to the decisions taken, as has been emphasized in previous sections. It is necessary to know how the fern became a problem in the first place, the habitats and terrain that are under threat and their ecological significance, as well as the extent of the problem to be dealt with and how rapidly its status may be changing. Comprehensive monitoring and survey work will be required to gain this information before decisions can be made as to appropriate management strategies and treatments.

The management of problem ferns may be either preemptive, to prevent problems arising, or retrospective where problems must be dealt with after the event. Prevention is better than cure in all cases concerning invasive species but early detection and treatment, as the most successful and economical management option, is not necessarily easy to put into effect.

Preemptive options for alien ferns can include outright import bans, legislation to control suppliers of ornamental plants, or declarations of prohibited noxious weeds. Few such controls are sufficiently timely. In most cases, the invasive potential of an alien species is at first unrecognized and the initial spread unnoticed until the species has established over a substantial area and has already become a problem. After introduction, public awareness programs may help prevent further recurrence of such escapes (e.g., Jacono, 2006; Anonymous, 2008a). However, prevention of further spread becomes extremely difficult for a species that can colonize remote locations by spores or, in aquatic environments, by fragmentation. Therefore, most problems created by invasive ferns are encountered retrospectively, requiring substantial control programs after the event. The sheer cost, scale and duration of such programs may also lead to delays in implementation, during which the problems worsen. In some cases (e.g., *Lygodium* or *Pteridium*), decades may pass before effective management is implemented, especially where novel control methods must be researched and tested.

A degree of preemptive management in new locations becomes feasible once the ecology of a fern and its responses to land management and control practices are sufficiently understood so as to be predictable. This is the case for *Pteridium* where the results of investigations from around the world over many years now permit better use of modern and traditional control methods. In a proactive approach to managing the spread of *Angiopteris evecta*, Christenhusz and Toivenen (2007) used distribution data from a number of sources to construct a model of its potential range based on climate and environmental variables. They found that *A. evecta* could easily spread in places where it was already invasive, including most of the Hawaiian Islands as well as large parts of Central America and southern Florida that were not too dry. They strongly recommended programs that minimize the impact of this popular ornamental plant in areas with suitable climate profiles. The difficult and urgent problems created by *Lygodium* species in Florida resulted in a program with a uniquely comprehensive, coordinated, regional approach to meet the challenges of control (Hutchinson *et al.*, 2006). This scheme was facilitated by the existence of organizations already well established in the region to combat massive invasions by alien seed plant species such as *Melaleuca quin-quenervia* (white bottlebrush tree) and *Schinus terebinthifolius* (Brazilian pepper).

Management programs for invasive species must also consider the depredations to the environment caused by that species and will need to include contingencies for vegetation recovery measures (Pakeman *et al.*, 1995). These measures may be as extended and as expensive as the problem-control programs and will often need to run concurrently.

8.5.1 Management strategies

Four basic strategies can be defined for the management of problem ferns (Robinson, 2007): (1) eradication; (2) containment; (3) releasing; and (4) long-term, low-level equilibrium stasis. Eradication measures may carry a high initial cost but this must be balanced against the potentially limitless duration of the other options. Apart from containment, the choice of strategy is, in most cases, dictated primarily by ecological and cultural requirements.

Eradication can provide a permanent solution for alien problem species at an early stage where a limited area is affected. Eradication becomes more difficult as the size of the affected area increases, especially for a species that has the ability to reinfest areas at random from spores that can lie dormant for several years. Eradication is however unacceptable for native problem ferns in most circumstances, and limited clearance measures may be proposed for well-defined, discrete areas but not obliteration from the landscape. Once initiated, any clearance program must be taken to completion, otherwise, as soon as control measures cease for any reason, the problem fern will almost certainly return to its dominant role.

Containment is a strategy that is used where areas of a problem fern must be left intact due to limited resources or other constraints. Containment requires occasional peripheral treatments to keep areas of the fern in check to prevent further encroach-ment. Regular site visits are required to detect advancing fronts that are incursions requiring priority treatment. Widely dispersed *Lygodium* colonies initiated by spores create challenges for containment in areas with poor access and where areas of sensitive habitat must be patrolled regularly to tackle new infestations before they become established. These activities may be termed "confinement" or "exclusion" depending on the context of the site. In many situations with a native problem species, such as *Pteridium* in the UK, the only alternatives to a perpetual containment program are to improve the competitiveness of the adjacent vegetation, clear the fern from the infested areas, or change the land use.

Releasing is a useful term borrowed from commercial forestry where *Pteridium* is controlled to prevent deleterious effects on newly planted trees (Biggin, 1982). When tall *Pteridium* leaves collapse on top of tree seedlings in the autumn, they can distort or break the young growth, especially with the added weight of winter snow. Releasing requires temporary control of *Pteridium* to permit the trees to grow tall enough to avoid damage. The concept is usefully extended to any situation where partial, temporary control is used to reduce the dominance of a problem fern to the direct benefit of underlying vegetation – usually targeted for conservation purposes. This strategy often applies to specific areas where a native problem fern cannot be eradicated because of the overall intrinsic benefits it offers as a nondominant component of the habitat (Table 8.3). *Pteridium* is a strong competitor with a tendency to dominate, so occasional releasing treatments will prove essential even on sites where its presence is vital to dependent conservation species. Releasing measures may be generalized (e.g., long-term, annual cutting regimes) or applied as specific, localized spot-treatments. Another option consists of rotational treatment of blocks with a cycle lasting several years so that optimal habitat conditions for the target conservation species are likely to occur at least somewhere on-site in a given year.

Long-term, low-level equilibrium stasis is a strategy that aims to maintain growth of the invasive fern at a low, acceptable level and to achieve a persistent and sustainable, natural equilibrium state, often using some form of biological control, as discussed in the next section. They are not cost free where added investment is required such as fencing to control grazing; and certainly not where extensive research may be necessary to develop an effective biocontrol agent.

The allocation of the above four management strategies will be determined by the ecological and infestation data from the initial surveys of a given site, the selected land-use options, the required vegetation recovery measures, the resources available, regulatory constraints and the feasible control technologies. Once these factors are known, attention can be paid to the treatment methods to be used.

8.5.2 Control options

Methods of controlling problem ferns can be categorized as physical, chemical or biological and sometimes a combination of such methods will be employed. Environmental control may represent a fourth option for aquatic ferns but this is rarely practical. In the case of *Salvinia* or *Azolla* species for example, drainage of the surface waters may be used to expose the weed to drought or to frost (Elder, 2004) but this treatment will also adversely affect the natural components of the ecosystem. Because the growth of *Salvinia* is reduced at pH >7.5 (Owens *et al.*, 2005) and increases with the nutrient content of the water (Madsen and Wersal, 2008), the control of agricultural run-off and sewage effluent to maintain low nutrient levels in the water may be helpful in suppressing a potential infestation to begin with, or potential recovery of the weed following control treatments. In practice, such manipulations of the aquatic environment may be hard to achieve, especially where contributing water flows may be irregular in quantity and quality.

For all control methods, discrimination in their manner of use remains essential to minimize "collateral damage" to nontarget species. In some cases, the scenic value of sites must be considered (e.g., where control of dominant *Dicranopteris linearis* or *Pteridium* may leave large, unattractive areas of persistent litter for many years). For native species that play an intrinsic, often useful, role within the landscape and its ecology, discrimination between sites remains essential and will exclude the use of biological control options that are not strictly local in effect. Conversely, an alien problem species may justifiably be subjected to all possible means aimed at selective annihilation within affected zones and where nonlocalized biological control measures may prove vital.

Physical control

Several methods of physical control have been used to treat problem fern infestations. Physical options for controlling terrestrial invasive species (e.g., *Pteridium*) have a place in "organic" and traditional farming situations (Bond *et al.*, 2006).

Plowing offers effective control of ferns, except when deep-rooted species (e.g., *Equisetum*) are involved, and therefore invasive ferns are rarely a problem on actively cultivated arable land. However, native ferns such as *Pteridium* can infest longer-term plantation crops such as tea (*Camellia sinensis*) in Southeast Asia (De Winter and Amoroso, 2003), a phenomenon also observed in Morocco and the Azores (R. C. Robinson, personal observation). Likewise, *Lygodium venustum* is a troublesome weed of vanilla plantations (*Vanilla planifolia*) in Mexico (Mehltreter, 2006).

Cutting is one option for physical control. For centuries, annual cutting of *Pteridium* using scythes was the traditional method of controlling extensive infestations in Europe. A variety of portable, tractor-operated or self-propelled mowers and brush-cutters now offer alternatives. Where high densities of dormant rhizome buds occur within entrenched patches, cutting may need to be repeated twice a year for over 20 years before such patches are likely to be killed (Le Duc *et al.*, 2007). Cutting the peripheral zone of a *Pteridium* stand just twice a year over several years will halt encroachment and thus offers an effective method of containment. Cutting cannot be used to eradicate *Pteridium*, except on very small areas, because the persistent regrowth eventually becomes too small to be dealt with by mechanized means. To control *Lygodium* growth, a technique called "poodle cutting" is used where the strands are cut 1.5 m above the ground to kill the top-growth, but such physical treatment alone is insufficient. Lower, uncut strands must then be treated with herbicide to prevent regrowth from the remaining strands and to minimize regrowth from the rhizomes in subsequent years. An extreme example of cutting has been used for the Australian tree fern *Sphaeropteris cooperi* in the Hawaiian Islands. It is physically controlled by felling the trunks, cutting off the terminal meristem and either chopping it up or placing it off the ground to prevent resprouting (Motooka *et al.*, 2003). Cutting treatments to control established aquatic ferns are usually inappropriate because of the difficulty in handling the huge masses of material concerned and the speed at which the ferns can regenerate in relation to the rate of cutting (McFarland *et al.*, 2004). Floating booms and other barriers to contain a floating fern will usually fail during storms and floods.

Crushing to control *Pteridium* and other terrestrial ferns may be undertaken using heavy, corrugated rollers, or baulks of timber, drawn by horses or vehicles. Crushing is faster and more repeatable than cutting because less care is needed in situations where cutting gear would otherwise suffer damage. Crushing may benefit conservation areas where *Pteridium* has become dominant because it appears to permit faster, initial recovery of other plants than cutting. Rapid regrowth of *Pteridium* occurs after cutting but not after crushing, possibly due to the release of apical dominance when the leaves are cut off. However, ultimately, repeated cutting is bound to be more effective than crushing.

Livestock can often effectively control *Pteridium* by trampling or rooting, especially in combination with other methods but, in all cases, there are risks that *Pteridium* may be consumed by the stock, with risks of acute poisoning in the short term or cancer in the longer term. Pigs, wild boar, cows, sheep, goats and horses are all susceptible to *Pteridium* and may die from its effects. The trampling and rooting effects of livestock represent a nonselective form of biological control that is both localized and concentrated in effect by fencing. Low-level equilibrium grazing (LEG) in which heavy animals such as horses or cattle are used for seasonal grazing of *Pteridium*-infested pastures can also be effective although lighter animals such as sheep are not useful for this strategy. More conventional forms of biological control are discussed below.

Burning land to rejuvenate vegetation on permanent grazing is inextricably linked to the spread of species such as *Pteridium* (Page, 1982). Burning encourages initial colonization by spore germination in areas not previously inhabited by the fern (Gliessman, 1978). Once the fern is established, repeated burning often encourages *Pteridium* (especially on heathland and moorland; Parry, 2003) by reducing the competition of other vegetation and releasing potash locked up in litter. *Dicranopteris linearis* appears to respond in a similar manner (Cohen *et al.*, 1995). Although burning of *Pteridium* litter may sometimes be useful in vegetation recovery programs, fire will not control *Pteridium* rhizomes. Ferns with shallow rhizomes, such as *D. linearis*, are susceptible to control by fire but will quickly recover unless the rhizomes are completely destroyed (Walker and Boneta, 1995). In Florida, wildfires caused by lightning are a means to natural habitat succession in many areas. Prescribed burns are therefore a useful means to promote rejuvenation of vegetation and to help exclude invasive species. However, the mats, walls and spires of dense *Lygodium* can carry forest fires to the tree canopy and thus are more destructive than wildfires that occur naturally only at ground level. The use of fire as an ecosystem management tool is thus limited when *Lygodium* is present and may actually increase the opportunities for spore dispersal (Hutchinson *et al.* 2006).

In general, with the exception of methods that disturb the soil, physical options tend to be less efficient at controlling invasive ferns than the chemical or biological control options that are discussed below. However, soil disturbance to bring seeds to the surface is an important means to promote natural vegetation recovery after control measures where dominant, invasive ferns such as *Pteridium* or *Dicranopteris linearis* have built up dense litter layers (Lowday and Marrs, 1992; Cohen *et al.*, 1995).

Chemical control

Although an irrational fear of pesticides persists in many quarters, the effects of problem ferns are sometimes so severe that they justify extreme control measures using chemical treatments on a scale that public concerns might inhibit in other circumstances. Herbicide application is thus an important control option for invasive ferns. Glyphosate, asulam and sulfonylurea herbicides are widely used to control ferns and examples of these and other chemicals, their characteristics, effects and limitations are summarized in Table 8.4. Also listed are some natural plant extracts used to control ferns in addition to synthetic chemicals. The natural products tend to be less effective and may have as many repercussions for other species in the environment as the use of a synthetic herbicide. For example, in rice fields, plant residues of *Parthenium hysteropherus* can have lethal allelopathic effects on *Salvinia* but unfortunately will also affect other species (Pandey, 1994). Trials undertaken in the UK using extracts of *Artemisia vulgaris* (mugwort) and A. *dracunculus* (tarragon) gave over 60% control of *Azolla filiculoides* in ten days (Oduro *et al.*, 2006) but such treatment alone will not eliminate the problem.

An innate aversion to synthetic herbicides in the EU (European Union) is itself a problem where asulam is necessary to control large areas of *Pteridium* within upland water catchments in the UK. The quantities of the herbicide entering drinking water supplies must be kept to a minimum and it sometimes proves necessary to limit spraying of asulam (Knapp, 2005) in order to meet the EU statutory limit for a pesticide in tap water ($0.1\,\mu g\ L^{-1}$). This limit is 100 times lower than that applied by the US Environmental Protection Agency and causes unnecessary restrictions on the use of asulam that may otherwise be justified as a means of reducing risks from the carcinogen ptaquiloside released by *Pteridium* in drinking water (see Section 8.2.1). By comparison, the European regulatory authorities show no concern over the presence of ptaquiloside that has been measured in drinking water at levels as high as $45\,\mu g\ L^{-1}$ (Rasmussen, 2003) whereas $0.005–0.016\,\mu g\ L^{-1}$ has been estimated as a tolerable maximum level for this carcinogen in drinking water (Rasmussen *et al.*, 2005). Asulam, a synthetic chemical that has no measurable animal toxicity still causes greater public

Table 8.4 *Some herbicides used to control problem ferns*

Chemical ingredient	Characteristics and limitations	Responsive problem ferns
Ammonium sulphamate	Total vegetation control. No replanting for 6–12 weeks.	*Equisetum* (eradication)
Artemisia extracts	Species native or cultivated in the UK	*Azolla filiculoides*
Asulam	Highly systemic, good selectivity. Suited to aerial application. Repeat spot-applications will kill the worst rhizomatous ferns. Once cleared for aquatic uses. No measurable animal toxicity.	*Pteridium* on moorland and pasture (± adjuvants), *Azolla filiculoides* *Equisetum arvense* (repeated applications)
Citronella oil	Extract of species of *Cymbopogon* grass. Not selective, affects invertebrates and heath vegetation.	*Pteridium* (standing leaves and following season's growth)
Dicamba and 2,4-D	Safe on grasslands.	*Sphaeropteris cooperi* (applied to individual shoot apices) *Equisetum* (with added trichlopyr)
Dichlobenil + amitrol	For situations with woody ornamentals, tree fruits, nurseries and shelterbelts, raspberries, cranberries and blueberries.	*Equisetum*
Fluridone	Systemic; broad-spectrum. Used for water weed control.	*Salvinia molesta*
Glyphosate	The most widely-used herbicide. Highly systemic, nonselective. Unsuited to aerial application unless as winter applications (*Lygodium*). Kills all green plants (except those with waxy leaves). Cleared for aquatic uses.	*Pteridium* (as dominant stands) *Pteridium* with mixed scrub in forestry *Lygodium* *Salvinia* (with adjuvants) and *Azolla* *Equisetum arvense* (repeated applications)
Imazapyr	Residual, broad-spectrum brushkiller.	*Pteridium* (temporary scrub control; pre-plant in forestry)

Table 8.4 (*cont.*)

Chemical ingredient	Characteristics and limitations	Responsive problem ferns
MCPA	Top-growth control only in grassland.	*Equisetum*
Orange oil	Repeat applications needed.	*Salvinia* (light infestations in Australia)
Paraquat and diquat	Contact-action. Highly effective in short-term. Hidden leaves will regrow. No long-term effect on rhizomatous species.	*Salvinia* *Azolla* Surfactants help to reduce buoyancy
Parthenium hysteropherus	Plant residues applied. Damages other paddy species.	*Salvinia* in rice paddy
Picloram	Residual action for total vegetation control in noncropped areas. Safe to grasses.	*Pteridium*
Salt (sodium chloride)	Limited to shallow waters due to huge amount of salt needed, adverse effects on other organisms in habitat.	*Azolla filiculoides* *Salvinia molesta*
Sulfonyl-ureas (metsulfuron-methyl, amidosulfuron, tribenuron-methyl, carfentrazone-ethyl, thifensulfuron-methyl)	Low dose but limited selectivity. Safe to grasses and some shrubs. Tribenuron-methyl and thifensulfuron-methyl as pretreatments at low doses may help to break dormancy of rhizome buds (e.g. in *Pteridium* and *Lygodium*).	*Pteridium* – further research needed to determine longer-term efficacy: amidosulfuron appears more effective than metsulfuron-methyl or tribenuron–methyl *Lygodium* – metsulfuron-methyl is mixed with glyphosate to improve short-term die-back of *Lygodium* *Salvinia* – carfentrazene-ethyl (repeated applications)
Terbutryn	Subsurface injection of granules.	*Azolla*
Triclopyr	Foliar-absorbed brushkiller with some systemic activity and limited selectivity. Partial effects on rhizomes with many dormant buds.	*Lygodium* (repeated applications) Not effective on *Cyathea* tree ferns
Triclopyr + 2,4-D + dicamba		*Equisetum* (repeated applications)

Source: The information is drawn from authors cited in the text and additionally Winfield (1988), Fagan *et al.* (1990), Lawrie and Butler (1995), West *et al.* (1995), Keary *et al.* (2000), Glomski and Getsinger (2006), Lockhart (2007).

Fig. 8.12 Helicopter spraying of grazing infested with *Pteridium*, Scotland.

concern than the hazardous, natural substances derived from *Pteridium* itself. This paradox was noted 20 years ago by Howell (1988).

Because the large tracts of uplands or wetlands affected by invasive ferns can often be inaccessible, aerial spraying (Fig. 8.12) proves essential and more cost effective in many situations than ground-based methods. Safety for nontarget species and their associated ecosystems will benefit from use of selective herbicides (such as asulam), appropriate application methods, accurately directed applications, or dormant-season applications. Pollution of surface waters, groundwaters and drinking-water catchments by herbicides must be kept to a minimum. Highly systemic herbicides such as asulam and glyphosate (Table 8.4) permit the use of large droplets that help to prevent spray drift into nontarget areas, even when applied by helicopter. To protect nontarget areas and reduce direct pollution of surface waters, effective buffer zones have been investigated for ground-based spraying using conventional nozzles (Marrs *et al.*, 1993) or for improved methods of helicopter spraying (Robinson *et al.*, 2000).

Pteridium spores will colonize newly cleared or disturbed land (see Section 8.2.1) but do not appear to recolonize land that has previously been occupied by *Pteridium* (R. C. Robinson, personal observation in the UK). Therefore, longstanding beds of native *Pteridium* can be controlled and even eliminated by a herbicide such as asulam. This result is fortunate because the mostly nonlocalized biocontrol options that are feasible for alien ferns are not acceptable for *Pteridium* or similar native problem ferns. It is significant that asulam does not appear to have been tested in the control of *Lygodium* in Florida where its selectivity to native species might be an improvement over the sulfonylureas recently introduced to complement the use of glyphosate (Hutchinson *et al.*, 2006) or glyphosate itself.

Fern species differ in their susceptibilities to a given herbicide such as asulam (Sheffield, 2002: Sheffield *et al.*, 2003). The native *Paesia scaberula* and tree ferns (several species of *Sphaeropteris* and *Dicksonia*) in New Zealand have survived aerially applied doses of asulam that are usually effective against *Pteridium* (R. C. Robinson, personal observation). *Sphaeropteris cooperi* in the Hawaiian Islands is susceptible to concentrated sprays of dicamba and 2,4-D applied directly on the stem terminals, whereas triclopyr, used frequently against *Lygodium* in the past, was less effective (Motooka *et al.*, 2003). *Equisetum arvense* has a reputation for being resistant to many foliar-applied herbicides and this reflects the effective protection provided by its waxy shoots, vigorous tubers which may create diversionary sinks, and extremely deep rhizomes.

Except for areas subject to total vegetation control, where nonselective doses of persistent, broad-spectrum herbicides may be used, a single herbicide treatment will rarely, if ever, kill a colony of a rhizomatous problem fern outright. Multiple follow-up treatments will usually be required for which planning over several years will be critical. Appropriate adjuvants (i.e., substances added to agricultural sprays to enhance effectiveness) can increase both the rate and degree of herbicide uptake and subsequent translocation within the plant. This approach can be significant when targeting the buds on extensive rhizome systems and can decrease the length of the follow-up period required.

Floating mats of *Salvinia* and *Azolla* are much more accessible to herbicide treatments than most terrestrial problem ferns, especially where aerial spraying is used. When used in the control of *Salvinia molesta*, surfactants (i.e., substances that reduce surface tension) permit escape of the air trapped between the trichomes (Plate 7B), thus causing a loss of buoyancy and increased exposure of the green leaves to any herbicide concentrated at the water surface (McFarland *et al.*, 2004). Van Oosterhout (2006) presents alternative, physical options for controlling aquatic ferns and their significant difficulties. Where possible, herbicides to control aquatic ferns should be applied before a dense surface cover has developed, not only to enhance efficacy but also to avoid the damaging effects on water quality when large masses of fern begin to die and decay in a short period after herbicide treatment.

Lygodium and *Marsilea* reproduce prolifically by spores. Despite successful elimination of a rhizomatous parent infestation of *Lygodium* using herbicide, if spores have been dispersed, a plethora of young sporophytes may subsequently appear within the controlled area and at new sites. A similar situation exists for the long-lived sporocarps of *Marsilea* species that can reinfest an area years after the rhizomatous parent colony has been dispatched. Because *Salvinia* and *Azolla* can spread by fragmentation, they can rapidly recolonize the entire area subject to control, if just one fragment survives herbicide treatment. All these reproductive traits mean that herbicides may be employed as essential stop-gap treatments to control such ferns, up to acceptable financial or environmental limits, but cannot guarantee substantial long-term control. In such cases, biological control becomes an essential alternative.

Biological control

Biological control of problem ferns conventionally refers to the release of a selective fungal or bacterial disease or herbivorous arthropod. Such biological controls can be effective means of reducing infestations of an alien problem fern to manageable levels over extensive areas. This approach is especially valuable in hill-land, woodland, wetland and aquatic situations that present diffi-cult access for other ground-based control methods and where aerial spraying may not always provide a solution. Although local extinctions of a problem fern are sometimes possible (e.g., *Azolla filiculoides* in South Africa using *Stenopelmus rufinasus*, McConnachie *et al.*, 2004), biocontrol usually achieves a long-term equilibrium stasis.

The trampling inherent in low-level equilibrium grazing also represents a form of biocontrol of *Pteridium* where, as appropriate for a native species, the effects can be localized by fencing. Nonlocalized biocontrol using arthropods or disease has no place in the control of native problem ferns when other rare or endangered, native species may be dependent either wholly or partially upon the target problem fern (see Section 8.2.1). This difficulty terminated research once undertaken on South African insect species with the potential to control *Pteridium* in Britain (Lawton, 1990; Fowler, 1993). However, in regions where *Pteridium* is newly established from spores after forest felling (e.g., in South America), and components of local ecology are unlikely to be dependent on the fern, the herbivorous South African insects may yet find potential uses. Furthermore, a sawfly from the Venezuelan Andes (*Aneugmenus merida*) is capable of inflicting severe damage on the forms of *Pteridium* found in that region (Calcagno-Pisarelli *et al.*, 2006).

Alien ferns do not usually create severe problems in their native environments where it must be assumed that herbivory, competition or disease keeps these ferns in check. These natural control organisms are likely candidates for use as biocontrol agents when such ferns become invasive in alien habitats. Considerable research is under way to discover such agents to control problem ferns in their nonnative locations. Potential control species have to be completely target specific and no threat to other nontarget vegetation in the problem areas. For example, biocontrol agents required to manage *Lygodium microphyllum* and *L. japonicum* in Florida will require complete safety for native *Lygodium* species, such as *L. palmatum* (found in the eastern USA from Florida northwards) and other noninvasive *Lygodium* species (e.g., *L. cubense*, *L. oligostachyum* and *L. volubile*) in the nearby West Indies. Similar concerns exist for some 40 other threatened or endangered fern species that are native to the region (Pemberton *et al.*, 2002b). Suitable biological methods to control problem *Lygodium* species are urgently required where neither physical nor herbicide treatments are capable of dealing with the large number of widely dispersed young sporophytes that can appear without warning in any niche available to wind-borne spores (Hutchinson *et al.*, 2006).

Once a nonlocalized biocontrol agent is released, there is no going back if unexpected effects occur. The relationships between a biocontrol agent and its targeted fern sometimes exist in a delicate balance. Ecological research on the efficacy and safety of biocontrol agents with respect to their target ferns may take many years. In that period, other control measures will be required to curtail an invasive alien species. Biocontrol programs do not always have predictable results and that is another reason for keeping alternative control methods on standby.

Biological control methods have a particular advantage against prolific alien aquatic ferns such as *Salvinia* and *Azolla* when mechanical or chemical control methods, designed to kill the weed in a short period, create problems of their own (Julien *et al.*, 2002). The spread of *Salvinia* may occasionally be so explosive that herbicide may be the only feasible control method in the short term. In such a situation, the decomposition of large amounts of dead fern material produced over a short period has serious consequences for water quality and the aquatic ecosystem. This is avoided by the much slower, sustained, season-long attack by a biological control agent (e.g., the weevil *Cyrtobagous salviniae*, Plate 7B, Julien *et al.*, 2002). Nevertheless, the low-level stasis achieved may ultimately be difficult to maintain and less predictable than that achieved by chemical control methods. An integrated approach, using different control methods is usually the best solution where sustainability is critical.

Mycoherbicides as biocontrol agents: Mycoherbicides are preparations of a fungus that are applied to cause disease levels that will control the vigor or reproduction of target plants. Relatively little progress has been made using mycoherbicides to control invasive ferns. Initial attempts to control *Pteridium* in Britain using such natural pathogens failed because of difficulties in developing a formulation with humectant (i.e., moisture-retaining) properties that ensured fungal spore germination and penetration of the fern leaves (Womack *et al.*, 1995).

Rust fungi are potentially useful as biocontrol agents because they have very narrow host ranges (see Chapter 7) such as *Puccinia lygodii*, a rust native to tropical South America, that has been found on *L. japonicum* in Louisiana (Rayamajhi *et al.*, 2005) and *Colletotrichum gloeosporioides* which has the potential to cause *c.* 50% dieback of *L. microphyllum*. Further opportunities to control *Lygodium* may reside in mycoherbicide preparations of *Myrothecium verrucaria*, a fungus native to Florida that is known to attack other alien species in the region (Jayachandran *et al.*, 2008). In India, conidial suspensions of the fungi *Phoma glomerata* or *Nigrospora sphaerica* are capable of suppressing the growth of *Salvinia molesta* (Kumar *et al.*, 2005). A species of *Dreschlera* has also been identified as a pathogen of *Salvinia* in Brazil (Muchovej, 1979; Muchovej and Kushalappa, 1979). *Azolla pinnata* is highly susceptible to several fungal pathogens such as *Myrothecium verrucaria*, *Rhizoctonia solani* and *Sclerotinia sclerotiorum* (De Winter and Amoroso, 2003). The last two are also serious diseases of potato and rice crops, respectively, which suggests that *Azolla* may sometimes also act as a reservoir for such diseases between crops.

Herbivorous arthropods as biocontrol agents: Within their native ranges, *Lygodium* species do not appear unduly invasive although they may be weedy in some areas. This suggests that substantial, natural herbivory may be taking place, as exemplified by *Octothrips lygodii*, a thrips which occurs widely in Southeast Asia and can severely damage *L. microphyllum* (De Winter and Amoroso, 2003). Representatives of about 20 herbivorous arthropod taxa have been collected for testing from *L. microphyllum*, *L. flexuosum*, *L. reticulatum* and *L. smithianum* in their native habitats in Asia and Africa. These include moths, thrips, mites and stem-borer beetles, as summarized briefly below based on Goolsby *et al.* (2003, 2004). Pyralid moths (*Austromusotima camptozonale* (syn. *Cataclysta*) and *Neomusotima conspurcatalis*) from Australia and Southeast Asia have been tested as *Lygodium* defoliants and the former was first released in Florida in 2005. Another candidate is the mite *Floracarus perrepae* that produces galls on the leaflets of *L. microphyllum* and causes curling of the margins of new leaflets. While feeding, the mite also transmits a fungal disease that causes further streaking and necrosis of the leaflets, a form of damage that has been noted in Southeast Asia as well as in Australia. Fungi isolated from the necrotic patches associated with mite damage included *Botryospheria* as a secondary pathogen. From Thailand, the leaf-mining beetle larvae of *Endelus bakerianus*, together with the leaf-feeding sawfly *Neostrombocerus abicomus*, may be more effective defoliants of *Lygodium* than the moth larvae already released in the USA. A pyralid stem-borer moth (*Ambia* sp.) is another prospect that damages the rachis just above the ground, either killing the rachis or disrupting its ability to transport water to a greater height. These species show promise if they can be bred in sufficient numbers. Because many of these herbivores of *Lygodium* are tropical examples, questions remain concerning their productivity in the cooler climate of subtropical Florida, despite a promising absence of natural predators and a good level of *Lygodium* specificity offering safety to native species.

Many herbivore species were investigated prior to biocontrol releases against the aquatic *Salvinia molesta* (McFarland *et al.*, 2004). *Cyrtobagous salviniae* (the Salvinia weevil), native to Brazil, Bolivia and Paraguay (Plate 7B), has proved highly effective in controlling *S. molesta* in 13 countries where this fern has invaded alien waters (Tipping and Center, 2003; Miller and Sheffield, 2004). Both the adults and the burrowing larvae of this weevil feed almost exclusively on the fern, can free surface waters from extreme infestations (Fig. 8.10a) and will usually hold the weed at about 1% of its initial coverage (Fig. 8.10b). The larvae feed destructively within all parts of the plants while the adults spend most of their time underneath the leaf lobes, feeding preferentially on new buds and can survive submergence due to air trapped on their bodies. The successful establishment of the weevils is highly dependent on nitrogen levels in waters infested with *Salvinia*. Application of urea to the plants can dramatically increase the initial weevil population buildup and subsequent levels of control. The closely related weevil *Cyrtobagous singularis* did not provide effective control of *Salvinia* (Julien *et al.*, 2002).

Paulinia acuminata (Salvinia grasshopper) will attack all species of *Salvinia* but is not specific to *Salvinia*. A significant reduction in *Salvinia* on Lake Kariba occurred two to three years after the release of *P. acuminata* in 1970 but this decline was attributed to nutrient stress rather than the effects of the insect (Julien *et al.*, 2002). Further trials have shown that larvae of *Samea multiplicalis* (Salvinia stem-borer or water-lettuce moth) may also present options to control *Salvinia minima*, although the moth has proved less successful in Australia for controlling *S. molesta* than *Cyrtobagous salviniae* (Anonymous, 2003).

Stenopelmus rufinasus, a weevil from North America, has been successfully used to control the aquatic fern *Azolla filiculoides*, which is an alien problem fern in South Africa (McConnachie *et al.*, 2004). The same weevil also shows promise against *A. filiculoides* in northern Iran (Filizadeh and Zolfinejad, 2006) and on infested waterways in the UK (Anonymous, 2004a, 2004b; Shaw *et al.*, 2006). In southern China, the aquatic larvae of a midge (*Polypedilum ivinoense*) can decimate a crop of *A. pinnata* in 3–5 days (De Winter and Amoroso, 2003). Larvae of *Chironomus* midges, *Pyralis* and *Nymphula* moths, the Azolla elephant beetle and several species of snails are known to attack *A. pinnata* in Asia and India. Where it is a problem fern in the Hawaiian Islands, *Azolla* is attacked by *Pomacea canaliculata* (Hawaiian apple snail) as well as by the moth larvae of *Agrotis ipsilon* (Lumpkin and Plunknett, 1980). As a cautionary note, the insects used for biocontrol of aquatic problem ferns will be susceptible to the pyrethroid insecticides that are sprayed regularly to control the carriers of malaria, bilharzia and tsetse fly within the same wetland habitats. Nevertheless, Kurugundia and Serumola (2007) showed that such spray programs may affect, but will not terminate, nearby biocontrol programs.

Fish as biocontrol agents: In the Colorado River network of the USA, infestations of *Salvinia molesta* have not proved as invasive as in other river systems, possibly due to grazing by *Oreochromis niloticus* (tilapia or Nile perch) that is ubiquitous in the Colorado system (McIntosh *et al.*, 2003). *Salvinia* is not very nutritious for these fish although they will feed on it, even when given other food sources. Grass carp feed on small infestations of *Azolla filiculoides* in the UK.

8.6 Examples of socioeconomic effects of problem ferns

Problem ferns rarely affect cultivated land and therefore receive little of the widespread attention that agricultural weeds receive in the context of crop production. This bias is clear from Alford and Backhaus (2005) where a conference on the introduction and spread of invasive species made no reference to invasive ferns. Similarly, Riches (2001) discussed only *Salvinia* and, briefly, *Azolla* within a select group of "the world's worst weeds". Taylor (1985) noted a universal tendency to downplay the issues associated with invasive *Pteridium*. A memorandum from the European Commission (Anonymous, 2008b) states that the damage caused by all invasive alien animal and plant species within the EU territories and the necessary control measures account for some US$12 billion annually (all monetary values in this section are converted to US dollars to facilitate comparisons), less than the annual alien plant control budget of Florida. *Azolla* was the only fern genus mentioned.

In the tropics, invasive ferns can affect poorer cultures to an extent that is difficult to assess in financial terms. For example, since first being reported as an alien fern in Sri Lanka in 1939, estimated costs of *Salvinia molesta* for 1987 in that country alone had reached $2.1 million, accounting for losses of rice paddy, fishing, power generation, transport and domestic water use as well as increased health costs and the costs of control measures (Julien *et al.*, 2002). The biocontrol program for *Azolla filiculoides* in South Africa was estimated at $206 million in 2000 (McConnachie *et al.*, 2003) with cost benefits outstripping control costs by 15:1 within 10 years. In any part of the world, most problem ferns affect aspects of ecology and conservation and these effects are difficult to quantify in financial terms until preventive or recovery measures are introduced. These measures are often disparate, ad-hoc and difficult to collate.

Of all the invasive ferns, *Pteridium* has received the most attention because of its direct economic impacts on human welfare, permanent grazing and forestry. The

UK offers ideal climatic conditions for the growth of *Pteridium* as an invasive fern. Virtually all the British land surface is modified by human interference, which creates ideal opportunities for encroachment. Britain is a crowded island with high land values where land infested with *Pteridium* carries approximately one third of the value of land that is *Pteridium*-free. The estimated 1.7 million ha of land occupied by, or under threat from, *Pteridium* (Pakeman *et al.*, 1996) represent an approximate total loss in capital value of at least $3.4 billion, before accounting for any loss of production, forestry or sporting interests.

Sheep farming in the UK uplands in recent years has been increasingly unprofitable, leading to a decline in associated land management. The current spread of *Pteridium* on to abandoned, upland sheep pastures is unprecedented and will be difficult to reverse if the land is ever required for grazing again. The permanent rough grazing occupied by *Pteridium* in Britain may represent an annual loss of production of some 2 million lambs (at 1 sheep ha^{-1}) or 250 000 calves (at 0.25 head ha^{-1}). Whether at a sale price of $60 per store lamb or $450 per yearling calf, this amounts to an estimated lost return of about $120 million year^{-1}.

The only significant, self-financing activity in the British uplands at present which is directly affected by *Pteridium* encroachment is the grouse shooting industry. It is estimated that about 1.5 million ha of heather moorland exist in the UK and that figure represents 75% of such vegetation in the world. On average, it is likely that some 10–15% of this heather moorland is *Pteridium* infested, even where the *Pteridium* is reasonably well managed for shooting purposes, amounting to some 225 000 ha of lost heather that would otherwise annually produce 112 500 pairs of grouse (0.5 pair ha^{-1}). A brace (pair) of driven grouse attracts a shooting receipt of about $280 (including the carcass value) so the lost income caused by *Pteridium* therefore amounts to about $31.5 million year^{-1}. This estimate does not include lost expenditure on accommodation and other necessities by sporting visitors that can make a significant contribution to the local economy. Not all heather land is managed for grouse but the above values give an estimate of the problem. If, instead, the capital value of productive grouse moor is considered at $12 000 per brace ($6000 ha^{-1}), compared with $2000 ha^{-1} for the rough grazing value of *Pteridium*-infested moor, then the loss in value of the 225 000 ha of lost heather on these moors amounts to $900 million. The costs of clearing this area of *Pteridium* at $2000 ha^{-1} to include 10 years of follow-up treatment and management of vegetation recovery for grouse, would amount to an additional $450 million.

The effect of increased tick-borne disease where tick habitat is provided by *Pteridium* (Sheaves and Brown, 1995a, 1995b) also brings a cost in both veterinary and medical terms when the effects of louping-ill virus, tick-borne fever, tick pyemia, redwater fever and Lyme disease are taken into account. Because Lyme disease and the effects of louping-ill virus in humans may not be diagnosed correctly at their early stages, the true cost of these potentially fatal diseases is difficult to establish. Another view of the financial consequences of *Pteridium* may be seen in the Azores dairy industry. The island of São Miguel is home to *c.* 50 000 highly bred Holstein dairy cattle worth about $2000 each while in milk. About 10 000 (20%) of these cattle are sent to slaughter annually, of which, until recently, as many as 1800 animals were rejected at the slaughterhouse due to bladder tumors and other cancers induced by consumption of *Pteridium* that had infested the edges of many of the pastures before corrective measures were introduced. The intrinsic value of the diseased animals, after rejection by inspection, amounted to *c.* $3.6 million year^{-1}. While a productive milk cow will produce *c.* 7500 L year^{-1} in the Azores, the loss in milk production caused by the *Pteridium*-induced cancers amounted to 13.5 million L year^{-1} with a value of about $11.4 million at the supermarket. The sum financial effects of lost production on this one island caused by *Pteridium* therefore amounted to well over $15 million year^{-1}. *Pteridium* impacts several other islands of the Azores archipelago that together produce 27% of the Portuguese milk quota (C. Pinto, personal communication). Similar losses of cattle are occurring in many other countries.

To further illustrate the financial consequences of environmental damage by invasive ferns, one may look at Florida where both *Lygodium microphyllum* and *L. japonicum* are capable of such extreme depredation of wildlife habitat that they are now the target of enormous resources aimed at control. In 2002, control of *L. microphyllum* with herbicides in areas with difficult access cost from $325 to $1250 ha^{-1} (Pemberton *et al.*, 2002a). By 2007, in Florida alone, the annual budget for *Lygodium* management rose to $6 million in order to protect an area exceeding 40 000 ha of unique wildlife habitat (D. C. Schmitz, personal communication). This included both herbicide applications and research into biocontrol methods. The sudden increase in expenditure after 2005 reflected damage from a total of eight hurricanes in 2004 and 2005 that destroyed the infestations of *Hydrilla verticillata*, an invasive aquatic seed plant of Florida lakes, and which released some of the $19 million formerly spent annually on *Hydrilla* control, for the control of *Lygodium* instead. These same hurricane seasons are also thought to have increased the spread of *Lygodium* by spores.

Just as with *Pteridium*, the motivation of different landowners to control *Lygodium* is a key issue. Socioeconomic factors are important where infestations affect property values, ecotourism, hunting leases, cattle grazing and timber production – all of which are incentives for private landowners to initiate and maintain control. Global warming is likely to further expand the range of *L. microphyllum* into northern Florida and even with the current level of effort, an infestation of central and southern Florida by 2014 of over 1 million ha may be an underestimate (Volin *et al.*, 2004). The above-mentioned level of funding for *Lygodium* compares with the $8.2 billion earmarked in 2004 for the overall Everglades restoration project, aimed at correcting the engineering and development intrusions that had reduced wading bird populations by 95%, endangered 68 plant and animal species and cut water flow through the glades by 70%. *Lygodium* has the potential to reverse much of this state-funded restoration work now under way in the Everglades. Because greater funding will be needed to fight the *Lygodium* epidemic using herbicides, sometimes inefficient and costly because of scale and access limitations, only biological control appears to offer a sustainable, long-term solution.

8.7 Conclusions and future prognoses

Invasive plants are one of the major factors that affect biodiversity, along with habitat change, climate change, overexploitation and pollution (Anonymous, 2008b). Within this context, there is a clear tendency among many scientists and the organizations they represent to overlook ferns as problems for ecology, the environment and human welfare. This oversight is partly because ferns rarely create issues for crop production that is the focus for huge resources aimed at feeding a hungry world. Problem ferns have much greater consequences for animal production as well as animal and human health but such effects are rarely treated as a priority. The scale of the problems posed by ferns around the world is much greater than many people would expect, perhaps partly due to the inertia that exists in addressing such issues. Only when problems created by ferns get demonstrably out of hand are efforts made to respond. By then it is usually too late to apply a timely, efficient solution.

8.7.1 Public awareness, the best preemptive control measure

De Winter and Amoroso (2003) proposed an annual trade value of $150–300 million for ornamental ferns, especially in *Nephrolepis*, *Asplenium* and *Adiantum*. They further noted that fern production in Florida during 1996 amounted to a wholesale value of $97 million, while 12 million *Nephrolepis* plants were produced by tissue culture in 1997. These data suggest that a modern census of the international trade in ferns and the risks it brings is long overdue. Botanical parks and gardens, the retail horticultural trade, TV gardening shows, the internet and other media continue to promote alien ferns without any apparent show of concern for the environmental risks that may result. Some outdated web sites continue to promote invasive ferns without any caution (e.g., *Sphaeropteris cooperi*, Gilman, 1999). Because alien problem ferns naturalize mostly by means of spores, the suppliers of these species and the public at large that may acquire them need to be made much more aware of the risks and the real problems caused by ferns. The trade in ferns and the employment it sustains is clearly important but rigorous legislation to control species that offer a threat, and more effective public awareness programs, appear essential to reduce the risk of further damaging escapes. The scale of the *Lygodium* invasion of Florida supports the adage that prevention is better than cure wherever such aggressively invasive species are concerned.

8.7.2 Pteridium

One of the outstanding challenges created by *Pteridium* is to quantify the environmental health risks of ptaquiloside and other carcinogens derived from inhaled spores, from drinking water contaminated with run-off from *Pteridium* beds and from milk derived from animals permitted to feed on *Pteridium*. Headway against the greatest problem posed by *Pteridium*, its widespread occupation of permanent pastures, will only be made when farm economics permit greater returns to the farmer that will justify the necessary control measures. The use of *Pteridium* as a biofuel may yet deserve more attention.

8.7.3 New hybrid ferns and other potential invasives

An outstanding, unknown consequence of problem ferns concerns the behavior of new hybrids being created between native and alien fern species. Apparently innocuous hybrids may prove to be aggressive colonists with the potential to create problems for sensitive habitats. Their ecological, environmental and economic impacts remain unknown. The same applies to many other species of alien ferns that have naturalized rapidly and widely (Table 8.1).

8.7.4 International cooperation in monitoring and research

Wilson (1996) noted that naturalized *L. japonicum* on the Island of Hawaii was not nearly as aggressive as in Florida. Mehltreter (2006) suggested that availability of water may be a key factor in the proliferation of *Lygodium* strands which may help to explain ecological differences that arise between regions. The lack of comprehensive understanding of such differences reflects the many questions that remain concerning invasive species and where answers may yet have a bearing on means of control. There is a clear need to maintain and extend international collaboration to improve the effectiveness and scope of legislative, chemical and biological management of problem ferns. At the same time, more widespread monitoring of naturalized ferns is required to flag and preempt invasive events at an early stage before they reach epidemic proportions (Kolar and Lodge, 2001). A deeper understanding of the ecological effects of many of the naturalized species is also required to safeguard ecosystems that may be at risk. Whereas close attention to accurate taxonomy resulted in a dramatic, international advance in the biological control of *Salvinia*, frequent misidentification of ferns and their numerous, confusing synonyms continue to hinder the study of invasive species. The list of problem ferns compiled in Table 8.1 represents a first attempt at such an undertaking and its scope is entirely provisional. Considerably more work will be required internationally to confirm the significance of many of the species listed and to recognize other species that have been overlooked.

8.7.5 *The* Lygodium *menace*

Areas such as Florida face insurmountable difficulties created by *Lygodium* if effective biological control methods cannot be introduced rapidly and successfully. The paleotropical species *L. microphyllum* is currently well established in shallow aquatic habitats or moist soils in the southern third of the Florida peninsula and, according to climatic analyses made by Pemberton *et al.* (2002a), it has the potential to spread north into Georgia and, by spores, west across the Gulf of Mexico to southern Louisiana and Texas and on through much of wet tropical America. It has already naturalized in Jamaica and Guyana, further threatening the Caribbean and eastern South America. When a well-funded program in the USA has difficulty in dealing with the scale of destruction caused by *Lygodium*, the potentially devastating effects of this alien fern in poorer tropical regions can only be imagined.

8.7.6 *Summary*

In conclusion, the examples of problem ferns discussed in this chapter suggest that the net cost of their effects around the world is enormous and likely to remain so. The costs to date have an unknown value and deserve further investigation so that the full effect of ferns on the global economy, human welfare and the environment might be recognized. If it were not already apparent, ferns deserve both reappraisal and respect when considering both their current and future effects on our own lives and the ecosystems of the planet.

Acknowledgements

The authors are indebted to several fern specialists and practitioners for their help in compiling the information in this review: Patrick Acock, Adrian Dyer, Richard Lewis, Heather McHaffie, Klaus Mehltreter, John Mickel, Robbin Moran, Chris Page, Robert Pemberton, Tom Ranker and Don Schmitz. We thank Robbin Moran for his permission to use line drawings of ferns adapted from Moran (2004). Permission to use photos from the Bugwood.org Network image archive and from the www.hear.org/starr/plants/images of Forest and Kim Starr is gratefully acknowledged.

References

Aguraiuja, R. and Wood, K. R. (2003). *Diellia mannii* (D. C. Eaton) Robins (Aspleniaceae) rediscovered in Hawai'i. *American Fern Journal*, **93**, 154–6.

Alford, D. V. and Backhaus, G. F. (eds.) (2005). *Introduction and Spread of Invasive Species*. BCPC Symposium Proceedings, 93. Alton, Hampshire, UK: British Crop Protection Council.

Alonso-Amelot, M. E., Castillo, U., Smith, B. L. and Lauren, D. R. (1996). Bracken ptaquiloside in milk. *Nature*, **382**, 587.

Alonso-Amelot, M. E., Oliveiros, A., Calcagno, M. P. and Arellano, E. (2001). Bracken adaptation mechanisms and xenobiotic chemistry. *Pure and Applied Chemistry*, **73**, 549–53.

Anonymous (1988). Bracken in Wales. In *Bracken in Wales: an Interagency Report on its Status in the Principality, with Recommendations*, ed. Senior Technical Officer's Group, Wales. Bangor, UK: Nature Conservancy Council, p. 16.

Anonymous (1996). *Horsetail (Equisetum): Weed Control Factsheet*. British Columbia Ministry of Agriculture and Lands, Canada. www.agf.gov.bc.ca/cropprot/hrsetail.htm. Viewed March 2009.

Anonymous (1999). *Noosa National Park Management Plan*. Queensland Parks and Wildlife Service, Australia. www.epa.qld.gov.au/register/p00207aa.pdf. Viewed September 2009.

Anonymous (2000). Mapele (*Cyrtandra cyaneoides*). In *Beacham's Guide to the Endangered Species of North America*. www.accessmylibrary.com/coms2/summary_0193–8521_ITM. Viewed March 2009.

Anonymous (2001). *Recovery Plan for the Giant Fern (Angiopteris evecta)*. NSW National Parks and Wildlife Service, Hurstville, Australia. www.environment.nsw.gov.au/resources/nature/approvedAngevect.pdf. Viewed March 2009.

Anonymous (2003). *Biological Control of Weeds in Texas: Salvinia Stem Borer*. Texas A & M University Department of Entomology, College Station, TX, USA. http://bc4weeds.tamu.edu/agents/salviniastemborer.html. Viewed March 2009.

Anonymous (2004a). *Azolla filiculoides, Water Fern*. Information Sheet 22. Centre for Ecology and Hydrology, Wallingford, UK. www.ceh.ac.uk/sections/wq/documents/22%20Azolla%20filiculoides.pdf. Viewed March 2009.

Anonymous (2004b). *Weevils of Mass Destruction*. British Waterways press release. www.british-waterways.org/newsroom/all-press-releases/display/id/1625. Viewed March 2009.

Anonymous (2005). *Salvinia molesta*. Global Invasive Species Database. Invasive Species Specialised Group. www.issg.org/database/species/impact_info.asp?si=569&fr=120&sts=&lang=EN. Viewed March 2009.

Anonymous (2006). *Twite (Carduelis flavirostris): Species Action Plan*. Northern Ireland Environment and Heritage Service. www.ni-environment.gov.uk/twiteniactionplanwebversionapril06.pdf. Viewed March 2009.

Anonymous (2007). *Rainforest Fernland: Threatened Native Vegetation Community Information Sheet*. Tasmania Forest Authority, Hobart, Australia. www.fpa.tas.gov.au/

fileadmin/user_upload/PDFs/Botany/rainforest_fernland_info_sheet_FE.pdf. Viewed March 2009.

Anonymous (2008a). *Weed of the Week: Mule's Foot Fern (Angiopteris evecta)*. Kauai Community Radio and Kauai Invasive Species Committee. www. hawaiiinvasivespecies.org/iscs/kisc/wow.html. Viewed March 2009.

Anonymous (2008b). Communication from the Commission to the Council, the European Parliament, the European Economic and Social Committee and the Committee of the Regions towards an EU strategy on invasive species. Com(2008) 789 final. European Council, Brussels. www.parliament.bg/pub/ECD/COM_2008_789_EN_ACTE_f.doc. Viewed March 2009.

Backshall, J. (1999). Managing bracken in the English uplands. *Enact*, **7**, 7–9.

Biggin, P. (1982). Forestry and bracken. *Proceedings of the Royal Society of Edinburgh*, **81B**, 19–27.

Bond, W., Davies, G. and Turner, R. (2006). *The Biology and Non-chemical Control of Bracken (Pteridium aquilinum (L.) Kuhn)*. Henry Doubleday Research Association (HDRA), Coventry, UK. www.gardenorganic.org.uk/organicweeds. Viewed March 2009.

Bondada, B. R. and Ma, L. Q. (2003). Tolerance of heavy metals in vascular plants: arsenic hyperaccumulation by Chinese brake fern (*Pteris vittata* L.). In *Pteridology in the New Millenium*, ed. S. Chandra and M. Srivastava. Dordrecht, The Netherlands: Kluwer Academic Publishers, pp. 397–420.

Borges, P. A. V., Cunha, R., Gabriel, R., *et al.* (eds.) (2005). *A List of the Terrestrial Fauna and Flora of the Azores*. Horta, Angra do Heroísmo and Ponta Delgada, Azores: Direcção Regional do Ambiente and Universidade dos Açores.

Boughey, A. S. (1963). The explosive development of floating vegetation on Lake Kariba. *Adansonia*, **3**, 49–61.

Brereton, T. M. and Warren, M. S. (1999). Ecology of the pearl bordered fritillary butterfly in Scotland and possible threats from bracken eradication measures in woodland grant schemes. In *Bracken Perceptions and Bracken Control in the British Uplands*. International Bracken Group Special Publication, 93, pp. 62–73.

Brinkhuis, H., Schouten, S., Collinson. M. E., *et al.* and the Expedition 302 Scientists. (2006). Episodic fresh surface waters in the Eocene Arctic Ocean. *Nature*, **441**, 606–9.

Brown, R. W. (1986). Bracken in the North York Moors: its ecological and amenity implications in national parks. In *Bracken: Ecology, Land Use and Control Technology*, ed. R. T. Smith and J. A. Taylor. Carnforth, UK: Parthenon, pp. 77–86.

Burn, A. (1988). Bracken and nature conservation. In *Bracken in Wales: an Interagency Report on its Status in the Principality, with Recommendations*, ed. Senior Technical Officer's Group, Wales. Bangor, UK: Nature Conservancy Council, Annex 4, pp. 47–55.

Calcagno-Pisarelli, M. P., Avila-Nunez, J. L., Otero, L. D., Silmi, S. and Naya, M. (2006). Bionomy of *Aneugmenus merida* (Hymenoptera: Tenthredinidae), a new natural enemy of bracken from the Venezuelan Andes. Paper presented at International Symposium on Intractable Weeds and Plant Invaders, Ponta Delgada, Azores, Portugal. Programme Abstracts, p. 38.

Callaghan, T. V. and Sheffield, E. (1985). *Pteridium aquilinum*: weed or resource? *Proceedings of the Royal Society of Edinburgh*, **93**, 461.

Callaghan, T. V., Lawson, G. J. L. and Scott, R. (1981). *Bracken as an Energy Crop: Current Scenarios and Future Research Requirements*. ITE Project 674 to the Department of Energy. Grange-over-Sands, UK: Merlewood Research Station, Institute of Terrestrial Ecology.

Caulton, E., Keddie, S. and Dyer, A. F. (1995). The incidence of airborne spores of bracken in the rooftop airstream over Edinburgh. In *Bracken: an Environmental Issue*, ed. R. T. Smith and J. A. Taylor. Aberystwyth, UK: International Bracken Group, Special Publication, 93, pp. 82–9.

Christenhusz, M. J. M. and Toivonen, T. K. (2007). Giants invading the tropics: the oriental vessel fern, *Angiopteris evecta* (Marattiaceae). *Biological Invasions*, **10**, 1215–28.

Cohen, A. L., Singhakumara, B. M. P. and Ashton, P. N. S. (1995). Releasing rain forest succession: a case study in the *Dicranopteris linearis* fernlands of Sri Lanka. *Restoration Ecology*, **3**, 261–70.

Davis, A. P. (1999). *The Genus Galanthus: A Botanical Magazine Monograph*. Portland, OR, USA: Timber Press.

Day, R. H., Cooper, B. A. and Blaha, R. F. (2003). Movement patterns of Hawaiian petrels and Newell's shearwaters on the Island of Hawaii. *Pacific Science*, **57**, 147–59.

Den Ouden, J. (2000). The role of bracken in forest dynamics. Unpublished Ph.D. thesis (2868), Wageningen University, The Netherlands.

De Winter, W. P., and Amoroso, V. B. (eds.) (2003). *Cryptogams: Ferns and Fern Allies*, Plant Resources of South-East Asia, 15. Leiden, The Netherlands: Backhuys.

Donnelly, E. (2003). Potential uses of bracken in organic agriculture in Scotland. Unpublished Ph.D. thesis, University of Aberdeen, Scotland.

Donnelly, E. (2004). A frond friend. *Organic Farming*, **81**, 30–1.

Durand, L. Z. and Goldstein, G. (2001). Photosynthesis, photoinhibition, and nitrogen use efficiency in native and invasive tree ferns in Hawaii. *Oecologia*, **126**, 345–54.

Dyatlov, S. and Vasilieva, T. (2008). *Salvinia natans*. www.grid.unep.ch/bsein/redbook/txt/salvinia.htm. Viewed March 2009.

Elder, H. (2004). *Giant Salvinia Infestation in Toledo Bend Poses Threat to Sam Rayburn*. Texas Parks and Wildlife Department, Austin, TX, USA. http://salvinia.er.usgs.gov/Salvinia_Advisory.pdf. Viewed March 2009.

Evans, I. A. (1986). The carcinogenic, mutagenic and teratogenic toxicity of bracken. In *Bracken: Ecology, Land Use and Control Technology*, ed. R. T. Smith and J. A. Taylor. Carnforth, UK: Parthenon, pp. 139–46.

Evans, W. C. (1986). The acute diseases caused by bracken in animals. In *Bracken: Ecology, Land Use and Control Technology*, ed. R. T. Smith and J. A. Taylor. Carnforth, UK: Parthenon, pp. 121–32.

Evans, W. C., Patel, M. C. and Kooby, Y. (1982). Acute bracken poisoning in homogastric and ruminant animals. *Proceedings of the Royal Society of Edinburgh*, **81B**, 29–64.

Fagan, R., McQuinn, A. D. and Mesch, P. (1990). Environmental and spray factors influencing control of bracken by boom spray applications of Brush-Off brush controller in Australia. In *Bracken Biology and Management*, ed. J. A. Thomson and R. T. Smith. Sydney, Australia: Australian Institute of Agricultural Science, Occasional Publication, 40, pp. 303–8.

Filizadeh, Y. and Zolfinejad, K. (2006). Evaluation of excessive growth of *Azolla* and its control in the north of Iran. Paper presented at International Symposium on Intractable Weeds and Plant Invaders, Ponta Delgada, Azores, Portugal. Programme Abstracts, p. 23.

Fowler, S. V. (1993). The potential for control of bracken in the UK using introduced herbivorous insects. *Pesticide Science*, **37**, 393–7.

Francesconi, K., Visoottiviseth, P., Sridokchan, W. and Goessler, W. (2002). Arsenic species in an arsenic hyperaccumulating fern, *Pityrogramma calomelanos*: a potential phytoremediator of arsenic-contaminated soils. *Science of the Total Environment*, **284**, 27–35.

Gilman, E. F. (1999). *Sphaeropteris cooperi: Australian Tree Fern*. Fact Sheet FPS-557. Gainsville, FL, USA: University of Florida, Cooperative Extension Service, Institute of Food and Agricultural Sciences.

Gliessman, S. R. (1978). The establishment of bracken following fire in tropical habitats. *American Fern Journal*, **68**, 41–4.

Glomski, L. A. M. and Getsinger, K. D. (2006). Carfentrazone-ethyl for control of Giant Salvinia. *Journal of Aquatic Plant Management*, **44**, 136–8.

Goolsby, J. A., Wright, D. and Pemberton, R. W. (2003). Exploratory surveys in Australia and Asia for natural enemies of Old World Climbing Fern, *Lygodium microphyllum*. *Biological Control*, **28**, 33–46.

Goolsby, J. A., Zonneveld, R. and Bourne, A. (2004). Pre-release assessment of impact on biomass production of an invasive weed, *Lygodium microphyllum*, by a potential biological control agent, *Floracarus perrepae*. *Environmental Entomology*, **33**, 997–1002.

Hartig, K. and Beck, E. (2003). The bracken fern dilemma in the Andes of southern Ecuador. *Ecotropica*, **9**, 3–13.

Hawaiian Ecosystems at Risk Project (2008). *Global Compendium of Weeds*. www.hear.org/gcw. Viewed March 2009.

Hendrix, S. D. (1977). The resistance of *Pteridium aquilinum* to insect attack by *Trichloplusia ni*. *Oecologia* **26**, 347–61.

Hirayama, T. (1979). Diet and cancer. *Nutrition and Cancer*, **1**, 67–81.

Holloway, S. M. (1994). Bracken stand characterization on the North York moors: a study of the rhizome and leaf system with regard to a large scale control programme. Unpublished Ph.D. thesis, University of Plymouth, UK.

Horning, D. S. (1983). A new fern record from the Snares, southern New Zealand. *New Zealand Journal Botany*, **21**, 205–8.

Hoshizaki, B. J. and Moran, R. C. (2001). *The Fern Grower's Manual*. Portland, OR, USA: Timber Press.

Howell, R. (1988). Bracken and water. In *Bracken in Wales: an Interagency Report on its Status in the Principality, with Recommendations*, ed. Senior Technical Officer's Group, Wales. Bangor, UK: Nature Conservancy Council, Annex 7, pp. 67–71.

Hubbard, C. E. (1968). *Grasses*, 2nd edn. Baltimore, MD, USA: Penguin Books.

Hussner, A. and Lösch, R. (2005). Alien aquatic plants in a thermally abnormal river and their assembly to neophyte-dominated macrophyte stands (River Erft, Northrhine-Westphalia). *Limnologica*, **35**, 18–30.

Hutchinson, J., Ferriter, A., Serbesoff-King, K., Langeland, K. and Rodgers, L. (eds.) (2006). *Old World Climbing Fern (Lygodium microphyllum) Management Plan for Florida*, 2nd edn. Florida Exotic Pest Plant Council *Lygodium* Task Force, West Palm Beach, FL, USA. www.fleppc.org/publications.htm. Viewed March 2009.

Ingrouille, M. (1992). *Diversity and Evolution of Land Plants*. London: Chapman and Hall.

Irvine, J. I. M., Burge, M. N. and McElwee, M. (1987). Association of *Phoma aquilina* and *Ascochyta pteridis* with curl-tip disease of bracken. *Annals of Applied Biology*, **110**, 25–31.

Jacono, C. C. (2002). *Four Leaf Unlucky? Non-indigenous Water Clovers*. US Geological Survey, Reston, VA, USA. http://nas.er.usgs.gov/taxgroup/plants/docs/marsilea/marsilea.html. Viewed March 2009.

Jacono, C. C. (2003). *Salvinia minima*. US Geological Survey, Reston, VA, USA. http://salvinia.er.usgs.gov/html/identification1.html. Viewed March 2009.

Jacono, C. C. (2006). *Giant Salvinia is an Aquatic Fern Prohibited in the United States by Federal Law*. A flyer from the US Geological Survey and the Mississippi Dept. of Marine Resources. www.dmr.state.ms.us/misc/species-of-concern/salvinia-flyer.pdf. Viewed March 2009.

Jayachandran, K., Shetty, K. G., Clarke, T. C., Smith, C. S., Miao, S. and Rodgers, L. (2008). Development of potential biological control agents for invasive plant species using native pathogens in south Florida. In *GEER 2008*. Naples, FL, USA: Greater Everglades Ecosystem Restoration, Abstracts, p. 194.

Jones, D. L. and Clemesha, S. C. (1980). *Australian Ferns and Fern Allies*. Sydney, Australia: A. H. and A. W. Reed.

Julien, M. H., Center, T. D. and Tipping, P. W. (2002). Floating Fern (*Salvinia*). In *Biological Control of Invasive Plants in the Eastern United States*, ed. R. Van Driesche, S. Lyon, B. Blossey, M. Hoddle and R. Reardon. Morgantown, WV, USA: USDA Forest Service, Publication, FHTET-2002–04, pp. 17–32.

Keary, I. P., Thomas, C. and Sheffield, E. (2000). The effects of the herbicide asulam on the gametophytes of *Pteridium aquilinum*, *Cryptogamma crispa* and *Dryopteris filix-mas*. *Annals of Botany*, **93** (Suppl. B), 47–51.

Kirk, A. A. (1977). The insect fauna of the weed *Pteridium aquilinum* in Papua New Guinea: a potential source of biological control agents. *Journal of the Australian Entomological Society*, **16**, 403–7.

Knapp, M. F. (2005). Diffuse pollution threats to groundwater: a UK water company perspective. *Quarterly Journal of Engineering Geology and Hydrology*, **38**, 39–51.

Kolar, C. S. and Lodge, D. M. (2001). Progress in invasion biology: predicting invaders. *Trends in Ecology and Evolution*, **16**, 199–204.

Kumar, P. S., Ramani, S. and Singh, S. P. (2005). Natural suppression of the aquatic weed, *Salvinia molesta*, by two previously unreported fungal pathogens. *Journal of Aquatic Plant Management*, **43**, 105–7.

Kuroda, A., Ikeda, S., Mukai, S. and Toyohara, G. (2006). Successive mapping of secondary pine forests affected by pine wilt disease and subsequent forest management in Miyajima Island, SW Japan. *Phytocoenologia*, **36**, 191–212.

Kurugundla, C. N., and Serumola, O. (2007). Impacts of deltamethrin spray on adults of the Giant Salvinia biocontrol agent, *Cyrtobagous salviniae*. *Journal of Aquatic Plant Management*, **45**, 124–9.

Lawrie, J. and Butler, R. C. (1995). Study of the low dose stimulation of buds and rhizome growth in bracken using a peaked logistic curve. *BCPC Weeds*, **3**, 1003–8.

Lawson, G. J., Callaghan, T. V. and Scott, R. (1986). Bracken as an energy resource. In *Bracken: Ecology, Land Use and Control Technology*, ed. R. T. Smith and J. A. Taylor. Carnforth, UK: Parthenon, pp. 239–47.

Lawton, J. H. (1990). Developments in the UK biological control programme for bracken. In *Bracken Biology and Management*, ed. J. A. Thomson and R. T. Smith. Sydney, Australia: Australian Institute of Agricultural Science, Occasional Publication, 93, pp. 309–14.

Le Duc, M. G., Pakeman, R. J. and Marrs, R. H. (2007). A restoration experiment on moorland infested by *Pteridium aquilinum*: plant species responses. *Agriculture, Ecosystems and Environment*, **93**, 53–9.

Lockhart, C. (2007). *Statewide Lygodium Treatment Site Evaluation Project*. Florida Natural Areas Inventory, Tallahassee, FL, USA. www.fnai.org. Viewed March 2009.

Lott, M. S., Volin, J. C., Pemberton, R. W. and Austin, D. F. (2003). The reproductive biology of the invasive ferns *Lygodium microphyllym* and *L. japonicum*: implications for invasive potential. *American Journal of Botany*, **93**, 1144–52.

Low, V. H. K. and Thomson. J. A. (1990). Cyanogenesis in Australian bracken (*Pteridium esculentum*): distribution of cyanogenic phenotypes and factors influencing activity of the cyanogenic glucosidase. In *Bracken Biology and Management*, ed. J. A. Thomson and R. T. Smith. Sydney, Australia. Australian Institute of Agricultural Science, Occasional Publication, 93, pp. 105–11.

Lowday, J. E. and Marrs, R. H. (1992). Control of bracken and the restoration of heathland. III. Bracken litter disturbance and heathland restoration. *Journal of Applied Ecology*, **29**, 212–7.

Lumpkin, T. A. and Plucknett, D. L. (1980). *Azolla*: botany, physiology and uses as a green manure. *Economic Botany*, **34**, 111–53.

Ma, L. Q., Komar, K. M., Tu, C., *et al.* (2001). A fern that hyperaccumulates arsenic. *Nature*, **409**, 579.

Ma, L. Q., Luongo, T., Srivastava, M. and Singh, N. (2006). *Contaminant Removal by Additional Ferns*, Official Gazette of the United States and Trademark Office Patents, June 27.

Mabey, R. (1998). *Flora Britannica: The Concise Edition*. London: Chatto and Windus.

Macdonald, A. (1887). *The Improvement of Hill Pasture Without Breaking It Up*. Transactions of the Highland and Agricultural Society of Scotland. www. electricscotland.com/agriculture/page72.htm. Viewed March 2009.

MacDonald, G., Sellers, B., Langeland, K., Duperron-Bond, T. and Ketterer-Guest, E. (2008). *Chinese Ladder Brake Fern (Pteris vittata): Invasive Species Management Plans for Florida*. Gainesville, FL, USA: Center for Aquatic and Invasive Plants, University of Florida.

MacLochlainn, C. (2004). *Farmland Birds on Cape Clear Island, West Cork*. Message held by ListServ archives for Heanet (Ireland's National Education and Research Network). http://listserv.heanet.ie/cgi-bin. Viewed March 2009.

Madsen, J. D. and Wersal, R. M. (2008). Growth regulation of *Salvinia molesta* by pH and available water column nutrients. *Journal of Freshwater Ecology*, **23**, 305–14.

Marliere, C. A., Wathern, P., Freitas, S. N., Castro, M. C. F. M. and Galvao, M. A. M. (2000). Bracken fern consumption and oesophageal and stomach cancer in the Ouro Preto region, Minas Gerais, Brazil. In *Bracken Fern: Toxicity, Biology and Control*, ed. J. A. Taylor and R. T. Smith. Aberystwyth, UK: International Bracken Group, Special Publication, 93, pp. 144–9.

Marinelli, J. (ed.) (2004). *Plant*. London: The Royal Botanic Gardens, Kew, and Dorling Kindersley.

Marrs, R. H., Frost, A. J., Plant, R. A. and Lunnis, P. (1993). Determination of buffer zones to protect seedlings of non-target plants from the effects of glyphosate spray drift. *Agriculture, Ecosystems and Environment*, **43**, 283–93.

Marrs, R. H. and Watt, A. S. (2006). Biological flora of the British Isles: *Pteridium aquilinum* (L.) Kuhn. *Journal of Ecology*, **93**, 1272–321.

Marrs, R. H., Le Duc, M. G., Mitchell, R. J., *et al.* (2000). The ecology of bracken: its role in succession and implications for control. *Annals of Botany*, **93** (Suppl. B), 3–15.

Mason, R. (2004). *Growing Native Plants: Platycerium bifurcatum*. Australian National Botanic Gardens, Australian National Herbarium, Canberra. www.anbg.gov.au/gnp/interns-2004/platycerium-bifurcatum.html. Viewed March 2009.

McConnachie, A. J., de Wit, M. P., Hill, M. P. and Byrne, M. J. (2003). Economic evaluation of the biological control of *Azolla filiculoides* in South Africa. *Biological Control*, **28**, 25–32.

McConnachie, A. J., Hill, M. P. and Byrne, M. J. (2004). Field assessment of a frond-feeding weevil, a successful control agent of red waterfern, *Azolla filiculoides*, in southern Africa. *Biological Control*, **29**, 326–31.

McElwee, M. and Burge, M. N. (1990). Lignification and associated disease resistance in bracken. In *Bracken Biology and Management*, ed. J. A. Thomson and R. T. Smith. Sydney, Australia. Australian Institute of Agricultural Science, Occasional Publication, 93, pp. 315–22.

McFarland, D. G., Nelson, L. S., Grodowitz, M. J., Smart, R. M. and Owens, C. S. (2004). *Salvinia molesta (Giant Salvinia) in the United States: A Review of Species Ecology and Approaches to Management*. US Army Engineer Research and Development Center, Vicksburg, MS, USA, Special Report, 04–02. www.erdc.usace.army.mil. Viewed March 2009.

McIntosh, D., King, C. and Fitzsimmons, K. (2003). *Tilapia* for biological control of Giant Salvinia. *Journal of Aquatic Plant Management*, **41**, 28–31.

Medeiros, A. C., Loope, L. L., Flynn, T., *et al.* (1992). Notes on the status of an invasive Australian tree fern (*Cyathea cooperi*) in Hawaiian rain forests. *American Fern Journal*, **82**, 27–33.

Mehltreter, K. (2006). Leaf phenology of the climbing fern *Lygodium venustum* in a semideciduous lowland forest on the Gulf of Mexico. *American Fern Journal*, **93**, 21–30.

Miller, J. and Sheffield, E. (2004). Controlling Kariba weed. *Biological Sciences Review*, **17**, 39–41.

Moran, R. C. (2004). *A Natural History of Ferns*. Portland, OR, USA: Timber Press.

Motooka, P., Castro, L., Nelson, D., Nagai, G. and Ching, L. (2003). *Weeds of Hawaii's Pastures and Natural Areas; An Identification and Management Guide*. College of Tropical Agriculture and Human Resources, University of Hawaii at Manoa. http://www.ctahr.hawaii.edu/invweed/WeedsHI/W_Sphaeropteris_cooperi pdf. Viewed March 2009.

Muchovej, J. J. (1979) *Dreschlera salviniae*: a new species from Brazil. *Transactions of the British Mycological Society*, **93**, 331–2.

Muchovej, J. J. and Kushalappa, A. C. (1979). *Dreschlera* leaf spot of *Salvinia auriculata*. *Plant Disease Reporter*, **63**, 154.

Negishi, J. N., Sidle, R. C., Noguchi, S., Nik, A. R. and Stanforth, R. (2006). Ecological roles of the roadside fern (*Dicranopteris curranii*) on logging road recovery in Peninsular Malaysia: preliminary results. *Forest Ecology and Management*, **224**, 176–86.

Nelson, G. (2000). *The Ferns of Florida: a Reference and Field Guide*. Sarasota, FL, USA: Pineapple Press.

Nicholls, D. (1999). *Ferns and Fern Allies in the Canberra Region: Histiopteris incisa, Bat's Wing Fern*. www.home.aone.net.au/~byzantium/ferns/descriptions/histiopteris/histiopt.html. Viewed March 2009.

Oates, M. (2004). The ecology of the pearl bordered fritillary in woodland. *British Wildlife*, **15**, 229–36.

Oduro, C., Newman, J. and Hatcher, P. E. (2006). Controlling *Azolla ficiluloides* using *Artemisia dracunculus* and *A. vulgaris* leaf extracts. Paper presented at International Symposium on Intractable Weeds and Plant Invaders, Ponta Delgada, Azores, Portugal. Programme Abstracts, p. 34.

Orr, D. (2008). *Marsilea villosa*. Center for Plant Conservation, St. Louis, MO, USA. www.centerforplantconservation.org/ASP. Viewed March 2009.

Ortega, F. J. M. (1994). Ethnobotanical notes on the bracken fern in the Amazon region and possible carcinogenic risks. *BioLlania*, **10**, 7–12.

Owens, C. S., Smart, R. M., Honnell, D. R. and Dick, G. O. (2005). Effects of pH on growth of *Salvinia molesta*. *Journal of Aquatic Plant Management*, **43**, 34–8.

Page, C. N. (1982). The history and spread of bracken in Britain. *Proceedings of the Royal Society of Edinburgh*, **81B**, 3–10.

Pakeman, R. J. and Marrs, R. H. (1992). The conservation value of bracken-dominated communities in the UK, and an assessment of the ecological impact of bracken expansion or its removal. *Biological Conservation*, **62**, 101–14.

Pakeman, R. J., Hill, M. O. and Marrs, R. H. (1995). Modelling vegetation succession after bracken control. *Journal of Environmental Management*, **43**, 29–39.

Pakeman, R. J., Marrs, R. H., Howard, D. C., Barr, C. J. and Fuller, R. M. (1996). The bracken problem in Great Britain, its present extent and future changes. *Applied Geography*, **16**, 65–86.

Palmer, D. D. (2003). *Hawaii's Ferns and Fern Allies*. Honolulu, HI, USA: University of Hawaii Press.

Pamukcu, A. M., Ertürk, E., Milli, U. and Bryan, G. T. (1978). Carcinogenic and mutagenic activities of milk from cows fed bracken fern. *Cancer Research*, **38**, 1556–60.

Pandey, D. K. (1994). Inhibition of salvinia (*Salvinia molesta* Mitchell) by parthenium (*Parthenium hysterophorus*). II. Relative effect of flower, leaf, stem, and root residue on *Salvinia* and paddy. *Journal of Chemical Ecology*, **20**, 3123–31.

Parry, J. (2003). *Living Landscapes: Heathland*. London: The National Trust.

Patrick, B. (1994). Antipodes island Lepidoptera. *Journal of the Royal Society of New Zealand*, **24**, 91–116.

Pemberton, R. W. and Ferriter, A. P. (1998). Old World Climbing Fern (*Lygodium microphyllum*), a dangerous invasive weed in Florida. *American Fern Journal*, **88**, 165–75.

Pemberton, R. W., Goolsby, J. A. and Wright, T. (2002a). Old World Climbing Fern. In *Biological Control of Invasive Plants in the Eastern United States*, ed. R. Van Driesche, S. Lyon, B. Blossey, M. Hoddle, M. and R. Reardon. Morgantown, WV, USA. USDA Forest Service, Publication, FHTET-2002–04, pp. 139–47.

Pemberton, R. W., Goolsby, J. A., Wright, T. and Buckingham, G. R. (2002b). *Getting on Top of Climbing Ferns*. CABI Biocontrol News and Information, 93. www.pestscience. com/Bni23–3/Gennews.htm. Viewed March 2009.

Pinto, C., Januário, T., Geraldes, M., *et al.* (2004). Bovine enzootic haematuria on São Miguel Island, Azores. In *Poisonous Plants and Related Toxins*, ed. T. Acamovic, C. S. Stewart, and T. W. Pennycott. Proceedings of the 6th International Symposium of Poisonous Plants, Glasgow, UK, 2001. Wallingford, UK: CAB International Publishing, pp. 564–74.

Pitman, R. and Webber, J. (1998). *Bracken as a Peat Alternative*. Forestry Authority Information Note. Edinburgh, UK: Forestry Commission

Proctor, G. R. (1989). Ferns of Puerto Rico and the Virgin Islands. *Memoirs of the New York Botanical Garden*, **53**, 1–389.

Rasmussen, L. H. (2003). Ptaquiloside: an environmental hazard? Occurrence and fate of a bracken toxin in terrestrial environments. Unpublished Ph.D. thesis, Royal Veterinary and Agriculture University, Copenhagen, Denmark.

Rasmussen, L. H., Jensen, L. S. and Hansen, H. C. B. (2003). Distribution of the carcinogenic terpene ptaquiloside in bracken fronds, rhizomes and litter in Denmark. *Journal of Chemical Ecology*, **29**, 771–8.

Rasmussen, L. H., Lauren, D. R. and Hansen, H. C. B. (2005). Sorption, degradation and mobility of ptaquiloside, a carcinogenic bracken constituent, in the soil environment. *Chemosphere*, **58**, 823–35.

Rayamajhi, M. B., Pemberton, R. W., Van, T. K. and Pratt, P. D. (2005). First report of infection of *Lygodium microphyllum* by *Puccinia lygodii*, a potential biocontrol agent of an invasive fern in Florida. *Plant Disease*, **89**, 110.

Richerson, M. M. and Jacono, C. C. (2005). *Salvinia molesta at Enchanted Lake, Kailua, on the Island of Oahu, Hawaii*. US Geological Survey. http://salvinia.er.usgs.gov/html/enchanted_lake.html. Viewed March 2009.

Riches, C. R. (ed.) (2001). *The World's Worst Weeds*. BCPC Symposium Proceedings 77. Alton, Hampshire, UK: British Crop Protection Council.

Robinson, R. C. (2007). Steps to more effective bracken management. *Aspects of Applied Biology*, **82**, 143–55.

Robinson, R. C. (2008). Bracken: a competitive stakeholder in the uplands. *Aspects of Applied Biology*, **85**, 7–84.

Robinson, R. C. and Page, C. N. (2000). Protection of non-target ferns during extensive spraying of bracken. In *Bracken Fern: Toxicity, Biology and Control*, ed. J. A. Taylor and R. T. Smith. Aberystwyth, UK: International Bracken Group, Special Publication, 93, pp. 163–74.

Robinson, R. C., Parsons, R. G., Barbe, G., Patel, P. T. and Murphy, S. (2000). Buffer zones for helicopter spraying of bracken. *Agriculture, Ecosystems and Environment*, **79**, 215–31.

Ruder, K. (2004). *Ferns Remove Arsenic From Soil And Water*. Genome News Network. www.middleport-future.com/cig/docs/library/gnn.phyto.pdf. Viewed March 2009.

Russell, A. E., Raich, J. W. and Vitousek, P. M. (1998). The ecology of the climbing fern *Dicranopteris linearis* on windward Mauna Loa, Hawaii. *Journal of Ecology*, **86**, 765–79.

Rymer, L. (1976). The history and ethnobotany of bracken. *Botanical Journal of the Linnean Society*, **73**, 151–76.

Sabbatini, M. R., Murphy, K. J. and Irigoyen, J. H. (1998). Vegetation–environment relationships in irrigation channel systems of southern Argentina. *Aquatic Botany*, **60**, 119–33.

Shaw, M. W. (1988). Ecology of bracken in relation to control methods. In *Bracken in Wales: an Interagency Report on its Status in the Principality, with Recommendations*, ed. Senior Technical Officer's Group, Wales. Bangor, UK: Nature Conservancy Council, Annex 2B, pp. 30–43.

Shaw, R. H., Reeder, R. H., Bacon, E. and Oduro, C. (2006). Biological control of a safe and effective alternative for the management of *Azolla filiculoides*. Paper presented at International Symposium on Intractable Weeds and Plant Invaders, Ponta Delgada, Azores, Portugal. Programme Abstracts, p. 24.

Sheaves, B. J. and Brown, R. W. (1995a). Densities of *Ixodes ricinus* ticks on moorland vegetation communities in the UK. *Experimental and Applied Acarology*, **19**, 489–97.

Sheaves, B. J. and Brown, R. W. (1995b). A zoonosis as a human health hazard in UK moorland recreational areas: a case study of Lyme disease. *Journal of Environmental Planning and Management*, **38**, 201–14.

Sheffield, E. (2002). Effects of asulam on non-target pteridophytes. *Fern Gazette*, **16**, 377–82.

Sheffield, E., Wolf, P. G. and Ranker, T. A. (1995). Genetic analysis of bracken in the Hawaiian Islands. In *Bracken: an Environmental Issue*, ed. R. T. Smith and J. A. Taylor. Aberystwyth, UK: International Bracken Group, Special Publication, 93, pp. 29–32.

Sheffield, E., Johns, M., Rumsey, F. J. and Rowntree, J. K. (2003). *An Investigation of the Effects of Low Doses of Asulox on Non-target Species*. English Nature Contracts EIT20–19–001 and 30–08–007.

Silvester, W. B. (1964). Forest regeneration problems in the Hunua range, Auckland. *Proceedings of the New Zealand Ecological Society*, **11**, 1–5.

Simán, S. E., Povey, A. C. and Sheffield, E. (1999). Human heath risks from fern spores? A review. *Fern Gazette* **15**, 275–87.

Simán, S. E., Povey, A. C., Ward, T. H., Margison, G. P. and Sheffield, E. (2000). Fern spore extracts can damage DNA. *British Journal of Cancer*, **83**, 69–73.

Smith, B. L., Lauren, D. L. and Prakash, A. S. (2000). Bracken fern: toxicity in animal and human health. In *Bracken Fern: Toxicity, Biology and Control*, ed. J. A. Taylor and R. T. Smith. Aberystwyth, UK: International Bracken Group, Special Publication, 93, pp. 76–85.

Somvanshi, R., Lauren, D. R., Smith, B. L., *et al.* (2006). Estimation of the fern toxin, ptaquiloside, in certain Indian ferns other than bracken. *Current Science*, **91**, 1547–52.

Srivastava, K. (2008). Conservation and management plans for *Angiopteris evecta*: an endangered species. *Southern Illinois University, Carbondale, USA, Ethnobotanical Leaflets*, **12**, 23–8. http://www.ethnoleaflets.com//leaflets/pteris.htm. Viewed March 2009.

Tan, R. (2001). Resam (*Dicranopteris linearis*). Mangrove and wetland wildlife at Sungei Buloh Wetlands Reserve, Singapore. www.naturia.per.sg/buloh/plants/resam.htm. Viewed March 2009.

Taylor, J. A. (1980). Bracken: an increasing problem and a threat to health. *Outlook on Agriculture*, **10**, 298–304.

Taylor, J. A. (1985). The bracken problem: a local hazard and global issue. In *Bracken: Ecology, Land Use and Control Technology*, ed. R. T. Smith and J. A. Taylor. Carnforth, UK: Parthenon, pp. 21–42.

Taylor, J. A. (1989). The British bracken problem. *Geography Review*, **2**, 7–11.

Taylor, J. A. (1990). The bracken problem: a global perspective. In *Bracken Biology and Management*, ed. J. A. Thomson and R. T. Smith. Sydney, Australia: Australian Institute of Agricultural Science, Occasional Publication, 93, pp. 3–19.

Taylor, J. E. and Thomson, J. A. (1990). Allelopathic activity of leaf run-off from *Pteridium esculentum*. In *Bracken Biology and Management*, ed. J. A. Thomson and R. T. Smith. Sydney, Australia: Australian Institute of Agricultural Science, Occasional Publication, 93, pp. 203–8.

Taylor, J. E. and Thomson, J. A. (1998). Bracken litter as mulch: glasshouse evaluation of phytotoxicity. *Australian Journal of Experimental Agriculture*, **38**, 161–9.

Tet-Vun, C. and Ismail, B. S. (2006). Field evidence for the allelopathic properties of *Dicranopteris linearis*. *Weed Biology and Management*, **6**, 59–67.

Thomson, J. A. (2004). Towards a taxonomic revision of *Pteridium* (Dennstaedtiaceae). *Telopea*, **10**, 798–803.

Tipping, P. W. and Center, T. D. (2003). *Cyrtobagous salviniae* (Coleoptera: Curulionidae) successfully overwinters in Texas and Louisiana. *Florida Entomologist*, **86**, 92–3.

Tyler, S. J. (1988). Birds and bracken in Wales. In *Bracken in Wales: an Interagency Report on its Status in the Principality, with Recommendations*, ed. Senior Technical Officer's Group, Wales. Bangor, UK: Nature Conservancy Council, Annex 5, pp. 56–62.

USDA, ARS, National Genetic Resources Program (2008). *Germplasm Resources Information Network (GRIN)*. National Germplasm Resources Laboratory, Beltsville, MD, USA. www.ars-grin.gov. Viewed March 2009.

USDI, US Geological Survey (2008). *Nonindigenous Aquatic Species*. http://nas.er.usgs. gov. Viewed March 2009.

Valier, K. (1995). *The Ferns of Hawaii*. Honolulu, HI, USA: University of Hawai'i Press.

Van Oosterhout, E. (2006). *Salvinia Control Manual: Management and Control Options for Salvinia molesta in Australia*. NSW Department of Primary Industries, Orange, Australia. www.dpi.nsw.gov.au. Viewed March 2009.

Veitch, B. (1990). Aspects of aboriginal use and manipulation of bracken fern. In *Bracken Biology and Management*, ed. J. A. Thomson and R. T. Smith. Sydney, Australia: Australian Institute of Agricultural Science, Occasional Publication, 93, pp. 215–26.

Villalobos-Salazar, J., Hernández, H., Meneses, A. and Salazar, G. (2000). Factors which may affect ptaquiloside levels in milk: effects of altitude, bracken fern growth stage and milk processing. In *Bracken Fern: Toxicity, Biology and Control*, ed. J. A. Taylor and R. T. Smith. Aberystwyth, UK: International Bracken Group, Special Publication, 93, pp. 68–74.

Volin, J. C., Lott, M. S., Muss, J. D. and Owen, D. (2004). Predicting rapid invasion of the Florida Everglades by Old World Climbing Fern (*Lygodium microphyllum*). *Diversity and Distributions*, **10**, 439–46.

Wagner, W. H., Jr. (1995). Evolution of Hawaiian ferns and fern allies in relation to their conservation status. *Pacific Science*, **49**, 31–41.

Walker, L. R. and Boneta, W. (1995). Plant and soil responses to fire on a fern-covered landslide in Puerto Rico. *Journal of Tropical Ecology*, **11**, 473–9.

Warren, M. S. (1993). Bracken: hot-spot for invertebrates. *Enact*, **1**, 6.

Warren, M. S. and Oates, M. R. (1995). The importance of bracken habitats to fritillary butterflies and their management for conservation. In *Bracken: an Environmental Issue*, ed. R. T. Smith and J. A. Taylor. Aberystwyth, UK: International Bracken Group, Special Publication, 93, pp. 178–81.

Watt, A. S. (1976). The ecological status of bracken. *Botanical Journal of the Linnean Society*, **73**, 217–39.

Wei, D.-S., Kuo, C.-M. and Pan, S.-M. (1997). Studies on the activity of superoxide dismutase in *Miscanthus floridulus* grown near a copper smelter in Taiwan. *Taiwania*, **42**, 189–200.

West, T. M., Lawrie, J. and Cromack, T. (1995). Responses of bracken and its understorey flora to some sulfonylurea herbicides and asulam. *BCPC Weeds*, **3**, 997–1002.

Wigglesworth, T., Parsons, M. and Warren, M. S. (2004). *Double Line (Mythimna turca) Fact Sheet*. Butterfly Conservation, UK. www.butterfly-conservation.org/uploads/ double_line.pdf. Viewed March 2009.

Williams, D. R. and Evans, R. A. (1959). Bracken *(Pteridium aquilinum)*: the effect of steaming on the nutritive value of bracken hay. *British Journal of Nutrition*, **93**, 129–36.

Wilkins, C. and Salter, L. (2003). Arsenic hyper-accumulation in ferns: a review. *Environmental Chemistry Group Bulletin of the Royal Society of Chemistry*, **July**, 8–10.

Wilson, D., Donaldson, L. J. and Sepai, O. (1998). Should we be frightened of bracken? A review of the evidence. *Journal of Epidemiology and Community Health*, **52**, 812–7.

Wilson, K. A. (1996). Alien ferns in Hawaii. *Pacific Science*, **50**, 127–41.

Winfield, R. J. (1988). Imazapyr for the control of bracken. *Aspects of Applied Biology*, **16**, 281–8.

Wolf, P. G., Haufler, C. H. and Sheffield, E. (1988). Maintenance of genetic variation in the clonal weed *Pteridium aquilinum* (bracken). *SAAS Bulletin of Biochemistry and Biotechnology*, **1**, 46–50.

Womack, J. G., Eccleston, G. M. and Burge, M. N. (1995). Progress in the development of a mycoherbicidal formulation for bracken control. In *Bracken: an Environmental Issue*, eds. R. T. Smith and J. A. Taylor. Aberystwyth, UK: International Bracken Group, Special Publication, 93, pp. 137–40.

Wynn, J. M., Small, J. L., Pakeman, R. J. and Sheffield, E. (2000). An assessment of genetic and environmental effects on sporangial development in bracken using a novel quantitative method. *Annals of Botany*, **85**, 113–5.

Xu, L. R. (1992). Bracken poisoning and enzootic haematuria in cattle in China. *Research in Veterinary Science*, **53**, 116–21.

Young, S. (1990). The horsetail's tale: a relic from the Paleozoic era. *New Scientist*, **1698**, 67.

Zakaria, Z. A., Abdul Ghani, Z. D. F., Raden Mohd. Nor, R. N. S., *et al.* (2006). Antinociceptive and anti-inflammatory activities of *Dicranopteris linearis* leaves: chloroform extract in experimental animals. *Yakugaku Zasshi*, **126**, 1197–203.

9

Fern conservation

KLAUS MEHLTRETER

Key points

1. Extant ferns and lycophytes are ecologically important and contribute 4% of the vascular plant diversity on Earth but currently face an unprecedented threat caused mainly by human disturbances such as fire or land use change. Few fern species benefit from these disturbances, and most become less abundant or locally extinct.
2. Current risk assessments for ferns are mainly based on abundance and geographic range. Risk assessments improve when additional ecological characteristics (e.g., habitat specificity, intrinsic biological factors, population dynamics and environmental disturbances) are considered.
3. By 2008 a global risk assessment had been completed for only 2% of the 11 000 species of ferns and lycophytes; 89% of these ferns were considered to be at risk. Global risk assessment of ferns is geographically biased toward Ecuador and China and taxonomically biased toward nine of c. 300 extant genera.
4. A mixed approach of habitat protection for hot spots of fern diversity and *in situ* protection of endangered fern species outside such hot spots is recommended as a management strategy for fern conservation. *Ex situ* cultivation of endangered fern species may supplement *in situ* protection efforts but should not replace them.

9.1 Introduction

Extant plants and animals likely represent only 1–2% of all organisms that have ever existed on our planet during the last 450 million years (May *et al.*, 1995) because extinction is common. Mass extinction events may have eliminated 75–95% of all living species during very short periods on the geological timescale, but 90–96% of all extinctions have occurred continuously during periods between these events (Raup, 1986). Total biodiversity on Earth returned to former levels within 1–8 million years through evolution of new species (radiation) following the three mass extinction events that occurred during the Permian (240 million years

Fern Ecology, ed. Klaus Mehltreter, Lawrence R. Walker and Joanne M. Sharpe. Published by Cambridge University Press. © Cambridge University Press 2010.

ago, Appendix C), Triassic (210 mya) and Cretaceous (65 mya; Rosenzweig, 1995). Ongoing plant extinction and speciation replaces on average 16–26% of the total diversity every one million years (Valentine *et al.*, 1991), and for a species the average duration, from origin to extinction, is about 5–10 million years (May *et al.*, 1995). If we apply these estimates of species turnover rates to the estimated count of 11 000 extant fern and lycophyte species (see Appendix A), then on average we would expect the natural extinction and speciation of one fern species every 500 years.

Ancestors of extant club-mosses, horsetails and some other ferns were the dominant life forms on Earth during the Carboniferous (360–290 mya), but most members of that first radiation, with the exception of the eusporangiate family Marattiaceae, died out during the mass extinction in the Permian. The origin of the other eusporangiate family, Ophioglossaceae (adder's tongue ferns), remains unclear because of a poor fossil record (Rothwell and Stockey, 2008). In the Mesozoic (225–65 mya) newly evolving gymnosperms (cycads, conifers and ginkgos) became the dominant vegetation (Sepkoski, 1992; Jablonski, 1995). Ferns experienced a second radiation between the Permian and the Jurassic, giving rise to the first extant leptosporangiate fern families (e.g., Cyatheaceae, Gleicheniaceae, Hymenophyllaceae and Osmundaceae, Rothwell and Stockey, 2008). In the Cretaceous, during a third radiation, all modern leptosporangiate fern families developed at the same time as the newly diversifying angiosperms, either in the dark understory (e.g., Tectariaceae) or in the canopy (e.g., Polypodiaceae) of angiosperms (Schneider *et al.*, 2004). For this reason, most extant fern families are not survivors of some pre-Cretaceous ancestral group.

During the last five decades, ferns have faced a new kind of threat as humans have caused the destruction of natural habitats and unparalleled species extinction (Pimm and Raven, 2000). Ferns that are able to adapt to fire, human disturbance (see Chapters 6 and 8) and agricultural ecosystems may thrive, but fern species of undisturbed habitats such as lowland and montane tropical forests are becoming endangered. Ferns that are collected from the wild because of ornamental or medicinal interest are also in trouble. At the same time, there is increasing human interest and concern for the conservation of ferns. If the discovery of new species of ferns continues at the same rate as it has for angiosperms (Prance, 2001), I expect at least 1000 new fern species to be found and described within the next 50 years, especially from floristically rich but poorly studied areas (e.g., Bolivia, Colombia, Indonesia, Malaysia).

The ecological importance of ferns is often underestimated, perhaps because most scientific research is performed in temperate climates where ferns are usually a minor element in the natural vegetation. The oldest and largest fern societies (British Pteridological Society founded in 1891 and American Fern Society in 1893) were established in temperate areas. Most research on ferns, especially in the tropics, is focused on taxonomy and molecular phylogenetics rather than ecology. This focus has led to remarkable advances in fern systematics over the last 30 years, concluding with several comprehensive floras (e.g., Moran and Riba, 1995; Mickel and Smith, 2004). These floras will aid new, interdisciplinary approaches to ecology and conservation based on modern classifications.

Although we recognize about 11 000 species of ferns and lycophytes in this book (see Appendix A), estimates as high as 13 025 (Baillie *et al.*, 2004) to 15 000 species (Roos, 1996) have been proposed. Ferns and lycophytes represent 3% of the vascular flora in Canada and the USA (Wagner and Smith, 1993), 5–10% in the continental tropics and 15–70% on tropical islands (see Chapter 2), where they can dominate in the lower canopy (Drake and Pratt, 2001) or the understory (Fig. 9.1). As colonizers and pioneers of open and disturbed habitats, ferns often modify habitat conditions during successional stages by building up leaf litter or stabilizing landslides with their dense network of rhizomes and roots, but may also inhibit future succession (see Chapters 6 and 8; Walker, 1994; Slocum *et al.*, 2004). Ferns interact with many animal species, especially herbivorous insects, some of which depend on specific fern species and could become extinct together with their hosts (see Chapter 7). Some ferns and lycophytes provide food, fertilizer and medicine (de Winter and Amoroso, 2003), or are ornamentals (Hoshizaki and Moran, 2001; Olsen, 2007). Recently, ferns have been used as indicator plants (Beukema and van Noordwijk, 2004) for soil characteristics or climates (Bickford and Laffan, 2006). Some species accumulate heavy metals, especially arsenic, and are now considered as potential organisms for phytoremediation (see Chapter 8). Most fern species have not yet been surveyed for their potential uses but Ehrlich and Ehrlich (1981) argue that even species without known direct or indirect economic use for humans should be protected.

In this chapter, I explore some of the basic ecological data needed to estimate the extinction risk for ferns and lycophytes: (1) abundance; (2) geographic range; (3) habitat specificity; (4) intrinsic biological factors; (5) population dynamics and genetic diversity; and (6) environmental disturbances. I then discuss how these ecological data are used by different methods of risk assessment and how these methods address situations where such data are currently unavailable. I review the current conservation status of ferns as reflected in the global Red List of the IUCN (International Union for Conservation of Nature and Natural Resources) and the legal actions on global trade in ferns that have been established through CITES (Convention on International Trade in Endangered Species). For species at risk of extinction, I describe management strategies for their conservation and then conclude with recommendations for future fern conservation priorities.

Fig. 9.1 In Reunion Island (Macaronesia, Indian Ocean) there are three species of *Cyathea*, which can (a) form part of the tree canopy, or (b) successfully establish in the understory of plantations of *Cryptomeria japonica*.

9.2 Ecological data needed for risk assessment

For a reliable risk assessment for a given species, we should gather data about at least the following six aspects of its ecology and biology: (1) abundance (number of individuals); (2) geographic range (and disjunctions, i.e., geographic separation of subpopulations); (3) habitat specificity (restrictions to specific substrates or abiotic conditions, e.g., soil pH, temperature range); (4) intrinsic biological factors (e.g., growth rates, dispersal or reproduction); (5) population dynamics and genetic diversity; and (6) environmental disturbances (e.g., fire, land use change, or introduction of invasive organisms). Episodic environmental disturbances (e.g., volcanic eruptions or inundations) that are impossible to predict, should be taken into account during the development of management strategies in order to provide species protection in several well-separated areas. A minimum threshold value to avoid species extinction is needed for all these characteristics. Because of the variety of biological and ecological traits exhibited by different ferns, risk levels for each species should be determined individually (Given, 1994). Current methods of risk assessment depend mainly on general empirical knowledge, driven by environmental activism and influenced by political interests because we do not know enough about either the partial impact of each of these six ecological aspects on species survival or how these ecological aspects interact.

9.2.1 Abundance

Population size is one critical measure of the status of a species but information about fern abundance is not sufficient to determine risk, especially for tropical species. Even if we could determine the exact abundance of a species, how many individuals are necessary to ensure the survival of a species?

In gathering data on local fern abundance, our goal should be the collection of sufficient quantitative data to make recommendations for decision makers. Most abundance data for ferns come from floras or herbarium labels where descriptive terms such as "frequent," "abundant," "common," "scattered" or "rare" are used. These qualitative terms lack clear definition and are consequently applied differently by each collector. Plant collectors should be encouraged to use standard quantitative approaches instead. I favor a numeric approach of counting the number of observed individuals within a defined area (e.g., density = individuals per $10\,\text{m}^2$, $100\,\text{m}^2$) or time (e.g., hours of exploration per collector). For quick assessments, a logarithmically ranked scale is helpful (e.g., 1, <10, <100, <1000, >10 000 individuals) and could be subdivided into more ranks (e.g., 1, <5, <10, <50, <100 individuals) if a higher resolution is required. If ferns and lycophytes form clones through vegetative propagation or have long, creeping rhizomes (e.g., in

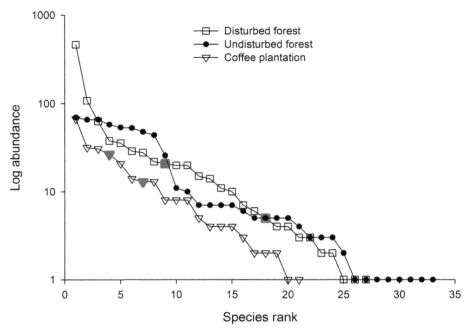

Fig. 9.2 Rank abundance curves of fern communities of three Mexican 1000 m²
plots in different vegetation types: an undisturbed montane forest, a disturbed
montane forest and an extensive coffee plantation. Dark gray symbols indicate
introduced species. (Data from Mehltreter, 2008a.)

Dennstaedtia) then individuals cannot be easily distinguished. Under these circumstances, one can instead count the number of stands, defined as the number of
clearly delimited patches (Sanford, 1968). There are two main advantages of
this quantitative approach: data can be directly compared and changes in local
abundance become apparent during repeated surveys of the same sites.

Some fern species are abundant while others are rare and a better understanding
of the ecological correlates that affect species abundance will benefit any conservation effort (Hubbell, 2001). In most plant communities, patterns of abundance for the different species become apparent at different scales. Ferns are
no exception. For example, logarithmic rank abundance curves for ferns were
constructed from three plots of 1000 m² each in Mexico, an undisturbed and a
disturbed montane forest as well as an extensive shade coffee plantation
(Fig. 9.2; Mehltreter, 2008a). Each curve was obtained by graphing the number
of individuals for each species ranked from the most abundant species on the
left to the rarest species on the right. Fern communities of all three vegetation
types showed the same pattern, with few abundant species, several more species
of intermediate abundance and most species represented by few individuals

(Mehltreter, 2008a). Although the three curves share the same general pattern, they also differ in several respects. The disturbed sites (forest and coffee plantations) had lower overall species diversity than the undisturbed forest, which results in a shorter right tail of the curves; the most dominant species were relatively more abundant, yielding a rising left tail on the curves. Both plots contained two introduced species (in gray) of intermediate abundance, while the undisturbed forest had none. Most fern species in the disturbed forest were more abundant compared with the undisturbed forest, perhaps because of increased light in the understory of the disturbed forest. In the coffee plantations, most species were of lower abundance, probably because of the effects of human impact. This example suggests that: (1) most fern species of the undisturbed forest are rare and become locally extinct after disturbance or conversion into a coffee plantation, and new fern communities that replace the original one are integrating introduced species even if they continue to follow the same general pattern of abundance; and (2) rare fern species of undisturbed forests are often represented by only one individual, which makes them vulnerable to local extinction because of accidental death or overcollecting (Mehltreter, 2008a). If these rare species are not represented in other communities within the region, they may not recover from local extinction and may disappear at regional scales.

9.2.2 *Geographical range*

Estimates of the geographical range of a given fern species are often based on herbarium specimen data. Although these estimates may be deficient at smaller scales they are often reasonable at larger scales (e.g., mountain ranges, national parks, provinces, or countries). However, in areas with poorly studied floras such as Botswana (Vega *et al.*, 1997), Bolivia (Smith *et al.*, 1999), Brazil (Ambrosio and De Melo, 2001) or Polynesia (Ranker *et al.*, 2005), botanists continue to report range extensions for fern species already known from other regions. In regions with detailed floristic treatments such as the continental USA, new records are mostly restricted to smaller, inconspicuous species, such as *Botryichum pallidum* (pale moonwort) in Maine (Gilman, 2004) or *Schizaea pennula* (ray fern) in Florida (Woodmansee and Sadle, 2005). By contrast, in regions that are less well studied, even large, conspicuous species such as *Acrostichum danaeifolium* (giant leather fern; León, 1990) and *Platycerium andinum* (staghorn fern) in Peru (Fernández and Vail, 2003) or *Angiopteris smithii* in Sulawesi (Camus and Pryor, 1996) have been overlooked. These range extensions either expand the geographic range of the species or fill in gaps between disjunct populations. For example, the extensive collections for *Flora Mesoamericana* (Moran and Riba,

1995) resulted in a reduction of the number of fern disjunctions between Mexico and Costa Rica (Mehltreter, 1995).

Endemics

If species are confined to a particular region they are called endemics and may be given special conservation priority in many countries. Sometimes paleoendemics are distinguished from neoendemics, although it is often difficult to classify species into one category or the other. Paleoendemic ferns are the last extant representatives of old and declining phylogenetic lineages, which may have been more widely distributed in the fossil record (see Cronk, 1992). Monotypic genera (containing only one species) may be examples of paleoendemics (e.g., *Thyrsopteris elegans* from the Juan-Fernández Islands; or *Anetium citrifolium* (matted fern) and *Hemidictyum marginatum* (large gully fern) confined to the neotropics; Tryon, 1989). Neoendemics are recently derived taxa, which often coexist with close relatives and have not yet spread because of their recent origin. For example, 5 of 12 species of *Argyrochosma* and 17 of 55 species of *Polypodium* are endemic to Mexico (Windham, 1987; Mickel and Smith, 2004). For conservation purposes, neoendemics and paleoendemics should be given the same priority, with no distinction made based on their evolutionary age. Because the term "endemism" is applied at different scales, endemic species do not necessarily have small ranges. For instance, endemic ferns of Mexico may be known only from the type collection (e.g., *Asplenium dianae*, *Athyrium tejeroi* and *Notholaena brachycaulis*) or may possess a range of several thousand square kilometers (e.g., *Adiantopsis seemannii*, *Anemia karwinskyana*) because Mexico is a relatively large country of nearly 2 000 000 km^2. The extent of the geographic range of *Bommeria elegans* is similar to that of *Cyathea ursina*, yet *B. elegans* is an endemic in Mexico while *C. ursina* is not an endemic in any country because it is distributed on the Central American land bridge from Guatemala in the North to Panama in the South, an area about one fourth the size of Mexico. For conservation purposes, geographic range should be expressed in km^2, although the term endemism should be retained as it may create social awareness and political responsibility at different governmental levels for safeguarding diversity.

Correlation between abundance and geographic range

There is no doubt that the abundance of a particular species is correlated with its geographical distribution, although the underlying mechanisms are not well understood (Hanski *et al.*, 1993). Rare species have restricted ranges and abundant species have wide ranges. However, rare species still can be locally abundant and abundant species can be scarce in some parts of their range (see Chapter 2). Abundant species are thought to be habitat generalists that survive in a wide

Table 9.1 *Assignment of all Mexican species of* Cheilanthes *to a modified table of Rabinowitz (1981) showing seven types of rarity (cells B–H) using a combination of two choices for each of three traits: geographic range (number of Mexican States), elevational range, and habitat specificity*

Most species are assigned to cells A (common species) and H (habitat-specific species with narrow ranges). In this example on ferns, Rabinowitz's original factor of local abundance is replaced by elevational range, because abundance data were unavailable.

Elevational range	Broad geographic range >5 states		Narrow geographic range ≤5 states	
	Generalist	Specialist	Generalist	Specialist
Broad >1000 m	18 (A)	10 (C)	6 (E)	6 (G)
Narrow <1000 m	0 (B)	2 (D)	4 (F)	14 (H)

Source: Data extracted from Mickel and Smith (2004).

variety of microenvironments, perhaps with variations in size, growth rate and reproduction at different sites. Rare species are often assumed to be habitat special-ists, and their rarity a consequence of their specialization. Rabinowitz (1981) showed different types of rarity using a table that considered the relationship of three traits: local abundance, geographic range and habitat specificity with two options for each (high/low). Of the eight possible combinations of the three traits (cells A–H, Table 9.1) only species with high values for all three traits (high abundance, broad range and habitat generalist) would not be at risk (cell A). However, species belonging to one of seven other categories can encounter an increasing degree of threat (cells B–H, Table 9.1) depending on how many of the three traits have the lower value (e.g., low local abundance, small geographic range and limited habitat tolerance).

Applying this general method of determining different types of rarity using com-parable traits for which data are available can be useful in establishing priorities for assessing risk in any group of organisms. For example, the genus *Cheilanthes* (lip fern) has 60 species in Mexico (Mickel and Smith, 2004) that can be categorized with a modified form of a Rabinowitz (1981) table. I summed the number of Mexican species of the genus *Cheilanthes* for each cell in the table, with habitat specialists defined as species occurring on one substrate type, generalists on two or more substrates and the geographic range measured arbitrarily as occurrence in more (or less) than five Mexican states. Because there were no data for local abundance, I replaced abundance with elevational range (difference between the lowest and highest elevation), a factor that proved to be correlated with the geographic distribution of

Mexican *Cheilanthes* (Fisher Exact test, $p<0.001$). Although local abundance and elevational range are not necessarily correlated, elevational range proved to be a useful trait for illustrating this method. The result (Table 9.1) demonstrates that species exist for each cell type, with one exception. There were no geographically wide-ranging, generalist species with a small elevational range, because these three categories almost exclude each other. Highest species counts appear in cells A (upper left corner, least risk) and H (lower right corner, most risk), indicating some correlation among the three chosen traits. Generalist species of Mexican *Cheilanthes* had broader elevational ranges (24 out of 28) than habitat specialists (16 out of 32, Fisher Exact test, $p<0.006$), but did not differ from habitat specialists in their geographic range (Fisher Exact test, $p=0.12$). Such a simple example may point out where detailed studies could provide interesting insights into how different traits are correlated with species rarity of ferns.

Ferns are most abundant in the tropics. Rapoport's rule (Rapoport, 1982) states that in lower latitudes (i.e., the tropics) species possess smaller ranges than in higher latitudes (i.e., temperate zones). Stevens (1989) assumes that Rapoport's rule is a consequence of the adaptation of temperate species to the strong annual climatic variation, which allows for survival across a wide range of similar climatic conditions. Tropical species experience a smaller range of annual climatic variation, and are not selected for tolerating extreme conditions, except at high altitudes. As a consequence, tropical species cannot survive under differing conditions and are therefore restricted to smaller geographic ranges (Stevens, 1989). Assuming that ferns are no exception to Rapoport's rule, we may conclude that tropical ferns occupy smaller geographic ranges than temperate ferns, making the former more vulnerable to extinction than the latter. In a similar way, terrestrial fern species have smaller elevational ranges and therefore may run higher extinction risks than epiphytic fern species (Bach *et al.*, 2007).

An approach commonly used for estimating species diversity may be applied to conservation management of habitats. Species–area curves correlate area with the number of observed species following the Arrhenius equation:

$$S = cA^z$$

where S is the number of species, A is area expressed in km^2, and c and z are constants that are based on environmental conditions and organisms. When drawn on a double logarithmic scale, the curves are linear ($\log S = \log c + z \log A$) with the slope z. Species curves are often drawn separately for continental areas and islands, and explain how species diversity increases with sampled area (or total area of islands). In contrast, Rosenzweig (1995) used these curves to estimate how many species could survive in a protected natural area if all surrounding habitat was destroyed. Rosenzweig (1995) used the island curve (with $z=0.25$ for the exponent in the

Arrhenius equation, a mean empirical value of several groups of organisms) to predict that the preservation of 5% of a defined area would save 47% of the original species diversity. The island curve is used because the destroyed area (95%) is empty and can no longer serve as a source area for species once they become extinct. Protection of 10% of natural areas (e.g., as approximately achieved in Costa Rica) may allow the survival of 56% of the original species diversity. If we collect more data on local fern abundance, we will be able to examine deviations from these general formulas and perhaps improve on such numerical approaches.

9.2.3 Habitat specificity

The use of the Arrhenius equation to predict the relative conservation of species diversity described above seems completely inadequate for habitat specialists. For these species, simple protection of a proportion of natural habitats is useless if the protected area does not happen to include their habitat. An example may illustrate the importance of understanding habitat specificity for species conservation. *Vandenboschia speciosa* (Killarney fern, syn. *Trichomanes speciosum*) is a tiny, endangered, filmy fern found in a few, scattered populations in the British Isles, France, Belgium, Germany, Italy, Spain, North Africa and the Macaronesian Islands (Stace, 1997). Many populations persist only as gametophytes, while sporophytes are extremely rare (Ratcliffe *et al.*, 1993; Vogel *et al.*, 1993; Rumsey *et al.*, 1996, 1998, 1999; Bennert, 1999). Loriot *et al.* (2006) investigated the populations in northwestern France, where the gametophytes of *V. speciosa* are restricted to microhabitats with low light, mostly northern exposure and acidic substrates. Their principal natural habitats are caves, rocky outcrops and woodland boulders, but two thirds of the populations thrive in human-made habitats, especially wells. Conservation efforts aim to create public awareness and to ask owners of these wells to protect the habitat (Loriot *et al.*, 2006). Without the application of a species-specific conservation management strategy, *V. speciosa* would have a high risk of extinction.

Some terrestrial fern species depend on particular soil characteristics (Tuomisto and Poulsen, 1996), just as epiphytic ferns (e.g., Plate 6D) may depend upon particular host tree characteristics (Mehltreter *et al.*, 2005) or microhabitat conditions (Gardette, 1996). Some habitats are so distinctive that ferns need specific adaptations in order to survive and grow. Ferns of riparian habitats are exposed to daily or seasonally changing water levels and currents, and may be heavily impacted by human-induced changes in the water regime or contamination (Boutin *et al.*, 2003). Lycophytes that colonize lakes such as some species of *Isoëtes* (quillworts, Plate 3D) may be heavily affected by changes in the chemical composition of the water (Vöge, 2004). A species that is a habitat specialist will always need an exceptional conservation strategy (Given, 1993). Nutrient cycles (see Chapter 4) and disturbances (see Chapter 6) can affect habitat conditions. Habitat specificity is discussed in more detail in Mehltreter (2008b).

9.2.4 Intrinsic biological factors

Intrinsic biological factors are defined here as all life history traits of a species that may have an influence on its abundance and may increase its risk of extinction (e.g., green spores, dark germination, large plant size). These traits are often called dimensionless or area-independent in the literature, because they are not directly correlated with geographic range. However, this distinction is misleading because some supposedly area-independent life history traits will have a direct effect on population dynamics and genetic diversity, both of which are mutually affected by geographic distribution and especially by geographic isolation. Ayensu (1981) emphasized that to facilitate conservation management, studies of basic plant biology should give priority to investigating reproduction and breeding systems in order to understand the importance of each trait as a risk factor for a species. It is also important to identify supposedly secure populations that primarily reproduce sexually and are genetically highly diverse, because these can be crucial for reintroduction projects (see Section 9.4.2). I now discuss some intrinsic biological factors of ferns that may cause declines in population size and make them more vulnerable to extinction.

Green spores

Most fern spores lack chlorophyll and are yellow, orange, brown or black. These spores are mostly viable for several months to several years. Long-lived spores may survive habitat disturbance (e.g., logging, fire) in the soil spore bank and may germinate after years of succession, when environmental conditions are appropriate again (Dyer, 1994). However, about 10% of fern species have green-colored spores because of their chlorophyll content (e.g., Equisetaceae, grammitid ferns, Hymenophyllaceae, Onocleaceae, Osmundaceae). For this reason, green spores can immediately germinate under favorable environmental conditions. If germination does not occur, the chlorophyll breaks down and mean spore viability is limited to about seven weeks (Lloyd and Klekowski, 1970). It seems likely that species with green spores may be strongly dependent on constantly humid microenvironments (e.g., tropical cloud forests), and more vulnerable to forest disturbance that decreases habitat humidity. The risk of such a short window of viability for ferns with green spores may have been overlooked, because few green-spored species are listed as endangered. It is possible that other traits, especially of the green-spored, epiphytic fern species, may compensate in part for this disadvantage. Sporophytes of most green-spored epiphytic fern species are small, increasing their chance of fitting into available microhabitats; they can become locally abundant and form large colonies. Gametophytes of green-spored ferns frequently produce gemmae (vegetative buds) that easily detach and allow for clonal colonization and dispersal. Although some green-spored species have limited ranges, they can be wide ranging (e.g., the epiphytic filmy fern *Hymenophyllum polyanthos* and terrestrial species of *Equisetum* and *Osmunda*) and have dispersed to remote volcanic islands such as Tristan da Cunha in the Atlantic Ocean (e.g., Hymenophyllaceae, Tryon, 1970).

Dark germination

Spore germination of most ferns and lycophytes requires light. Thus, their spores germinate on the substrate surface where the green gametophytes then develop. In contrast, all spores of Lycopodiaceae (with the exception of *Lycopodiella*, bog club-moss), Ophioglossaceae and Psilotaceae germinate belowground, in the dark. After spore germination, the development of subterranean gametophytes is comparatively slow, because it depends on a saprophytic association with mycorrhizal fungi. In conservation studies on German ferns and lycophytes, dark germination could only be induced in the laboratory for two of 11 species known for their dark germination under field conditions (Bennert and Danzebrink, 1996; Bennert, 1999). These results could reflect the importance of interactions with mycorrhizal fungi in the natural environment during spore germination.

It was assumed that subterranean gametophytes are too distant from each other for intergametophytic fertilization. Thus, it was expected that their bisexual gametophytes reproduce mainly by intragametophytic selfing and that their populations would suffer subsequent loss of genetic diversity. This hypothesis was supported for *Botrychium virginianum* (common grape fern) that has a high percentage of intragametophytic selfing (Soltis and Soltis, 1986). However, this hypothesis has been rejected for *Lycopodium complanatum* (syn. *Diphasiastrum complanatum*, northern ground cedar), another species with subterranean gametophytes but high intergametophytic fertilization (Soltis and Soltis, 1989). The latter species may possess gametophyte populations growing in larger densities to allow for outcrossing or may produce antheridiogens, hormones released by female gametophytes that cause nearby immature gametophytes to become male, thus promoting intergametophytic fertilization (Schneller, 2008). Clearly, a better understanding of the longevity, density and breeding systems of natural gametophyte populations is required in order to explain the observed patterns between dark germination and species conservation status. In conclusion, many species with dark germination may be rare or threatened (e.g., all Lycopodiaceae and Ophioglossaceae in Germany, Bennert, 1999), because their slow gametophyte development that depends on mycorrhizal fungi makes them potentially more vulnerable to frequent habitat changes or disturbances. Future research should focus on environmental conditions that favor the development of these species *in situ* (in their natural habitats) and should look for successful methods for their cultivation *ex situ* (outside their natural habitats such as in botanical gardens).

Photosynthesis

Photosynthetic rates of ferns and lycophytes vary widely among species and habitats (Table 9.2). On average, most ferns have lower mean water conduction and net assimilation rates than seed plants, especially in sunny environments (Brodribb *et al.*, 2005, 2007). Although slow growth rates might be considered as an obvious competitive disadvantage, there is no evidence that low net assimilation rates correlate with high extinction risk of ferns. Other factors may compensate for the apparent disadvantage of lower photosynthetic rates in ferns, making them successful competitors against coexisting seed plants. *Dicranopteris linearis*, a scrambling fern in Hawaii, has lower net assimilation rates than the codominant tree, *Metrosideros polymorpha*, but dominates the aboveground net primary productivity in some habitats, perhaps because of its high phosphorus use efficiency (Russell *et al.*, 1998; see Chapters 4 and 6). Bannister and Wildish (1982) suggested that ferns compensate for low carbon assimilation rates with low dark respiration rates (resulting in low light compensation points), allowing for net photosynthetic gains even in the dark understory. Under low light conditions, ferns reach values of leaf hydraulic conductivity similar to those of seed plants (Brodribb and Holbrook, 2004). In conclusion, it should not be expected that fern species are endangered because of their lower photosynthetic efficiency. Fern abundance in some forest understories clearly demonstrates their competitive success against seed plants under low light conditions, whereas in high light environments ferns can be strong competitors by other traits such as allelopathy (see Chapters 6 and 8).

Plant age at maturity and sporophyte size

Slow-growing fern species (e.g., many tree ferns as well as dark germinating species, see above) often need more time to develop from a spore into a reproductively mature plant. In habitats with frequent natural or human disturbances, this implies a proportionally higher extinction risk, because individual sporophytes in a fern population cannot reach maturity between two subsequent disturbance events. Most fern species require several years to reach maturity. For example *Polystichum tripteron* in Japan needs 11 years (Sato, 1990) and the tree fern *Alsophila firma* in Mexico 12 years to reach maturity (Mehltreter and

Table 9.2 *Mean maximum net CO_2 assimilation (A_{max}) for ferns and lycophytes*

Species	A_{max} (μmol m^{-2} s^{-1})	Country	Reference
Asplenium cuspidatum*	2.6	Mexico	Hietz and Briones, 2001
Bolbitis portoricensis	2.8	Costa Rica	Brodribb *et al.*, 2007
Botrychium lunaria	11.0	Germany	Giers *et al.*, 1996
Cibotium sp.	5.8 (3.6)	Hawaii	Durand and Goldstein, 2001
Cryptogramma crispa	14.0	Germany	Giers *et al.*, 1996
Dicranopteris linearis	2.9–5.0	Hawaii	Russell *et al.*, 1998
Dicksonia antarctica	10.8	Tasmania	Hunt *et al.*, 2002
Diphasiastrum complanatum	7.0	Germany	Heiser *et al.*, 1996
Diphasiastrum tristachyum	10.0	Germany	Heiser *et al.*, 1996
Dryopteris cristatum	8.5	Germany	Giers *et al.*, 1996
Dryopteris oreades*	11.0	Germany	Benemann, 1996
Elaphoglossum glaucum*	5.2	Mexico	Hietz and Briones, 2001
Elaphoglossum petiolatum*	2.6	Mexico	Hietz and Briones, 2001
Lygodium venustum	5.4	Costa Rica	Brodribb *et al.*, 2007
Matteuccia struthiopteris	4.7	Germany	Giers *et al.*, 1996
Osmunda regalis	12.0	Germany	Giers *et al.*, 1996
Phlebodium areolatum*	3.2	Mexico	Hietz and Briones, 2001
Pleopeltis mexicana*	4.4	Mexico	Hietz and Briones, 2001
Pleopeltis plebeia*	4.8	Mexico	Hietz and Briones, 2001
Polypodium puberulum*	2.5	Mexico	Hietz and Briones, 2001
Polypodium triseriale	3.2	Costa Rica	Brodribb *et al.*, 2007
Polystichum braunii*	4.4–6.4	Germany	Benemann, 1996
Pteris altissima	5.4	Costa Rica	Brodribb *et al.*, 2007
Selaginella pallescens	1.7	Costa Rica	Brodribb *et al.*, 2007
Sphaeropteris cooperi	11.2 (7.1)	Hawaii	Durand and Goldstein, 2001
Tectaria confluens	2.1	Costa Rica	Brodribb *et al.*, 2007
Thelypteris palustris*	11.8	Germany	Benemann, 1996
Tmesipteris obliqua	1.1	Tasmania	Brodribb *et al.*, 2007
Trichomanes bucinatum*	0.6	Mexico	Hietz and Briones, 2001

* Values determined under laboratory conditions. Separate measurements made in the shade are noted within parentheses.

García-Franco, 2008), while other fern species such as *Macrothelypteris torresi-ana* in Mexico can reproduce within several months at very early developmental stages, when their leaves are shorter than 0.3 m, and long before the plant produces leaves of their maximum length of about 1.5 m (K. Mehltreter, personal observation).

Larger organisms may have a higher probability of extinction than smaller ones. Larger organisms have the advantage of being less vulnerable to interspecific competition, but they need more space and fewer individuals will fit into a protected area, keeping their populations small. Humans also tend to overexploit larger organisms such as tree ferns because their harvest easily yields benefits. Fortunately, large species are also the first to receive attention by conservation managers because they represent a significant proportion of the biomass, and may contribute substantially to the local economy.

If large fern species are for some reason exposed to a higher extinction risk than small ferns, one would expect fewer fern species to be large (e.g., leaf length >2 m) than small or midsized. Hutchinson and MacArthur (1959) demonstrated that there are more intermediate-sized animal species than smaller or larger ones. In ferns, midsized species with a mean leaf length of 0.5–1.5 m are also the most numerous. Some families such as Blechnaceae include representatives of a large range of leaf lengths from 0.1 m to over 3 m, but most species are midsized. Several families do have mostly small species, such as Hymenophyllaceae, Schizaeaceae and the water fern families Salviniaceae and Marsileaceae. Only four of the 37 fern families (Appendices A, B; Smith *et al.*, 2006) comprise mainly species of large adult size (i.e., tree ferns): Culcitaceae, Cibotiaceae, Cyatheaceae and Dicksoniaceae, although there are a few trunk forming or large-leaved representatives in other families (see Chapter 3, Table 3.2; e.g., Blechnaceae).

Blechnum werckleanum (Fig. 9.3a) is one of the largest species in the genus, forming a trunk 1 m tall with leaves 1.1–1.7 m long (Moran and Riba, 1995). In 2007, I observed only one large, mature plant and a dozen small, sterile ones in the Tapantí Reserve in Costa Rica. At the same site, medium-sized species such as *B. schiedeanum*, and smaller species such as *B. appendiculatum* (Plate 5D) were both abundant. One large individual of *B. werckleanum* may produce several million spores but apparently few sporophytes establish successfully in the immediate neighborhood of the mother plant. *Blechnum werckleanum* perhaps requires some degree of disturbance to reproduce from spores, but changes driven by succession quickly alter environmental conditions before young sporophytes may achieve reproductive maturity.

Blechnum auratum (syn. *B. buchtienii;* A. R. Smith, personal communication; Fig. 9.3b) is another large species with trunks up to 2 m high and leaves 0.5–1.2 m

Fig. 9.3 *Blechnum* species of varying sizes and habitats achieving different local abundances: (a) *Blechnum werckleanum*, very large trunk-forming, rare species of the Costa Rican cloud forest. (b) *B. auratum* (syn. *B. buchtienii*) trunk-forming, codominant species of the bogs in Costa Rican páramo.

long (Moran and Riba, 1995) that achieves high local abundances at páramo sites in Costa Rica. Why does *B. auratum* occur in exceptionally large numbers compared with the low abundance of *B. werckleanum*? The páramo vegetation is prone to regular fires, caused by natural events such as lightning or by human activities. The leaves and trunk surface of *B. auratum* burn like the rest of the vegetation, but the growing tip must be well protected as it has the ability to produce new leaves from the apex of the bare trunks (Horn, 1988, 1989). In this way, areas with fire disturbance are dominated by fire-resistant species such as *B. auratum* and the bamboo *Chusquea subtessellata*. Interestingly, fire resistance in *B. auratum* may not be a primary adaptation, because they are most common in bogs, where they become codominant. Consequently, I hypothesize that *B. auratum* may have spread from the bogs into the surrounding habitats where it exhibits a competitive advantage over otherwise dominant trees and shrubs because of its fire resistance.

Small size is not necessarily a guarantee for survival either as indicated by the midsize peak of body size frequencies in animal species (Hutchinson and MacArthur, 1959). For example, *Blechnum stoloniferum* in Mexico is one of the smaller species of the genus with leaves only 0.1–0.35 m long (Mickel and Smith, 2004). Despite its capacity to produce stolons, it is a relatively rare species, especially in comparison with *B. appendiculatum* (Plate 5D).

These contrasting examples of growth and abundance of different species of *Blechnum* indicate that it may be difficult to find any general correlations between the size of fern species and their abundance or extinction risk because of additional, often unique species traits with opposite effects. However, it is probably worthwhile to take a closer look at these types of potential correlations, especially regarding other factors that are related to species size such as population dynamics.

9.2.5 *Population dynamics and genetic diversity*

Shorter-lived organisms are more likely to have fluctuating population sizes (Pimm, 1991; Lawton, 1995) than longer-lived species, because short-lived species can quickly build up populations of many individuals during favorable times but may quickly die back to a few individuals when conditions become unfavorable. Because nearly all fern species are perennial and some species are very long lived, natural population dynamics should not be expected to be a major extinction risk, but human activities such as habitat destruction and fragmentation may cause their populations to collapse to just a few individuals.

There may be a significant loss of genetic diversity in isolated populations that decrease in size to a small number of individuals. Isolated populations of four fern species in Switzerland showed little or no isozymic variation within populations (*Asplenium ruta-muraria* (wall rue), *A. septentrionale* (forked spleenwort), *A. viride* (green spleenwort) and *Polypodium vulgare* (common polypody)), but all seemed to be viable because they did not suffer from inbreeding depression (i.e., reduced fitness because of the expression of harmful or lethal plant traits; Schneller and Holderegger, 1996). Because these species can successfully disperse over long distances, they are not endangered as long as their isolated populations remain viable. In contrast, *Adiantum reniforme* var. *sinense* is endangered in China because of partial submersion of its natural habitat after the construction of the Three Gorge Dam on the Yangtze River, and because of overcollecting due to its use as a medicinal herb. It possesses highly polymorphic microsatellite loci (i.e., a measure of genetic diversity) that suggest outbreeding. However, heterozygotic deficiencies indicate reduced gene flow between the isolated small populations. This species may become extinct if genetic variation cannot be maintained by establishing viable populations elsewhere (Kang *et al.*, 2006).

Even more threatened by land use change is *Archangiopteris itoi* (Marattiaceae), which is endemic to Taiwan. The last known population of 18 individuals possesses very low genetic variation (Hsu *et al.*, 2000). In this case, the continuous loss of original habitat required that viable populations be established and maintained elsewhere. Because spore germination and gameto-phyte development are very slow in this species, with only 1% of gametophytes producing sporophytes after 2.5 years, an alternative propagation method is needed. Stipule buds are common in the Marattiaceae although they rarely develop into sporophytes in their natural habitat (Sharpe and Jernstedt, 1991). In cultivation, however, plantlets sprouted from 90% of the stipule buds after only seven months. This method is now the most successful propagation strategy for species in the Marattiaceae (Chiou *et al.*, 2006), though it does not resolve the problems of low genetic variation within these species.

Some fern species are considered to be threatened in one country where they are at the margins of their geographical range, whereas in a neighboring country they may be in the central part of their range and maintain viable populations. The geographic range of *Dryopteris arguta* (coastal wood fern) extends from Baja California (Mexico) and the western USA into British Columbia (Canada). Although common in the southern part of its range, at its northern limit in Canada there are only 18 extant populations on 16 sites (Jamison and Douglas, 1998). From a global point of view, this species is not endangered. So why should we be concerned about protecting these peripheral populations? Peripheral populations often diverge genetically and morphologically from central populations, partly because of their spatial isolation and partly because they may encounter increasingly unfavorable abiotic or biotic conditions for their survival. Peripheral and central populations may even occur in different microhabitats (Lesica and Allendorf, 1995). Consequently, conservation of peripheral populations maintains high genetic variability as well as securing potential sites for future evolution of new ecotypes and species. This last argument is sometimes overlooked when disjunct and isolated populations are neglected within conservation management plans that typically give preference to the largest populations in central areas or several smaller adjacent and inter-connected habitat islands (Diamond, 1975, 1986; Whittaker, 1998).

9.2.6 Environmental disturbances

Environmental disturbances can be caused by natural phenomena (e.g., earth-quakes, fire, storms, volcanism) or human impact. Direct human impact is probably the most important cause of recent extinctions, because most plants are not pre-adapted against adverse human impact factors (e.g., habitat destruction, fertilizers, pesticides). Greuter (1995) compared the number of extinctions during the last century in areas with Mediterranean climates. He found that the Mediterranean, with its ancient human influence, had the lowest extinction rates (0.13% of plant species extinct) while southwestern Australia, a more recently impacted area had the highest extinction rates (0.66% of plant species extinct). Greuter (1995) assumed that undocumented, large-scale extinctions had already reduced plant diversity in the Mediterranean in earlier times, selectively leaving more extant species resistant to human civilization.

At present, fire often results from human impacts, but fire occurred naturally before human evolution and civilization. A high percentage of fire-resistant ferns have been observed in Africa (Kornás, 1977, 1978), the continent with the least diverse tropical fern flora (see Chapter 2). In a study of forested habitats in Australia, tree fern populations recovered faster after forest fires than after logging (Ough, 2001; Ough and Murphy, 2004). Additionally, the success of *Pteridium* (bracken), one of the most cosmopolitan ferns, might be based on fire resistance. It is protected against fire because of subterranean rhizomes, but also quickly recovers after fires from the spore bank (see Chapters 6 and 8; Gliessman, 1978). Several species of Gleicheniaceae successfully colonize landslides, but do not survive fires; however, they invade burned landslides from unaffected populations on the margins (Walker and Boneta, 1995).

Most endangered fern species are threatened by a combination of several human activities (e.g., overexploitation, contamination, introduction of invasive species, urban development). By far the most important threat comes from habitat destruction, fragmentation and land use change due to conversion from forest into agricultural lands (Sukopp and Trautmann, 1981; Paciencia and Prado, 2005). Deforestation is particularly damaging because the trees provide shade for understory ferns as well as substrate for epiphytic ferns. A few fern species are endangered by commercial sale. For example, the ornamental *Dicksonia antarctica* (soft tree fern) is harvested in Tasmania to supply the Australian and European market (Unwin and Hunt, 1996). *Dicksonia antarctica* can still be harvested, because only the four New World species of *Dicksonia* (*D. berteriana, D. externa, D. sellowiana* and *D. stuebelii*) are internationally protected by CITES (2008), but not the approximately 17 species from the Old World. *Dicksonia antarctica* is not included on the Australia EPBC Act List of threatened species (Department of the Environment, Water, Heritage and the Arts, 2009), but its sustainable harvest, transport and trading is regulated by a fern management plan (Department of the Environment, Water, Heritage and the Arts, 2007). This management plan foresees no or limited harvest of natural populations of *D. antarctica*, but instead encourages their commercial cultivation and harvest in the understory of *Eucalyptus* plantations. This type of management plan encourages land owners to participate in species conservation because they are still allowed to harvest species under legally controlled conditions.

Human impacts on ferns are multiple. Arcand and Ranker (2008) discuss additional examples of human impact and other threats to ferns and lycophytes, while

Ehrlich and Ehrlich (1981) and Given (1994) give a comprehensive overview of human impacts on native plants in general and examples of their conservation management.

9.2.7 Summary

I have discussed six aspects of the biology of a fern species that may be correlated with the risk of extinction. To fully understand the ecology of a species, knowledge of all six topics is necessary but the data are only available for few species because fern conservation is clearly not a current research priority. In the next section, I will address the practical challenges of using these data to assess extinction risk for fern species and the problems that emerge when data are missing.

9.3 Risk assessment

The IUCN is the leading organization for species conservation on a worldwide scale. It applies five main risk criteria, which are in part based on the six ecological aspects just discussed in order to assign species to one of eight risk categories. The IUCN publishes their results in the annually updated global Red List of Threatened Species (IUCN, 2008) and promotes the compilation of regional Red Lists through governmental agencies and nongovernmental organizations by providing guidelines to apply Red List criteria at regional levels (IUCN, 2003). In this section, I will present an overview of current issues and problems with the global and regional Red Lists for ferns and lycophytes and will discuss the role of CITES in fern conservation.

9.3.1 IUCN Red List criteria

The IUCN currently defines five main criteria of risk (A–E) (Table 9.3; IUCN Standards and Petitions Working Group, 2008), three of which (A, C and D) address total population size and decline. One criterion (B) represents the reduction and fragmentation or fluctuation of geographic range, and another criterion (E) represents a quantitative analysis (e.g., population viability analysis, Brigham and Schwartz, 2003) that estimates probability of extinction. Thus, the first four IUCN criteria (A–D) focus on the first two ecological aspects: abundance (see Section 9.2.1) and geographic range (see Section 9.2.2). The last criterion (E) asks for a quantitative analysis and has not yet been used for ferns (IUCN, 2008). Consequently, IUCN criteria do not rely directly on the four remaining ecological aspects (see Sections 9.2.3–9.2.6), but use them indirectly as explanatory factors for the changes in abundance and range of a species. For this reason, we should

Table 9.3 *Simplified summary of the major requirements of the five IUCN criteria (A–E) for each of three risk categories*
A species that meets all requirements of one criterion (e.g., B2 or C1) can be assigned to the corresponding threatened category.

	Threatened categories		
Criterion	Critically endangered (CR)	Endangered (EN)	Vulnerable (VU)
A. Population reduction: declines measured over the longer of 10 years or 3 generations			
A1. and if causes of reduction are reversible, understood, *and* ceased	>90%	>70%	>50%
A2–A4. and if causes of reduction may not be reversible, understood, *or* may not have ceased	>80%	>50%	>30%
B. Geographic range			
B1. Extent of occurrence (EOO)	$<100\,\mathrm{km}^2$	$<5000\,\mathrm{km}^2$	$<20\,000\,\mathrm{km}^2$
B2. Area of occupancy (AOO)	$<10\,\mathrm{km}^2$	$<500\,\mathrm{km}^2$	$<2000\,\mathrm{km}^2$
C. Small population size and decline.			
C1–C2. Number of mature individuals	<250	<2500	<10 000
D. Very small or restricted population. Number of mature individuals	<50	<250	<1000
E. Quantitative analysis indicating probability of extinction in the wild within 100 years maximum	>50% in 10 years or 3 generations	>20% in 20 years or 5 generations	>10% in 100 years

Source: After IUCN Standards and Petitions Working Group (2008).

promote the study of the correlation of ecological aspects with species abundance and species ranges to understand the underlying causes of species rarity.

Risk criteria are used to set limits that define different risk assessment categories. The IUCN Standards and Petitions Working Group (2008) fixed the minimum number of mature individuals (criteria C and D) for the three threatened categories (Table 9.3) for practical reasons, based on empirical data and the quest for international agreement across a wide range of biological fields. However, the number of individuals is unknown for most fern species. For this reason, criterion B1, which is based on extent of occurrence (EOO), seems to be a more practical approach, because species ranges are commonly included in floras (e.g., Mickel and Smith, 2004). Extent of occurrence is a measure of the spatial spread of areas currently occupied by a species and is defined as the area (in km^2) of the smallest polygon without internal

(a) 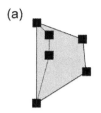 (b) 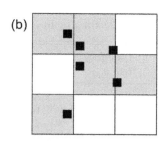 (c)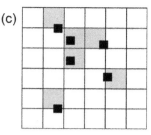

Fig. 9.4 Example for the determination of the (a) extent of occurrence and (b–c) area of occupancy of a species at two different scales. (a) Extent of occurrence is defined as the area of the smallest polygon (in gray) without internal angles >180° that contains all sites of occurrence. The smaller inner polygon delimited by six points does not accomplish this requirement because there are two internal angles >180°. (b–c) Area of occupancy is the sum of occupied squares (in gray) and is scale dependent. In this example, the area of occupancy in (b) is 3.33 times larger at lower resolution than in (c), where one grid square is four times smaller. (Adapted from IUCN, 2001).

angles >180° that contains all subpopulations (Fig. 9.4a, Table 9.3). The extent of occurrence for any species can be calculated with computer software (geographical information systems) if geographically referenced specimen locations are available. The alternative geographic range evaluation method of IUCN (criterion B2) requires the knowledge of the area of occupancy (AOO), defined as the area actually occupied by the species within its extent of occurrence. The area of occupancy may be available for well-studied species with very restricted distribution ranges, whereas for all other species the measure of the area of occupancy is difficult to achieve. Moreover, the area of occupancy is scale dependent (i.e., the result increases at larger scales), because it includes increasingly more uninhabited areas (Fig. 9.4b, c; Hartley and Kunin, 2003). In conclusion, I recommend that the risk assessment of ferns be based on criterion B1, the extent of occurrence, as it is the most accessible and standardized approach, until more detailed information for the other criteria becomes available. Because IUCN guidelines require that a species that fulfills a single criterion (with several subcriteria) be assigned to a risk category on the Red List of Threatened Species (IUCN Standards and Petitions Working Group, 2008), this recommendation could accelerate the assessment of the conservation status of ferns by scientists and knowledgeable amateurs.

9.3.2 *Risk categories of the IUCN Red List of Threatened Species*

Global risk category assignment

The most accepted international standard for risk assessment is that of the IUCN Species Survival Commission and it uses nine official categories in decreasing order

Table 9.4 *Number of fern and lycophyte species in each Red List category*

Evaluation	Data	Category type	Category name	Category code	Lyco-phytes	Ferns	Total
Evaluated	Adequate	Extinct	Extinct	EX	0	3	3
			Extinct in the wild	EW	0	0	0
		Threatened	Critically endangered	CR	3	29	32
			Endangered	EN	2	37	39
			Vulnerable	VU	10	58	68
		Near threatened		NT	2	14	16
		Least concern		LC	1	7	8
	Deficient	Deficient	Data deficient	DD	0	45	45
	Total	Evaluated		18	193 (1.3%)	211 (2.1%)	(2.0%)
Not evaluated	Total	Not evaluated		NE	~1300	~9000	~10 300

Species of the categories LC and NE are not considered as red listed.
Source: After IUCN (2008).

of extinction risk to assign the conservation status at species level (IUCN, 2001, 2008; Table 9.4). All species that are assigned to the seven categories EX, EW, CR, EN, VU, NT and DD are considered "red listed" (IUCN, 2008). Species that are assigned to the remaining two categories (LC and NE) are not red listed, either because they have been evaluated but are not at risk (i.e., least concern, LC) or they still have not been evaluated (NE).

There are few ferns and lycophytes in the two extinction categories (EX and EW) for several reasons. First, detailed studies over long periods are required to declare a species extinct; even then extinction is difficult or impossible to prove. Exceptions are endemics on small islands, where the chance of rediscovery is much reduced. Not surprisingly, two of the three fern species listed as extinct, *Anogramma ascensionis* and *Dryopteris ascensionis*, are from Ascension Island, a remote volcanic island in the Southern Atlantic with a surface area of only 88 km^2. However, the third fern species for which extinction status has been approved is *Adiantum lianxianense* from mainland China (China Plant Specialist Group, 2004). Second, for a large number of rare species (e.g., species known only from the type collection), it is very difficult to collect the evidence of their extinction. Finally, undescribed fern species may have never been encountered because of their rarity or because they occur in remote, inaccessible areas. Such species may become extinct before they have been discovered.

We need to give high priority to the risk assessment of known, rare species, and study more intensively the ferns in highly diverse, tropical hot spots. In these hot spots, >1500 endemic vascular plant species and >70% of the original natural vegetation has been lost (Myers, 1988). The 25 plant hot spots worldwide defined by Myers *et al.* (2000) are mostly identical for ferns (see Chapter 2) and seed plants (e.g., Brazilian Atlantic rain forest, Madagascar, Philippines, tropical Andes). However, in some seed plant hot spots, fern endemism is considerably lower. For example, less than 0.5% of all ferns worldwide (45 of 11 000 fern species; Schelpe and Anthony, 1986), but >2.0% of all seed plants (Mittermeier *et al.*, 2005) are endemic to the Cape region in South Africa. West Australia has <0.1% of ferns worldwide as endemics, while 1% of the World's seed plant flora is restricted to Southwest Australia alone (Flora of Australia Online, 2009).

The 2008 IUCN Red List of Threatened Species shows that the conservation status has only been evaluated for 2% of ferns worldwide, of which 74.9% have been assessed as extinct, threatened, or near threatened (Table 9.4). It is our incomplete knowledge of the abundance and ecology of ferns that impedes risk assessment. For this reason, the criteria used for risk assessment (see Table 9.3) of threatened fern species (IUCN, 2008) are mainly B1 (82 times), A2 (19) and D2 (17). Criterion B1, extent of occurrence (EOO), has been the most commonly used criterion to classify ferns at risk because this information is currently the most available for ferns (see Section 9.2.1) and its application in the future could result in a faster risk assessment of the remaining 98% of fern species. In addition to the very small number of species evaluated, the current global IUCN Red List of ferns and lycophytes are subject to a strong taxonomic and geographic bias. Of all 211 evaluated species, 141 species belong to only nine genera for which more than five species have been assessed: *Asplenium* (8 of 720 species), *Blechnum* (9/200), *Cyathea* (8/185), *Diplazium* (17/400), *Elaphoglossum* (50/700), *Huperzia* (12/400), *Hymenophyllum* (11/250), *Polypodium* (11/45) and *Thelypteris* (15/280). The geographical bias is even stronger with 168 risk assessments of fern species from Ecuador (*c.* 13% of its fern flora), 25 from China (*c.* 1%), 13 from Saint Helena (including its dependencies Ascension Island and Tristan da Cunha, *c.* 25%), 4 from Hawaii and 1 from Yemen. Because of the low number and large taxonomic and geographical bias of available risk assessments, we cannot yet estimate the real threat for ferns and lycophytes worldwide, but have to look at regional risk assessments for additional insights.

Regional risk assessment

Red List criteria have been more widely applied to ferns at regional than at global levels (IUCN, 2003). Some regional Red Lists have already evaluated all fern species at a country-wide level (see Table 9.5). Regionally red-listed species may not be

Table 9.5 *Number of fern and lycophyte species of selected regional Red Lists*

Country	EX	CR	EN	VU	NT	LC	DD	Total	Reference
Australia	7	2	12	14	nd	nd	0	nd	Department of the Environment, Water, Heritage and the Arts, 2009
Bolivia	0	1	12	107	0	1047	1	1168	Kessler *et al.*, 2006
Germany	1	3	17	11	7	43	1	83	Korneck *et al.*, 1996
Great Britain	1	1	1	7	6	42	0	nd	Cheffings and Farrell, 2005
Japan	8	62	37	32	6	nd	nd	nd	J-IBIS, 1997
Luxemburg	5	2	4	4	9	nd	1	nd	Colling, 2005
São Tomé and Príncipe	0	8	3	11	nd	nd	nd	nd	Figueiredo and Gascoigne, 2001
Pitcairn Islands	0	3	3	8	6	11	1	32	Kingston and Waldren, 2002

Categories follow IUCN (2008). Total species numbers are only given when unthreatened species (LC) were also evaluated. EX, Regionally Extinct, CR, Critically Endangered, EN, Endangered, VU, Vulnerable, NT, Near Threatened, LC, Least Concern (Not Threatened), DD, Data Deficient, nd = not determined.

endangered globally, as species that are rare in one country may be still common in other countries. However, for species that are endemic to one country, the risk assessment category could be directly integrated into the global Red List after the Red List authorities have accepted the regional assessments submitted by specialists. For instance, Figueiredo and Gascoigne (2001) evaluated the conservation status of 22 ferns and lycophytes of the Islands of São Tomé and Príncipe in the Gulf of Guinea, which included 13 endemic species (Table 9.5). The regional risk assessment of these endemic species (6 CR, 1 EN and 6 VU) should be directly relevant for their global evaluation. Another example is the regional Red List in preparation for Bolivia (Kessler *et al.*, 2006) that classifies about 10% of the species (121 out of 1168 species) as threatened at some level. Most of these species are endemic to Bolivia and are known from fewer than 10 collections and some are new species currently being described or known only from the type collection (36%). If these endemic Bolivian species cannot be found in the neighboring countries within a reasonable number of years, they might be proposed for classification as globally threatened. However, few regional assessments are submitted to IUCN Red List authorities, most likely because their global evaluation requires evidence that these species are missing in other regions where there may be a lack of funding and expertise for risk assessment. Consequently the global Red List of ferns and lycophytes is increasing slowly.

Newly described species present another problem, especially if they are only known from the type collection (e.g., 23 newly described species from Mexico,

2.3% of the total fern flora). Should these species immediately be classified as threatened? How long should we look out for more collections of these species and wait until we can confirm that they are threatened or that they have become extinct? The IUCN Standards and Petitions Working Group (2008) now suggests that these species will be considered case by case after their status has been widely reviewed and accepted by experts. Because case studies and individual evaluations are time consuming, this process is slow. I would recommend that these presumably endangered species be quickly accepted in a preliminary category (e.g., DD) until more information is available and their final conservation status can be determined.

9.3.3 CITES

Species that are considered overexploited are protected by CITES, the Convention on International Trade in Endangered Species of Wild Fauna and Flora, which was enacted July 1975 with 80 participating countries. Currently, 174 member countries have signed CITES. All 28 417 plant species listed in the three appendices (CITES, 2008) require permission for their international trade, but not for their local commerce. Tree ferns (all Cyatheaceae as a family, four species of Dicksoniaceae from the New World and *Cibotium barometz* from Southeast Asia) are the fern species protected by CITES. In contrast, the global IUCN (2008) Red List considers only seven tree fern species as endangered, a clear underestimation of the threat to tree ferns, because only 9 of about 600 species have been evaluated so far.

Living tree ferns are traded infrequently as ornamental plants, but their trunks, when covered by a thick root mantle, gained popularity as a substrate for cultivation of epiphytes and especially orchids in the 1970s. CITES first restricted the trade of tree fern trunks in 1977. After protection of living tree ferns by CITES in 2000, international interest in tree fern products decreased and started to focus on sustainable alternative materials (e.g., cocos fiber). Demand and exploitation continued in Tasmania with the native *Dicksonia antarctica* (not included in CITES) until the Australian government regulated the sustainable harvest, transport and trading in a tree fern management plan (Department of the Environment, Water, Heritage and the Arts, 2007) that still allows some national and international trade of the Australian tree ferns. Because most ferns of horticultural interest can be grown easily from spores or from sterile tissue culture, and both are exempted from CITES for international trade, there should be no need for intensive extraction of plants from their natural habitats. However, some system of accountability for commercial fern growers, especially in tropical countries, may be required to avoid nursery sales of plants collected in the wild rather than cultivated in the greenhouse.

9.4 Management strategies for fern conservation

Fern species may be protected within their native range by conservation of their natural habitat (*in situ*) or by cultivation of individual species in botanical gardens or other horticultural institutions either by growing them from spores, vegetative propagation or tissue culture (*ex situ*). I will first discuss *ex situ* conservation, because its various methodologies increase our understanding of species reproduction, although it should never be considered as a substitute for *in situ* conservation.

9.4.1 Ex situ *conservation*

Ecological field knowledge is very limited for most endangered species, therefore *ex situ* cultivation should be used in concert with *in situ* habitat protection (Sutherland *et al.*, 2002). Rare species may still maintain a soil spore bank (viable spores that are present in the upper soil) at their natural sites (see Chapter 3), making possible the collection of viable spores to subsequently induce germination in the laboratory (Dyer, 1994). About 50% of fern species of Great Britain produce spore banks (Dyer and Lindsay, 1992, 1996). Even in tropical environments, some species form spore banks. Ranal (2003) found a dozen fern species successfully growing from soil spore banks in Brazil. Modern propagation techniques aim for species conservation in less space and use cryopreservation (storage in liquid nitrogen at $-196\,°C$) of spores (Ballesteros *et al.*, 2006), gametophytes and sporophytes or in vitro cultures (living plant tissues kept in glass tubes under sterile conditions) at temperatures of $15–30\,°C$ depending on the species (Pence, 2002, 2008). Spore viability varies among species from days to decades, mostly depending on air humidity and temperature. The longer spores are stored, the fewer germinate (and at increasingly slower rates). Comparative studies between wet and dry storage yielded contradictory results and differed among species. Wet storage worked better at temperatures above $5\,°C$ and dry storage below $5\,°C$, the latter also reducing contamination (Quintanilla *et al.*, 2002). In general, green spores lose germination capability considerably faster than nongreen spores. Several research institutions and botanical gardens worldwide have successfully propagated many fern species from in vitro cultures, from spores, gametophytes or meristematic tissues of sporophytes such as from the root tip or rhizome apex (Fernández and Revilla, 2003), yielding sporophytes within 3–4 months (e.g., *Drynaria quercifolia*, oak leaf fern, Hegde and D'Souza, 1996). Tissue culture is the most efficient method to satisfy the increasing demand for ornamental ferns without damaging natural populations and provides a modern tool for the conservation of endangered fern species. However, *in situ* management strategies should always be considered as the first choice to prevent fern species from becoming endangered in the wild, rather than trying to avoid species loss at the last moment.

9.4.2 In situ *management strategies*

In situ management strategies comprise habitat protection (e.g., cloud forests) and species protection (e.g., Red List species). Habitat protection should be always given priority when ferns are concentrated in hot spots (e.g., tropical mountainous islands), and species protection may be chosen for species with narrow ranges that do not correspond with these hot spots. There are few examples for *in situ* management of endangered fern populations, although in Singapore *Dipteris conjugata* stands are controlled and freed of secondary overgrowth by other ferns (e.g., Gleicheniaceae) in order to guarantee optimum growth conditions (Chan, 2002).

Once an endangered species has been successfully cultivated *ex situ*, several reintroductory steps can be attempted. Rae (1999) differentiated between reinforcement (i.e., returning *ex situ* propagated plants to their original site), reintroduction (i.e., bringing cultivated plants back to historic sites where the species has become extinct) and introduction (i.e., introducing plants into new sites in protected areas with similar habitat conditions, but still within the geographic range of the species). Because reintroduction projects are expensive and their success cannot be guaranteed, habitat conservation should be always attempted first to avoid local extinction in the wild (Rae, 1999).

All types of reintroduction projects need careful planning, as illustrated by the management program for the rare British fern *Woodsia ilvensis* (rusty cliff fern; Lusby *et al.*, 2002). Reintroduced plant material may cross-fertilize and genetically impact nearby natural populations by contributing less-adapted genes. For this reason, Lusby *et al.* (2002) selected spore material from different locations to raise plants for reintroduction. If the new sites were close to a natural population (e.g., 6 km), plants were raised from spores of this same population. If the reintroduction site was far from natural populations (e.g., 70 km), they introduced plants grown from spores of several different origins in order to guarantee the highest possible genetic diversity. Reintroduction sites were selected using historical plant records, and plants were translocated to microhabitats inaccessible to grazing animals during spring and autumn, when climatic conditions were most favorable. Plants were watered artificially during the first summer and monitored at least twice a year. After one year, 90% of the plants had survived and some had even produced spores (Lusby *et al.*, 2002). Long-term monitoring is essential to improve future reintroductions and to ensure self-sustaining populations. In Poland, Zenkteler (2002) had to mix spores from three distant populations of *Osmunda regalis* (royal fern) to produce heterozygous sporophytes because the genetic load of single populations resulted in lethal homozygous individuals. Both studies found that introducing larger plants resulted in higher survival rates than smaller plants. These two examples illustrate how complex reintroduction projects may become and that each species may require a unique approach.

9.5 Conclusions

The successful conservation of ferns and lycophytes requires a number of actions such as improving knowledge of their abundance and distribution patterns and conducting field studies to investigate habitat specificity, population dynamics, genetic diversity and any intrinsic factors or environmental disturbances that affect their survival. This basic information for fern species requires an increase in the number of studies conducted at larger spatial and temporal scales (Franklin *et al.*, 1990). Better ecological information, combined with faster risk assessments of ferns and lycophytes at regional and global levels, will facilitate production of essential risk management plans. Contemporary *in situ* plant conservation strategies should always focus on a mixed approach of species and habitat protection. Habitat protection may refer to microhabitats for specialized species (e.g., epiphytes specific to tree ferns) or hot spots for fern diversity such as mid-elevation cloud forests and tropical oceanic islands that are subject to heavy destruction. The propagation of endangered ferns in tissue culture is an additional tool for conservation that also can satisfy the demand for ornamental ferns without damaging natural populations.

Acknowledgements

I thank Tom Ranker, Joanne M. Sharpe, Alan R. Smith and Lawrence R. Walker for corrections and helpful comments on the manuscript. I also acknowledge support from the Instituto de Ecología, A.C., Xalapa, Veracruz, Mexico, CONACYT-SEMARNAT (2002-C01–0194) and CONACYT-SEP (2003-C02–43082).

References

Ambrosio, S. T. and De Melo, N. F. (2001). New records of pteridophytes in the semi-arid region of Brazil. *American Fern Journal*, **91**, 227–9.
Arcand, N. N. and Ranker, T. (2008). Conservation Biology. In *Biology and Evolution of Ferns and Lycophytes*, ed. T. A. Ranker and C. H. Haufler. Cambridge, UK: Cambridge University Press, pp. 257–83.
Ayensu, E. S. (1981). Assessment of threatened plant species in the United States. In *The Biological Aspects of Rare Plant Conservation*, ed. H. Synge. New York: John Wiley, pp. 19–58.

Bach, K., Kessler, M. and Gradstein, S. R. (2007). A simulation approach to determine statistical significance of species turnover peaks in a species-rich tropical cloud forest. *Diversity and Distributions*, **13**, 863–70.

Baillie, J. E. M., Hilton-Taylor, C. and Stuart, S. N. (eds.) (2004). *2004 IUCN Red List of Threatened Species. A Global Species Assessment*. Gland, Switzerland, and Cambridge, UK: IUCN.

Ballesteros, D., Estrelles, E. and Ibars, A. M. (2006). Responses of pteridophyte spores to ultrafreezing temperatures for long-term conservation in germplasm banks. *Fern Gazette*, **17**, 293–302.

Bannister, P. and Wildish, K. L. (1982). Light compensation points and specific leaf areas in some New Zealand ferns. *New Zealand Journal of Botany*, **20**, 421–4.

Benemann, J. (1996). Freilanduntersuchungen zum Gaswechsel ausgewählter gefährdeter Farne. Unpublished Diploma thesis, Ruhr-Universität Bochum, Germany.

Bennert, H. W. (1999). *Die seltenen und gefährdeten Farnpflanzen Deutschlands. Biologie, Verbeitung, Schutz*. Bonn: Bundesamt für Naturschutz.

Bennert, H. W. and Danzebrink, B. (1996). Spore germination of endangered pteridophytes in Germany. *Verhandlungen der Gesellschaft für Ökologie*, **26**, 197–207.

Beukema, H. and van Noordwijk, M. (2004). Terrestrial pteridophytes as indicators of a forest-like environment in rubber production systems in the lowlands of Jambi, Sumatra. *Agriculture, Ecosystems and Environment*, **104**, 63–73.

Bickford, S. A. and Laffan, S. W. (2006). Multi-extent analysis of the relationship between pteridophyte species richness and climate. *Global Ecology and Biogeography*, **15**, 588–601.

Boutin, C., Jobin, B. and Belanger, L. (2003). Importance of riparian habitats to flora conservation in farming landscapes of southern Quebec, Canada. *Agriculture, Ecosystems and Environment*, **94**, 73–87.

Brigham, C. A. and Schwartz, M. A. (eds.) (2003). *Population Viability in Plants: Conservation, Management and Modeling of Rare Plants*. Ecological Studies, 165. Berlin: Springer-Verlag.

Brodribb, T. J. and Holbrook, N. M. (2004). Stomatal protection against hydraulic failure: a comparison of coexisting ferns and angiosperms. *New Phytologist*, **162**, 663–70.

Brodribb, T. J., Holbrook, N. M., Zwienickii, M. A. and Palma, B. (2005). Leaf hydraulic capacity in ferns, conifers and angiosperms: impacts on photosynthetic maxima. *New Phytologist*, **165**, 839–46.

Brodribb, T. J., Field, T. S. and Jordan, G. J. (2007). Leaf maximum photosynthetic rate and venation are linked by hydraulics. *Plant Physiology*, **144**, 1890–8.

Camus, J. M. and Pryor, K. V. (1996). The pteridophyte flora of Sulawesi, Indonesia. In *Pteridology in Perspective*, ed. J. M. Camus, M. Gibby and R. J. Johns. Kew, UK: Royal Botanic Gardens, pp. 169–170.

Chan, L. (2002). Plant conservation initiatives in Singapore. In *Plant Conservation in the Tropics: Perspectives and Practice*, ed. M. Maunder, C. Clubbe, C. Hankamer and M. Groves. Kew, UK: Royal Botanic Gardens, pp. 171–83.

Cheffings, C. M. and Farrell, L. (eds.) (2005). *The Vascular Plant Red Data List for Great Britain*. Species Status, 7. Peterborough: Joint Nature Conservation Committee.

China Plant Specialist Group (2004). *Adiantum lianxianense*. In *2008 IUCN Red List of Threatened Species*, ed. IUCN. www.iucnredlist.org. Viewed January 2009.

Chiou, W. L., Huang, Y. M. and Chen, C. M. (2006). Conservation of two endangered ferns, *Archangiopteris somai* and *A. itoi* (Marattiaceae, Pteridophyta), by propagation from stipules. *Fern Gazette*, **17**, 271–78.

CITES (2008). *Convention on International Trade in Endangered Species of Wild Fauna and Flora.* www.cites.org. Viewed January 2009.

Colling, G. (2005). Red list of the vascular plants of Luxembourg. *Ferrantia*, **42**, 1–77.

Cronk, Q. C. B. (1992). Relict floras of Atlantic islands: patterns assessed. *Biological Journal of the Linnean Society*, **46**, 91–103.

De Winter, W. P. and Amoroso, V. B. (eds.) (2003). *Cryptogams: Ferns and Fern Allies*, Plant Resources of South-East Asia, 15. Leiden, The Netherlands: Backhuys.

Department of the Environment, Water, Heritage and the Arts (2007). *Tree Fern Management Plan for the Sustainable Harvesting, Transporting or Trading of Dicksonia antarctica in Tasmania.* Australian Government, Canberra. www.environment.gov.au/biodiversity/trade-use/sources/management-plans/treefern-tas/pubs/treefern-tas.pdf. Viewed January 2009.

Department of the Environment, Water, Heritage and the Arts (2009). *EPBC Act List of Threatened Flora.* Australian Government, Canberra. www.environment.gov.au/cgi-bin/sprat/public/sprat.pl. Viewed January 2009.

Diamond, J. M. (1975). The island dilemma: lessons of modern biogeographic studies for the design of natural preserves. *Biological Conservation*, **7**, 129–46.

Diamond, J. M. (1986). The design of a nature reserve system for Indonesian New Guinea. In *Conservation Biology: the Science of Scarcity and Diversity*, ed. M. Soulé. Sunderland, MA, USA: Sinauer Associates.

Drake, D. R. and Pratt, L. W. (2001). Seedling mortality in Hawaiian rain forest: the role of small-scale physical disturbance. *Biotropica*, **33**, 319–23.

Durand, L. Z. and Goldstein, G. (2001). Photosynthesis, photoinhibition, and nitrogen use efficiency in native and invasive tree ferns in Hawaii. *Oecologia*, **126**, 345–54.

Dyer, A. F. (1994). Natural soil spore banks: can they be used to retrieve lost ferns? *Biodiversity and Conservation*, **3**, 160–75.

Dyer, A. F. and Lindsay, S. (1992). Soil spore banks of temperate ferns. *American Fern Journal*, **82**, 89–123.

Dyer, A. F. and Lindsay, S. (1996). Soil spore banks: a new resource for conservation. In *Pteridology in Perspective*, ed. J. M. Camus, M. Gibby and R. J. Johns. Kew, UK: Royal Botanic Gardens, pp. 153–60.

Ehrlich, P. and Ehrlich, A. (1981). *Extinction.* New York: Random House.

Fernández, H. and Revilla, M. A. (2003). *In vitro* culture of ornamental ferns. *Plant Cell, Tissue and Organ Culture*, **73**, 1–13.

Fernández, R. and Vail, R. (2003). New records for *Platycerium andinum* Baker in Peru. *American Fern Journal*, **93**, 160–3.

Figueiredo, E. and Gascoigne, A. (2001). Conservation of pteridophytes in São Tomé e Príncipe (Gulf of Guinea). *Biodiversity and Conservation*, **10**, 45–68.

Flora of Australia Online (2009). *Australian Biological Resources Study, Canberra.* www.environment.gov.au/biodiversity/abrs/online-resources/flora/main/index.html. Viewed January 2009.

Franklin, J. F., Bledsoe, C. S. and Callahan, J. T. (1990). Contributions of the Long-Term Ecological Research Program. *BioScience*, **40**, 509–23.

Gardette, E. (1996). Microhabitats of epiphytic fern communities in large lowland rain forest plots in Sumatra. In *Pteridology in Perspective*, ed. J. M. Camus, M. Gibby and R. J. Johns. Kew, UK: Royal Botanic Gardens, pp. 655–8.

Giers, A., Heiser, T. and Bennert, H. W. (1996). Freilanduntersuchungen zum Gaswechsel einheimischer Farnarten. *Verhandlungen der Gesellschaft für Ökologie*, **25**, 277–85.

Gilman, A. V. (2004). *Botrychium pallidum* newly discovered in Maine. *American Fern Journal*, **94**, 155–6.

Given, D. R. (1993). Changing aspects of endemism and endangerment in Pteridophyta. *Journal of Biogeography*, **20**, 293–302.

Given, D. R. (1994). *Principles and Practice of Plant Conservation*. Portland, OR, USA: Timber Press.

Gliessman, S. R. (1978). The establishment of bracken following fire in tropical habitats. *American Fern Journal*, **68**, 41–4.

Greuter, W. (1995). Extinctions in Mediterranean areas. In *Extinction Rates*, ed. J. H. Lawton and R. M. May. Oxford, UK: Oxford University Press, pp. 88–97.

Hanski, I., Kouki, J. and Halkka, A. (1993). Three explanations of the positive relationship between distribution and abundance of species. In *Species Diversity in Ecological Communities: Historical and Geographical Perspectives*, ed. R. Ricklefs and S. Schluter. Chicago, IL, USA: University of Chicago Press, pp. 108–16.

Hartley, S. and Kunin, W. A. (2003). Scale dependency of rarity, extinction risk, and conservation priority. *Conservation Biology*, **17**, 1559–70.

Hegde, S. and D'Souza, L. (1996). *In vitro* propagation of *Drynaria quercifolia*, an endangered fern. In *Pteridology in Perspective*, ed. J. M. Camus, M. Gibby and R. J. Johns. Kew, UK: Royal Botanic Gardens, pp. 171.

Heiser, T., Giers, A. and Bennert, H. W. (1996). *In situ* gas exchange measurements and adaptations to the light regime in three species of *Lycopodium*. In *Pteridology in Perspective*, ed. J. M. Camus, M. Gibby and R. J. Johns. Kew, UK: Royal Botanic Gardens, pp. 599–610.

Hietz, P. and Briones, O. (2001). Photosynthesis, chlorophyll fluorescence and within-canopy distribution of epiphytic ferns in a Mexican cloud forest. *Plant Biology*, **3**, 279–87.

Horn, S. P. (1988). Effect of burning on a montane mire in the Cordillera de Talamanca, Costa Rica. *Brenesia*, **30**, 81–92.

Horn, S. P. (1989). Post-fire vegetation development in the Costa Rica páramos. *Madroño*, **36**, 93–114.

Hoshizaki, B. J. and Moran, R. C. (2001). *The Fern Grower's Manual*, revised and expanded edition. Portland, OR, USA: Timber Press.

Hsu, T.-W., Moore, S.-J. and Chiang, T.-Y. (2000). Low RAPD polymorphism in *Archangiopteris itoi*, a rare and endemic fern in Taiwan. *Botanical Bulletin of Academia Sinica*, **41**, 15–8.

Hubbell, S. P. (2001). *The Unified Neutral Theory of Biodiversity and Biogeography*. Princeton, NJ, USA: Princeton University Press.

Hunt, M. A., Davidson, N. J., Unwin, G. L. and Close, D. C. (2002). Ecophysiology of the Soft Tree Fern *Dicksonia antarctica* Labill. *Austral Ecology*, **27**, 360–8.

Hutchinson, G. E. and MacArthur, R. H. (1959). A theoretical ecological model of size distributions among species of animals. *American Naturalist*, **93**, 117–25.

IUCN (2001). *IUCN Red List Categories and Criteria: Version 3.1*. Gland, Switzerland and Cambridge, UK: IUCN Species Survival Commission.

IUCN (2003). *Guidelines for Application of IUCN Red List Criteria at Regional Levels: Version 3.0*. Gland, Switzerland and Cambridge, UK: IUCN Species Survival Commission.

IUCN (2008). *2008 IUCN Red List of Threatened Species*. www.iucnredlist.org. Viewed January 2009.

IUCN Standards and Petitions Working Group (2008). *Guidelines for Using the IUCN Red List Categories and Criteria. Version 7.0*. Prepared by the Standards and Petitions Working Group of the IUCN SSC Biodiversity Assessments Sub-Committee in August 2008. intranet.iucn.org/webfiles/doc/SSC/RedList/RedListGuidelines.pdf. Viewed January 2009.

Jablonski, D. (1995). Extinctions in the fossil record. In *Extinction Rates*, ed. J. H. Lawton and R. M. May. Oxford, UK: Oxford University Press, pp. 25–44.

Jamison, J. A. and Douglas, G. W. (1998). Status of the coastal wood fern, *Dryopteris arguta* (Dryopteridaceae) in Canada. *Canadian Field-Naturalist*, **112**, 284–8.

J-IBIS (1997). *Japan Integrated Biodiversity Information System*. Ministry of the Environment. www.biodic.go.jp/english/rdb/rdb_f. Viewed January 2009.

Kang, M., Pan, L., Yao, X. and Huang, H. (2006). Development and characterization of polymorphic microsatellite loci in endangered fern *Adiantum reniforme* var. *sinense*. *Conservation Genetics*, **7**, 807–10.

Kessler, M., Betz, L. and Roedde, S. (2006). *Red List of the Pteridophytes of Bolivia*. www. fernsofbolivia.uni-goettingen.de. Viewed January 2009.

Kingston, N. and Waldren, S. (2002). A conservation assessment of the pteridophyte flora of the Pitcairn Islands. *Fern Gazette*, **16**, 404–10.

Kornás, J. (1977). Life-forms and seasonal patterns in the pteridophytes in Zambia. *Acta Societatis Botanicorum Poloniae*, **46**, 668–90.

Kornás, J. (1978). Fire-resistance in the pteridophytes of Zambia. *Fern Gazette*, **11**, 373–84.

Korneck, D., Schnittler, M. and Vollmer, I. (1996). Rote Liste der Farn- und Blütenpflanzen (Pteridophyta et Spermatophyta) Deutschlands. *Schriftenreihe für Vegetationskunde*, **28**, 21–187.

Lawton, J. H. (1995). Population dynamic principles. In *Extinction Rates*, ed. J. H. Lawton and R. M. May. Oxford, UK: Oxford University Press, pp. 147–63.

León, B. (1990). New localities for *Acrostichum danaeifolium* in Peru. *American Fern Journal*, **80**, 113–14.

Lesica, P. and Allendorf, F. W. (1995). When are peripheral populations valuable for conservation? *Conservation Biology*, **9**, 753–60.

Lloyd, R. M. and Klekowski, E. J. (1970). Spore germination and viability in Pteridophyta: evolutionary significance of chlorophyllous spores. *Biotropica*, **2**, 129–37.

Loriot, S., Magnanon, S. and Deslandes, E. (2006). *Trichomanes speciosum* (Hymenophyllaceae: Pteridophyta) in northwestern France. *Fern Gazette*, **17**, 333–49.

Lusby, P., Lindsay, S. and Dyer, A. F. (2002). Principles, practice and problems of conserving the rare British fern *Woodsia ilvensis* (L.) R. Br. *Fern Gazette*, **16**, 350–5.

May, R. M., Lawton, J. H. and Stork, N. E. (1995). Assessing extinction rates. In *Extinction Rates*, ed. J. H. Lawton and R. M. May. Oxford, UK: Oxford University Press, pp. 1–24.

Mehltreter, K. (1995). Species richness and geographical distribution of montane pteridophytes of Costa Rica, Central America. *Feddes Repertorium*, **106**, 563–84.

Mehltreter, K. (2008a). Helechos. In *Agroecosistemas Cafetaleros de Veracruz: Biodiversidad, Manejo y Conservación*, ed. R. H. Manson, V. Hernández-Ortiz, S. Gallina and K. Mehltreter. Mexico City: Instituto de Ecología, A. C. (INECOL) e Instituto Nacional de Ecología (INE), pp. 83–93.

Mehltreter, K. (2008b). Phenology and habitat specificity in tropical ferns. In *Biology and Evolution of Ferns and Lycophytes*, ed. T. A. Ranker and C. H. Haufler. Cambridge, UK: Cambridge University Press, pp. 201–21.

Mehltreter, K. and García-Franco, J. G. (2008). Leaf phenology and trunk growth of the deciduous tree fern *Alsophila firma* (Baker) D.S. Conant in a lower montane Mexican forest. *American Fern Journal*, **98**, 1–13.

Mehltreter, K., Flores-Palacios, A. and García-Franco, J. G. (2005). Host preferences of vascular trunk epiphytes in a cloud forest of Veracruz, México. *Journal of Tropical Ecology*, **21**, 651–60.

Mickel, J. T. and Smith, A. R. (2004). *The Pteridophytes of Mexico*. New York: The New York Botanical Garden.

Mittermeier, R. A., Gil, P. R., Hoffmann, M., *et al.* (2005). *Hotspots Revisited: Earth's Biologically Richest and Most Endangered Terrestrial Ecoregions*. Mexico: CEMEX.

Moran, R. C. and Riba, R. (eds.) (1995). *Psilotaceae a Salviniaceae*, Vol. 1. of *Flora Mesoamericana*, ed. G. Davidse, M. S. Souza and S. Knapp. Ciudad Universitaria, Mexico: Universidad Nacional Autónoma de México.

Myers, N. (1988). Threatened biotas: 'hotspots' in tropical forests. *The Environmentalist*, **8**, 1–20.

Myers, N., Mittermeier, R. A., Mittermeier, C. G., da Fonseca, G. A. B. and Kent, J. (2000). Biodiversity hotspots for conservation priorities. *Nature*, **403**, 853–8.

Olsen, S. (2007). *Encyclopedia of Garden Ferns*. Portland, OR, USA: Timber Press.

Ough, K. (2001). Regeneration of wet forest flora a decade after clear-felling or wildfire – is there a difference? *Australian Journal of Botany*, **49**, 645–64.

Ough, K. and Murphy, A. (2004). Decline in tree-fern abundance after clearfell harvesting. *Forest Ecology and Management*, **199**, 153–63.

Paciencia, M. L. B. and Prado, J. (2005). Effects of forest fragmentation on pteridophyte diversity in a tropical rain forest in Brazil. *Plant Ecology*, **180**, 87–104.

Pence, V. C. (2002). Cryopreservation and *in vitro* methods for *ex situ* conservation of pteridophytes. *Fern Gazette*, **16**, 362–8.

Pence, V. C. (2008). *Ex situ* conservation of ferns and lycophytes: approaches and techniques. In *Biology and Evolution of Ferns and Lycophytes*, ed. T. A. Ranker and C. H. Haufler. Cambridge, UK: Cambridge University Press, pp. 284–300.

Pimm, S. L. (1991). *The Balance of Nature? Ecological Issues in the Conservation of Species and Communities*. Chicago, IL, USA: The University of Chicago Press.

Pimm, S. L. and Raven, P. (2000). Extinction by numbers. *Nature*, **403**, 843–5.

Prance, G. T. (2001). Discovering the plant world. *Taxon*, **50**, 345–59.

Quintanilla, L. G., Amigo, J., Pangua, E. and Pajarón, S. (2002). Effect of storage method on spore viability in five globally threatened fern species. *Annals of Botany*, **90**, 461–7.

Rabinowitz, D. (1981). Seven forms of rarity. In *The Biological Aspects of Rare Plant Conservation*, ed. H. Synge. New York: John Wiley, pp. 205–17.

Rae, D. (1999). Conservation of natural populations. In *A Color Atlas of Plant Propagation and Conservation*, ed. B. G. Bowes. New York: The New York Botanical Garden Press, pp. 177–86.

Ranal, M. A. (2003). Soil spore bank of ferns in a gallery forest of the ecological station of Panga, Uberlândia, MG, Brazil. *American Fern Journal*, **93**, 97–115.

Ranker, T. A., Trapp, P. G., Smith, A. R., Moran, R. C. and Parris, B. S. (2005). New records of lycophytes and ferns from Moorea, French Polynesia. *American Fern Journal*, **95**, 126–7.

Rapoport, E. H. (1982). *Aerography: Geographical Strategies of Species*. Oxford, UK: Pergamon Press.

Ratcliffe, D. A., Birks, J. B. and Birks, H. H. (1993). The ecology and conservation of the Killarney fern *Trichomanes speciosum* Willd. in Britain and Ireland. *Biological Conservation*, **66**, 231–47.

Raup, D. M. (1986). Biological extinction in Earth history. *Science*, **231**, 1528–33.

Roos, M. (1996). Mapping the world's pteridophyte diversity: systematics and floras. In *Pteridology in Perspective*, ed. J. M. Camus, M. Gibby and R. J. Johns. Kew, UK: Royal Botanic Gardens, pp. 29–42.

Rosenzweig, M. L. (1995). *Species Diversity in Space and Time*. Cambridge, UK: Cambridge University Press.

Rothwell, G. W. and Stockey, R. A. (2008). Phylogeny and evolution of ferns: a paleontological perspective. In *Biology and Evolution of Ferns and Lycophytes*, ed. T. A. Ranker and C. H. Haufler. Cambridge, UK: Cambridge University Press, pp. 332–66.

Rumsey, F. J., Russell, S. J., Ji, J., Barrett, J. A. and Gibby, M. (1996). Genetic variation in the endangered filmy fern *Trichomanes speciosum*. In *Pteridology in Perspective*, ed. J. M. Camus, M. Gibby and R. J. Johns. Kew, UK: Royal Botanic Gardens, pp. 161–5.

Rumsey, F. J., Jermy, A. C. and Sheffield, E. (1998). The independent gametophyte stage of *Trichomanes speciosum* Willd. (Hymenophyllaceae) the Killarney fern and its distribution on the British Isles. *Watsonia*, **22**, 1–19.

Rumsey, F. J., Vogel, J. C., Russell, S. J., Barrett, J. A. and Gibby, M. (1999). Population structure and conservation biology of the endangered fern *Trichomanes speciosum* Willd. (Hymenophyllaceae) at its northern distributional limit. *Biological Journal of the Linnean Society*, **66**, 333–44.

Russell, A. E., Raich, J. W. and Vitousek, P. M. (1998). The ecology of the climbing fern *Dicranopteris linearis* on windward Mauna Loa, Hawaii. *Journal of Ecology*, **86**, 765–79.

Sanford, W. W. (1968). Distribution of epiphytic orchids in semideciduous tropical forest in southern Nigeria. *Journal of Ecology*, **56**, 697–705.

Sato, T. (1990). Estimation of chronological age for sporophyte maturation in three semi-evergreen ferns in Hokkaido. *Ecological Research*, **5**, 55–62.

Schelpe, E. A. C. L. E. and Anthony, N. C. (1986). Pteridophyta. In *Flora of Southern Africa*, ed. O. A. Leistner. Pretoria, Republic of South Africa: Botanical Research Institute.

Schneider, H., Schuettpelz, E., Pryer, K. M., Cranfill, R., Magallón, S. and Lupia, R. (2004). Ferns diversified in the shadow of angiosperms. *Nature*, **428**, 553–7.

Schneller, J. J. (2008). Antheridiogens. In *Biology and Evolution of Ferns and Lycophytes*, ed. T. A. Ranker and C. H. Haufler. Cambridge, UK: Cambridge University Press, pp. 134–58.

Schneller, J. J. and Holderegger, R. (1996). Genetic variation in small, isolated fern populations. *Journal of Vegetation Science*, **7**, 113–20.

Sepkoski, J. J., Jr. (1992). Phylogenetic and ecologic patterns in the Phanerozoic history of marine biodiversity. In *Systematics, Ecology and the Biodiversity Crisis*, ed. N. Eldridge. New York: Columbia University Press, pp. 77–100.

Sharpe, J. and Jernstedt, J. A. (1991). Stipular bud development in *Danaea wendlandii* (Marattiaceae). *American Fern Journal*, **81**, 119–27.

Slocum, M. G., Aide, T. M., Zimmerman, J. K. and Navarro, L. (2004). Natural regeneration of subtropical montane forest after clearing fern thickets in the Dominican Republic. *Journal of Tropical Ecology*, **20**, 483–6.

Smith, A. R., Kessler, M. and Gonzales, J. (1999). New records of pteridophytes from Bolivia. *American Fern Journal*, **89**, 244–66.

Smith, A. R., Pryer, K. M., Schuettpelz, E., *et al.* (2006). A classification for extant ferns. *Taxon*, **55**, 705–31.

Soltis, D. E. and Soltis, P. S. (1986). Electrophoretic evidence for inbreeding in the fern *Botrychium virginianum* (Ophioglossaceae). *American Journal of Botany*, **73**, 248–56.

Soltis, D. E. and Soltis, P. S. (1989). Polyploidy, breeding systems, and genetic differentiation in homosporous pteridophytes. In *Isozymes in Plant Biology*, ed. D. E. Soltis and P. S. Soltis. Portland, OR, USA: Dioscorides Press, pp. 241–58.

Stace, C. (1997). *New Flora of the British Isles*. Cambridge, UK: Cambridge University Press.

Stevens, G. C. (1989). The latitudinal gradient in geographical range: how so many species coexist in the tropics. *American Naturalist*, **133**, 240–56.

Sukopp, H. and Trautmann, W. (1981). Causes of the decline of threatened plants in the Federal Republic of Germany. In *The Biological Aspects of Rare Plant Conservation*, ed. H. Synge. New York: John Wiley, pp. 113–16.

Sutherland, T., Blackmore, P., Ndam, N. and Nkefor, J. (2002). Conservation through cultivation: the work of the Limbe Botanic Garden, Cameroon. In *Plant Conservation in the Tropics: Perspectives and Practice*, ed. M. Maunder, C. Clubbe, C. Hankamer and M. Groves. Kew, UK: Royal Botanic Gardens, pp. 395–419.

Tryon, R. M. (1970). Development and evolution of fern floras of oceanic islands. *Biotropica*, **2**, 76–84.

Tryon, R. M. (1989). Pteridophytes. In *Tropical Rain Forest Ecosystems: Biogeographical and Ecological Studies*. Vol. 14B of *Ecosystems of the World*, ed. H. Lieth and M. J. A. Werger. Amsterdam, The Netherlands: Elsevier, pp. 327–38.

Tuomisto, H. and Poulsen, A. D. (1996). Influence of edaphic specialization on pteridophyte distribution in neotropical rain forests. *Journal of Biogeography*, **23**, 283–93.

Unwin, G. L. and Hunt, M. A. (1996). Conservation and management of soft tree fern *Dicksonia antarctica* in relation to commercial forestry and horticulture. In *Pteridology in Perspective*, ed. J. M. Camus, M. Gibby and R. J. Johns. Kew, UK: Royal Botanic Gardens, pp. 125–37.

Valentine, J. W., Yiffney, B. H. and Sepkowski, J. J., Jr. (1991). Evolutionary dynamics of plants and animals: a comparative approach. *Palaios*, **6**, 81–8.

Vega, E., Caudales, R. and Sánchez-Pérez, A. (1997). New records of pteridophytes for Botswana. *American Fern Journal*, **87**, 127–30.

Vöge, M. (2004). Non-destructive assessing and monitoring of populations of *Isoëtes lacustris* L. *Limnologica,* **34**, 147–53.

Vogel, J. C., Jessen, S., Gibby, M., Jermy, A. C. and Ellis, L. (1993). Gametophytes of *Trichomanes speciosum* (Hymenophyllaceae: Pteridophyta) in Central Europe. *Fern Gazette*, **14**, 227–32.

Wagner, W. H., Jr. and Smith, A. R. (1993). Pteridophytes of North America. In *Introduction*, Vol. 1 of *Flora of North America North of Mexico*, ed. Flora of North America Editorial Committee. New York and Oxford, UK: Oxford University Press, pp. 247–66.

Walker, L. R. (1994). Effects of fern thickets on woodland development on landslides in Puerto Rico. *Journal of Vegetation Science*, **5**, 525–32.

Walker, L. R. and Boneta, W. (1995). Plant and soil responses to fire on a fern-covered landslide in Puerto Rico. *Journal of Tropical Ecology*, **11**, 473–9.

Whittaker, R. J. (1998). *Island Biogeography: Ecology, Evolution and Conservation*. Oxford, UK: Oxford University Press.

Windham, M. D. (1987). *Argyrochosma*, a new genus of cheilanthoid ferns. *American Fern Journal*, **77**, 37–41.

Woodmansee, S. W. and Sadle, J. L. (2005). New occurrences of *Schizaea pennula* Sw. in Florida. *American Fern Journal*, **95**, 84–7.

Zenkteler, E. K. (2002). *Ex situ* breeding and reintroduction of *Osmunda regalis* L. in Poland. *Fern Gazette*, **16**, 371–6.

10

Current and future directions in fern ecology

LAWRENCE R. WALKER, KLAUS MEHLTRETER
AND JOANNE M. SHARPE

10.1 Introduction

The ecology of ferns is a rapidly growing discipline that offers new and exciting insights into general ecological principles and applications. Progress has been made in studying fern biogeography, population dynamics, natural resource use, disturbance responses, species interactions and links with humans (Table 10.1). In this concluding chapter, we explore the lessons learned about each of these topics and how they clarify the ecological role of ferns. We then raise some unanswered questions that might become the foci for future research on fern ecology and improve the integration of ferns into general studies of ecology.

Ferns (and lycophytes) differ from seed plants in fundamental ways. Ferns have a different evolutionary background, phenology, nutrient acquisition patterns, adaptations to xeric environments, responses to disturbance, interactions with fungi and animals and invasion patterns that provide an excellent contrast to seed plants. However, ferns also share fundamental similarities with seed plants, especially herbaceous perennials. Ferns have similar physiological pathways of energy capture and nutrient distribution and share some common traits such as colonization abilities, habitat specificity, leaf function, growth patterns, vegetative propagation, population dynamics, species interactions (e.g., shading) and mycorrhizal infection. Examining differences and similarities between ferns and seed plants is one useful approach to the rapidly expanding field of fern ecology.

10.2 Biogeography: dispersal, habitats and diversity

Ferns have only one potentially long distance dispersal phase (spores) while seed plants have two (pollen and seeds). The ecological implications of these contrasting dispersal modes are unclear. The one-step dispersal of ferns might suggest fewer dispersal bottlenecks than for seed plants with two steps. In addition, the ancestors

Fern Ecology, ed. Klaus Mehltreter, Lawrence R. Walker and Joanne M. Sharpe. Published by Cambridge University Press. © Cambridge University Press 2010.

Table 10.1 *Selected topics in fern ecology with examples of current research and possible areas for expansion of research*

Topic	Subtopic	Current research	Areas for research expansion	Chapters	Representative citations
Evolution	Phylogeography	Clade maintenance	Population gene flow	1, 2, 9	Kato, 1993; Wolf et al., 2001
Biogeography	Dispersal	Richness patterns	Beta diversity	1, 2, 6, 9	Smith et al., 2006; Soria-Auza and Kessler, 2008
	Habitats	Description	Adaptive mechanisms	1, 2, 5, 9	Page, 1979; Tryon and Tryon, 1982
	Diversity	Links to site fertility	Global documentation	2, 4	Tuomisto and Poulsen, 2000; Hemp, 2006
Populations	Growth	Belowground structures	Tropical ferns	1, 3, 6	Greer and McCarthy, 2000; Johnson-Groh et al., 2002; Whittier, 2006
	Reproduction	Spore production	Long-term gradient work	1, 3	Mehltreter and Palacios-Rios, 2003
	Life cycles	Pathways, exceptions	Triggers for stage transitions	1, 3, 6	Mehltreter and García-Franco, 2008
	Structure	Stage classes	Transition matrices	3	Bremer, 2004, 2007
	Phenology	Seasonal patterns	Annual patterns, disturbance	3	Sharpe, 2005; Mehltreter and García-Franco, 2008
Resources	Nutrients	Stoichiometry	Decomposition, recycling	2, 4, 6	Russell et al., 1998; Amatangelo and Vitousek, 2008
	Water	Drought tolerance	Genetic basis	5	Carlquist and Schneider, 2001; Proctor and Tuba, 2002
	Light	Photoinhibition, shading	Competitive consequences	5, 6	George and Bazzaz, 1999a; Coomes et al., 2005
Disturbance	Hurricanes	Differential responses	Experimental tree falls	6	Cooper-Ellis et al., 1999; Halleck et al., 2004
	Landslides	Colonization dynamics	Landscape carbon dynamics	6	Kessler, 1999; Walker and Shiels, 2008

Table 10.1 (cont.)

Topic	Subtopic	Current research	Areas for research expansion	Chapters	Representative citations
	Fire	Resistance and recovery	Effects on invasive species	6, 9	Alonso-Amelot and Rodulfo-Baechler, 1996; Atkinson, 2004
Species interactions	Competition	Rates	Allelopathy	5, 6, 7	Gliessman and Muller, 1978; George and Bazzaz, 1999a,b
	Herbivory	Specialization	Deterrents	7, 8	Hegnauer, 1962, 1986; Hendrix, 1980
	Mycorrhizae	Extent and type of infection	Impact on competition	4, 7	Iqbal et al., 1981; Gemma et al., 1992; Moteetee et al., 1996
	Ants	Spore dispersal	Nutrient implications	4, 7	Tryon, 1985
Temporal dynamics	Succession	Thickets, long-term plots	Functional diversity	3, 6	Walker et al., 1996; Slocum et al., 2004
Human links	Conservation	Bioindicators	Hot spot protection	2, 9	Beukema and van Noordwijk, 2004; Bickford and Laffan, 2006
	Invasives	Biocontrol, management	Ecological and economic impacts	8	Durand and Goldstein, 2001; Miller and Sheffield, 2004; Christenhusz and Toivonen, 2008
	Medicine	Carcinogen detectors	Novel compounds	9	Simán et al., 2000; Villalobos-Salazar et al., 2000
	Restoration	Heavy metal tolerance	Phytoremediation	6, 9	Marrs et al., 2000; Tu and Ma, 2005

of extant ferns have been dominant earlier in evolutionary history than seed plants (Schneider *et al.*, 2004), allowing more time to disperse to and evolve in a variety of habitats. However, these potential advantages for ferns are perhaps offset by the need for an aqueous film as an environment for fertilization. Fern species are on average more widely distributed than species of seed plants (see Chapter 2), although fern dominance is partly linked to specific types of habitat.

The habitats that support the highest number of fern species are humid tropical cloud forests. Fern species richness does not reach high values in regions of high seed plant richness such as Mediterranean-type climates but can exceed seed plant richness on some remote, oceanic islands. With increasing latitude, fern species richness drops off more precipitously than that of seed plants, but epiphytic ferns do reach further into cold and dry tropical climates than seed plant epiphytes (see Chapter 2). Fern species richness is highest on nutrient-rich soils but ferns make their greatest proportional contribution to overall plant biomass on nutrient-poor soils such as those found in both arctic and alpine tundra or in early successional soils (see Chapter 4). Fern distribution and abundance are also modified by historical factors. The relative paucity of ferns in Africa may be due to limited refugia for ferns on moist tropical mountains during periods of global drought (Aldasoro *et al.*, 2004).

There are several unsolved issues about fern ecology that future biogeographical research could address. For example, how many more species of ferns exist in the world and where will they be found? More surveys are urgently needed because some authors estimate that there are still several thousand more fern species to discover beyond the *c.* 11 000 currently described (see Chapter 2) and the expansion of human populations is probably destroying fern habitats faster than new ferns are being identified (see Chapter 9). In addition, too little is known about gametophyte ecology and the importance of gametophyte habitat requirements for fern establishment (see Section 10.3). Addressing these issues will help clarify the historical basis for current distributions.

10.3 Population dynamics: growth, reproduction and life cycles

Ferns have evolved a large number of growth forms, from small, aquatic or epiphytic ferns with single leaves, to vine-like ferns with indeterminate growth, to sturdy tree ferns with trunks and leaves that can reach several meters in length. Ferns survive in remarkably diverse habitats from deserts to boreal forests and have many ecological roles (Table 10.2). However, no annual ferns have been described (perhaps with the exception of *Anogramma leptophylla* or some species of *Marsilea*) and ferns do not appear to be parasitic on other plants. The absence of annual ferns might make ferns less competitive with seed plants in habitats with

Table 10.2 *Key ecological roles of ferns*

The indicated number of sets of chromosomes may vary.

Fern life stage	Direct effects	Indirect consequences
Spore (1n)	Source of food for birds, insects	Long distance dispersal
Gametophyte (1n)	Competitors with mosses	Establishment in moist, low-light forest understories
Sporophyte (2n)	Colonizers after disturbance; competitors with seed plants; critical epiphyte substrate; drought tolerators; efficient nutrient accumulators; facilitators and inhibitors of succession; providers of food for birds, insects, humans; source of forage habitat, nest material; invaders of native ecosystems; slope stabilizers	Aesthetic values; contribution to biodiversity; soil formation

Sources come from several chapters of this book as well as Arcand and Ranker (2008).

resource pulses but more competitive in habitats with consistently low resource availability. Ferns that are parasitic may not have evolved because fern secondary biochemistry and vascular anatomy might have been too different from that of seed plants to support the development of haustoria on the host plant. However, some parasitic seed plants (e.g., *Cuscuta*; dodder) parasitize ferns and are even involved in gene transfer to ferns such as *Botrychium virginianum* (Davis *et al.*, 2005).

The reproductive ecology of ferns in some ways resembles that of both mosses and seed plants but in some ways is also unique. We present some questions about the ecological implications of four characteristics of fern reproduction. First, fern spores are much smaller dispersal units than the seeds and fruits of seed plants. Are there key advantages of smallness for dispersal into microhabitats suitable for gametophyte development? Second, fern gametophytes and sporophytes exist independently. Which generation tolerates more extreme environmental conditions and limits the species range? Similarly, how is the ploidy level of the generations connected to their morphological complexity? Apogamous gametophytes may be diploid or triploid, but there is no evidence of haploid sporophytes in nature. Third, what are the ecological consequences of the two requirements for sperm to fertilize the eggs: water for movement and a close neighbor to swim to? Does this latter requirement favor dense stands of adult sporophytes or are they thinned out to solitary plants because of intraspecific competition? Finally, seed plants require a small, specific target for their pollen but seeds can disperse more randomly. Ferns, in contrast, can have random spore dispersal yet require fairly specific habitats in order to have effective gametophyte development. How do these variations in dispersal dynamics impact the patterns of fern distribution and aggregation compared with those of seed plants?

Studies of fern life cycles have helped to clarify patterns of spore production, spore germination, initiation of sexual reproduction of gametophytes, leaf production, leaf growth and leaf life spans. However, there is a need to identify ecological triggers for the transition between each life history stage, perhaps with an experimental approach. Identifying an individual fern plant is not easy because many ferns, like many seed plants, are able to clone, making population parameters difficult to assess (Hamilton, 1992). The determination of the age of fern individuals is as challenging as aging any other long-lived herbaceous perennial, although tree fern ages can be approximated in most cases (Schmitt and Windisch, 2006; Mehltreter and García-Franco, 2008).

Future studies of the population dynamics of both fern generations can help to distinguish between optimal and marginal habitats for growth and reproduction. Development of standardized categories for life history transition matrices for fern populations can not only facilitate comparisons with other fern populations, but also with seed plant populations. Multiyear studies with careful monitoring will help quantify many unknowns including the variation in individual growth rates and responses to unpredictable disturbances. Development of standardized methods to measure ferns must account for variable architecture, growth rates and spore production phenology.

10.4 Ecophysiology

10.4.1 Nutrients

The gametophyte role in nutrient cycling is poorly understood but there are a number of recent studies on how fern sporophytes respond to nutrients in their environment and how ferns, in turn, alter ecosystem nutrient dynamics. The roots of most ferns are infected with mycorrhizal fungi that aid in nutrient uptake. The often elongated rhizomes might also be involved in nutrient uptake or sequestration, but there is as yet little evidence to support this idea. Some leaves have water-absorbing scales, but the significance of these scales, especially for nutrient absorption, is unknown. A small number of fern species also have symbiotic relationships with ants or nitrogen-fixing cyanobacteria (see Section 10.6). How these relationships alter the growth, population dynamics or competitive balances of such ferns is not yet understood.

Locally, fern species richness rises with increasing soil fertility, although there are exceptions (e.g., from lowland rain forest in Ecuador, Tuomisto *et al.*, 2002). Higher availability of epiphytic microsites might account for some of this increase.

Richness could also vary along soil fertility gradients, possibly augmented by increases due to disturbance, changes in light regime or species interactions (e.g., competition, herbivory). Ferns likely make their greatest contribution to total plant biomass on infertile soils that characterize many types of recently disturbed sites.

Fern species can also have diverse responses to soil nutrients, particularly soil cations, suggesting that some fern species might be used as indicators of soil fertility. For example, fern leaves from three forested ecosystems have lower leaf calcium concentrations than adjacent seed plants, at least in low calcium environments (see Chapter 4). Perhaps this is due to a lack of the pectin that binds cell walls together in seed plants (Manton, 1950; Page 2002a). Does low leaf calcium imply that evolution of ferns occurred in low calcium environments (Amatangelo and Vitousek, 2008); or are ferns competitively dominant over seed plants in low calcium environments? In contrast, some ferns grow better in calcium-rich than calcium-poor environments, so studying fern nutrient ecology along gradients of soil calcium should be very informative. Fern-dominated ecosystems may have reduced levels of calcium cycling, particularly because fern leaves may be slow to decompose relative to leaves of most seed plants. Finally, can the capability of some fern species to hyperaccumulate aluminum and arsenic extend to other soil characteristics and toxic substances and be exploited for restoration or agricultural uses? Clearly, there are many interesting questions about fern nutrient dynamics that remain unresolved. Perhaps the answers to some of these questions will alter our fundamental understanding of nutrient cycling in plants.

10.4.2 Water

Contrary to common perception, ferns can tolerate dry habitats. Approximately 2% of all ferns grow in arid environments such as deserts or on the surfaces of rocks. Ferns have a large range of strategies to cope with drought and are arguably as well adapted to tolerating drought as are seed plants. Fern sporophytes that tolerate drought do so through pigments, antioxidants, leaf curling and (rarely) CAM photosynthesis. Ferns avoid desiccation by reducing water loss with cuticles and scales and by being drought deciduous. Many fern gametophytes are also highly tolerant of drought (Watkins *et al.*, 2007). Despite many physiological studies about ferns in dry habitats (see Chapter 5), the ecological implications of such adaptations are unclear. What role do ferns have in dry environments? Do they contribute substantially to the entrapment of dispersing seeds; do they affect competitive balances within plant communities, or alter herbivore dynamics, nutrient cycling or soil formation? Despite their remarkable adaptations, it is likely that xerophytic ferns only occasionally reach high enough densities or biomass in dry habitats to have a major impact on ecosystem functions.

10.4.3 *Light*

There is a wide range of light tolerance among fern species that allows some species to adapt to such high-light environments as landslides, cliffs or forest gaps, and other species to such low-light environments as forest understories or cave entrances. In contrast, intolerance of shade by ferns on landslides in Puerto Rico leads to their replacement by trees (Walker *et al.*, 1996). Understory ferns can be important light filters in temperate forests and indirectly impact seedling establishment and eventual composition of canopy-forming trees. Coexistence of four *Cyathea* tree fern species within some New Zealand forests may be due to both variable shade tolerance and different responses to disturbance (N. Bystriakova, personal communication). Fern responses to light represent a largely untapped but important research area.

10.5 Disturbance responses

Some ferns are no strangers to disturbance. These ferns can be frequent colonizers following disturbances such as hurricanes, landslides, fire, floods and volcanoes, and also can become abundant when humans have disrupted existing vegetation (e.g., pavement, mine tailings, road cuts and abandoned agricultural fields, see Chapter 6). These ferns are generally not the same as those found in more stable habitats such as forest understories. Rapid, long distance spore dispersal by wind to disturbed sites is clearly an advantage over the slower wind or animal dispersal of larger, heavier seeds (or fruits). Despite the presence of spores in high-altitude wind streams, most spore (and seed) dispersal is local, declining rapidly with distance from the source. Therefore, neither ferns nor seed plants rapidly invade the centers of large disturbances. Some ferns are so linked with disturbance that their regeneration can depend on it (Page, 2002b). However, the disturbance responses of many ferns are unknown, creating potential problems when previously undamaged fern habitat is disturbed by human activities.

Once established on a site, some ferns can compete effectively with seed plants for nutrients, water and light. Ferns in the Gleicheniaceae (scrambling ferns) are very successful on disturbed sites throughout the tropics and dominate resources through growth of dense thickets. Other disturbance-adapted ferns such as *Pteridium* (bracken) are allelopathic and often dominate disturbed areas in temperate regions. Dense growth of thicket-forming ferns can impede successional transitions but most colonizing ferns are shade intolerant. Therefore, when canopy species become established, the fern colonizers are replaced by shade-tolerant understory plants. Initial fern dominance, followed eventually by replacement by shade-tolerant species, has many potential implications for habitat restoration that have not been adequately examined. For example, slope stability can be achieved by promoting the growth of fern thickets, although biodiversity may be reduced (Walker *et al.*, 2009). Studying fern responses to various disturbances will lead to a better understanding of both natural fern replacements during succession and the potential role of ferns in land management.

Some fern species can tolerate such extremely stressful environments as toxic mine tailings (Tu and Ma, 2005) or radioactivity (Odum and Pigeon, 1970). The mechanisms and consequences of such stress tolerances have not been examined except in the context of phytoremediation of toxic sites through such mechanisms as hyperaccumulation of heavy metals (see Section 10.7).

10.6 Species interactions

Ferns interact with other organisms in their environment in many ways. Ferns involved in epiphytic relationships (ferns growing on other plants or plants growing on ferns) do not significantly impact their host. Ferns compete with seed plants, are infested by fungal parasites and eaten by herbivores. Finally, ferns often have a mutualistic relationship with mycorrhizal fungi.

Ferns often compete with seed plants for nutrient, water and light resources. Ferns and grasses may compete strongly with each other because they have a similar network of fine, adventitious roots. Dense layers of ferns in forest understories may also influence the regeneration of trees by competing successfully against tree seedlings, thus ultimately influencing canopy composition. Studying competitive relationships in a successional context can be helpful because it elucidates ecological principles of dispersal, competition, facilitation and community assembly.

Most fern herbivores are arthropods. Rates of fern herbivory by arthropods may be similar to those on seed plants (*c.* 5–15% of all seed plant and fern leaves typically show some damage) although some researchers maintain that ferns are less likely to experience damage than seed plants. Many arthropod herbivores that specialize on ferns appear to have a long, coevolutionary history with ferns, so most ferns, like most seed plants, have developed chemical defenses to herbivores. Mammals do not favor fern leaves in their diet (Forsyth *et al.*, 2005), but wild pigs do consume the starchy trunks of tree ferns in Hawaii (see Chapter 7; Diong, 1982; Arcand, 2007).

Mutualistic relationships with mycorrhizal fungi are just as common for ferns as for seed plants with >70% of ferns infected (see Chapter 7). Gametophytes as well as sporophytes support mycorrhizal fungi. In fact, ferns with subterranean gametophytes depend on mycorrhizal fungi that in turn infect a photosynthetic host. Fern mutualisms with ants cover a range of levels of interdependence between the fern and the ant. Some ferns supply nectar rewards and even hollowed out rhizomes for ant nests and, in turn, receive protection from potential competing plants or herbivores.

10.7 Ferns and humans

Humans are increasingly disrupting fern populations and fern habitats by clearing and burning forests and adding fertilizers and pesticides to croplands that then can alter fern habitats downslope. Overcollecting for the horticultural trade has also disrupted fern populations. Humans have also introduced some ferns to new habitats, sometimes creating problems with invasive fern species. However, many ferns are also beneficial to humans, so we have a multifaceted relationship with them.

Many ferns that are useful or even endangered in their original habitat can pose problems for humans when transported around the world, often as ornamentals. Many such introduced ferns escape from cultivation, tolerate high-light conditions and spread quickly, thus outcompeting native plants, sometimes with devastating consequences for local ecosystems. Successful fern invaders include the Australian *Lygodium microphyllum* (small-leaved climbing fern, Plate 8D), which can engulf trees in Florida, and the Australian tree fern *Sphaeropteris cooperi* that replaces native tree ferns species of *Cibotium* in Hawaii (Fig. 6.7d, e). Other invaders include the shade-tolerant *Angiopteris evecta* (mule's foot fern, Plate 4D) that invades stream banks in rain forests, several species of *Nephrolepis* (sword fern) that invade disturbed habitats worldwide, and several aquatic ferns (e.g., *Salvinia molesta*; giant salvinia) that clog inland waterways (Fig. 8.10). Species of *Pteridium* (Fig. 8.2) and ferns in the Gleicheniaceae (Figs. 6.8, 8.7, 8.8) invade disturbed areas while species of *Equisetum* (horsetails, Plate 4C) can dominate wetlands. Management of problem ferns may involve eradication or the release of competitors to stabilize the problem ferns at low population levels. Chemical or biological controls and physical removal can be used with each of these management strategies, but can be costly.

Several ferns are beneficial to humans (see Chapter 8). We use such species as the nitrogen-fixing *Azolla* (mosquito fern) as fertilizer in rice paddies; the South African *Rumohra adiantiformis* (leatherleaf fern) in the florist trade; the tropical *Nephrolepis exaltata* (Boston fern) as an ornamental around the house or garden; and the lycophyte *Huperzia serrata* (Chinese fir-moss) in Asian medicines. The trunks of tree ferns were once used for growing orchids and making handicrafts but are now protected (see Chapter 9). Mutation rates in some fern rhizomes have been used to detect carcinogens from toxins in the environment. An exciting and expanding use of ferns is for phytoremediation because some ferns hyperaccumulate heavy metals such as As, Cd, Cr, Cu, Ni or Pb (see Chapter 8, Plate 8B). Ferns capable of tolerating sites contaminated by heavy metals may become a source of genes to modify crop plants. Much research remains to determine how best to utilize ferns in genetic research and ecological restoration of toxic sites.

Conservation of ferns serves many purposes, including preserving habitat and populations of rare ferns, ensuring the continuation of key ecological roles of ferns such as providing food for specialist herbivores (Table 10.2), and protecting ferns that humans use directly for food, fertilizer or medicine. Some ferns have been identified as indicator plants for particular soil characteristics, land use, disturbance regimes or climates. Humans will likely find many more uses of ferns in the future. However, ferns are still so poorly studied that we do not even know how many ferns are rare or endangered.

Extinction of a few fern species has been documented (e.g., two from the remote Ascension Island and one from China, IUCN, 2008). Habitat destruction of tropical forests may have already driven many fern species to extinction, particularly endemic species with narrow ecological niches. Intrinsic features of some ferns that might predispose them to extinction include short-term viability of green spores; slow growth of subterraneous, mycotrophic gametophytes; lower water conduction; and lower net assimilation rates than seed plants (see Chapter 9). The large size of tree ferns may lead to overexploitation by humans. Low genetic diversity appears to be a concern for the few fern species that have been identified as threatened. Conservation priorities are to protect endemic ferns, fern habitats and hot spots (regions with high fern diversity).

10.8 Unresolved questions in fern ecology

Many topics have been covered in this book on fern ecology (Table 10.1) and it is clear that ferns, where present, are intimately integrated within their ecosystems (Table 10.2). However, there are many unresolved aspects of fern ecology that will direct future research. We present some questions on nine topics that intrigue us and that, when addressed, will provide a much more robust foundation for understanding fern ecology than we have today. Ideally, addressing these questions will also help integrate the study of ferns into general ecological studies.

10.8.1 Correlates of fern diversity

Several seminal studies have begun to clarify patterns of fern species diversity (see Chapter 2) and explore possible correlations with environmental factors such as substrate fertility, topography and isolation. However, what ultimately controls patterns of fern diversity? A better understanding of the reproductive processes and the genetic diversity within and among taxa will help unravel fern phylogenies and aid in conservation decisions. Studies of diversity within and between communities provide comparisons across landscape scales. Comparisons of the patterns and controls of fern and seed plant biodiversity will help clarify how different fern communities are structured. Finally, as undiscovered species of ferns are found, our current understanding of fern habitat requirements and biodiversity will certainly improve.

10.8.2 Correlates of fern abundance

As with any group of plants, at local scales (plots <1 ha) few fern species are very abundant, some more are intermediate in abundance and most are represented by few individuals, whether abundance is measured by biomass, density or some other measure. Which ecological traits account for the remarkable success of certain fern genera (e.g., *Pteridium*), families (e.g., Gleicheniaceae) and orders (e.g., Cyatheales)? Does clonal growth or allelopathy play a role in these success stories? Conversely, what determines the rarity of other ferns? Are the factors that control fern abundance any different from those that control seed plant abundance?

10.8.3 Controls over gametophyte growth and development

The gametophyte, as an independent plant, is unique to ferns, mosses and some algae but has not received much attention from plant ecologists (but see Farrar *et al.*, 2008). Understanding what controls gametophyte growth and development is central to fern ecology. Are habitats optimal for gametophytes different from those favoring sporophytes? For example, it appears that desiccation tolerance for a gametophyte may or may not resemble that of its sporophyte (see Chapter 5). Is establishment of new sporophytes most likely in disturbed habitats where high light and low competition with seed plants favor gametophyte growth? Under what conditions can gametophytes develop in the same habitat as sporophytes? Parallels with seed germination versus seed plant habitat requirements are obvious, but what are the different controls that limit the success of a seedling versus that of a gametophyte?

10.8.4 Spore size and dispersal

Seed size varies widely with interesting ecological consequences from the energy requirements to produce a seed, to dispersal trajectories of small versus large seeds, to the degree of dependence on resources in the germination environment and eventual consequences for competition for resources. Fern spores, however, have remained uniformly small. What are the consequences of spore size in ferns? Are the consequences for ferns similar to those for small-seeded plants such as orchids (e.g., minimal parental investment, the possibility of long distance dispersal, and few internal resources)? Or is dispersal more effective for ferns because it takes place at the unicellular stage (spore) which can be smaller than the multicellular dispersal unit of seed plants?

10.8.5 Controls over life history transitions

What factors (environmental and intrinsic) control fern transitions between each stage of its life history, from spore production to spore germination to gameto-phyte growth to gamete production to fertilization and subsequent sporophyte growth? An understanding of what controls these transitions and how those controls vary among fern species would be the ultimate challenge for fern ecologists.

10.8.6 Ecological niches

How does the variety of niches that ferns occupy differ from those of seed plants? Are there potential explanations to be found in the different life cycles of ferns and seed plants? Fern gametophytes grow inconspicuously in often low-light niches (e.g., under rocks or in crevices within the root mantle of tree fern trunks) while seed plants often germinate under high-light conditions. Fern spor-ophytes and adult seed plants, however, often share an affinity for high-light niches such as the branches of forest trees (occupied mainly by rosette-forming ferns in the Old World and rosette-forming bromeliads in the New World). However, fern sporophytes generally compete most successfully with seed plants in shady understories.

10.8.7 Interactions

Ferns have a lower number of interactions with animals than seed plants, because they lack flowers, seeds and fruits. However, there is no convincing evidence that ferns and seed plants differ in the number of interactions with mycorrhizal fungi, parasitic fungi or leaf-feeding herbivores. Because ferns possess less diverse and somewhat different biochemical defense mechanisms than seed plants, one might expect that generalist species would be capable of feeding on ferns and that some insect species would specialize on ferns. How many insect species specialize on ferns and did these coevolve with ferns or did they switch from seed plant or bryophyte hosts? Do ferns of older lineages interact more often with older insect groups? How many fern herbivores remain to be discovered? Finally, do insect species that currently feed on pollen grains also consume fern spores of similar size or do extant insects of lineages that formerly fed on fern spores now feed on pollen of seed plants?

10.8.8 Roles in community and ecosystem development

How does the presence of ferns alter various ecosystem functions, disturbance regimes and successional trajectories? As ferns are not dominant players in all ecosystems, the answers may lie in large-scale comparisons of ecosystems with and without ferns. On some remote islands where ferns constitute 30–70% of the vascular flora (Plate 1A, B) or in open habitats where ferns constitute >90% of the biomass, do they facilitate or inhibit seed plant colonization compared with similar environments with fewer ferns? We suspect that ferns could be replaced in many early successional systems by functionally equivalent seed plants such as grasses with only moderate changes in ongoing disturbances (e.g., erosion), community assembly and successional trajectories. However, the role of ferns in nutrient uptake and sequestration is sufficiently unique that if ferns were excluded from some disturbed habitats (e.g., landslides, toxic mine tailings) the return of substantial vegetative cover might be delayed. Comparative analyses and perhaps experiments could be undertaken to address such questions.

10.8.9 Human uses of ferns

The ornamental fern industry has achieved worldwide success in a global economy. What further discoveries can be expected to increase the economic role of ferns? Although there are currently no natural by-products of ferns being used in Western pharmacopeias, there are many compounds found in ferns that are considered useful in traditional herbal medicines. To date, only *Azolla* has been used in an agricultural context as a manure and animal feed, but perhaps there is a role for utilization of other ferns (e.g., *Pteridium*) as a biofuel or animal fodder.

10.9 Conclusions

10.9.1 The importance of ferns

One goal in producing this book was to illustrate how ferns are unique in the plant world. Their uniqueness is often what attracts people to study ferns. However, the perception that ferns are unique can sometimes lead to the omission of ferns from ecological and botanical studies. This omission, perhaps unintentionally reinforced by several generations of scientists, perpetuates a widespread ignorance about fern ecology. Therefore, another goal we had was to illustrate how ferns should become a routine part of integrative studies of plant ecology. The majority of ferns interact with their environment in much the same way as do herbaceous perennials while tree ferns are quite comparable to palms (no secondary growth, branching or tap root and trunks formed by leaf bases). Therefore, standardized measurements for seed plants can generally include ferns. With an increased awareness of ferns, fern ecology will hopefully become more integrated into general studies of plant ecology and ferns will achieve coverage in ecology textbooks proportional to their importance. At the same time, fern-focused research can continue to explore the wonderful oddities that ferns present.

10.9.2 The future of ferns

Ferns delight us with their aesthetic qualities and intrigue us with their unusual shapes and life histories. When examined more closely, ferns also bear strong resemblances to seed plants. This awareness of and engagement with ferns by humans bodes well for the survival of ferns. However, the high diversity of ferns on tropical mountains, on remote islands and in forest understories worldwide makes ferns susceptible to the increasing human impacts on those ecosystems. Fern hot spots should be identified and conservation of rare ferns should become a priority for conservation advocacy groups. The prohibition of international trade in living tree ferns by CITES in 2000 (see Chapter 9) was an important step toward appropriate recognition of the vulnerability of ferns when wild populations are harvested without sustainable management. Cultivation and propagation can help the recovery of some populations and assist restoration efforts but these approaches are not sufficient to replace conservation of natural areas. Because ferns are dynamic participants in many habitats, represent ancient and modern lineages of plants, and simultaneously resemble and differ from other plants, we hope and expect that the study of fern ecology will be a high priority for future generations.

Acknowledgements

Joanne Sharpe and Lawrence Walker acknowledge support from the Luquillo Long-Term Ecological Research Program in Puerto Rico funded by the U.S. National Science Foundation. Klaus Mehltreter acknowledges support from the Instituto de Ecología, A. C., Xalapa, Veracruz, Mexico, CONACYT-SEMARNAT (2002-C01–0194) and CONACYT-SEP (2003-C02–43082). We thank Bruce Clarkson, David Coomes, Michael Kessler, Jill Rapson and George Yatskievych for helpful reviews of the manuscript.

References

Aldasoro, J. J., Cabezas, F. and Aedo, C. (2004). Diversity and distribution of ferns in sub-Saharan Africa, Madagascar and some islands of the South Atlantic. *Journal of Biogeography*, **31**, 1579–604.

Alonso-Amelot, M. E. and Rodulfo-Baechler, S. (1996). Comparative spatial distribution, size, biomass and growth rate of two varieties of bracken fern (*Pteridium aquilinum* (L.) Kuhn) in a neotropical montane habitat. *Vegetatio*, **125**, 137–47.

Amatangelo, K. L. and Vitousek, P. M. (2008). Stoichiometry of ferns in Hawaii: implications for nutrient cycling. *Oecologia*, **157**, 619–27.

Arcand, N. N. (2007). Population structure of the Hawaiian tree fern *Cibotium chamissoi* across intact and degraded forests, Oahu, Hawaii. Unpublished Master thesis, University of Hawaii, Oahu.

Arcand, N. N. and Ranker, T. A. (2008). Conservation biology. In *Biology and Evolution of Ferns and Lycophytes*, ed. T. A. Ranker and C. H. Haufler. Cambridge, UK: Cambridge University Press, pp. 257–83.

Atkinson, I. A. E. (2004). Successional processes induced by fires on the northern offshore islands of New Zealand. *New Zealand Journal of Ecology*, **28**, 181–93.

Beukema, H. and van Noordwijk, M. (2004). Terrestrial pteridophytes as indicators of a forest-like environment in rubber production systems in the lowlands of Jambi, Sumatra. *Agriculture, Ecosystems and Environment*, **104**, 63–73.

Bickford, S. A. and Laffan, S. W. (2006). Multi-extent analysis of the relationship between pteridophyte species richness and climate. *Global Ecology and Biogeography*, **15**, 588–601.

Bremer, P. (2004). On the ecology and demography of a terrestrial population of *Asplenium trichomanes* (Aspleniaceae: Pteridophyta) in the Netherlands. *Fern Gazette*, **17**, 85–96.

Bremer, R. (2007). Frost and forest stand effects on population dynamics of *Asplenium scolopendrium* L. in the colonization of a former sea-floor by ferns. Unpublished Ph.D. thesis. Wageningen University, Wageningen, The Netherlands.

Carlquist, S. and Schneider, E. L. (2001). Vessels in ferns: structural, ecological and evolutionary significance. *American Journal of Botany*, **88**, 1–13.

Christenhusz, M. J. M. and Toivonen, T. K. (2008). Giants invading the tropics: the oriental vessel fern, *Angiopteris evecta* (Marattiaceae). *Biological Invasions*, **10**, 1215–28.

Coomes, D. A., Allen, R. B., Bentley, W. A., *et al.* (2005). The hare, the tortoise and the crocodile: the ecology of angiosperm dominance, conifer persistence and fern filtering. *Journal of Ecology*, **93**, 918–35.

Cooper-Ellis, S., Foster, D. R., Carlton, G. and Lezberg, A. (1999). Forest response to catastrophic wind: results from an experimental hurricane. *Ecology*, **80**, 2683–96.

Davis, C. C., Anderson, W. R. and Wurdack, K. J. (2005). Gene transfer from a parasitic flowering plant to a fern. *Proceedings of the Royal Society, Series B*, **272**, 2237–42.

Diong, C. H. (1982). Population biology and management of the feral pig (*Sus scrofa* L.) in Kipahulu Valley, Maui. Unpublished Ph.D. thesis, University of Hawaii, Oahu.

Durand, L. Z. and Goldstein, G. (2001). Photosynthesis, photoinhibition, and nitrogen use efficiency in native and invasive tree ferns in Hawaii. *Oecologia*, **126**, 345–54.

Farrar, D. R., Dassler, C., Watkins, J. E., Jr. and Skelton, C. (2008). Gametophyte ecology. In *Biology and Evolution of Ferns and Lycophytes*, ed. T. A. Ranker and C. H. Haufler. Cambridge, UK: Cambridge University Press, pp. 222–56.

Forsyth, D. M., Richardson, S. J. and Menchenton, K. (2005). Foliar fibre diet selection by invasive red deer *Cervus elaphus* in a New Zealand temperate forest. *Functional Ecology*, **19**, 495–504.

Gemma, J. N., Koske, R. E. and Flynn, T. (1992). Mycorrhizae in Hawaiian pteridophytes: occurrence and evolutionary significance. *American Journal of Botany*, **79**, 843–52.

George, L. O. and Bazzaz, F. A. (1999a). The fern understory as an ecological filter: emergence and establishment of canopy-tree seedlings. *Ecology*, **80**, 833–45.

George, L. O. and Bazzaz, F. A. (1999b). The fern understory as an ecological filter: growth and survival of canopy-tree seedlings. *Ecology*, **80**, 846–56.

Gliessman, S. R. and Muller, C. H. (1978). The allelopathic mechanisms of dominance in bracken (*Pteridium aquilinum*) in southern California. *Journal of Chemical Ecology*, **4**, 337–62.

Greer, G. K. and McCarthy, B. C. (2000). Patterns of growth and reproduction in a natural population of the fern *Polystichum acrostichoides*. *American Fern Journal*, **90**, 60–76.

Halleck, L. F., Sharpe, J. M. and Zou, Z. (2004). Understorey fern responses to post-hurricane fertilization and debris removal in a Puerto Rican rain forest. *Journal of Tropical Ecology*, **20**, 173–81.

Hamilton, R. G. (1992). Allozyme variation and ramet distribution in two species of athyrioid ferns. *Plant Species Biology*, **7**, 69–75.

Hegnauer, R. (1962). *Chemotaxonomie der Pflanzen, Bd. I*. Basel, Switzerland: Birkhäuser.

Hegnauer, R. (1986). *Chemotaxonomie der Pflanzen, Bd. VII*. Basel, Switzerland: Birkhäuser.

Hendrix, S. D. (1980). An evolutionary and ecological perspective of the insect fauna of ferns. *American Naturalist*, **115**, 171–96.

Hemp, A. (2006). Continuum or zonation? Altitudinal diversity patterns in the forests of Mt. Kilimanjaro. *Plant Ecology*, **184**, 27–42.

Iqbal, S. H., Yousaf, M. and Younus, M. (1981). A field survey of mycorrhizal associations in ferns of Pakistan. *New Phytologist*, **87**, 69–79.

IUCN (2008). *2008 IUCN Red List of Threatened Species*. www.iucnredlist.org. Viewed 21 January 2009.

Johnson-Groh, C., Riedel, C., Schoessler, L. and Skogen, K. (2002). Belowground distribution and abundance of *Botrychium* gametophytes and juvenile sporophytes. *American Fern Journal*, **92**, 80–92.

Kato, M. (1993). Biogeography of ferns: dispersal and vicariance. *Journal of Biogeography*, **20**, 265–74.

Kessler, M. (1999). Plant species richness and endemism during natural landslide succession in a perhumid montane forest in the Bolivian Andes. *Ecotropica*, **4**, 123–36.

Manton, I. (1950). *Problems of Cytology and Evolution in the Pteridophyta*. Cambridge, UK: Cambridge University Press.

Marrs, R. H., Le Duc, M. G., Mitchell, R. J., *et al.* (2000). The ecology of bracken: its role in succession and implications for its control. *Annals of Botany*, **85** (Suppl. B), 3–15.

Mehltreter, K. and García-Franco, J. G. (2008). Leaf phenology and trunk growth of the deciduous tree fern *Alsophila firma* (Baker) D. S. Conant in a lower montane Mexican forest. *American Fern Journal*, **98**, 1–13.

Mehltreter, K. and Palacios-Rios, M. (2003). Phenological studies of *Acrostichum danaeifolium* (Pteridaceae, Pteridophyta) at a mangrove site on the Gulf of Mexico. *Journal of Tropical Ecology*, **19**, 155–62.

Miller, J. and Sheffield, E. (2004). Controlling Kariba weed. *Biological Sciences Review*, **17**, 39–41.

Moteetee, A., Duckett, J. G. and Russell, A. J. (1996). Mycorrhizas in the fern of Lesotho. In *Pteridology in Perspective*, ed. J. M. Camus, M. Gibby and R. J. Johns. Kew, UK: Royal Botanic Gardens, pp. 621–31.

Odum, H. T. and Pigeon, R. F. (eds.) (1970). *A Tropical Rainforest: a Study of Irradiation and Ecology at El Verde, Puerto Rico*. Washington, D.C., USA: Division of Technical Information, U.S. Atomic Energy Commission.

Page, C. N. (1979). Experimental aspects of fern ecology. In *The Experimental Biology of Ferns*, ed. A. F. Dyer. London: Academic Press, pp. 552–89.

Page, C. N. (2002a). Ecological strategies in fern evolution, a neopteridological overview. *Review of Palaeobotany and Palynology*, **119**, 1–33.

Page, C. N. (2002b). The role of natural disturbance regimes in pteridophyte conservation management. *Fern Gazette*, **16**, 284–9.

Proctor, M. C. F. and Tuba, Z. (2002). Poikilohydry and homoihydry: antithesis or spectrum of possibilities? *New Phytologist*, **156**, 327–49.

Russell, A. E., Raich, J. W. and Vitousek, P. M. (1998). The ecology of the climbing fern *Dicranopteris linearis* on windward Mauna Loa, Hawai'i. *Journal of Ecology*, **86**, 765–79.

Schmitt, J. L. and Windisch, P. G. (2006). Growth rates and age estimates of *Alsophila setosa* Kaulf. in southern Brazil. *American Fern Journal*, **96**, 103–11.

Schneider, H., Schuettpelz, E., Pryer, K. M., *et al.* (2004). Ferns diversified in the shadow of angiosperms. *Nature*, **428**, 553–7.

Sharpe, J. M. (2005). Temporal variation in sporophyte fertility in *Dryopteris intermedia* and *Polystichum acrostichoides* (Dryopteridaceae: Pteridophyta). *Fern Gazette*, **17**, 223–34.

Simán, S. E., Povey, A. C., Ward, T. H., Margison, G. P. and Sheffield, E. (2000). Fern spore extracts can damage DNA. *British Journal of Cancer*, **83**, 69–73.

Slocum, M. G., Aide, T. M., Zimmerman, J. K. and Navarro, L. (2004). Natural regeneration of subtropical montane forest after clearing fern thickets in the Dominican Republic. *Journal of Tropical Ecology*, **20**, 483–6.

Smith, A. R., Pryer, K. M., Schuettpelz, E., *et al.* (2006). A classification for extant ferns. *Taxon*, **55**, 705–31.

Soria-Auza, R. W. and Kessler, M. (2008). The influence of sampling intensity on the perception of the spatial distribution of tropical diversity and endemism, a case study of ferns from Bolivia. *Diversity and Distributions*, **14**, 123–30.

Tryon, A. F. (1985). Spores of myrmecophytic ferns. *Proceedings of the Royal Society of Edinburgh*, **86B**, 105–10.

Tryon, R. M. and Tryon, A. F. (1982). *Ferns and Allied Plants with Special Reference to Tropical America*. New York: Springer-Verlag.

Tu, C. and Ma, L. Q. (2005). Effects of arsenic on concentration and distribution of nutrients in the fronds of the arsenic hyperaccumulator *Pteris vittata* L. *Environmental Pollution*, **135**, 333–40.

Tuomisto, H. and Poulsen, A. D. (2000). Pteridophyte diversity and species composition in four Amazonian rain forests. *Journal of Vegetation Science*, **11**, 383–396.

Tuomisto, H., Ruokolainen, K., Poulsen, A. D., *et al.* (2002). Distribution and diversity of pteridophytes and Melastomataceae along edaphic gradients in Yasuni National Park, Ecuadorian Amazonia. *Biotropica*, **34**, 516–33.

Villalobos-Salazar, J., Hernández, H., Meneses, A. and Salazar, G. (2000). Factors which may affect ptaquiloside levels in milk: effects of altitude, bracken fern growth stage and milk processing. In *Bracken Fern: Toxicity, Biology & Control*, ed. J. A. Taylor and R. T. Smith. Manchester, UK: International Bracken Group, Special Publication, pp. 68–74.

Walker, L. R. and Shiels, A. B. (2008). Post-disturbance erosion impacts carbon fluxes and plant succession on recent tropical landslides. *Plant and Soil*, **313**, 205–16.

Walker, L. R., Zarin, D. J., Fetcher, N., Myster, R. W. and Johnson, A. H. (1996). Ecosystem development and plant succession on landslides in the Caribbean. *Biotropica*, **28**, 566–76.

Walker, L. R., Velázquez, E. and Shiels, A. B. (2009). Ecological succession as a restoration tool on landslides. *Plant and Soil*, **324**, 157–68.

Watkins, J. E., Mack, M. C., Sinclair, T. R. and Mulkey, S. S. (2007). Ecological and evolutionary consequences of desiccation tolerance in tropical fern gametophytes. *New Phytologist*, **176**, 708–17.

Whittier, D. (2006). Gametophytes of four tropical, terrestrial *Huperzia* species (Lycopodiaceae). *American Fern Journal*, **96**, 54–61.

Wolf, P. G., Schneider, H. and Ranker, T. A. (2001). Geographic distributions of homosporous ferns: does dispersal obscure evidence of vicariance? *Journal of Biogeography*, **28**, 263–70.

Appendix A

Classification system of ferns and lycophytes

Classification system of ferns and lycophytes with key distinctive characters of classes, orders and families, with exception of monotypic taxa. Numbers of genera and species are given between parentheses

See Appendix B for an alphabetical list of genera and Fig. 1.1 for a phylogeny of the orders.

Classification	Key distinctive characters	Genera
LYCOPHYTES (5/1380) CLASS LYCOPODIOPSIDA	Small single-veined leaves (i.e., microphylls), single sporangia on adaxial leaf bases of sporophylls, sporophylls often grouped to a strobilus, eusporangiate	
Order Isoëtales	Heterosporous, leaves grass-like, secondary growth, aquatic or terrestrial and seasonal, sometimes CAM photosynthesis, ligules	
Isoëtaceae (1/150)		Isoëtes
Order Lycopodiales	Homosporous, terrestrial or epiphytic, leaves isophyllous, with gemmae in Huperzia, gametophytes often subterraneous and mycotrophic	
Lycopodiaceae (3/480)		Huperzia, Lycopodium, Lycopodiella
Order Selaginellales	Heterosporous, leaves mostly anisophyllous in four rows, mostly terrestrial or epipetric, rhizophores, ligules	
Selaginellaceae (1/750)		Selaginella
FERNS (282 /9462–9573+) CLASS PSILOTOPSIDA	>1000 spores per sporangium, eusporangiate, lacking an annulus, subterranean gametophytes	
Order Ophioglossales	Vernation not circinate, without root hairs, spores trilete	
Ophioglossaceae (3/87–92)		Botrychium, Helminthostachys, Ophioglossum
Order Psilotales	Without roots, leaves reduced, synangia of 2–3 sporangia, spores monolete	
Psilotaceae (2/12)		Psilotum, Tmesipteris

CLASS EQUISETOPSIDA

Order Equisetales	> 1000 spores per sporangium, green spores with elaters, stems branched with leaves whorled and fused into sheaths, sporangia lacking an annulus, borne on peltate sporangiophores	
Equisetaceae (1/15)		*Equisetum*

CLASS MARATTIOPSIDA

Order Marattiales	> 1000 spores per sporangium, eusporangiate, large, fleshy leaves with stipules, mucilage canals, swollen pulvinae, synangia of several sporangia	
Marattiaceae (6/91–111)		*Angiopteris, Christensenia, Danaea, Eupodium, Marattia, Ptisana*

CLASS POLYPODIOPSIDA — Vernation circinate, leptosporangiate, < 512 spores per sporangium

Order Osmundales	128–512 spores per sporangium, stipules green, trilete spores, leaves dimorphic or with distinct fertile pinnae	
Osmundaceae (4/19)		*Leptopteris, Osmunda, Osmundastrum, Todea*
Order Hymenophyllales	Gametophytes often with gemmae, indusia tubular or bivalvate, green, trilete spores, lamina one cell thick, without stomata	
Hymenophyllaceae (9/429+)		*Abrodictyum, Callistopteris, Cephalomanes, Crepidomanes, Didymoglossum, Hymenophyllum, Polyphlebium, Trichomanes, Vandenboschia*
Order Gleicheniales	Heterogenous order with families grouped together because of molecular and antheridial characteristics, roots with 3–5 protoxylem poles	
Gleicheniaceae (6/138)	Scramblers, indeterminate, pseudodichotomously branched leaves with axillary buds, >128 spores per sporangium	*Dicranopteris, Diplopterygium, Gleichenella, Gleichenia, Sticherus, Stromatopteris*
Dipteridaceae (2/11)	64–128 spores per sporangium	*Cheiropleuria, Dipteris*
Matoniaceae (2/4)	Blades flabellate	*Matonia, Phanerososrus*

Order Schizaeales

	No well defined sori, 128–256 spores per sporangium, sporangia with transverse, subapical, continuous annulus	
Lygodiaceae (1/25)	Climbers, indeterminate leaves with axillary buds	*Lygodium*
Anemiaceae (1/120)	Sporangia on 2 to several, erect, basal pinnae	*Anemia*
Schizaeaceae (2/35–40)	Simple grass-like or fan-shaped leaves	*Actinostachys, Schizaea*

Order Salviniales

	Water ferns, heterosporous, endosporic gametophytes	
Marsileaceae (3/54–77)	Rooting aquatic ferns, sori in bean-shaped sporocarps, 0, 2 or 4 leaflets	*Marsilea, Pilularia, Regnellidium*
Salviniaceae (2/16)	Floating aquatic ferns, some N-fixers, small leaves	*Azolla, Salvinia*

Order Cyatheales

	Mostly trunk-forming tree ferns, trilete spores	
Thyrsopteridaceae (1/1)	Monospecific, restricted to Juan-Fernandez Island (Chile)	*Thyrsopteris*
Loxomataceae (2/2)	Long-creeping rhizomes, marginal sori	*Loxoma, Loxsomopsis*
Culcitaceae (1/2)	Marginal sori	*Culcita*
Plagiogyriaceae (1/15–20)	Trunkless, without scales and hairs	*Plagiogyria*
Cibotiaceae (1/10)	Hairy tree ferns, marginal sori	*Cibotium*
Cyatheaceae (5/558)	Scaly tree ferns with abaxial sori	*Alsophila, Cyathea, Gymnosphaera, Hymenophyllopsis, Sphaeropteris*
Dicksoniaceae (3/28)	Hairy tree ferns, sori abaxial or marginal	*Calochlaena, Dicksonia, Lophosoria*
Metaxyaceae (1/2)	Trunkless, sori abaxial, exindusiate	*Metaxya*

Order Polypodiales

	Vertical annulus interrupted by sporangial stalk, sori abaxial	
Lindsaeaceae (8/199–200)	Indusia open outwardly	*Cystodium, Lindsaea, Lonchitis, Odontosoria, Ormoloma, Sphenomeris, Tapeinidium, Xyropteris*
Saccolomataceae (1/12)	Indusia marginal	*Saccoloma*
Dennstaedtiaceae (11/189)	Long-creeping rhizomes, often epipetiolar buds, often scramblers and species of forest margins and disturbed habitats	*Blotiella, Coptidipteris, Dennstaedtia, Histiopteris, Hypolepis, Leptolepia, Microlepia, Monachosorum, Oenotrichia, Paesia, Pteridium*

Family (genera/species)	Characters	Genera
Pteridaceae (53/1104–1110+)	Sori mostly marginal with false indusia, or along veins and exindusiate, or acrostichoid in *Acrostichum*, indeterminate leaves (e.g., *Jamesonia*), xerophytes (e.g., *Cheilanthes*), farina (e.g., *Notholaena*), green spores (e.g., *Vittaria*)	*Acrostichum, Actiniopteris, Adiantopsis, Adiantum, Aleuritopteris, Ananthacorus, Anetium, Anogramma, Antrophyum, Argyrochosma, Aspidotis, Astrolepis, Austrogramme, Bommeria, Ceratopteris, Cerosora, Cheilanthes, Cheiloplecton, Coniogramme, Cosentinia, Cryptogramma, Doryopteris, Eriosorus, Haplopteris, Hecistopteris, Hemionitis, Jamesonia, Llavea, Mildella, Monogramma, Nephopteris, Notholaena, Ochropteris, Onychium, Paraceterach, Parahemionitis, Pellaea, Pellaeopsis, Pentagramma, Pityrogramma, Platyloma, Platyzoma, Polytaenium, Pteris, Pterozonium, Radiovittaria, Rheopteris, Scoliosorus, Syngramma, Taenitis, Trachypteris, Vaginularia, Vittaria*
Aspleniaceae (2/745) Thelypteridaceae (5/933)	Clathrate rhizome scales, linear sori along veins Petioles with 2 vascular bundles, sori abaxial, round to oblong, mostly terrestrial	*Asplenium, Hymenasplenium* *Cyclosorus, Macrothelypteris, Phegopteris, Pseudophegopteris, Thelypteris*
Woodsiaceae (15/735)	Petioles with 2 vascular bundles, mostly terrestrial	*Acystopteris, Athyrium, Cheilanthopsis, Cornopteris, Cystopteris, Deparia, Diplaziopsis, Diplazium, Gymnocarpium, Hemidictyum, Homalosorus, Protowoodsia, Pseudocystopteris, Rhachidosorus, Woodsia*
Blechnaceae (9/244)	Sori along midvein, indusia open inwardly	*Blechnum, Brainea, Doodia, Pteridoblechnum, Sadleria, Salpichlaena, Steenisioblechnum, Stenchlaena, Woodwardia*
Onocleaceae (4/5)	Green spores, leaves strongly dimorphic	*Matteuccia, Onoclea, Onocleopsis, Pentarhizidium*
Dryopteridaceae (33/1663–1685+)	Petioles with 3 or more vascular bundles, sori mostly round to oblong, but acrostichoid in *Elaphoglossum*	*Acrophorus, Adenoderris, Arachniodes, Ataxipteris, Bolbitis, Coveniella, Ctenitis, Cyclodium, Cyrtogonellum, Cyrtomium, Didymochlaena, Dryopolystichum, Dryopsis, Dryopteris, Elaphoglossum, Hypodematium, Lastreopsis, Leucostegia, Lomagramma, Maxonia, Megalastrum, Oenotrichia, Olfersia, Peranema,*

Family (genera/species)	Description	Genera
Lomariopsidaceae (4/71)	Creeping to climbing rhizomes, some hemiepiphytes	*Phanerophlebia, Polybotrya, Polystichopsis, Polystichum, Rewattsia, Rumohra, Stenolepia, Stigmatopteris, Teratophyllum*
Tectariaceae (10/294–297)	Petioles with 3 or more vascular bundles, terrestrial, veins often areolate with included veins	*Cyclopeltis, Lomariopsis, Nephrolepis, Thysanosoria* *Aenigmopteris, Arthropteris, Heterogonium, Hypoderris, Pleocnemia, Psammiosorus, Psomiocarpa, Pteridrys, Tectaria, Triplophyllum*
Oleandraceae (1/40)	Leaves articulate, abscising leaving phyllopodia	*Oleandra*
Davalliaceae (5/665)	Rhizomes long-creeping, petioles abscising at base, epiphytic and epipetric	*Araiostegia, Davallia, Davallodes, Humata, Wibelia*
Polypodiaceae (62/1487–1508+)	Sori round to oblong, exindusiate, leaves abscising often leaving phyllopodia, green spores (e.g., *Grammitis*), mostly epiphytic and epipetric, veins often areolate with included veins	*Acosorus, Adenophorus, Aglaomorpha, Arthromeris, Belvisia, Calymmodon, Campyloneurum, Caobangia, Ceradenia, Christiopteris, Chrysogrammitis, Cochlidium, Colysis, Ctenopterella, Dasygrammitis, Dictymia, Drymotaenium, Drynaria, Enterosora, Goniophlebium, Grammitis, Gymnogrammitis, Kontumia, Lecanopteris, Lellingeria, Lemmaphyllum, Lepidogrammitis, Lepisorus, Leptochilus, Lomaphlebia, Lomagramme, Luisma, Melpomene, Microgramma, Micropolypodium, Microsorum, Neocheiropteris, Niphidium, Oreogrammitis, Pecluma, Phlebodium, Phymatopteris, Phymatosorus, Platycerium, Pleopeltis, Pleurosoriopsis, Podosorus, Polypodiodes, Polypodium, Prosaptia, Pyrrosia, Radiogrammitis, Scleroglossum, Selliguea, Serpocaulon, Synammia, Terpsichore, Themelium, Thylacopteris, Tomophyllum, Xiphopterella, Zygophlebia*

Source: From Kramer *et al.* (1995), Smith *et al.* (2008). Suprageneric classification of ferns follows Smith *et al.* (2008) and suprageneric nomenclature of lycophytes follows Hoogland and Reveal (2005). Generic classification and the estimates of genus and species counts in parentheses were provided by A. R. Smith.

References

Hoogland, R. D. and Reveal, J. L. (2005). Index Nominum Familiarum Plantarum Vascularium. *The Botanical Review*, **71**, 1–291.

Kramer, K. U., Schneller, J. J. and Wollenweber, E. (1995). *Farne und Farnverwandte*. Stuttgart: Georg Thieme Verlag.

Smith, A. R., Pryer, K. M., Schuettpelz, E., *et al.* (2008). Fern classification. In *Biology and Evolution of Ferns and Lycophytes*, ed. T. A. Ranker and C. H. Haufler. Cambridge, UK: Cambridge University Press, pp. 417–67.

Appendix B

Index to genera of ferns and lycophytes in alphabetical order

ALAN R. SMITH

Index to genera of ferns and lycophytes and their family placement. Families listed below are those accepted by Smith *et al.* (2006, 2008). Numbers in parentheses correspond to family numbers assigned by Smith *et al.* (2006), and are indicated for each accepted genus. All accepted genera (but not all synonyms) in Pichi Sermolli (1977), Ching (1978), and Kramer (1990) are accounted for. Newly described or recircumscribed genera since 1990 are also included. Accepted names are in roman, synonyms are in italics; for both accepted names and synonyms, the number of species (sometimes approximate) is given, except for synonyms whose circumscription varies sufficiently (or is unclear) such that this number would be relatively meaningless. An asterisk (*) indicates genera likely or soon to undergo redefinition or inclusion in another accepted genus, based on existing morphological and molecular data. Superscript 1 (^{1}) indicates genera in which species circumscription requires more study before limits are clear. Sources for numbers of species include Copeland (1947), Kramer (1990), and several recent publications, e.g., Ebihara *et al.* (2006), as well as some unpublished information by Smith. IPNI (The International Plant Names Index: http://www.ipni.org/) has also been consulted for many genera, in order to incorporate recently described species in the totals.

For more complete references documenting the family level classification, see Smith *et al.* (2006, 2008). A few accepted names given herein are not validly published, as, for example, some in Thelypteridaceae (see Smith, in Kramer, 1990) and are indicated by superscript 2 (^{2}).

References to family letters and numbers

Lycophytes: Lycopodiaceae (A), Selaginellaceae (B), Isoëtaceae (C).

Ferns (Monilophytes): Ophioglossaceae (1), Psilotaceae (2), Equisetaceae (3), Marattiaceae (4), Osmundaceae (5), Hymenophyllaceae (6), Gleicheniaceae (7), Dipteridaceae (8), Matoniaceae (9), Lygodiaceae (10), Anemiaceae (11), Schizaeaceae (12), Marsileaceae (13), Salviniaceae (14), Thyrsopteridaceae (15),

Loxomataceae (16), Culcitaceae (17), Plagiogyriaceae (18), Cibotiaceae (19), Cyatheaceae (20), Dicksoniaceae (21), Metaxyaceae (22), Lindsaeaceae (23), Saccolomataceae (24), Dennstaedtiaceae (25), Pteridaceae (26), Aspleniaceae (27), Thelypteridaceae (28), Woodsiaceae (29), Blechnaceae (30), Onocleaceae (31), Dryopteridaceae (32), Lomariopsidaceae (33), Tectariaceae (34), Oleandraceae (35), Davalliaceae (36), Polypodiaceae (37).

Genera

Abacopteris	= Cyclosorus subg. Abacopteris
Abrodictyum (6):	25+ spp. (Ebihara *et al.*, 2006)
Acarpacrium	= Trichomanes
Achomanes	= Trichomanes
Acropelta: 1 sp.	= Polystichum
Acrophorus (32):	2 spp.
Acrorumohra: 7 spp.	= Dryopteris
Acrosorus (37):	*c.* 7 spp.
Acrostichum (26):	3 spp.
Actiniopteris (26):	5 spp.
Actinostachys* (12):	*c.* 15–20 spp.
Acystopteris (29):	2 spp.
Adenoderris (32):	2 spp.
Adenophorus (37):	10 spp.
Adiantopsis (26):	*c.* 20 spp. (Barker and Hickey, 2006)
Adiantum (26):	*c.* 200 spp.
Aenigmopteris (34):	5 spp.
Afropteris: 2 sp.	= Pteris
Aglaomorpha (37):	32 spp.
Aleuritopteris* (26):	*c.* 45 spp. (Schuettpelz *et al.*, 2007)
Allantodia: *c.* 100 spp.	= Diplazium
Alsophila (20):	*c.* 235 spp.
Amauropelta: *c.* 200 spp.	= Thelypteris subg. Amauropelta
Ampelopteris: 1 sp.	= Cyclosorus subg. Ampelopteris
Amphiblestra: 1 sp.	= Tectaria
Amphidesmium	= Metaxya
Amphineuron: 12 spp.	= Cyclosorus subg. Amphineuron[2]
Amphipteron: 5 spp.	= Hymenophyllum subg. Hymeno-phyllum (Ebihara *et al.*, 2006)
Amphoradenium	= Adenophorus

Ananthacorus (26): 1 sp.
Anapausia = Bolbitis
Anapeltis: *c*. 10 spp. = Microgramma
Anarthropteris: 1 sp. = Loxogramme
Anchistea: 1 sp. = Woodwardia
Anemia (11): *c*. 120 spp.
Anetium (26): 1 sp.
Angiopteris[1] (4): *c*. 10–30 spp. [more than 200 described] (Murdock, 2007)

Anisocampium: 2 spp. = Athyrium
Anogramma (26): *c*. 2 spp. (Nakazato and Gastony, 2003)
Anopteris: 1 sp. = Pteris
Antigramma: *c*. 3 spp. = Asplenium
Antrophyum (26): *c*. 20 spp. (Crane, 1997)
Apteropteris: 2 spp. = Hymenophyllum subg. Sphaerocionium (Ebihara *et al.*, 2006)

Arachniodes (32): 50–70 spp.
Araiostegia = Davallodes (Tsutsumi and Kato, 2005; Kato and Tsutsumi, 2008, Tsutsumi *et al.*, 2008)

Araiostegiella (36): 3 spp. (Kato and Tsutsumi, 2008)
Archangiopteris: *c*. 5 spp. = Angiopteris
Arcypteris = Pleocnemia
Argyrochosma (26): 18+ spp.
Arthrobotrya: 3 spp. = Teratophyllum
Arthromeris[1] (37): *c*. 10–15 spp.
Arthropteris (34): 12–15 spp.
Aspidotis (26): 4 spp.
Aspleniopsis: 2 or 3 spp. = Austrogramme
Asplenium (27): *c*. 720 spp.
×*Asplenoceterach* [*Asplenium* × *Ceterach*]: 3 sterile hybrids and fertile allopolyploid spp. = Asplenium
×*Asplenophyllitis* [*Asplenium* × *Phyllitis*]: 7 sterile hybrids and fertile allopolyploid spp. = Asplenium
×*Asplenosorus* [*Asplenium* × *Camptosorus*]: 13 sterile hybrids and fertile allopolyploid spp. = Asplenium

Astrolepis (26):	5 spp.
Atalopteris: 3 spp.	= Ctenitis
Ataxipteris (32):	2 spp.
Athyriopsis: *c.* 13 spp.	= Deparia sect. Athyriopsis
Athyrium (29):	*c.* 180 spp.
Austrogramme (26):	5 spp.
Azolla (14):	6 spp.
Belvisia (37):	8 spp.
Blechnidium: 1 sp.	= Blechnum
Blechnum* (30):	*c.* 200 spp.
Blotiella (25):	*c.* 18 spp.
Bolbitis (32):	*c.* 50 spp.
Bommeria (26):	4 spp.
Boniniella: 1 sp.	= Hymenasplenium (Kato *et al.*, 1990)
Botrychium (1):	*c.* 60 spp.
Botrypus: 2 or 3 spp.	= Botrychium
Brainea (30):	1 sp.
Bryodesma: *c.* 45 spp.	= Selaginella
Buesia: *c.* 5 spp.	= Hymenophyllum subg. Hymeno-phyllum (Ebihara *et al.*, 2006)
Callipteris	= Diplazium
Callistopteris (6):	*c.* 5 spp. (Ebihara *et al.*, 2006)
Calochlaena (21):	5 spp.
Calymella	= Gleichenia
Calymmodon (37):	*c.* 30 spp.
Camptodium: 2 spp.	= Tectaria
Camptosorus: 2 spp.	= Asplenium
Campyloneurum (37):	*c.* 52 spp.
Caobangia (37):	1 sp.
Cardiomanes: 1 sp.	= Hymenophyllum subg. Cardiomanes (Ebihara *et al.*, 2006)
Cassebeera	= Doryopteris
Cephalomanes (6):	*c.* 4 spp. (Ebihara *et al.*, 2006)
Ceradenia (37):	70+ spp.
Ceratopteris (26):	3 spp.
Cerosora (26):	3 spp.
Ceterach	= Asplenium
Ceterachopsis: *c.* 5 spp.	= Asplenium
Cheilanthes* (26):	*c.* 150 spp.

Cheilanthopsis (29):	4 spp.
Cheiloplecton (26): 2 or 3 spp.	= Notholaena (Rothfels *et al.*, 2008)
Cheilosoria: *c*. 10 spp.	= Cheilanthes?
Cheiroglossa: 1 sp.	= Ophioglossum
Cheiropleuria (8):	3 spp. (Kato *et al.*, 2001)
Chieniopteris: 3 spp.	= Woodwardia
Chingia: 22 spp.	= Cyclosorus subg. Chingia[2]
Chlamydogramme: 2 spp.	= Tectaria
Choristosoria: 1 sp.	= Pellaeopsis
Christella: *c*. 50 spp.	= Cyclosorus subg. Cyclosoriopsis
Christensenia (4):	2 spp.
Christiopteris (37):	2 spp.
Chrysochosma	= Notholaena
Chrysogrammitis (37):	2 spp.
Cibotium (19):	*c*. 10 spp.
Cionidium: 1 sp.	= Tectaria
Cnemidaria: *c*. 27 spp.	= Cyathea
Cochlidium (37):	16 spp.
Colysis (37):	*c*. 30 spp.
Coniogramme (26):	12+ spp.
Copelandiopteris: 3 sp.	= Pteris
Coptidipteris (25):	1 sp.
Cornopteris (29):	9 spp.
Coryphopteris: 47 spp.	= Thelypteris subg. Coryphopteris
Cosentinia (26):	1 sp.
Costaricia: 1 sp.	= Dennstaedtia
Coveniella (32):	1 sp.
Craspedaria	= Microgramma
Craspedodictyum: 2 spp.	= Syngramma
Craspedophyllum: *c*. 3 spp.	= Hymenophyllum subg. Hymeno-phyllum (Ebihara *et al.*, 2006)
Craspedosorus: 1 sp.	= Cyclosorus subg. Stegnogramma
Crepidium	= Crepidomanes
Crepidomanes (6):	30+ spp. (Ebihara *et al.*, 2006)
Crepidopteris: 1 sp.	= Crepidomanes sect. Crepidium (Ebihara *et al.*, 2006)
Crypsinopsis: *c*. 5 spp.	= Selliguea
Crypsinus	= Selliguea
Cryptogramma (26):	8–11 spp.

Appendix B

Dictyodroma: 2 or 3 spp.   = Deparia

Dictyoxiphium: 1 sp.   = Tectaria

Didymochlaena (32):   1 or 2 spp. (Ebihara *et al.*, 2006)

Didymoglossum (6):   *c.* 35 spp.

Diellia: 6 spp.   = Asplenium

Diphasiastrum: *c.* 20 spp.   = Lycopodium sect. Complanata (Øllgaard, 1989)

Diphasium: *c.* 6 spp.   = Lycopodium sect. Diphasium (Øllgaard, 1989)

Diplaziopsis (29):   *c.* 3 spp.

Diplazium (29):   *c.* 400 spp.

Diploblechnum: 1 sp.   = Blechnum

Diplopterygium (7):   *c.* 25 spp.

Diplora: *c.* 5 spp.   = Asplenium

Dipteris (8):   *c.* 8 spp.

Doodia (30):   *c.* 12 spp.

Doryopteris (26):   *c.* 50 spp.

Dracoglossum: 2 spp.   = Tectaria (Christenhusz, 2007b)

Drymoglossum: 6 sp.   = Pyrrosia

Drymotaenium (37):   1 sp.

Drynaria (37):   *c.* 15 spp.

Drynariopsis: 1 sp.   = Aglaomorpha

Dryoathyrium: *c.* 20 spp.   = Deparia sect. Dryoathyrium

Dryopolystichum (32):   1 sp.

Dryopsis (32):   25 spp.

Dryopteris (32):   *c.* 225 spp.

Dryostachyum: 1 sp.   = Aglaomorpha

Edanyoa: 1 sp.   = Bolbitis

Egenolfia c. 13 spp.   = Bolbitis

Elaphoglossum (32):   700+ spp.

Emodiopteris: 2 spp.   = Dennstaedtia

Enterosora (37):   12 spp.

Equisetum (3):   15 spp.

×*Eriosonia*: 3+ hybrids   = Eriosorus × Jamesonia hybrids

Eriosorus* (26):   *c.* 30 spp.

Eschatogramme   = Pleopeltis

Eupectinatum   = Hymenophyllum

Eupodium (4):   2 spp. (Murdock, 2007, 2008a, 2008b)

Fadyenia: 1 sp.   = Tectaria

Feea: 5 spp. = Trichomanes subg. Feea (Ebihara *et al.*, 2006)

Flabellata = Hymenophyllum subg. Sphaerocionium

Fourniera = Sphaeropteris

Glaphyropteridopsis: 4 spp. = Cyclosorus subg. Glaphyropterid-opsis[2]

Glaphyropteris = Cyclosorus subg. Steiropteris[2]

Gleichenella (7): 1 sp.

Gleichenia (7): *c.* 10 spp.

Gleicheniastrum = Gleichenia

Glyphotaenium = Enterosora

Goniophlebium (37): *c.* 15 spp.

Goniopteris: *c.* 75 spp. = Cyclosorus subg. Goniopteris[2]

Gonocormus: 1+ sp. = Crepidomanes sect. Gonocormus (Ebihara *et al.*, 2006)

Grammatopteridium: 1 sp. = Selliguea

Grammitis* (37): *c.* 30 spp., polyphyletic, includes elements to be segregated as new genera, by B. Parris; Grammitis s.s. is *c.* 15 spp.

Gymnocarpium (29): 6 spp.

Gymnogramma = Hemionitis

Gymnogrammitis (37): 1 sp.

Gymnopteris: 2 spp. = Hemionitis

Gymnosphaera (20): *c.* 30 spp. (Korall *et al.*, 2007)

Haplodictyum: 5 spp. = Cyclosorus subg. Abacopteris

Haplopteris (26): *c.* 25 spp.

Hecistopteris (26): 3 spp.

Helminthostachys (1): 1 sp.

Hemicyatheon: 2 spp. = Hymenophyllum subg. Hymenophyllum (Ebihara *et al.*, 2006)

Hemidictyum (29): 1 sp.

Hemigramma: 9 spp. = Tectaria

Hemionanthes: 1 sp. = Cheilanthes s.l.

Hemionitis (26): 7 spp.

Hemipteris = Pteris

Hemitelia = Cyathea

Heterogonium (34): 22 spp.

Hicriopteris: 1 sp. = Dicranopteris

Hippochaete: 7 spp. = Equisetum subg. Hippochaete

Histiopteris (25):   *c.* 7 spp.
Holcosorus: 2 spp.   = Selliguea
Holodictyum: 1 sp.   = Asplenium
Holostachyum: 1 sp.   = Aglaomorpha
Holttumiella: 1 sp.   = Taenitis
Homalosorus (29):   2 spp. (Kato and Darnaedi, 1988)
Homoëtes: 1 sp.   = Trichomanes subg. Trichomanes
                                      (Ebihara *et al.*, 2006)

Humata (36):   *c.* 25 spp. (Tsutsumi *et al.*, 2008)
Humblotiella: 1 sp.   = Lindsaea
Huperzia (A):   *c.* 400 sp.
Hyalotricha: 1 sp.   = Campyloneurum
Hyalotrichopteris: 1 sp.   = Campyloneurum
Hymenasplenium (27):   *c.* 25 spp.
Hymenocystis: 1 or 2 spp.   = Woodsia
Hymenoglossum: 3 sp.   = Hymenophyllum subg. Hymeno-
                                      glossum (Ebihara *et al.*, 2006)

Hymenophyllopsis (20):   8 spp.
Hymenophyllum (6):   *c.* 250 spp. (Ebihara *et al.*, 2006)
Hymenostachys: 2 spp.   = Trichomanes subg. Feea (Ebihara
                                      et al., 2006)

Hypodematium (32):   4 spp.
Hypoderris (34):   1 sp.
Hypolepis[1] (25):   *c.* 40 spp.
Idiopteris: 1 sp.   = Pteris
Isoëtes (C):   *c.* 150 spp.
Isoloma   = Lindsaea
Jamesonia (26):   18 spp.
Japanobotrychium: 1 or 2 spp.   = Botrychium
Kontumia (37):   1 sp.
Kuniwatsukia: 1 sp.   = Athyrium
Lacostea: 5 spp.   = Trichomanes subg. Lacostea (Ebihara
                                      et al., 2006)

Lacosteopsis: 2+ spp.   = Vandenboschia subg. Lacosteopsis
                                      (Ebihara *et al.*, 2006)

Lastrea: *c.* 3 spp.   = Thelypteris subg. Lastrea
Lastreopsis (32):   *c.* 40 spp.
Lecanium: 1 sp.   = Didymoglossum subg. Didymoglossum
                                      (Ebihara *et al.*, 2006)

Lycopodium (A):   *c.* 40 spp.
Lygodium (10):   *c.* 25 spp.
Macroglena: 12+ spp.   = Abrodictyum subg. Pachychaetum
                                         (Ebihara *et al.*, 2006)

Macroglossum: 1 or 2 spp.   = Angiopteris
Macrothelypteris (28):   *c.* 10 spp.
Mankyua (1):   1 sp.
Marattia (4):   *c.* 7 spp. (Murdock, 2007, 2008b)
Marginaria   = Pleopeltis
Marginariopsis: 1 sp.   = Pleopeltis
Marsilea (13):   50–70 spp.
Matonia (9):   2 spp. (Kato, 1993)
Matteuccia (31):   1 sp.
Maxonia (32):   1 sp.
Mecodium: *c.* 35+ spp.   = Hymenophyllum subg. Mecodium
                                         (Ebihara *et al.*, 2006)

Megalastrum[1] (32):   *c.* 50+ spp.
Melpomene (37):   *c.* 30 spp.
Meniscium: *c.* 22 spp.   = Cyclosorus subg. Meniscium[2]
Menisorus: 1 sp.   = Cyclosorus subg. Menisorus[2]
Meringium   = Hymenophyllum subg. Hymenophyl-
                                         lum (Ebihara *et al.*, 2006)

Merinthosorus: 1 sp.   = Aglaomorpha
Mesophlebion: 19 spp.   = Cyclosorus subg. Mesophlebion[2]
Mesopteris: 1 sp.   = Cyclosorus subg. Amphineuron
Metapolypodium: 1 sp.   = Polypodiodes (Lu and Li, 2006)
Metathelypteris: *c.* 14 spp.   = Thelypteris subg. Metathelypteris
Metaxya (22):   2 spp.
Microchlaena   = Athyrium
Microgonium: *c.* 11 spp.   = Didymoglossum subg. Microgonium
                                         (Ebihara *et al.*, 2006)

Microgramma (37):   *c.* 30 spp.
Microlepia[1] (25):   *c.* 45 spp.
Microphlebodium: 2 spp.   = Pleopeltis
Micropolypodium (37):   *c.* 30 spp.
Microsorum* (37):   *c.* 60 spp.
Microstaphyla: 1 sp.   = Elaphoglossum
Microtrichomanes: 10 spp.   = Hymenophyllum subg. Sphaerocionium
(polyphyletic)   (Ebihara *et al.*, 2006)

Mildella* (26):	9 spp.
Mohria: 7 spp.	= Anemia
Monachosorum (25):	*c.* 3 spp.
Monogramma: *c.* 2 spp.	= nested within Haplopteris (Ruhfel *et al.*, 2008); Monogramma is older, suggesting the desirability of conserving Haplopteris
Monomelangium: 1 sp.	= Diplazium
Mortoniopteris: 1 sp.	= Trichomanes subg. Trichomanes (Ebihara *et al.*, 2006)
Myriodon: 1(2) spp.	= Hymenophyllum subg. Hymenophyllum (Ebihara *et al.*, 2006)
Myriopteris	= Cheilanthes
Myrmecopteris: 5 spp.	= Lecanopteris
Nannothelypteris: 7 spp.	= Cyclosorus subg. Abacopteris
Negripteris: 3 spp.	= Aleuritopteris
Nematopteris: 2 spp.	= Scleroglossum
Neoathyrium: 1 sp.	= Cornopteris
Neocheiropteris (37):	15–20 spp.
Neolepisorus: *c.* 10 spp.	= Neocheiropteris
Neottopteris: *c.* 20 spp.	= Asplenium
Nephelea: 22 spp.	= Alsophila
Nephopteris (26):	1 sp.
Nephrolepis (33):	20 spp.
Nesopteris: 4+ spp.	= Crepidomanes subg. Nesopteris (Ebihara *et al.*, 2006)
Neurocallis: 1 sp.	= Pteris (Schuettpelz *et al.*, 2007)
Neurodium: 2 spp.	= Pleopeltis
Neuromanes: 1 sp.	= Trichomanes subg. Trichomanes (Ebihara *et al.*, 2006)
Neurophyllum	= Trichomanes
Neurosoria: 1 sp.	= Cheilanthes
Niphidium (37):	10 spp.
Notholaena (26):	*c.* 34 spp.
Nothoperanema: 5 spp.	= Dryopteris
Ochropteris (26):	1 sp.
Odontomanes: 1 sp.	= Trichomanes subg. Trichomanes (Ebihara *et al.*, 2006)
Odontosoria* (23):	*c.* 15 spp.

Oenotrichia (25):   2 spp.
Oenotrichia p.p. (32):   1 sp. (Tindale and Roy, 2002)
Oleandra (35):   *c.* 40 spp.
Oleandropsis: 1 sp.   = Selliguea
Olfersia (32):   2 spp.
Onoclea (31):   1 sp.
Onocleopsis (31):   1 sp.
Onychium (26):   *c.* 8 spp.
Ophioderma: 1(4) spp.   = Ophioglossum
Ophioglossum (1):   25–30 spp.
Oreogrammitis (37):   110+ spp. (Parris, 2007)
Oreopteris: 3 spp.   = Thelypteris subg. Lastrea
Ormoloma (23):   1 or 2 spp.
Ormopteris: c/ 7 spp.   = Doryopteris
Orthiopteris   = Saccoloma
Osmunda (5):   *c.* 10 spp. (Metzgar *et al.*, 2008)
Osmundastrum (5):   1 sp. (Metzgar *et al.*, 2008)
Pachychaetum   = Abrodictyum subg. Pachychaetum
Pachypleuria   = Humata
Paesia (25):   *c.* 12 spp.
Palhinhaea: *c.* 30 spp.   = Lycopodiella sect. Campylostachys (Øllgaard, 1989)

Paltonium: 1 sp.   = Neurodium
Papuapteris: 1 sp.   = Polystichum
Paraceterach (26):   7 spp.
Paradavallodes: 4 spp.   = Davallodes (Tsutsumi and Kato, 2005; Tsutsumi *et al.*, 2008)

Paragramma   = Lepisorus
Paragymnopteris   = Parahemionitis (26): 1 sp.
Paraleptochilus: 3 spp.   = Colysis
Parasorus: 1 sp.   = Davallia
Parathelypteris: *c.* 15 spp.   = Thelypteris subg. Parathelypteris
Pecluma* (37):   *c.* 40 spp.
Pelazoneuron: *c.* 25   = Cyclosorus subg. Pelazoneuron[2]
Pellaea* (26):   *c.* 25 spp.
Pellaeopsis* (26):   *c.* 10 spp.
Peltapteris: 3 spp.   = Elaphoglossum
Pentagramma (26):   2 spp.
Pentarhizidium (31):   2 spp.

Peranema (32): 3 or 4 spp.
Pessopteris: 9 spp. = Niphidium
Phanerophlebia (32): 9 spp.
Phanerophlebiopsis: *c.* 4 or 5 spp. = Arachniodes
Phanerosorus (9): 2 spp.
Phegopteris (28): 3 spp.
Phlebiophyllum = Polyphlebium
Phlebodium (37): 3 spp.
Phlegmariurus = Huperzia
Photinopteris: 1 sp. = Aglaomorpha
Phyllitis: *c.* 4 spp. = Asplenium
Phylloglossum = Huperzia
Phymatodes = Phymatosorus
Phymatopsis = Phymatopteris
Phymatopteris (37): *c.* 60 spp.
Phymatosorus* (37): *c.* 15–20 spp.
Pilularia (13): 3–6 spp.
Pityrogramma (26): *c.* 22 spp.
Plagiogyria (18): 15–20 spp.
Platycerium (37): 15 spp.
Platygyria: 5 spp. = Lepisorus
Platyloma (26): 5 spp.
Platytaenia = Taenitis
Platyzoma (26): 1 sp.
Plecosorus: 1 sp. = Polystichum
Plenasium = Osmunda
Pleocnemia (34): 19 spp.
Pleopeltis* (37): *c.* 90 spp.
Plesioneuron: *c.* 45 spp. = Cyclosorus subg. Plesioneuron[2]
×*Pleuroderris*: 1 sp. = Tectaria hybrid
Pleuromanes: 5 spp. = Hymenophyllum subg. Pleuromanes (Ebihara *et al.*, 2006)

Pleurosoriopsis (37): 1 sp.
Pleurosorus: 3(4) spp. = Asplenium
Pneumatopteris: *c.* 80 spp. = Cyclosorus subg. Pneumatopteris
Podosorus (37): 1 sp.
Polybotrya (32): 35 spp.
Polyphlebium (6): *c.* 15 spp. (Ebihara *et al.*, 2006)
Polypodiastrum: *c.* 7 spp. = Polypodiodes (Lu and Li, 2006)

Polypodiodes (37):	*c.* 25 spp.
Polypodiopsis	= Selliguea
Polypodiopteris: 3 spp.	= Selliguea
Polypodium (37):	*c.* 45 spp.
Polystichopsis (32):	6 spp.
Polystichum (32):	*c.* 200 spp.
Polytaenium (26):	8–10 spp.
Pronephrium: *c.* 72 spp.	= Cyclosorus subg. Abacopteris
Prosaptia (37):	*c.* 35 spp.
Protomarattia: 1 sp.	= Angiopteris
Protowoodsia (29):	1 sp.
Psammiosorus (34):	1 sp.
Pseudocolysis: 1 sp.	= Pleopeltis
Pseudocyclosorus: *c.* 12 spp.	= Cyclosorus subg. Pseudocyclosorus
Pseudocystopteris (29):	*c.* 19 spp.
Pseudodrynaria: 1 sp.	= Aglaomorpha
Pseudolycopodiella: *c.* 10 spp.	= Lycopodiella sect. Caroliniana (Øllgaard, 1989)
Pseudophegopteris (28):	*c.* 20 spp.
Pseudotectaria: 6 spp.:	= Tectaria
Psilotum (2):	2 spp.
Psomiocarpa (34):	1 sp.
Pteridanetium	= Anetium
Pteridium (25):	*c.* 5 spp.
Pteridoblechnum (30):	1 sp.
Pteridrys (34):	8 spp.
Pteris* (26):	*c.* 250 spp.
Pteromanes: 1 sp.	= Trichomanes subg. Trichomanes
Pteropsis	= Pyrrosia
Pterozonium (26):	13 spp.
Ptilopteris: 1 sp.	= Monachorosum
Ptisana (4):	*c.* 20 spp. (Murdock, 2007, 2008a, 2008b)
Ptychophyllum	= Hymenophyllum
Pycnodoria	= Pteris
Pycnoloma: 3 spp.	= Selliguea
Pyrrosia (37):	*c.* 65 spp.
Quercifilix: 1 sp.	= Tectaria
Radiogrammitis (37):	28 spp. (Parris, 2007)

Radiovittaria (26):   *c.* 10 spp.

Ragatelus   = Trichomanes

Reediella: 1+ sp.   = Crepidomanes sect. Crepidium (Ebihara *et al.*, 2006)

Regnellidium (13):   1 sp.

Revwattsia (32):   1 sp.

Rhachidosorus (29):   8 spp.

Rheopteris (26):   1 sp. (Ruhfel *et al.*, 2008)

Rhizoglossum: 1 sp.   = Ophioglossum

Rhipidopteris   = Elaphoglossum

Rosenstockia: 1 sp.   = Hymenophyllum subg. Hymenophyllum (Ebihara *et al.*, 2006)

Rumohra (32):   7 spp.

Saccoloma (24):   *c.* 12 spp.

Sadleria (30):   6 spp.

Saffordia: 1 sp.   = Trachypteris

Sagenia: *c.* 65 spp.   = Tectaria (Holttum, 1991)

Salpichlaena (30):   *c.* 2 spp.

Salvinia (14):   *c.* 10 spp.

Sambirania: 2 spp.   = Lindsaea

Saxiglossum: 1 sp.   = Pyrrosia

Sceptridium: 25–30 spp.   = Botrychium

Schaffneria: 1 sp.   = Asplenium

Schellolepis   = Goniophlebium

Schizaea (12):   *c.* 20 spp.

Schizocaena   = Sphaeropteris

Schizolepton: 1 sp.   = Taenitis

Schizoloma   = Lindsaea

Schizostege   = Pteris

Scleroglossum (37):   *c.* 8 spp.

Scoliosorus (26):   3 spp.

Scyphularia: 3 spp.   = Davallia s.l.

Selaginella (B):   *c.* 750 spp.

Selenodesmium: *c.* 15 spp.   = Abrodictyum subg. Pachychaetum (Ebihara *et al.*, 2006)

Selliguea* (37):   *c.* 55 spp.

Serpocaulon (37):   *c.* 45–50 spp.

Serpyllopsis: 1 sp.   = Hymenophyllum subg. Hymenophyllum (Ebihara *et al.*, 2006)

Sinephropteris: 1 sp.	= Asplenium
Sinopteris: *c.* 2 spp.	= Aleuritopteris (Schuettpelz *et al.*, 2007)
Solanopteris: 5 spp.	= Microgramma
Sorolepidium: 2 spp.	= Polystichum
Soromanes: 1 sp.	= Polybotrya
Sphaerocionium: *c.* 70 spp.	= Hymenophyllum subg. Sphaerocionium (Ebihara *et al.*, 2006)
Sphaeropteris (20):	*c.* 100 spp.
Sphaerostephanos: *c.* 175 spp.	= Cyclosorus subg. Sphaerostephanos
Sphenomeris (23):	*c.* 12 spp.
Steenisioblechnum (30):	1 sp.
Stegnogramma: 15 spp.	= Cyclosorus subg. Stegnogramma[2]
Steiropteris: 24 spp.	= Cyclosorus subg. Steiropteris[2]
Stenochlaena (30):	6 spp.
Stenolepia (32):	1 sp.
Stenoloma	= Odontosoria
Stenosemia: *c.* 5 spp.	= Tectaria
Sticherus (7):	*c.* 90 spp.
Stigmatopteris (32):	23 spp.
Stromatopteris (7):	1 sp.
Struthiopteris: *c.* 10 spp.	= Blechnum
Stylites: 1 sp.	= Isoëtes
Synammia (37):	3 spp.
Syngramma (26):	*c.* 15 spp.
Syngrammatopteris: 8 spp.	= Pterozonium
Taenitis (26):	*c.* 15 spp.
Tapeinidium (23):	17 spp.
Tectaria (34):	*c.* 200 spp.
Tectaridium: 1 sp.	= Tectaria
Teratophyllum (32):	9 spp.
Terpsichore (37):	*c.* 60 spp.
Thamnopteris: *c.* 20 spp.	= Asplenium
Thayeria: 1 sp.	= Aglaomorpha
Thelypteris* (28):	*c.* 280 spp.
Themelium (37):	18 spp.
Thylacopteris (37):	1 sp.
Thyrsopteris (15):	1 sp.
Thysanosoria (33):	1 sp.

Tmesipteris (2):	*c*. 10 spp.
Todea (5):	2 spp.
Tomophyllum (37):	22 spp.
Toxopteris: 4 spp.	= Syngramma
Triblemma: 2 spp.	= Deparia
Trachypteris (26):	3 spp.
Trichipteris	= Cyathea
Tricholepidium: *c*. 7 spp.	= Microsorum
Trichomanes (6):	*c*. 50 spp. (Ebihara *et al*., 2006)
Trichoneuron: 1 sp.	= Lastreopsis
Trichopteris	= Cyathea
Trigonophyllum: *c*. 2 spp.	= Trichomanes subg. Trichomanes (Ebihara *et al*., 2006)
Trigonospora: 9 spp.	= Cyclosorus subg. Trigonospora
Triplophyllum (34):	*c*. 25 spp.
Trismeria: 1 sp.	= Pityrogramma
Trogostolon: 1 sp.	= Davallia sect. Trogostolon (Kato and Tsutsumi, 2008)
Tryonella: 2 spp.	= Doryopteris
Urostachys	= Huperzia (Øllgaard, 1989)
Vaginularia (26):	*c*. 6 spp. (Ruhfel *et al*., 2008)
Vandenboschia (6):	15+ spp. (Ebihara *et al*., 2006)
Vittaria (26):	6 spp.
Weatherbya: 1 sp.	= Lemmaphyllum
Wibelia (36):	*c*. 8 spp. (Tsutsumi and Kato, 2005, Kato and Tsutsumi, 2008; Tsutsumi *et al*., 2008)
Woodsia[1] (29):	*c*. 40 spp.
Woodwardia (30):	*c*. 15 spp.
Xiphopterella (37):	6 spp.
Xiphopteris	= Cochlidium
Xyropteris (23):	1 sp.
Zygophlebia (37):	16 spp.

References

Barker, M. S. and Hickey, R. J. (2006). A taxonomic revision of Caribbean *Adiantopsis* (Pteridaceae). *Annals of the Missouri Botanical Garden*, **93**, 371–401.

Ching, R. C. (1978). The Chinese fern families and genera: systematic arrangement and historical origin. *Acta Phytotaxonomica Sinica*, **16**, 16–37.

Christenhusz, M. J. M. (2007a). Evolutionary history and taxonomy of neotropical marattioid ferns. Unpublished Ph.D. thesis, University of Turku, Finland.

Christenhusz, M. J. M. (2007b). *Dracoglossum*, a new neotropical fern genus (Pteridophyta). *Thaiszia*, **17**, 1–10.

Copeland, E. B. (1947). *Genera Filicum*. Waltham, MA, USA: Chronica Botanica.

Crane, E. H. (1997). A revised circumscription of the genera of the family Vittariaceae. *Systematic Botany*, **22**, 509–17.

Ebihara, A., Dubuisson, J.-Y., Iwatsuki, K., Hennequin, S. and Ito, M. (2006). A taxonomic revision of Hymenophyllaceae. *Blumea*, **51**, 221–80.

Holttum, R. E. (1991). *Tectaria* group. In *Flora Malesiana*, Ser. II *Pteridophyta* 2(1), ed. H. P. Nooteboom. Leiden, The Netherlands: Rijksherbarium/Hortus Botanicus, pp. 1–133.

Kato, M. (1993). A taxonomic study of the genus *Matonia* (Matoniaceae). *Blumea*, **38**, 167–72.

Kato, M. and Darnaedi, D. (1988). Taxonomic and phytogeographic relationships of *Diplazium flavoviride, D. pycnocarpon*, and *Diplaziopsis. American Fern Journal*, **78**, 77–85.

Kato, M. and Tsutsumi, C. (2008). Generic classification of Davalliaceae. *Acta Phytotaxonomica et Geobotanica*, **59**, 1–14.

Kato, M., Nakato, N., Akiyama, S. and Iwatsuki, K. (1990). The systematic position of *Asplenium cardiophyllum* (Aspleniaceae). *Botanical Magazine Tokyo*, **103**, 461–8.

Kato, M., Yatabe, Y., Sahashi, N. and Murakami, N. (2001). Taxonomic studies of *Cheiropleuria* (Dipteridaceae). *Blumea*, **46**, 513–25.

Korall, P., Conant, D. S., Metzgar, J. S., Schneider, H. and Pryer, K. M. (2007). A molecular phylogeny of scaly tree ferns (Cyatheaceae). *American Journal of Botany*, **94**, 873–86.

Kramer, K. U. (1990). Pteridophytes. In *Pteridophytes and Gymnosperms*, Vol. 1 of *The Families and Genera of Vascular Plants*, ed. K. U. Kramer and P. S. Green. Berlin: Springer-Verlag, pp. 1–277 + i–xiii.

Lu, S.-G. and Li, L.-X. (2006). Phylogenetic position of the monotypic genus *Metapolypodium* Ching endemic to Asia: evidence from chloroplast DNA sequences of *rbc*L gene and *rps4-trn*S region. *Acta Phytotaxonomica Sinica*, **44**, 494–502.

Metzgar, J. S., Skog, J. E., Zimmer, E. A. and Pryer, K. M. (2008). The paraphyly of *Osmunda* is confirmed by phylogenetic analyses of seven plastid loci. *Systematic Botany*, **33**, 31–6.

Murdock, A. G. (2007). Systematics and molecular evolution of marattioid ferns (Marattiaceae). Ph.D. thesis, University of California, Berkeley.

Murdock, A. G. (2008a). Phylogeny of marrattioid ferns (Marattiaceae): inferring a root in the absence of a closely related outgroup. *American Journal of Botany*, **95**, 626–41.

Murdock, A. G. (2008b). A taxonomic revision of the eusporangiate fern family Marattiaceae, with description of a new genus *Ptisana. Taxon*, **57**, 737–55.

Nakazato, T. and Gastony, G. J. (2003). Molecular phylogenetics of *Anogramma* species and related genera (Pteridaceae: Taenitidoideae). *Systematic Botany*, **28**, 490–502.

Øllgaard, B. (1989). Index of the Lycopodiaceae. *Biologiske Skrifter*, **34**, 1–135.

Parris, B. S. (2007). Five new genera and three new species of Grammitidaceae (Filicales) and the re-establishment of *Oreogrammitis. Gardens' Bulletin Singapore*, **58**, 233–74.

Pichi Sermolli, R. E. G. (1977). Tentamen pteridophytorum genera in taxonomicum ordinem redigendi. *Webbia*, **31**, 313–512.

Rothfels, C. J., Windham, M. D., Grusz, A. L., Gastony, G. J. and Pryer, K. M. (2008). Toward a monophyletic *Notholaena* (Pteridaceae): resolving patterns of evolutionary convergence in xeric-adapted ferns. *Taxon*, **57**, 712–24.

Rouhan G., Hanks, J. G., McClelland, D. and Moran, R. C. (2007). Preliminary phylogenetic analysis of the fern genus *Lomariopsis* (Lomariopsidaceae). *Brittonia*, **69**, 115–28.

Ruhfel, B., Lindsay, S. and Davis, C. C. (2008). Phylogenetic placement of *Rheopteris* and the polyphyly of *Monogramma* (Pteridaceae s.l.): evidence from *rbcL* sequence data. *Systematic Botany*, **331**, 37–43.

Schuettpelz, E., Schneider, H., Huiet, L., Windham, M. D. and Pryer, K. M. (2007). A molecular phylogeny of the fern family Pteridaceae: assessing overall relationships and the affinities of previously unsampled genera. *Molecular Phylogenetics and Evolution*, **44**, 1172–85.

Shi, L. and Zhang, X.-C. (1999a). Taxonomy of the fern genus *Leptochilus* Kaulf. (Polypodiaceae). *Acta Phytotaxonomica Sinica*, **37**, 145–52.

Shi, L. and Zhang, X.-C. (1999b). Taxonomic studies of the fern genus *Lepidomicrosorum* Ching et Shing (Polypodiaceae) from China and neighboring regions. *Acta Phytotaxonomica Sinica*, **37**, 509–22.

Smith, A. R., Pryer, K. M., Schuettpelz, E., *et al.* (2006). A classification for extant ferns. *Taxon*, **55**, 705–31.

Smith, A. R., Pryer, K. M., Schuettpelz, E., *et al.* (2008). Fern classification. In *Biology and Evolution of Ferns and Lycophytes*, ed. T. A. Ranker and C. H. Haufler. Cambridge, UK: Cambridge University Press, pp. 419–69.

Tindale, M. D. and Roy, S. K. (2002). A cytotaxonomic survey of the Pteridophyta of Australia. *Australian Systematic Botany*, **15**, 839–937.

Tsutsumi, C. and Kato, M. (2005). Molecular phylogenetic study on Davalliaceae. *Fern Gazette*, **17**, 147–62.

Tsutsumi, C., Zhang, X.-C. and Kato, M. (2008). Molecular phylogeny of Davalliaceae and implications for generic classification. *Systematic Botany*, **33**, 44–8.

Appendix C

Geological timescale

Era	Period	Epoch	Time (millions of years ago)
Cenozoic			
	Quaternary		
		Recent	5000 years to present
		Pleistocene	1.5–0.005
	Tertiary		
	Neogene	Pliocene	5.3–1.5
		Miocene	23–5.3
	Palaeogene	Oligocene	34–23
		Eocene	53–34
		Palaeocene	65–53
Mesozoic			
	Cretaceous		135–65
	Jurassic		205–135
	Triassic		250–205
Palaeozoic			
	Permian		290–250
	Carboniferous		355–290
	Devonian		410–355
	Silurian		438–410
	Ordovician		510–438
	Cambrian		570–510
Proterozoic			2500–570
Archaean			4000–2500
Hadean			>4000

Adapted with permission from Camus *et al.* (1996).

Reference

Camus, J. M., Gibby, M. and Johns, R. J. (eds.) (1996). *Pteridology in Perspective*. Whitstable, UK: Whitstable Litho. Ltd

Glossary

μm	micrometer or micron (1×10^{-6} m)
ABA	abscisic acid, plant hormone involved in leaf abscission
abaxial	lower surface of a leaf (*compare with* adaxial)
abiotic	nonliving; not caused or induced by organisms
abscission	dropping of leaves or other plant parts after a separation layer has been formed
acrostichoid	sporangia covering the entire lower leaf surface
adaxial	upper surface of a leaf (*compare with* abaxial)
adjuvant	additive to enhance herbicide or pesticide effectiveness
adventitious root	root that originates from aboveground tissue such as rhizome or leaf rather than laterally from the root; aerial root
aerenchyma	parenchymatic plant tissues with large intercellular spaces for aeration
age structure	composition of a population of individuals of different age classes
alien species	species that occurs outside its natural geographic range as a result of intentional or accidental dispersal (often by humans)
alkaloid	plant compound characterized by its alkaline reaction (e.g., nicotine, caffeine)
allelopathy	chemical inhibition of the growth of one organism by another
allopatry	speciation that occurs when plants in a geographically isolated population evolve into a new species

alpha diversity number of species within a single study site or plant community

alternation of generations reproductive life cycle that requires the alternation between two morphologically different generations of the same species: a gametophyte (usually haploid) and a sporophyte (usually diploid)

anisophyllous rhizomes that bear leaves of different size but similar shape (*compare with* isophyllous)

annual plant plant that completes its life cycle in less than one year

annulus ring or cluster of thick-walled cells on a sporangium that opens and expels the spores, typical of leptosporangiate ferns

anoxic without oxygen

antagonism inhibition of one species by the action of another

antheraxanthin carotenoid, an accessory photosynthetic pigment in the xanthophyll cycle that dissipates excess light energy

antheridiogen plant hormones produced by female gametophytes that induce (1) formation of antheridia on another gametophyte or (2) belowground germination of some spores

antheridium structure on a gametophyte that produces male **(pl. antheridia)** gametes (sperm)

anthocyanin pigments of red, violet or blue color

anthracnosis soil-borne, fungal disease of ferns and seed plants

antioxidant molecule that prevents oxidation of other molecules, e.g. vitamin E

AOO area of occupancy, sum of the areas (km^2) actually occupied by a species within the larger outlines of its geographic range

apical dominance prevention of growth of lateral meristems by the apical meristem, often with help of plant hormones

apogamy asexual development of a sporophyte from vegetative cells of the gametophyte rather than from a fertilized egg cell

arbuscular mycorrhiza group of fungi that grow inside the cortex cells and intercellular spaces of roots; *see also* endomycorrhizae

arbuscule	branched tree-like structure that promotes nutrient exchange between endomycorrhizal fungi and their host plant
archegonium (pl. archegonia)	structure on a gametophyte that produces a female gamete (egg)
areolate	netted leaf vein pattern that outlines polygons (areoles)
Arrhenius equation	formula used to model species–area relationships
arthropod	animal in the Phylum Arthropoda (e.g., insects, spiders)
ascomycete	class of fungi (Ascomycetes) bearing spores in a cup-like structure called an ascus (e.g., mildews, yeasts)
aseasonal	without apparent temporal pattern related to climate or photoperiod
aseptate	lacking crosswalls in the hyphae of fungi
assimilation rate	rate at which carbon dioxide is taken up by a leaf
basidiomycete	class of fungi (Basidiomycetes) bearing spores on stalk-like structure called a basidium (e.g., mushrooms, smuts, rusts)
beta diversity	difference in species composition along a gradient from one community to another
bioassay	measurement of the biological response to a treatment (e.g., nutrient)
biocontrol	biological control of pests by the introduction, preservation or facilitation of natural predators, parasites or other enemies
biogeography	study of the underlying mechanisms of geographical distributions of living organisms
bottleneck	loss of genetic diversity in very small populations that increases inbreeding and genetic drift
brackish	water of intermediate salinity between freshwater and seawater
bud	in ferns, a secondary meristematic tissue that develops into a new fern plant
calcicole	plant that grows in calcium-rich soil
CAM	*see* crassulacean acid metabolism
capillary	narrow, elongate tube that promotes water conduction

carcinogenic	chemical compounds able to cause cancer
carotenoid	accessory photosynthetic pigment
catabolize	degrading metabolic process that liberates energy
cation	positively charged ion
cavitation	breakage of the water column in xylem capillaries
chain hopping	successive colonization of islands from their closest neighbor island within an island chain (e.g., Lesser Antilles or Hawaii)
cheilanthoid	mostly drought-resistant fern belonging to the subfamily Cheilanthoideae within the family Pteridaceae
chlorophyllous	containing chlorophyll, the green pigment for photosynthesis in plants
chlorosis	failure of plant tissues to produce regular amounts of chlorophyll
chronosequence	set of communities of different ages since disturbance, assumed to represent a single community over time
CITES	Convention on International Trade in Endangered Species that regulates international trade in plants and animals
clade	group of taxa sharing a common ancestor
clathrate	feature of plant scales composed of cells with dark cell walls and clear central portions that appear lattice-like
clavate	club-shaped, an elongated structure gradually thickening toward the apex
climbing fern	fern that climbs over other vegetation with help of a slender, elongate rhizome or twining leaves (e.g., *Lygodium*, *Salpichlaena*)
clone	population consisting of potentially independent but genetically identical ramets
cohort	group of individuals in a population that share a developmental history
community assembly	process by which ecological communities are created through repeated invasions of species
conidium (pl. conidia)	wind dispersed spore of fungi
coppice	dense grove of small trees or shrubs that has grown from suckers and sprouts, not seeds; maintained by periodic cutting

coprolite	fossilized fecal material
cordiform	heart-shaped
coriaceous	of leathery texture
crassulacean acid metabolism	carbon dioxide fixation pathway that improves water economy by fixing carbon dioxide at night and photosynthesis during the day with closed stomata (e.g., in cacti, some ferns)
cross-fertilization	outcrossing; fertilization between gametophytes that developed from spores of two different sporophytes
crozier	young fern leaf that is coiled in a spiral pattern; fiddlehead
cryptogam	plant that reproduces by spores (e.g., alga, lichen, moss, fern)
cuticle	waxy layer on the outer epidermal cell wall of a plant
cyanobacteria	blue-green algae; photosynthetic organisms capable of nitrogen fixation
cytology	cell biology
dark germination	spore germination that does not require light; may be induced by antheridiogen
deciduous	seasonal abscission or loss of leaves
decumbent	reclining or lying on the ground, but with an ascending terminal part
de-epoxidation	chemical reaction to avoid damage in plants caused by excess light
demography	study of changes in population size and structure through time
density	number of individuals per unit area
derived	used for a character of an extant species that was not present in the ancestral stock; advanced
desiccation	water loss during the process of drying
determinate growth	limited growth of leaves (or plants) that ceases after differentiation and complete development of its tissues (*compare with* indeterminate growth)
detritivore	organism feeding on fragmented, dead organic matter
diaspore	dispersal unit of a plant (e.g., seed or spore)
dichotomous	forked into two parts

dimorphism	presence of two morphologically distinct forms of leaves (e.g., sterile and fertile) on the same fern species
diploid	condition of having two sets of chromosomes: often abbreviated as 2n (*compare with* haploid)
disjunct distribution	range that consists of geographically separate parts
disturbance	relatively abrupt event (e.g., fire, storm) causing loss of biomass or structure
disturbance regime	characteristics of disturbance occurring in a given ecosystem; generally described by intensity, size or frequency
divergence time	period of acquisition of dissimilar characters by organisms of the same clade
domatia	ant houses; hollow plant organs in which ants live
ECM	*see* ectomycorrhiza
ecosystem	a group of organisms and their interactions with the physical environment in a specified place and time
ecotype	one of several populations of a species from different habitats or locations that possesses a similar set of genetically based adaptations
ectendomycorrhiza (pl. ectendomycorrhizae)	mycorrhiza that is formed by septate Ascomycota and is intermediate between an endomycorrhiza and an ectomycorrhiza
ectomycorrhiza (pl. ectomycorrhizae)	mycorrhiza in which the fungal hyphae forms a sheath around the root of the host plant and penetrates the intercellular spaces of the root cortex, but not into the root cells (*compare with* endomycorrhiza)
edaphic	pertaining to soils
elater	strap-shaped band attached to a spore that may move with changes in humidity
embolus (pl. emboli)	breakage in a water column due to entry of air
endemism	taxon confined to a single, limited geographical area; paleoendemism, last extant representatives of old and declining phylogenetic lineage; neoendemism, recently derived taxa that have not yet spread because of recent origin

endomycorrhiza (pl. **endomycorrhizae**)	internal mycorrhiza with fungal hyphae that penetrate the cells of the root cortex but do not form a sheath around the root (*compare with* ectomycorrhiza)
endophytic	living within another plant
endosporic	within the spore wall; refers to gametophytes of heterosporous ferns that develop within the spore wall
ENM	*see* endomycorrhiza
EOO	extent of occurrence, geographic range (km^2) or extent of areas currently occupied by a species (*compare with* AOO)
ephemeral	short-lived; e.g., refers to short-lived fertile leaves of dimorphic fern species
epilithic	growing on rocks; epipetric
epiphyte	plant that grows on another plant and is not rooted in the ground
ericoid	pertaining to the plant order Ericales (e.g., heaths): refers also to characters similar to those observed in Ericales (e.g., ericoid mycorrhiza)
EU	European Union
eusporangium (**eusporangiate**)	sporangium with a wall composed of two or more cell layers (*compare with* leptosporangium)
evapotranspiration	transfer of water to the atmosphere by evaporation from humid surfaces plus transpiration by plants
evergreen	plant that is never barren of foliage as leaves are shed and replaced gradually
ex situ	not in its natural habitat; e.g., refers to species conservation measures such as propagation in laboratories or green houses (*compare with in situ*)
exindusiate	without indusium; refers to a sorus that lacks the covering structure of the indusium
exosporic	outside the spore wall; refers to gametophytes of homosporous ferns that develop outside the spore wall
exudate	substance that oozes or diffuses out of a plant
facilitation	positive interactions among plants; a process by which early pioneer species increase the establishment and survival of later successional species

facultative — not obligatory; can refer to the ability to grow under either of two opposing conditions

farina — meal-like, usually white or yellow powder of flavonoids on the leaf surface of some cheilanthoid ferns

fertile — in ferns, referring to leaves or plants that bear spores

fertilization — fusion of a male and a female gamete to form a zygote

flabellate — flattened with a rounded apex; fan-shaped

flush — in ferns, simultaneous emergence of leaves from a single plant

frass — fecal matter or other fine animal debris serving as evidence of herbivore activity

frond — megaphyll; leaf of ferns, often highly divided and of large size

fungicide — chemical substances used to kill or control fungi

gall — abnormal outgrowth of plant tissues that can be caused by various parasites (e.g., fungi, bacteria, insects, mites)

gametes — sex cells (egg and sperm) that fuse during fertilization

gametophyte — small sexual generation of ferns that grows from a spore and produces gametes

gemma (pl. gemmae) — bud for vegetative reproduction of gametophytes of some fern species

generalist — animal that feeds on a wide variety of plants or plant that is tolerant of a wide range of habitats (*compare with* specialist)

genet — sexually produced, genetically unique plant individual

genetic drift — change in gene frequencies due to random sampling effects, especially important in small populations (*see* bottleneck)

genetic fragmentation — loss of genetic connectivity between populations of a species as a result of geographical, ecological or reproductive isolation

genome — entire genetic information composed of all chromosomes of an individual cell of a species

gradient	regularly increasing or decreasing change in an environmental factor
grammitid	diverse group of mostly epiphytic, tropical ferns in the family Polypodiaceae that are more closely related to *Grammitis* than to other ferns in the family
gravitropism	turning or growth movement by a plant in response to gravity
habit	general appearance and growth form of a plant species
habitat	natural environment in which an organism lives
haploid	having just one set of chromosomes; a condition usually characteristic of the gametophyte generation in ferns; often abbreviated as 1n (*compare with* diploid)
haplotype	haploid genotype; combinations of DNA sequences of different genes on the same chromosome of a species
Hartig net	network of ectomycorrhizal fungal hyphae between the root cells of the host plant
haustorium (pl. haustoria)	specialized plant organ of parasitic plants to attach, penetrate and transfer water and solutes from a host plant
heathland	vegetation type dominated by low-growing shrubs of the family Ericaceae
hemiepiphyte	plant that spends only part of its life cycle as an epiphyte, germinating on the ground then growing toward and up a tree, or germinating as epiphyte and developing roots that reach the ground.
hemimetabolous	insects with incomplete metamorphosis: each larval stage gradually resembles more the final adult stage; a pupal stage is missing (*compare with* homometabolous)
hemiparasitic	plants that parasitize the xylem of their hosts for water but are capable of photosynthesis for nutrition
heteroblasty	morphological change of leaf shape and dissection at different stages of plant development
heteroecy	development of parasitic fungi on two alternating hosts

heterospory	producing two kinds of spores (i.e., male microspores and female megaspores)
heterotroph	organism that cannot synthesize its own food (e.g., animals, bacteria, fungi) but depends on organic substances produced by autotrophic organisms (e.g., plants)
homometabolous	insects with complete metamorphosis with four stages, egg, larvae, pupa and adult imago (*compare with* hemimetabolous)
homospory	producing only one kind of spore
homozygous	having identical parental genes
hot spot	threatened, species-rich region with unique edaphic, climatic or historical conditions
hydathode	enlarged vein tip on the upper leaf surface that often secretes water and diluted salts
hydraulic conductance	property of vascular plants, soil or rock; to move water through porous media
hydrenchyma	water-storing tissue
hygroscopic	readily absorbing and retaining moisture from the atmosphere
hymenophylloid	filmy fern in the Hymenophyllaceae family that has a common ancestor and is more closely related to *Hymenophyllum* than to *Trichomanes*
hyperaccumulator	plant that accumulates up to 30% of their dry mass in various elements (e.g., arsenic, cadmium, cobalt, manganese, nickel, selenium, zinc)
hypha (pl. hyphae)	branching filamentous structure of fungi
in situ	in its natural habitat; e.g., refers to species conservation measures (*compare with ex situ*)
in vitro	applies to the culture of fern species in a laboratory using spores, gametophytes or meristematic tissues of sporophytes
inbreeding depression	reduced fitness of inbred plants because of the expression of harmful or lethal traits
indeterminate growth	growth that continues throughout the life span of an individual (or leaf) such that size and age are roughly correlated (*compare with* determinate growth)
indicator species	a species that indicates by its presence the existence of particular environmental conditions

indument	epidermal outgrowth such as hairs, scales or glands on a plant structure
indusium (pl. indusia)	thin, membranous structure that covers the sorus
intragametic selfing	fertilization of egg by resident sperm on a bisexual gametophyte
intrinsic biological factors	all life history traits of a species
introduction	establishment of a species in a new site that may (e.g., conservation plantings) or may not (e.g., alien species) be within the current geographic range of a species
island curve	steeper species area curve for islands in comparison with continental areas
isophyllous	rhizomes that bear leaves that are similar in shape and size (*compare with* anisophyllous)
isotope	atoms of the same element (e.g., carbon) but differing in atomic mass (e.g., ^{12}C and ^{13}C)
isozyme	variant of the same enzyme catalyzing identical reactions but differing in their biochemical properties
IUCN	International Union for Conservation of Nature and Natural Resources
lamina	expanded green part of a fern leaf without the petiole
leaf primordium	embryonic stage of a leaf in the apical meristem
leaflet	primary division of a compound leaf; pinna
leptokurtic	probability distribution with higher central peak and flatter tails than the normal distribution; refers to the probability of spore dispersal from its mother plant
leptosporangium	stalked, thin-walled (by one cell layer) sporangium derived from a single epidermal cell
life cycle graph	diagram showing the size, age or stage classes of a population with arrows indicating transitions among classes (loop diagram)
life history	schedule of birth, mortality and growth of individuals in a population
life table	probability table of the mortality rate of a cohort of individuals at each age
light compensation point	light level at which photosynthetic gain matches respiratory loss of carbon dioxide

lignified sclerenchymatic plates supporting tissues of dead cells with lignin in the cell walls, providing stiffness to the tree fern trunk

lignin complex polymeric macromolecule of secondary cell walls in plants, providing strength and resistance against most microorganisms

lineage line of common descent; a group of species sharing descent from a common ancestor

lithophyte plant growing on rock

litter dead plant material on the substrate surface that is not completely decomposed

lycophyte species of the families Lycopodiaceae, Selaginellaceae, and Isoëtaceae and their fossil relatives; characterized by simple entire leaves with a single vein (microphylls) and a single sporangium borne on the upper leaf surface or the leaf axil

macroecology study of the relationship between organisms and their environment at large spatial scales

macromolecule molecule with large molecular mass (e.g., nucleic acid, protein, carbohydrate, lipid)

malate salt or ester of malic acid

marcescent having withered leaves that persist on a plant (e.g., forming a skirt of dead leaves in tree ferns)

MDE *see* mid-domain effect

megapascal unit of pressure

megaphyll large, multiveined leaf, characteristic of ferns and seed plants with leaf gaps in the central vascular system (*compare with* microphyll)

megaspore larger female spore of heterosporous plants (*compare with* microspore)

meiosis cell division during which the chromosome number per cell is halved; in ferns, meiosis gives rise to spores

meristem embryonic tissue of undifferentiated plant cells that can undergo cell division and produce other tissues; the apical meristem is located at the tip of a growing stem or branch

mesic habitat with moderate water balance; moist (*compare with* xeric)

mesophyll photosynthetic tissue between the upper and lower leaf epidermis

metamorphosis	marked structural transformation of an organism during the transition from larval or pupal stage to adult stage
microhabitat	environmental conditions in the immediate surroundings of an organism; a small specialized habitat
microlepidoptera	artificial and controversial grouping of small, supposedly more primitive moths in comparison with the larger and less primitive macrolepidoptera
microphyll	small, single-veined, characteristic of the lycophytes, not associated with a leaf gap in the central vascular system (*compare with* megaphyll)
microspore	smaller male spore of heterosporous plants (*compare with* megaspore)
mid-domain effect	model in which species' altitudinal ranges are randomly placed, resulting in hump-shaped curve with highest species richness at mid-elevation and increasingly lower species richness at the lower and upper elevational limits
modular	having a structure made up of repeated units (e.g., branches)
monolete	bean-shaped spore with one linear aperture, where spore germinates (*compare with* trilete)
monophyletic	describes a group of organisms that consists of an ancestor and all its descendants
monotypic genus	a genus containing only one species
mucilage	slimy or sticky fluid that may be exuded by some plants
mutualism	symbiosis; a relationship in which both organisms benefit
mycoherbicide	herbicide created from fungi
mycorrhiza (pl. mycorrhizae)	symbiosis between a fungus and the root of a host plant
mycotrophy	the condition where a plant receives its nutrients from a fungus
myrmecophilous	plants thriving in association with ants
native species	species that grows within its natural range
nectary	a gland that secretes nectar containing sugars and amino acids

net primary productivity   amount of light energy fixed by plants through photosynthesis minus respiratory losses

neutralism   association between two species in which both are unaffected by the presence of the other

niche   range of ecological conditions in which an organism can exist; ecological role of an organism

niche assembly model   a model that predicts how ecological factors and species traits determine the occupation of niches by different species

niche breadth hypothesis   a hypothesis that predicts that the niches of species are smaller with increasing amount of available environmental resources because of specialization

niche partitioning   process by which natural selection drives competing species into different patterns of resource use or different niches

niche position hypothesis   predicts that specialists that depend on relatively rare resources may be too scarce in low energy areas to maintain viable populations

NPP   net primary productivity

nucleotides   basic subunits of nucleic acids (DNA, RNA)

nutrients   substances needed for plant life such as growth and reproduction (e.g., phosphorus, nitrogen, copper, zinc)

nutrient cycling   flow of nutrients through an ecosystem

nutrient resorption   process by which nutrients are absorbed by a plant during leaf senescence

NV   number of veins; a method of determining developmental status of a sporophyte based on a count of the veins extending from the midrib of sequentially-produced leaves

ontogeny   course of development of an individual from birth to maturity and death

organic material   decaying and decomposed material in the soil that comes from living organisms

osmotic potential   potential of water molecules to move through the semi-permeable cell membrane into the plant cells because of the higher concentration of the solutes within the cells; turgor pressure that will

	be reached at osmotic equilibrium between a solution and pure water
palustrin	alkaloid that deters herbivory in some plants such as horsetails
páramo	a discontinuous, neotropical ecosystem of high mountains, located between the upper forest line (about 3400 m above sea level) and the permanent snow line (about 5000 m)
parasitism	intimate relationship between two organisms in which the heterotrophic parasite lives at the expense of the host
parenchyma	ground tissue of living, thin-walled plant cells
patch	specified area within a habitat
pectin	insoluble substance that is part of the non-woody parts of terrestrial plants; helps bind cells together
PEP-C	*see* phospho-enol-pyruvate carboxylase; enzyme that catalyzes carbon dioxide fixation of CAM plants
perennial	plant that completes its life cycle in more than one year
PET	potential evapotranspiration
petiole	stalk of a leaf; stipe
phenology	study of the relationship between periodic life history events (e.g., initiation of growth, spore production, dormancy) and environmental factors (e.g., temperature, precipitation)
phenotypic plasticity	range of structural and functional responses of an organism to environmental changes
phospho-enol-pyruvate carboxylase	PEP-carboxylase; enzyme that catalyzes carbon dioxide fixation on phospho-enol-pyruvate during photosynthesis of CAM plants
photochemical efficiency	measure of the proportion of photosynthetic active radiation used for photosynthesis
photoinhibition	decline in photosynthesis upon excessive light exposure and absorption which damages the photosystems
photoperiodism	response of an organism to changes in day length or to a light–dark cycle

phyllopodium (pl. phyllopodia)	stump-like extension of the rhizome to which the leaves are attached and at which leaves are abscised
phylogeny	evolutionary history of lineages of organisms through time
phylogeography	study of the historical processes to explain the current biogeographical distribution of plants
phytoalexins	chemical compounds of plants that are produced in response to an attacking organism and in order to inhibit further damage
phytoecdysone	insect-molting hormone produced by some plants as antiherbivore defense
phytohormone	plant hormone (e.g., abscisic acid, gibberellic acid, antheridiogen, ethylene)
phytoremediation	use of green plants to remove, contain or render harmless environmental contaminants
pinna (pl. pinnae)	leaflet; primary division of a pinnate or more divided leaf
pinnate	having a feather-like arrangement of the leaflets along both sides of a single midvein or rachis
pinnule	secondary division of a divided, compoundly pinnate leaf
poikilohydric	capacity of plants or plant parts to dry out without losing their capacity to function upon rehydration
polyploidy	possession of more than two sets of chromosomes within a cell
population dynamics	changes in the number of individuals, genetic composition, structure, or spatial dispersion of a population
population structure	relative frequency of individuals in each class of different size, age or stage
primary succession	succession in which plants colonize surfaces with little or no prior soil development (e.g., lava, sand, landslide, floodplain)
propagule	vegetative or reproductive unit of dispersal such as a spore or a seed
prothallus	gametophyte of a fern

pseudodichotomous	seemingly equal branching pattern but which did not develop from equal divisions of the growing tip (e.g., the branching of the rachis of scrambling ferns which bears a bud in each apparent division point)
ptaquiloside	carcinogenic chemical compound of bracken that is poisonous to animals
pteridologist	person who studies seedless vascular plants (i.e., ferns and lycophytes)
pulvinus (pl. pulvini)	joint-like structure at petiole or pinnule bases of a leaf containing cells that change turgor pressure in response to environmental conditions
rachis	prolongation of the petiole along the lamina; midrib of a leaf; midvein
ramet	vegetatively produced plant individual that may detach from the genetically identical parent plant
recalcitrant litter	plant litter that decomposes slowly
recumbent	spreading across the surface or over other vegetation (e.g., scrambling fern, *Dicranopteris*)
Red List	list of threatened species that is derived from a risk assessment; The Red List uses IUCN criteria
redox reaction	chemical reaction in which some atoms are oxidized (i.e., lose electrons) while others are reduced (i.e., gain electrons)
refugium (pl. refugia)	isolated habitat that retains species that were once widespread or that has escaped major environmental changes of a larger region or both
reinforcement	introduction of plants that have been propagated *ex situ* from plants of the same site in order to increase their population size
reintroduction	re-establishing populations of a species in a historical site where that species had gone extinct
rejuvenation	production of leaves with juvenile characteristics following a period when leaves of adult form have been produced
reproductive effort	fraction of resources allocated to reproduction at a given age or stage
rheophyte	flood-adapted, rooted plant that lives in running freshwater

rhizoid   elongated, hair-like structure that performs root-like functions

rhizome   horizontal or ascending stem of ferns growing at or below the soil surface, anchored to the soil by its roots

richness   number of species in a defined area

risk assessment   use of standardized criteria to estimate the probability that species may go extinct worldwide or in a particular area

rosette fern   fern with an ascending or erect, compact rhizome with leaves arranged in a funnel-like rosette

rubisco   RuBP carboxylase/oxygenase; the enzyme catalyzing primary carbon dioxide fixation in plants with C3-photosynthesis

ruderal   plant species that grow well in habitats disturbed by human activities such as roadsides, trails and agricultural land

rust fungus (pl. fungi)   parasitic fungus with alternation of hosts, named for the orange, powdery spore masses it produces on its host plant

safe site   favorable location for fern spore germination, gametophyte growth and establishment of a sporophyte

saprophyte   plant that obtains nutrients from dead or decaying organic matter

scale   spatial or ecological context of a data set; small, flat outgrowth of the epidermal layer that is two or more cells wide at its base (*compare with trichome*)

sclerenchyma   supporting plant tissue containing elongated fibers or sclereids, both with thick, lignified cell walls

scrambling fern   fern with leaves of indeterminate growth or dormant leaf buds and that spreads across the soil surface and over other low-growing vegetation

secondary growth   growth in diameter of stems by means of a secondary meristematic cambium that produces wood (xylem) and bast (phloem)

secondary metabolite	chemical compound that serves a wide variety of functions but not necessary for growth and plant development
seral	referring to a plant community that is part of a successional sequence or sere
sesquiterpene glycoside	carcinogenic compound found in some ferns
soral crypsis	protective covering of microlepidoptera that mimics a fern sorus
sorus (pl. sori)	cluster of sporangia
specialist	herbivore species that feeds on one or several host plant species to which it is often adapted by specific traits such as the ability to detoxify their chemical defensive compounds (*compare with* generalist); plant that is limited to very specific habitat conditions
species–area curve	mathematical function that shows how the number of species increases with larger sample area
sporangium (pl. sporangia)	multicellular structure in which fern spores are produced
spore	one-celled, wind-dispersed propagule of ferns, product of meiosis within the fern sporangium, capable of germinating and developing into a gametophyte
spore bank	viable spores that are buried in the soil at a specific location
sporeling	young sporophyte that is known to have developed sexually from fertilization on a gametophyte
sporocarp	hard, round or bean-shaped, sporangia-bearing structure of the heterosporous, aquatic fern families Marsileaeceae and Salviniaceae
sporophyll	fern leaf that bears sporangia; fertile leaf
sporophyte	spore-producing fern generation
stand	all of the plants in an area; a group of ferns or fern leaves growing together
sterile	a sporophyte without fertile leaves; leaves without sporangia; a gametophyte that does not produce sperm or egg
stipule	appendage at either side of petiole base, especially in the family Marattiaceae

stoichiometry	relative proportion of different elements
stolon	elongated, horizontal stem that forms new plants at its end; runner
stress tolerant	ability to survive unfavorable conditions of low resources or extreme environments or both
strobilus (pl. strobili)	cone of spore-bearing leaves inserted directly on the stem of a lycophyte
succession	sequential change in species composition or other ecosystem characteristic
summer-green	fern species with leaves that die at the end of the growing season
survivorship curve	plot of the relative number of survivors of a cohort of individuals against time
sustainable	capable to reproduce and to survive by utilization of renewable resources
symbiosis	intimate relationship between two organisms interacting to their mutual benefit
sympatry	closely related species with overlapping species ranges
synangium (pl. synangia)	structure formed by the fusion of two or more sporangia and divided internally into compartments (locules)
synaptospory	postulated spore dispersal in groups promoted by entangled filaments such as in *Equisetum* or by spore sculpture
taeniafuge	chemical compound for expelling tapeworms
tannin	phenolic compound that reduces the digestibility of plant tissues
tectonic	pertaining to the movement of the plates that comprise the Earth's crust
tephra	fragmented volcanic materials ejected from a volcano during an eruption and transported through the air
tetraploid	sporophyte having four sets of homologous chromosomes
thiaminese	enzyme that breaks down thiamine, a component of the essential vitamin B1 complex, causing vitamin deficiency

thicket	dense, often monospecific vegetation that covers and shades the entire ground, and can deter colonization by other species
thylakoid membrane	double membrane in a chloroplast that bears photosynthetic pigments
tocopherol	lipophilic, organic compound with antioxidant vitamin E activity
tolerance	endurance of unfavorable environmental conditions
tracheid	type of elongated cell that forms from water-conducting tissue in primitive vascular plants
trichomanoid	filmy fern in the Hymenophyllaceae family that has a common ancestor and is more closely related to *Trichomanes* than to *Hymenophyllum*
trichome	plant hair that is only one cell wide at its base (*compare with* scale)
trilete	tetrahedral to nearly globose spore with three apertures at its proximal side, where spore germinates (*compare with* monolete)
turgescent	turgid; water-filled cells or tissues providing structural support to plant organs
turgor pressure	outwardly directed pressure within a cell resulting from the absorption of water
turnover time	measure of how rapidly materials are completely replaced within a system
type collection	designated herbarium specimen that serves as reference for a species' name
ultramafic soil	soil derived from rocks of volcanic origin that is rich in magnesium and iron
unimodal	frequency distribution having a single mode or peak; refers to the size or age classes of a population
vegetative reproduction	production of a new, independent plant by establishment and growth of a ramet (e.g., a stolon, leaf bud, etc.)
velamen radicum	several layers of dead root cells that absorb water
venation	arrangement or pattern of veins on a leaf
vernation	arrangement of the leaf parts during early leaf development; in circinate vernation the leaf is rolled at its apex

vesicle	lipid-storing swelling of hyphae of endomycorrhizal fungi
vessel	xylem tube composed of several stacked cells that have fused together by breaking down the walls between them
vicariance	originally contiguous populations that are split by geological or climatic change into separate populations
violaxanthin	carotenoid that acts as an accessory photosynthetic pigment in the xanthophyll cycle to dissipate excess light energy
vittarioid	fern in the Pteridaceae family that shares a common ancestor and is more closely related to *Vittaria* than to other genera in the family
winter-green	fern species that keep green leaves during winter and replace them in spring
xanthophyll cycle	chemical transformation of a number of carotenoid molecules in the chloroplast to avoid photodamage by excess light levels
xeric	habitat with short water supply; dry (*compare with* mesic)
xerophyte	plant adapted to dry habitats
zeaxanthin	a carotenoid that acts as an accessory photosynthetic pigment in the xanthophyll cycle to dissipate excess light energy

Index

Numbers in *italics* point to b/w figures, numbers in **bold** refer to color plate numbers.

Lightning Source UK Ltd.
Milton Keynes UK
UKOW07f1539050617

302711UK00005B/17/P